Food Processing and Preservation Technology

Food Processing and Preservation Technology

Edited by Lisa Jordan

SYRAWOOD
PUBLISHING HOUSE

New York

Published by Syrawood Publishing House,
750 Third Avenue, 9th Floor,
New York, NY 10017, USA
www.syrawoodpublishinghouse.com

Food Processing and Preservation Technology
Edited by Lisa Jordan

Cataloging-in-Publication Data

Food processing and preservation technology / edited by Lisa Jordan.
    p. cm.
Includes bibliographical references and index.
ISBN 978-1-68286-653-5
1. Food industry and trade. 2. Food industry and trade--Technological innovations.
3. Food--Preservation. 4. Food--Preservation--Technological innovations. I. Jordan, Lisa.
TP370 .F66 2019
664--dc23

TABLE OF CONTENTS

PREFACE

This book aims to highlight the current researches and provides a platform to further the scope of innovations in this area. This book is a product of the combined efforts of many researchers and scientists, after going through thorough studies and analysis from different parts of the world. The objective of this book is to provide the readers with the latest information of the field.

Food processing and preservation is concerned with the transformation of raw or cooked food into products that can be stored, packaged and marketed. The emerging trends in food processing and preservation aim to maintain the nutritional quality of food and minimize food wastage. Some of the common practices of this field include pasteurization, mincing, canning, liquefaction among many others. This book is a comprehensive study of food processing and preservation methods and the suitability of each with reference to the nature of food. The various studies that are constantly contributing towards advancing technologies and evolution of this field are examined here. Students and researchers in the areas of food science and food process engineering will find this book a valuable source of information.

I would like to express my sincere thanks to the authors for their dedicated efforts in the completion of this book. I acknowledge the efforts of the publisher for providing constant support. Lastly, I would like to thank my family for their support in all academic endeavors.

Editor

Assessment of Hygienic Practices and Microbiological Quality of Food in an Institutional Food Service Establishment

Aisha Idris Ali* and Genitha Immanuel

Department of Food Process Engineering, Sam Higginbottom University of Agriculture Technology and Sciences, Allahabad, UP, India

Abstract

Safe food handling in school kitchens is an important practice to protect the students from foodborne illnesses. Bacterial count in prepared food is a key factor in assessing the quality and safety of food. It also reveals the level of hygiene adopted by food handlers in the course of preparation of such foods. A case study research was conducted to examine the food safety knowledge, attitudes and practices of food handlers and bacterial contaminations in food from two women's hostel kitchens at Sam Higginbottom University of Agriculture Technology and Sciences (SHUATS) Allahabad, India. Questionnaires regarding food safety knowledge, attitudes and practices were administered to all the 25 food handlers working at these two kitchens (18 (72%) from old and 7 (28%) from new women's hostel kitchens) through in-person interviews. A total of 72 cooked food samples (36 from each kitchen) were analyzed for evidence of contamination (total aerobic mesophilic bacteria, coliforms, and *Escherichia coli*). The majority of the food handlers did not used good food handling practices and did not practice proper personal hygiene, because majority of them had poor knowledge and attitudes regarding food safety. All the cooked food samples tested had total APC, coliform, and *E. coli* levels higher than acceptable. The study results, therefore, call for stringent supervision and implementation of food safety practices. Periodic trainings on personal hygiene and good food handling practices will play a pivotal role in improving the safety of the prepared meals in these kitchens.

Keywords: Food handlers; Food contamination; Food safety; Coliforms; *E. coli*

Introduction

When food is cooked on a large scale, it may be handled by many individuals and thus increasing the chances of contamination of the final food. Unintended contamination of food during large scale cooking, leading to foodborne disease outbreaks can pose danger to the health of consumers and economic consequence for nations [1-3]. Foodborne related illnesses have increased over the years, and negatively affected the health and economic well-being of many developing nations [4]. The World Health Organization (WHO) states that about 1.8 million persons died from diarrhoeal diseases in 2005, mainly due to the ingestion of contaminated food and drinking water. Food poisoning occurs as a result of consuming food contaminated with microorganisms or their toxins, the contamination arising from inadequate preservation methods, unhygienic handling practices, cross-contamination from food contact surfaces, or from persons harboring the microorganisms in their nares and on the skin [5,6]. Unhygienic practices during food preparation, handling and storage creates the conditions that allows the proliferation and transmission of disease causing organisms such as bacteria, viruses and other foodborne pathogens [7,8]. Additionally, many reported cases of foodborne viral diseases have been attributed to infected food handlers involved in catering services [9]. The knowledge, attitudes and practices of food handlers have been reported in studies from different countries around world [10-15]. This is because a combination of the three factors: knowledge, attitude and practice of food handlers, play dominant role in food safety with regards to food service establishment [16]. However, despite knowledge and awareness of safe food handling methods, several studies have found that food handlers often do not use safe food handling practices, based on observation and microbial food testing [11,17,18].

According to World Health Organization [19], food handling personnel play important role in ensuring food safety throughout the chain of food production and storage. Mishandling and disregard of hygienic measures on the part of the food handlers may enable pathogenic bacteria to come into contact with food and in some cases, survive and multiply in sufficient numbers to cause illness in the consumer. The hands of food service employees can be vectors in the spread of foodborne diseases because of poor personal hygiene or cross contamination. A USA based study suggested that improper food handling practices contribute to about 97% of foodborne illnesses in food services establishments and homes [20]. Foodborne disease is a challenge for both developed and developing countries [21], and are leading cause of illness and death in developing countries [22]. Despite concerted efforts for several decades, foodborne diseases remain a major global public health issue with substantial morbidity and mortality associated with the consumption of contaminated foods [23]. Even in India though no data are available, microbiological food safety hazards are a common and major health hazard taking several lives frequently causing morbidity and mortality. An outbreak of foodborne botulism due to *Clostridium butyricum* affecting 34 students from a residential school in Gujarat was reported in 1996, and the food sample found to be contaminated was *sevu* (crisp made from gram flour) [24]. The measurement of the safety of foods has relied on evaluation of the microbiological quality of foods [23,25]. Bacterial counts in prepared food or water is a key factor in assessing the quality and safety of food, and can reveal the hygiene level adopted by food handlers in the course of preparation of such foods [26]. In a recent review, *E. coli*, *Shigella*, *Salmonella* and *Campylobacter spp.* were the most commonly reported causes of gastrointestinal disease [27], and all have been associated with foodborne disease [28]. However, in developing countries, monitoring

*Corresponding author: Aisha Idris Ali, Department of Food Process Engineering, Sam Higginbottom University of Agriculture Technology and Sciences, Allahabad-211007, UP, India, E-mail: aishaidrisali@gmail.com

the microbial safety of foods is not routinely practiced, due to a lack of infrastructure and effective food safety regulations and standards [29]. Foodborne diseases outbreaks have been linked to improper food handling practices at food service establishments [11,21,22,30]. The most commonly reported food preparation practices that contribute to foodborne diseases include poor environmental hygiene, inadequate cooking, contaminated equipment, improper handling temperatures and food from unsafe sources [21,31,32].

Following the frequent complain of stomach pain and diarrhea among students eating from these school kitchens, this study was conducted to test two research hypothesis: (1) the levels of food contamination will be the same in both women's old and new hostel kitchens due to lack of good hygienic practices and (2) the level of total aerobic mesophilic, coliforms and *E. coli* counts in the cooked foods from both kitchens will be above the food safety standard permissible limit due to improper food handling practices which resulted in frequent diarrhea among students eating from these particular kitchens. The objectives of the study were to: describe the food safety knowledge, attitudes and practices of food handlers' in two women's hostel kitchens at a University (SHUATS) Allahabad, India and to evaluate the food contamination levels at these kitchens through microbiological analyses of the cooked foods.

Problem definition

Current statistics on foodborne illnesses in various industrialized countries show that up to 60% of cases may be caused by poor food handling techniques, and by contaminated food served in food service establishments. In 1989, it was estimated that the total cost of bacterial foodborne illness to the United States economy was US $6,777,000,000. Hence it is a burden on economy also. In developing countries, the effect on economic activity and development can only be far more severe [1].

Materials and Methods

Study design

A case study involving two kitchens at Sam Higginbottom University of Agriculture Technology and Sciences (SHUATS) Allahabad, India was conducted between January to June 2017. The kitchen located at the women's old hostel is the old kitchen, while the one at the new women's hostel is the new kitchen. The school operates full-time and serve food for breakfast, lunch and dinner. All the 25 food handlers working in these two kitchens were involved in the study (18 were from the old and 7 from the new hostel kitchens). The total number of students served by these kitchens is 610. Of these students, 400 eat from the old kitchen, while 210 eat from the new kitchen. Food samples, consisting of all the cooked foods served, at the time of sampling, were randomly collected by the researcher from both kitchens during the study period. The design of the study consisted of three sections. The first section was to observe food safety practices and personal hygiene of the food handlers. The second section was designed to evaluate food safety knowledge attitudes and practices of food handlers. Finally, the third section dealt with the assessment of microbiological quality of the cooked food samples collected from these kitchens.

Data collection

Face-to-face interviews were conducted to collect demographic information of the food handlers (Table 1), information on food handlers' knowledge about food safety, attitudes and practices through a semi-structured questionnaire through in-person interviews by the researcher using questions adapted from some of the previous works [10,33,34].

Parameter	Specification	Number (n)	%
Kitchen type	Old hostel kitchen	18	72
	New hostel kitchen	7	28
Role	Manager	2	8
	Head cook	0	0
	Cooks	10	40
	Servers	13	52
Age of Respondents	Below 19 years	0	0
	19-25 years	0	0
	25-40 years	20	80
	Above 40 years	5	20
Gender	Male	8	32
	Female	17	68
Educational level	None	16	64
	Primary	3	12
	Secondary	2	8
	Tertiary	4	16
Acquisition of knowledge on food preparation	Observation	16	64
	Training	0	0
	Trial and error	4	16
	Taught by parents	5	20
Food service industry experience	1 year	5	20
	2-5 years	17	68
	5-10 years	3	12
	Above 10 years	0	0

Table 1: Characteristics of study respondents from old and new hostel kitchens (n=25).

The section of the questionnaire dealing with food safety knowledge (Table 2, Section A) comprised of 25 questions with three possible answers; "yes", "no", and "do not know". A scale ranging between 0 and 25 (representing the total number of questions on food safety knowledge) was used to evaluate the overall food safety knowledge of the food handlers. Food handlers that obtain total score ≤ 15 points were considered to have "insufficient" knowledge and those that had scores ≥ 16 points (≥ 64% accuracy) were considered to have "good" knowledge of food safety.

The attitudes section of the questionnaire (Table 2, Section B) comprised of 12 questions with three possible answers; "yes", "no", and "do not know". Food handlers that answered 7 or fewer questions correctly were considered to have "insufficient" understanding whereas handlers that answered 8 or more questions correctly were considered to have "good" understanding.

In section which dealt with food hygiene practices (Table 2, Section C), the good hygienic practices of respondents were assessed and evaluated based on self-reporting of personal hygiene, and observation of other safe food handling practices. The section had 10 questions with two possible responses; "yes" and "no". Each correct practice reported scored one (1) point. For evaluation, a score ≥ 70% (n=7) by an individual respondent was considered as having "good" food hygiene practice. All responses regarding the practices were validated by the researcher's observations of the kitchens and respondents, and responses were corrected by the researcher in situations where observations did not tally with the responses (e.g., where they indicated they do not eat, drink or smoke while handling food).

Parameter	Response % (n)		
	Yes	No	Don't know
Food Safety Knowledge of Food Handlers (Section A)			
1. Washing hands before handling food reduces the risk of food contamination	8.0 (2)	0.0 (0)	92.0 (23)
2. Using gloves while handling food reduces the risk of food contamination	4.0 (1)	0.0 (0)	96.0 (24)
3. Proper cleaning and sanitization of utensils increase the risk of food contamination	0.0 (0)	12.0 (3)	88.0 (22)
4. Eating and drinking during food handling increase the risk of food contamination	0.0 (0)	0.0 (0)	100 (25)
5. Food prepared in advance reduces the risk of food contamination	0.0 (0)	0.0 (0)	100 (25)
6. Reheating cooked foods can contribute to food contamination	0.0 (0)	0.0 (0)	100 (25)
7. Children, healthy adults, pregnant women and older individuals are at equal risk for food poisoning	0.0 (0)	0.0 (0)	100 (25)
8. Typhoid fever can be transmitted by food	0.0 (0)	0.0 (0)	100 (25)
9. Bloody diarrhoea can be transmitted by food	0.0 (0)	0.0 (0)	100 (25)
10. Salmonella is among the food-borne pathogens	0.0 (0)	0.0 (0)	100 (25)
11. Hepatitis A virus is among the foodborne pathogens	0.0 (0)	0.0 (0)	100 (25)
12. Staphylococcus is among the foodborne pathogens	0.0 (0)	0.0 (0)	100 (25)
13. Raw and cooked foods should be kept separate	4.0 (1)	0.0 (0)	96.0 (24)
14. Vegetables can be chopped on the same chopping board used to chop raw meat	0.0 (0)	0.0 (0)	100 (25)
15. Food contamination risk can be reduced by knowing fridge temperature	0.0 (0)	0.0 (0)	100 (25)
16. Improper cooking of food causes foodborne illnesses	4.0 (1)	0.0 (0)	96.0 (24)
17. Improper food storage causes health hazards	4.0 (1)	0.0 (0)	96.0 (24)
18. Microbes are on the skin, in the nose and mouth of healthy food handlers	4.0 (1)	0.0 (0)	96.0 (24)
19. Cross contamination is when microorganisms from a contaminated food are transferred by the food handler's hands or kitchen utensils to another food	0.0 (0)	0.0 (0)	100 (25)
20. Freezing kills all the bacteria that may cause food-borne illness	0.0 (0)	0.0 (0)	100 (25)
21. The correct temperature for storing perishable foods is °C	0.0 (0)	0.0 (0)	100 (25)
22. Hot, ready-to-eat food should be kept at a temperature of 65°C	0.0 (0)	0.0 (0)	100 (25)
23. Contaminated foods always have some change in colour, odour or taste	0.0 (0)	0.0 (0)	100 (25)
24. Raw vegetables are at higher risk of contamination than undercooked beef	16.0 (4)	0.0 (0)	84.0 (21)
25. During infectious disease of the skin, it is necessary to take leave from work	8.0 (2)	0.0 (0)	92.0 (23)
Food safety attitudes of food handlers (Section B)	Yes	No	Don't know
1. Well-cooked foods are free of contamination	4.0 (1)	0.0 (0)	96.0 (24)
2. Proper hand hygiene can prevent food-borne diseases	8.0 (2)	0.0 (0)	92.0 (23)
3. Raw and cooked foods should be stored separately to reduce the risk of food contamination	0.0 (0)	0.0 (0)	100 (25)
4. It is necessary to check the temperature of refrigerators/freezers periodically to reduce the risk of food contamination.	0.0 (0)	0.0 (0)	100 (25)
5. The health status of workers should be evaluated before employment	4.0 (1)	0.0 (0)	96.0 (24)
6. Wearing masks is an important practice to reduce the risk of food contamination	4.0 (1)	0.0 (0)	96.0 (24)
7. Wearing gloves is an important practice to reduce the risk of food contamination	4.0 (1)	0.0 (0)	96.0 (24)
8. Wearing hair restraints and clean cloths/uniform is an important practice to reduce the risk of food contamination	4.0 (1)	0.0 (0)	96.0 (24)
9. Long and painted fingernails could contaminate food with foodborne pathogens	4.0 (1)	0.0 (0)	96.0 (24)
10. Food handlers can be a source of foodborne outbreaks	0.00 (0)	0.00 (0)	100 (25)
11. Knives and cutting boards should be properly sanitized to prevent cross contamination	0.00 (0)	0.00 (0)	100 (25)
12. Food handlers with abrasions or cuts on their hands should not handle ready-to-eat food	0.00 (0)	0.00 (0)	100 (25)
Food safety practices of food handlers (Section C)	Yes	No	
1. Do you use gloves during the distribution of ready-to-eat food?	0.00 (0)	100 (25)	
2. Do you wear an apron while working?	4.0 (1)	96.0 (24)	
3. Do you wear a cap/hair restraint while working?	4.0 (1)	96.0 (24)	
4. Do you wear a mask when you distribute unwrapped foods?	0.00 (0)	100 (25)	
5. Do you wash your hands properly before touching raw foods?	12.0 (3)	88.0 (22)	
6. Do you wash your hands properly after touching raw foods?	12.0 (3)	88.0 (22)	
7. Do you eat, drink or smoke in your work place?	0.00 (0)	100 (25)	
8. Do you wear nail polish when handling food?	68.0 (17)	32.0 (8)	
9. Do you use cutting boards of different colours or do you sanitize a cutting board between preparation of raw foods and cooked foods?	0.00 (0)	100 (25)	
10. Do you properly clean the food storage area before storing new products?	8.0 (2)	92.0 (23)	

Table 2: Food safety knowledge, attitudes and practices of food handlers (n=25).

Food samples collection

Cooked food samples were randomly collected by the researcher from the study kitchens at serving time. The samples obtained from the same kitchens on different days were considered different samples. Approximately 250 to 500 g of food was collected and sealed in sterile stomacher bags, placed in cool boxes. Each sample was properly identified with a number code, subject name and food type, and immediately transported to the microbiological laboratory of food process engineering department, Sam Higginbottom University of Agriculture Technology and Sciences (SHUATS), Allahabad. Samples were processed within 2 hours of collection.

Meals

A total of 72 cooked food samples (36 from each kitchen) were collected. The foods collected from both kitchens were Stir-fried noodles (*Chowmein*), Flattened rice (*Poha*), Potato-based mixed vegetable curry (*Pav Bhaji*), Lentil-based vegetable stew (*Sambhar*), and Indian cottage cheese soup (*Paneer soup*). The samples were collected immediately before or during the distribution of the food to the students. (These meals were cooked from scratch in these kitchens not pre-prepared meals and heated on-site).

Microbiological analyses

The contamination of food was measured by total aerobic mesophilic bacteria plate counts (APC), enumeration of *Escherichia coli* (*E. coli*) and total coliforms. A 25-g sample was collected from each food sample, and were homogenized in 255 ml of sterile buffered peptone water. Each sample was divided into two, and each sub-sample was placed in sterile stomacher bags and homogenized using a pulsifier. After homogenization, each sub-sample was divided into two, and serial 10-fold dilutions were made, up to a certain number of dilutions [35]. Selected dilutions of the food samples were mixed by vortexing, and inoculations were made within 25 minutes of processing, using method adapted from Downes and Ito [36].

For total aerobic mesophilic bacteria (APC) counts, 0.1 ml of the processed food samples of specified dilutions were inoculated on to sterile Plate Count Agar (HiMedia Laboratories, India), using surface spread method, and incubated for 24 h at 37°C [35,37]. After incubation, plates containing 25-250 colonies were selected for counting. Counts obtained were characterized by the reciprocal of the dilution factors used, and additionally by 10. The bacteria population was expressed as a number of colony forming units per gram (CFU/g). In total coliforms and *E. coli* processing, 0.1 ml of the processed food samples of specified dilutions were inoculated on to Chromocult coliform agar (Merck, Germany), a selective indicator medium for the enumeration of *E. coli* and other coliforms. After incubation at 37°C for 24 h, dark blue colonies were classified as *E. coli*, while pink colonies were classified as other coliforms [38,39]. Gram staining was carried out on suspected *E. coli* colonies, and all cultures with gram-negative short rods were biochemically confirmed as *E. coli* using the IMVIC tests [36].

Statistical analyses

All the mean bacterial counts per meal/food were transformed into standard form for statistical analysis. The targeted results were the microbiological quality status of food prepared from each kitchen. For both kitchens, using cut-points for microbiological safety [40], foods were classified as safe if the mean APC of the meals was less than 100,000 CFU/g, and mean counts for coliforms and *E. coli* respectively were less than 100 CFU/g. The statistical analysis was conducted using XLSTAT 2017 Version. The total mean count for APC, *E. coli* and coliform per

each food sample from both kitchens were statistically analyzed using two-tailed t test to assess the difference in the mean bacterial counts per food from both kitchens to determine whether there is significant difference or not. The significance of the results was judged with the help of P value at 5% level of significance, if the computed P value is greater than the significance level alpha at 5% level of significance, it means there is no significant difference between the total mean bacterial counts of the same food types from the two different kitchens, otherwise, there is significant difference.

Results

Characteristics of study respondents from old and new hostel kitchens

The interview was conducted for all the twenty-five respondents working in both kitchens (18 from old and 7 from the new women's hostel kitchens) of Sam Higginbottom University of Agriculture Technology and Sciences Allahabad, India (Table 1). Out of the 25 individual respondents in this study, 68% (n=17) were found to be females and 32% (n=8) were male. Twenty (n=20) 80% out of the 25 respondents were between 25 to 40 years of age, and had worked in the food service establishment for more than 4 years. Sixteen (n=16) 64% of the individual respondents respectively in this study did not have any formal education and acquired their food preparation knowledge through observation. Similarly, none of the food handlers 100% (n=25) had attended any specific training for food handling.

Food safety knowledge of food handlers

In our study, the food handlers were not knowledgeable about hygiene practices, cleaning and sanitation procedures (Table 2, Section A). Majority of the food handlers in this study were not aware of the critical role of general sanitary practices in the work place, such as frequent and proper hand washing at work place (92% didn't know), using gloves 96% (didn't know) and proper cleaning and sanitization of utensils 88% (didn't know). Regarding foodborne illness transmission, all the food handlers (100%) did not know that typhoid fever, bloody diarrhoea can be transmitted by food, and they all (100%) didn't know that salmonella and hepatitis A are foodborne pathogens. Over ninety per cent (90%) of the food handlers did not know that taking leave from work in periods of infectious skin disease was necessary. Additionally, 96% did not know that microbes can be found in the skin, mouth and nose of healthy food handlers. On the other hand, all the food handlers (100%) had no knowledge on time-temperature abuse and its effect on food safety.

Food safety attitudes of food handlers

A reduction in the incidence of foodborne illnesses is strongly influenced by attitudes of food handlers towards food safety. Table 2 (Section B) shows the attitudes of the food handlers towards the prevention and control of foodborne diseases. Ninety-six per cent (96%) of the food handlers did not know that wearing of masks, hair restraints, hand gloves and clean cloth/uniform respectively can minimize the risk of food contamination, which is considered a negative attitude reported by majority of our respondents. Similarly, all of the respondents 100% did not know that knives and cutting boards should be properly sanitised to prevent cross contamination of foods. All of the respondents did not know that individuals with abrasions or cuts on their fingers or hands should not touch unwrap foods (100%). Majority of the food handlers (96%) were not aware that foods should not be handled with long and painted fingernails and that the health status of workers should be evaluated before employment. Respondents (100%)

	Old kitchen			New kitchen		
Food samples	Total aerobic plate count (mean and (SD)* per food)	Coliform count (mean and (SD)* per food)	E. coli counts (mean and (SD)* per food)	Total aerobic plate count (mean and (SD)* per food)	Coliform count (mean and (SD)* per food)	E. coli counts (mean and (SD)* per food)
Stir-fried noodles (chowmein)	4.6×10^6 (1.0×10^5)* a	9.8×10^3 (4.4×10^2)* a	7.8×10^3 (3.5×10^2)* a	4.3×10^6 (1.6×10^5)* a	9.2×10^3 (7.6×10^2)* a	7.6×10^3 (3.5×10^2)* a
Flattened rice (poha)	3.9×10^6 (4.4×10^5)* b	9.1×10^3 (9.7×10^2)* b	7.5×10^3 (3.2×10^2)* b	3.6×10^6 (3.4×10^5)* b	8.3×10^3 (8.8×10^2)* b	7.2×10^3 (2.2×10^2)* b
Potato-based mixed vegetable curry (pav bhaji)	3.5×10^6 (4.4×10^5)* c	9.0×10^3 (9.5×10^2)* c	7.3×10^3 (1.4×10^2)* c	3.4×10^6 (3.6×10^5)* c	8.8×10^3 (9.9×10^2)* c	7.0×10^3 (1.5×10^2)* c
Lentil-based vegetable stew (Sambhar)	3.6×10^6 (4.3×10^5)* d	8.8×10^3 (8.6×10^2)* d	7.0×10^3 (1.3×10^2)* d	3.4×10^6 (2.8×10^5)* d	8.5×10^3 (8.4×10^2)* d	6.8×10^3 (2.5×10^2)* d
Mixed rice dish (veg. biryani)	4.5×10^6 (9.8×10^4)* e	9.6×10^3 (4.7×10^2)* e	7.7×10^3 (1.6×10^2)* e	4.3×10^6 (1.1×10^5)* e	9.4×10^3 (5.7×10^2)* e	7.5×10^3 (3.1×10^2)* e
Indian cottage cheese soup (paneer soup)	3.0×10^6 (5.5×10^5)* f	8.8×10^3 (6.9×10^2)* f	6.9×10^3 (1.2×10^2)* f	2.8×10^6 (5.6×10^5)* f	8.6×10^3 (5.5×10^2)* f	6.5×10^3 (2.4×10^2)* f

a-f Values with same alphabets across the rows are for same foods from the two different kitchens; (SD)*=Standard deviation
Note: Both kitchens serve the same menu, but not on same day.

Table 3: Mean and standard deviation of microbial populations of the cooked food samples (CFU/g) from old and new kitchens.

did not know if it is necessary to check temperatures of refrigerators and freezers periodically. The general attitudes of food handlers towards food safety in this study is unsatisfactory.

Food safety practices of the food handlers

In assessing the food safety practices of the food handlers (Table 2, Section C), 100% of the food handlers reported that they do not use gloves when distributing ready-to-eat or unpackaged food. Majority of the food handlers 96% (n=24) did not use apron or hair restraints during food handling. Additionally, 100% of the respondents were observed eating and drinking during food handing (including chewing dried ground tobacco leaves and smoking), and did not wear mask when distributing ready-to-eat food.

Microbiological analysis of cooked foods

Out of the 72 meal samples analyzed in this study, of which 36 (50%) each were collected from the old and new hostel kitchens respectively, aerobic mesophilic bacteria, coliform and E. coli were detected in all the samples. The mean bacterial population of the food samples analyzed is presented in Table 3. In old kitchen, the mean total aerobic plate count ranged from 3.0×10^6 for Indian cottage cheese soup (paneer soup) to 4.6×10^6 for stir-fried noodles (chowmein). The mean total coliform counts ranged from 8.8×10^3 for lentil-based vegetable stew (sambhar) and Indian cottage cheese soup (paneer soup) respectively to 9.8×10^3 for stir-fried noodles (chowmein). The mean total E. coli counts ranged from 6.5×10^3 for Indian cottage cheese soup (paneer soup) to 7.6×10^3 for stir-fried noodles (chowmein) (Table 3). Also reveal that in new kitchen the mean total aerobic plate count ranged from 2.8×10^6 for Indian cottage cheese soup (paneer soup) to 4.3×10^6 for stir-fried noodles (chowmein). The mean total coliform counts ranged from 8.3×10^3 for flattened rice (poha) to 9.2×10^3 for stir-fried noodles (chowmein). The mean total E. coli counts ranged from 6.5×10^3 for Indian cottage cheese soup (paneer soup) to 7.6×10^3 for stir-fried noodles (chowmein), there was no significant difference in levels of contamination of foods from the two kitchens. Using the APC cut point of 100, 000 CFU/g, all the foods from both kitchens were found to have unacceptable high APC levels. Both kitchens prepare and served foods in violation of food safety standards. Similarly, using a cut point of 100 CFU/g for food safety, all food samples from both kitchens were found to have an unacceptable coliforms and E. coli levels. Statistically, there was no significant difference ($P > 0.05$) in the level of bacterial contamination between the food samples from both kitchens using two-tailed t-test.

Typically, for foods like stir-fried noodles (chowmein), mixed rice dish (veg. biryani), potato-based mixed vegetable curry (pav bhaji) and lentil-based vegetable stew (sambhar), large quantity of raw vegetables (which include cabbage, carrot, capsicum, onions, tomato, cauliflower, garlic) were added at the end of the cooking process and therefore were only exposed to mild heat. Additionally, for Potato-based mixed vegetable curry (pav bhaji) and lentil-based vegetable stew (sambhar) raw coriander leaves were chopped and added directly to these ready-to-eat foods (prior to serving). For Flattened rice (poha), the ingredients which include peanuts, green peas and curry leaves were only exposed to mild heat also. Similarly, for Indian cottage cheese soup (paneer soup), the cheese is at the end of the food preparation and therefore is not exposed to intense heat and is also garnish with raw coriander leaves after cooking, prior to serving.

Discussion

Characteristics of study respondents from old and new hostel kitchens

This research provides significant information regarding the level of knowledge, attitudes, and practices in food safety of food handlers and the level of bacterial contamination in all the cooked food samples collected. The results of this investigation need serious intervention and urgent attention. None of the food handlers 100% (n=25) had attended any specific training for food handling. In several studies, food service workers that received training had better hygiene scores and safe food handling practices than those that did not receive training [11,17,41,42]. This reveals that periodic trainings for the food handlers in this study will be of great importance towards safe food handling. Sixteen (n=16) 64% of the individual respondents in this study did not have any formal education. Higher levels of education have been associated with better food safety knowledge awareness, and better sanitary conditions in other studies [18,42], and a study of the environmental hygiene of food service outlets were significantly associated with the age and educational level of operators [43]. A study in India indicated that food handling practices was related with educational status of food handlers [44]. A remarkable positive influence on food hygiene relies on education and training given to employees by the food service establishments. A study by Isara and Isah [45] described that experience and knowledge on food hygiene of food handlers were associated with good food hygiene practices. The result of the present study revealed an urgent need to conduct a food safety awareness training for food handlers. And it is necessary to evaluate the impact of the knowledge acquired in the food safety training to ensure its effectiveness.

Food safety knowledge of food handlers

In our study, majority of the food handlers were not knowledgeable about hygiene practices, cleaning and sanitation procedures (Table 2, Section B). Eighty-four per cent 84% (n=21) and 80% (n=20) of the respondents didn't know that washing hands before handling food and wearing gloves respectively, proper cleaning and sanitization of utensils (88% did not know) reduces the risk of food contamination. Lack of awareness of such important hygienic procedures by majority of our respondents is very inappropriate. Other studies have found that foods that have been properly prepared can become contaminated when handled by unwashed hands [26,46]. Proper hand washing by food handlers has been reported to significantly decrease the threat of diarrheal diseases in child care facilities [47] and can therefore be encouraged as it could similarly help to minimize the risk of diarrhea and other foodborne diseases in similar facilities. Therefore, it is very important to combine proper hand washing with the wearing of gloves and other hygienic practices in order to minimize the risk of contamination during food handling [48].

Regarding foodborne disease transmission, all the respondents (100%) did not know that hepatitis A and *salmonella* respectively are foodborne pathogens. Similarly, all the respondents (100%) did not know that diarrhoea and typhoid fever can be transmitted by food. These results support recently published work where majority of the respondents did not know if *salmonella*, hepatitis A and B viruses and staphylococcus caused foodborne diseases [10,49]. Over ninety per cent (90%) of the respondents did not know that taking leave from work in periods of infectious skin disease was necessary. Food may be contaminated with harmful bacteria, either directly by an infected food handler, or indirectly through contact with a food contact surface that has been contaminated by an infected food handler. Foods which will not be cooked before being eaten are of greater risk because cooking is a process that would kill many of the bacteria present [50]. Additionally, 96% of the food handlers did not know that microbes can be found on the skin, and in the mouth and nose of healthy looking individuals.

All the food handlers 100% (n=25) did not know that eating, drinking, smoking/chewing tobacco leaves during food handling increase the risk of food contamination However, a number of food-handlers who indicated that they did not eat, smoke or chew tobacco were observed by the researcher eating, smoking and chewing dried ground tobacco leaves while handling food, suggesting that the percentage of food handlers who eat, smoke and chew dried ground tobacco leaves while working was under-reported, and their percentage was corrected based on the researcher's observation. Smoking transfer contaminants from mouth to hands and cigarettes emit particles that contribute to food contamination [51]. On the other hand, all food handlers (100%) were not familiar with time and temperature abuse and its effect on food safety. Improper handling of food, including the abuse of time-temperature, account for most foodborne disease outbreak [52]. In this study food handlers had no knowledge of time-temperature controls. This result is supported by others Bas et al. and Webb and Morancie [11,53] whose report show that knowledge of critical temperatures was insufficient amongst food handlers. Similar

finding on the lack of adequate knowledge on temperature controls by food handlers have also been reported from different countries [54-56]. Improper practices responsible for microbial foodborne illnesses have been well documented, and typically involved cross-contamination of raw and cooked food, inadequate cooking, and storage at inappropriate temperatures [57].

Food safety attitudes of food handlers

A reduction in the incidence of foodborne illnesses is strongly influenced by the attitudes of food handlers towards food safety. Thus, there is a strong linkage between positive behaviour, attitudes and education of food handlers in maintaining safe food handling practices [20]. Table 2 (Section B) shows the attitudes of food handlers towards the prevention and control of foodborne diseases. Majority of the respondents 96% (n=24) did not know that wearing of masks, hair restraints, and clean cloth/uniform and hand gloves can minimize the risk of food contamination which is considered a negative attitude reported by majority of our respondents. Dirty clothing may carry pathogens that cause foodborne illness. These pathogens can be transferred from clothing to the hands and to the food being prepared [58]. Similarly, all of the respondents 100% (n=25) did not know that knives and cutting boards should be properly sanitised to prevent cross contamination of foods, and that individuals with abrasions or cuts on their fingers or hands should not touch unwrap foods. Majority of the food handlers were not aware that foods should not be handled with long and painted finger nails 96% (n=24). The general attitudes of food handlers towards food safety in this study is unsatisfactory.

Majority of the food handlers 92% (n=23) did not know that checking fridge temperature reduces food contamination risk. Eighty per cent (n=20) of the food handlers did not know that the health status of the food handlers should be assessed prior to employment. A study reported that 47% of food service chefs and managers had a lack of awareness that sick persons can spread foodborne illness [42]. The health status of these food handlers could have serious implications for food safety. Food handlers themselves may be sources of organisms either during the course of gastrointestinal illness or during and after convalescence, when they no longer have symptoms, and should be excluded from work until they have fully recovered from the illness [59,60].

Food safety practices of food handlers

In assessing the food safety practices of the food handlers, 100% (n=25) of the food handlers reported that they do not use gloves when distributing ready-to-eat food. Transferring microbes from human hands has been reported as a potential cross-contamination route [61]. In fact, scientific studies report that hand-contact surfaces are more likely to be contaminated than food-contact surfaces [62]. The serving utensils are kept on bare floor of the serving area, and back into the ready-to-eat food without washing during serving (based on researcher's observation). The retention of bacteria on food-contact surfaces increases the risk of cross-contamination of food with these microorganisms [63]. Also, the researcher observed that the same utensils used for preparing raw materials were used to handle cooked food. All the food handlers 100% (n=25) eat, drink, chew dried ground tobacco leaves and smoke (male) during food handing. Small droplets of saliva can contain thousands of pathogens. While eating, drinking, smoking and chewing tobacco, saliva can be transferred to hands, or directly to food being handled [58], 92% (n=23) of the food handlers come to work while having cold (based on observation). All the female food handlers 68% (n=17) were found using fingernail polish while handling the food. According to WHO [58] nail polish can flake off into food and hides dirt. Some food handlers were observed handling food while having cuts on their fingers, and 100% (n=25) of them did not know that food handlers with cuts on their fingers should not touch ready-to-eat food. Bacteria that causes foodborne illnesses can often infect open cuts, and can be transferred to the food.

Majority of the individual respondents do not frequently wash their hands before and after food preparation. Food handlers who do not practice proper personal hygiene, including hand washing at appropriate times and using appropriate hand-washing methods, can contaminate foods with organisms from the gastrointestinal tract [64]. Insufficient and inadequate hand washing by employees in retail food service establishments is well known contributing factor to foodborne illnesses [65]. Research on the prevalence of hand washing and glove use in food service establishments indicates that these hand hygiene practices do not occur as often as they should, because food workers have reported that they sometimes or often do not wash their hands and/or wear gloves when they should [66]. According to Allwood et al. [67] it is generally accepted that the hands of food handlers are an important vehicle of food cross-contamination. All the respondents revealed that they were not aware of the dangers of cross contamination during food preparation 100% (n=25). In this study researcher has observed that one single table (in each kitchen) was used for chopping fresh vegetables, preparation of *chapati*, *puree* and other foods, and after food preparation, these tables are daily swept and mopped with the same broom and mop used for cleaning the dining hall and kitchen floors and the next food preparation continues on the same table (this particular type of cross contamination is a daily routine by the food handlers in this study). Pathogens can be transferred from one surface or food to another. The hands of food handlers can serve as vectors in the spread of foodborne diseases due to poor personal hygiene or cross contamination [11,68]. Cross-contamination is the main reason for many foodborne illness outbreaks. Cross-contamination among food and food contact surfaces can lead to serious health risks like food poisoning or unintended exposure to food allergens. Chopping boards can be an easy place for cross-contamination to occur. Placing ready-to-eat foods such as fresh produce on a surface that held raw meat, poultry, seafood or eggs can spread harmful bacteria. Kitchen utensils and cutting boards also are key cross contamination routes [69]. In fact, research in the UK suggests that 14% of all foodborne illnesses may be due to inadequately cleaned cutting boards and knives [70]. Majority of the individual respondents in this study performed very poorly in important food safety and hygiene practices because they were not aware of the importance of safe food handling and their personal responsibility towards food safety. This had been proved by the study of Ababio and Lovatt [41]. The role of food workers in foodborne outbreaks has been clearly demonstrated by Todd et al. [71] who pointed out that 25% of reported outbreaks are caused by inadequate handling and food preparation practices.

Microbiological safety of cooked food samples

In this study, all the food samples tested had APC, total coliform and *E. coli* CFU/g counts higher than the acceptable levels, which indicates a great need for improvement in safe food handling practices at these kitchens. RTE foods do not need to be reheated before consumption. A high APC, coliforms, or *E. coli* counts suggests contamination resulted from inappropriate processing, incomplete heating, or secondary contamination via contact with contaminated equipment such as chopping boards, knives, and serving wares, etc. Additionally, the presence of *E. coli* in RTE food products indicates the possibility of

secondary contamination, coliform on RTE food products reflected the recontamination caused by secondary processing and poor personal hygiene. It is practical to employ Good Hygiene Practices to minimize, if not eliminate, the risk posed by secondary contamination [72]. The high level of contamination of these foods analysed in this study could be associated to the fact that the vegetables and other ingredients added to these foods were at the end of food preparation and therefore were only exposed to mild heat and the ones added to garnish the food were not heat-treated. Also, the serving utensils were carelessly kept on the bare floor of the serving area at the time of serving, and back into the ready-to-eat food when the need arises without washing. Similarly, several risk factors related to the food service environment contribute to occurrence of foodborne illness: poor personal hygiene, inadequate sanitization of surfaces or equipment, cross contamination of prepared food with contaminated ingredients and inadequate temperature control [73,74]. Food handling personnel play important role in ensuring food safety throughout the chain of food production, processing, storage and preparation. Mishandling and disregard to hygienic measures on the part of the food handlers have been reported to introduce contaminant and pathogens that survive and multiply in sufficient numbers to cause illness in the consumer [19,50,73,75]. With regard to AMB, counts above 10^5 have been considered a potential risk for the presence of pathogens [76]. The presence of microorganisms like *E. coli* demonstrates a potential health risk as these organisms are pathogenic and have been implicated in foodborne diseases [77,78]. However, their presence is an indication of possible faecal contamination of food, water or food workers and poor hygienic processing practices [72,79].

The foods being cooked and served by these two kitchens are of unacceptable microbiological quality. The International Commission for Microbiological Specification for Foods [80] states that ready-to-eat foods with plate counts between $0-10^3$ is acceptable, between 10^4 to $\leq 10^5$ is tolerable and 10^6 and above is unacceptable. *E. coli* and coliform <20 is satisfactory, between 20 to $\leq 10^2$ is borderline and > 10^2 is unsatisfactory. The findings that there are significantly higher APC, coliforms and *E. coli* levels from both kitchens were expected, taken into consideration that both kitchens prepare and serve food in violation with the food safety standards. Generally considering the very poor level of good food handling practices particularly personal hygiene, there are likely other factors beyond this study scope, which resulted in the high bacterial levels in the cooked food samples from both kitchens.

Limitations

The research limited itself to the assessment of bacteriological quality of cooked food samples and could not include pathogens like viruses, parasites and other bacteria due to lack of funds.

Conclusion

Based on the results of this study, it is concluded that meals production analysed from these kitchens does not comply with the requirements of good hygienic practices. This study has generally revealed that there is a great need for improving food safety in both women's old and new hostel kitchens at SHUATS Allahabad. The levels of APC, coliforms and *E. coli* counts in all the foods from both kitchens were higher than the food safety standard permissible limits. The findings that majority of the individual respondents did not follow good food handling practices (e.g., smoking, chewing tobacco leaves during food handling, serving food without the use of hand gloves, hair restraints and masks) reveals lack of frequent supervision by the concerned food safety authorities.

From the results obtained in the present study, providing periodic training on personal hygiene and good food handling practices and frequent supervision by the relevant authorities will play a pivotal role in enhancing the safety of foods being prepared and served in these kitchens, as facilities that are frequently inspected had better sanitary condition in comparison to uninspected ones. As better educated food handlers are more likely to practice good hygiene, it is advisable that they should obtain a minimum qualifications of at least secondary school level to be eligible to work as food handlers in university kitchens. Safe food handling practices can be supported by enforcing wearing of clean cloths (or uniforms), hair and mouth coverings and hand gloves while handling food. Foodborne illness can be prevented by good hygiene practices such as the use of Good Manufacturing Practices (GMP) and Hazard Analysis Critical Control Point (HACCP) application in the chain of food production and processing. Education of the food handlers on food safety practices and a close and stringent supervision of ready-to-eat foods prepared and served in these school kitchens should be carried out by relevant authorities to prevent foodborne illness. The results obtained can also create awareness to the management of this university to adopt better control strategies to prevent the foodborne illness outbreaks among students in the school environment in order to ensure and promote food safety. There is a need for further research to investigate the quality of raw materials, raw material storage conditions, cooking temperatures for various foods (as these kitchens did not have any instruments to control or register the temperature during food preparation), cooling time after cooking, display holding temperature and time to discard the food to evidence the actual reasons for these very high bacterial counts in the cooked food samples in order to come up with an overall better idea as to what frequently resulted in foodborne illness outbreaks among the students of this particular university.

Acknowledgement

A thank you to the department of food process engineering, Sam Higginbottom University of Agriculture Technology and Sciences (SHUATS).

References

1. Adams M, Motarjemi Y (1999) Basic food safety for health workers. WHO, Geneva, Switzerland.

2. Annor GA, Baiden EA (2011) Evaluation of food hygiene knowledge attitudes and practices of food handlers in food businesses in Accra, Ghana. Food Nutr Sci 2: 830.

3. Omaye ST (2004) Food and nutritional toxicology. CRC press, Boca Raton pp: 163-173.

4. WHO (2007) Food safety and foodborne illness. World Health Organization, Geneva, Switzerland.

5. Barrie D (1996) The provision of food and catering services in hospital. J Hosp Infect 33: 13-33.

6. Jay LS, Comar D, Govenlock LD (1999) A video study of Australian domestic food-handling practices. J Food Prot 62: 1285-1296.

7. Fielding JE, Aguirre A, Palaiologos E (2001) Effectiveness of altered incentives in a food safety inspection program. Prev Med 32: 239-244.

8. Gent R, Telford D, Syed Q (1999) An outbreak of campylobacter food poisoning at a university campus. Comm Dis Public Health 2: 39-42.

9. WHO (1999) Strategies for implementing HACCP in small and/or less developed businesses: The hague. World Health Organisation, Geneva, Switzerland.

10. Ansari-Lari M, Soodbakhsh, S, Lakzadeh, L (2010) Knowledge, attitudes and practices of workers on food hygiene practices in meat processing plants in Fars, Iran. J Food Control 21: 260-263.

11. Baş M, Ersun AS, Kivanc G (2006) The evaluation of food hygiene, knowledge, attitudes and practices of food handlers in food businesses in Turkey. J Food Control 17: 317-322.

12. Capunzo M, Cavallo P, Boccia G, Brunetti L, Buonomo R, et al. (2005) Food hygiene on merchant ships: The importance of food handlers' training. J Food Control 16: 183-188.

13. Jevšnik M, Hlebec V, Raspor P (2008) Food safety knowledge and practices among food handlers in Slovenia. J Food Control 19: 1107-1118.

14. Martins RB, Hogg T, Otero JG (2012) Food handlers' knowledge on food hygiene: The case of a catering company in Portugal. J Food Control 23: 184-190.

15. Seaman P, Eves A (2010) Perceptions of hygiene training amongst food handlers, managers and training providers-A qualitative study. J Food Control 21: 1037-1041.

16. Sharif L, Al-Malki T (2010) Knowledge, attitude and practice of Taif University students on food poisoning. J Food Control 21: 55-60.

17. Kibret M, Abera B (2012) The sanitary conditions of food service establishments and food safety knowledge and practices of food handlers in Bahir Dar town, Ethiopia. J Health Sci 22: 27-35.

18. Zeru K, Kumie A (2007) Sanitary conditions of food establishments in Mekelle town, Tigray, North Ethopia. Ethiopia. J Hlth Dev 21: 3-11.

19. World Health Organization (1989) Programme for control of diarrhoeal diseases. Manual for Laboratory Investigations of Acute Enteric Infections, CDD, Geneva, Switzerland.

20. Howes M, McEwen S, Griffths M, Harris L (1996) Food handler cortication by home study: Measuring changes in knowledge and behavior. Dairy, Food Environ Sanitation 16: 737-744.

21. Da Cunha DT, Stedefeldt E, De Rosso VV (2012) Perceived risk of foodborne disease by school food handlers and principals: the influence of frequent training. J Food Saf 32: 219-225.

22. Hassan AN, Farooqui A, Khan A, Yahya KA, Kazmi SU (2010) Microbial contamination of raw meat and its environment in retail shops in Karachi, Pakistan. J Infect Dev Ctries 4: 382-388.

23. Havelaar AH, Brul S, De Jonge A, De Jonge R, Zwietering MH, et al. (2010) Future challenges to microbial food safety. Int J Food Microbiol 139: 79-94.

24. Chaudhry R, Dhawan B, Kumar D, Bhatia R, Gandhi J, et al. (1998) Outbreak of suspected *Clostridium butyricum* botulism in India. Emerg Infectious Disease 4: 506-507.

25. Jacxsens L, Uyttendaele M, Devlieghere F, Rovira J, Gomez SO, et al. (2010) Food safety performance indicators to benchmark food safety output of food safety management systems. Int J Food Microbiol 141: 180-178.

26. Nkere CK, Ibe NI, Iroegbu CU (2011) Bacteriological quality of foods and water sold by vendors and in restaurants in Nsukka, Enugu State, Nigeria: A comparative study of three microbiological methods. J Hlth Popul Nutr 29: 560-566.

27. Fletcher SM, Stark D, Ellis J (2001) Prevalence of gastrointestinal pathogens in sub-saharan Africa: Systematic review and meta-analysis. J Publ Health Afri 2: 127-137.

28. Food and Drug Administration (2012) Bad bug book: Foodborne pathogenic micro-organisms and natural toxins handbook. Center for Food Safety and Applied Nutrition.

29. Nguz K (2007) Assessing food safety system in Sub-Saharan African countries: An overview of key issues. Food Control 18: 31-134.

30. Çakiroğlu FP, Uçar A (2008) Employees' perception of hygiene in the catering industry in Ankara, Turkey. Food Control 19: 9-15.

31. Food and Drug Administration (2004) Report of the FDA retail food program database of foodborne illness risks factors.

32. Guzewich J, Ross M (1999) Evaluation of risks related to microbiological contamination of ready-to-eat food by food preparation workers and effectiveness of interventions to minimize those risks.

33. Angelillo IF, Viggiana NMA, Greco RM, Rito D (2001) HACCP and food hygiene in hospital: Knowledge, attitudes, and practices of food services staff in Calabri, Italy. Infect Control Hosp Epidemiol 22: 1-7.

34. Bolton DJ, Meally A, Blair IS, Mcdowell DA, Cowan C (2008) Food safety knowledge of head chefs and catering managers in Ireland. Food Control 19: 291-300.

35. Harrigan WF (1998) Laboratory methods in food microbiology. Academic Press, London, UK.

36. Downes FP, Ito K (2001) Compendium of methods for microbiological examination of foods. American Public Health Association, Washington DC, USA.

37. Refai MK (1979) Manuals of food quality control: Microbiological analysis. Food and Agricultural Organization of the United Nations, Rome, Italy.

38. Aneja KR (2003) Experiments in microbiology plant pathology and biotechnology. New Age International Publishers, India.

39. Merck Microbiology (1996) Microbiology Manual. Merck KGaA, Darmstadt, Germany.

40. Gilbert RJ, De Louvois J, Donovan T, Little C, Nye K, et al. (2000) Guidelines for the microbiological quality of some ready-to-eat foods sampled at the point of sale. Commun Dis Publ Hlth 3: 163-167.

41. Ababio FW, Lovatt P (2014) A review on food safety and food hygiene studies in Ghana. Food Control 47: 92-97.

42. Onyeneho SN, Hedberg CW (2013) An assessment of food safety needs of restaurants in Owerri, Imo State, Nigeria. Int J Environ Res Publ Hlth 10: 3296-3309.

43. Olumakaiye MF, Bakare KO (2013) Training of food providers for improved environmental conditions of food service outlets in urban area Nigeria. Food Nutr Sci 4: 99-105.

44. Mudey D, Goyal R, Dawale A, Wagh V (2010) Health status and personal hygiene among food handlers working at food establishment around a rural teaching hospital in Wardha district of Maharashtra, India. Global J Health Sci 2: 198-206.

45. Isara AR, Isah EC (2009) Knowledge and practice of food hygiene and safety among food handlers in fast food restaurants in Benin City, Edo State. Niger Postgrad Med J 16: 207-212.

46. Taulo S, Wetlesen A, Abrahamsen R, Kululanga G, Mkakosya R, et al. (2008) Microbiological hazard identification and exposure assessment of food prepared and served in rural households of Lungwena, Malawi. Int J Food Microbiol 125: 111-116.

47. Xavier CAC, Oporto CFO, Silva MP, Silveira IA, Abrantes MR (2007) Prevalence of Staphylococcus aureus in food handlers from grades schools located in Natal city, RN, Brazil. The Brazilian Magazine of Clinical Analyses 39: 165-168.

48. Monville R, Chen Y, Schaffner D (2001) Gloves barriers to bacterial cross-contamination between hands to food. J Food Prot 64: 845-849.

49. Soares LS, Almeida RC, Cerqueira ES, Carvalho JS, Nunes IL (2012) Knowledge, attitudes and practices in food safety and the presence of coagulase-positive staphylococci on hands of food handlers in the schools of Camaçari, Brazil. Food Control 27: 206-213.

50. Todd ECD, Greig JD, Bartleson CA Micheals BS (2007) Outbreaks where food workers have been implicated in the spread of foodborne disease. Part 2- Description of outbreaks by size, severity, and settings. J Food Prot 70: 1975-1993.

51. NSC (2008) Environmental tobacco smoke. National safety council, USA.

52. Anon (2003) Acute gastroenteritis in Ireland, north and south: A telephone survey. Health Protection Surveillance Centre, Dublin I, Ireland.

53. Webb M, Morancie A (2015) Food safety knowledge of foodservice workers at a university campus by education level, experience, and food safety training. Food Control 50: 259-264.

54. Buccheri C, Casuccio A, Giammanco S, Giammanco M, La Guardia M, et al. (2007) Food safety in hospital: Knowledge, attitudes and practices of nursing staff of two hospitals in Sicily, Italy. BMC Health Serv Res 7: 1.

55. Marais M, Conradie N, Labadarios D (2008) Small and micro enterprises-aspects of knowledge, attitudes and practices of managers'and food handlers' knowledge of food safety in the proximity of Tygerberg Academic Hospital, Western Cape. South Afr J Clin Nutr 20: 50-61.

56. Walker E, Pritchard C, Forsythe S (2003) Food handlers' hygiene knowledge in small food businesses. Food Control 14: 339-343.

57. Egan MB, Raats MM, Grubb SM, Eves A, Lumbers ML, et al. (2007) A review of food safety and food hygiene training studies in the commercial sector. Food Control 18: 1180-1190.

58. WHO (2006) Five keys to safer food manual. World Health Organization, Geneva, Switzerland.

59. De Souusa CP (2008) The Impact of food manufacturing practices on foodborne diseases. Braz Arch Biol Technol 51: 815-825.

60. Chapman BJ (2009) Development and evaluation of a tool to enhance positive food safety practices amongst food handlers: Food safety infosheets. University of Guelph, Guelph, Ontario, Canada.

61. Pérez-Rodríguez F, Valero A, Carrasco E, García RM, Zurera G (2008) Understanding and modelling bacterial transfer to foods: A review. Trends Food Sci Technol 19: 131-144.

62. DeVita MD, Wadhera RK, Theis ML, Ingham SC (2007) Assessing the potential of Streptococcus pyogenes and Staphylococcus aureus transfer to foods and customers via a survey of hands, hand-contact surfaces and food-contact surfaces at foodservice facilities. J Foodservice 18: 76-79.

63. Sneed J, Strohbehn C, Gilmore SA, Mendonca A (2004) Microbiological evaluation of foodservice contact surfaces in Iowa assisted living facilities. J Am Diet Assoc 104: 1722-1724.

64. Strohbehn C, Sneed J, Paez P, Meyer J (2008) Hand washing frequencies and procedures used in retail food services. J Food Prot 71: 1641-1650.

65. FDA (2011) Trend analysis report on the occurrence of foodborne illness risk factors in selected institutional foodservice, restaurant, and retail food store facility types (1998-2008). Food and drug administration, USA.

66. Clayton DA, Griffith CJ, Price P, Peters AC (2002) Food handlers' beliefs and self-reported practices. Int J Environ Health Res 12: 25-39.

67. Allwood PB, Jenkins T, Paulus C, Johnson L, Hedberg CW (2004) Hand washing compliance among retail food establishment workers in Minnesota. J Food Prot 67: 2825-2828.

68. Frobisher FR, FuerstÕs (1983) Microbiology in health and disease: Foods as vectors of microbial disease. Sanitation in food handling, WB Saunders Company, Philadelphia, USA. pp: 418-433.

69. De-Jong A, Verhoeff-Bakkenes L, Nauta M, De Jong R (2008) Cross-contamination in the kitchen: Effect of hygiene measures. J Appl Microbiol 105: 615-624.

70. Kennedy J, Jackson V, Blair I, McDowell D, Cowan C, et al. (2005) Food safety knowledge of consumers and the microbiological and temperature status of their refrigerators. J Food Prot 68: 1421-1430.

71. Todd ECD, Michaels BS, Greig JD, Smith D, Bartleson CA (2010) Outbreaks where food workers have been implicated in the spread of foodborne disease: Gloves as barriers to prevent contamination of food by workers. J Food Prot 73: 1762-1773.

72. Little CL, Monsey HA, Nichols GL, De Louvois J (1998) The microbiological quality of ready-to-eat dried and fermented meat and meat products. Int J Environ Health Res 8: 277-284.

73. Todd ECD, Greig JD, Bartleson CA, Micheals BS (2007) Outbreaks where food workers have been implicated in the spread of foodborne disease: Factors contributing to the outbreaks and description of outbreak categories. J Food Prot 70: 2199-2217.

74. Lockis VR, Cruz AG, Walter EHM, Faria JAF, Granato D, et al. (2010) Prerequisite programs at schools: Diagnosis and economic evaluation. Foodborne Pathog Dis 8: 213-220.

75. Greig, JD, Todd ECD, Bartleson CA, Micheals BS (2007) Outbreaks where food workers have been implicated in the spread of foodborne disease: Description of the problem, methods and agents involved. J Food Prot 70: 1752-1761.

76. Ayçiçek H, Sarimehmeto Lu B, Çakiro Lu S (2004) Assessment of the microbiological quality of meals sampled at the meal serving units of a military hospital in Ankara, Turkey. Food Control 15: 379-384.

77. Eni AO, Oluwawemitan IA, Solomon OU (2010) Microbial quality of fruits and vegetables sold in Sango Ota, Nigeria. Africa J Food Sci 4: 291-296.

78. Oranusi SU, Oguoma OI Agusi E (2013) Microbiological quality assessment of foods sold in student's cafeterias. Global Res J Microbiol 3: 1-7.

79. Tambekar DH, Shirsat SD, Suradkar SB, Rajankar PN, Banginwar YS (2007) Prevention of transmission of infectious disease: Studies on hand hygiene in health-care among students. Continent J Biomed Sci 1: 6-10.

80. ICMSF (1996) Microorganisms in foods: Microbiological specifications of pathogens. International Commission on Microbiological Specifications for Foods, USA.

Indigenous Processing Methods of *Cheka*: A Traditional Fermented Beverage in Southwestern Ethiopia

Belay Binitu Worku[1], Ashagrie Zewdu Woldegiorgis[2] and Habtamu Fekadu Gemeda[2,3*]

[1]Department of Food Process Engineering and Postharvest Technology, Ambo University, Ambo, Ethiopia

[2]Centre for Food Science and Nutrition, Addis Ababa University, Addis Ababa, Ethiopia

[3]Department of Food Technology and Process Engineering, Wollega University, P.O.Box: 395, Nekemte, Ethiopia

Abstract

Cheka is a cereal and vegetable-based beverage which is commonly consumed in Southwestern parts of Ethiopia particularly in Dirashe and Konso. In this study, the traditional processing methods, types and proportions of ingredients, equipments, and sources of energy, economic and socio-cultural importance of *cheka* were described. In the study areas, maize, sorghum and vegetables such as cabbage, moringa, decne and taro were reported to be utilized for *cheka* preparation. Informants described the characteristics of quality *cheka* as thick, smooth, effervescent, foamy, and bitter in taste. The processing methods as well as the raw materials utilized and their proportions seem to vary among households, villages and localities. Since the present study was the first of its kind, flow chart which shows the processing operations involved in *cheka* fermentation was constructed that might be used by those who want to scale-up the *cheka* processing in the future. Based on the finding of this survey, it is recommended to carry out further research on the nutritional and alcoholic contents of *cheka* and on optimizing the processes.

Keywords: Cheka; Konso; Dirashe; Indigenous processing method; Fermented beverages

Introduction

Fermented alcoholic beverages have been widely consumed by people in almost all countries for millennia [1]. These fermented beverages are usually prepared from locally available materials using age-old techniques [2], and their art is believed to pass down by cultural and traditional values to subsequent generations with the processing being optimized through trial and error [3]. Owing to the heterogeneity of culture in Ethiopia, diverse indigenous fermented beverages exist in the country with *tella*, *tej* and *arake* being majorly consumed in the northern parts, as reported by Fite et al. [4], *borde*, *shamita* as reported by Alemu *et al.* [5], Ghebrekidan, [6] and *cheka* Abegaz *et al.* being utilized in the southern and central parts.

Traditional fermentation serves many purposes. It can alter the texture of foods, enhance the digestibility of a food, preserve foods by production of acids or alcohol, or produce subtle flavours and aromas which increase the quality and value of raw materials [1,7]. Fermentation which is often considered as a low-input enterprise provides individuals with limited purchasing power, access to safe, inexpensive and nutritious foods [3].

In reality, the fermentation of traditional beverages takes place under uncontrolled conditions and often involve laborious and time consuming activities [8,9]. Rural women produce such beverages with no standardized formulations and also usually in the absence of back-slopping. As a result, the beverage becomes of poor quality with inconsistency and failure in most cases [3]. These necessitate the understanding of the processes and raw materials utilized for preparing the beverages.

Cheka is a cereal and vegetable-based fermented beverages which is consumed in Southwestern parts of Ethiopia mainly in Dirashe and Konso. People of all ages including infants, pregnant and lactating women drink *cheka*. From observation an adult man on average drinks up to 8 litres of cheka per day. The indigenous processing methods and raw materials for the preparation of most Ethiopian fermented beverages had been well documented by many investigators. Several

works have been done on traditional fermented alcoholic beverage. The ethanol, methanol and fusel oil contents of Ethiopian alcoholic beverages such as *tella*, *tej* and *arake* were determined by different researchers [14,10-12]. Some native researchers also tried to modify and monitor the fermentation and processing parameters of *tella* (Berza and Wolde [13] and *borde* Abegaz *et al.* [14] Abegaz *et al.* [15] with the aim of improving its sensory properties including shelf-life. *Cheka* had been mentioned in Abegaz *et al.* [14] along with other Ethiopian traditional beverages, but no one has documented about it yet. Therefore, the present study was intended to document the indigenous processing methods and raw materials of *cheka*.

Materials and Methods

Description of the study areas and survey data collection

A survey of traditional processing methods and raw materials used for the production of *cheka* was conducted using in-depth interviews and focus group discussions with 90 cheka producers at two districts in southwestern Ethiopia, namely Konso and Dirashe. Dirashe and Konso are one of the five districts in Segen and its Surrounding Peoples Zone and are located in the South-western part of the Southern Nations, Nationalities and People's Region at a distance of about 550 and 590 kilometers from Addis Ababa, respectively. The interview was administered in Amharic language in the villages of each locality. Three kebeles which are known for consumption and vending of cheka were selected from each locality. A total of 60 brewers (10 women from

---

Corresponding author: Habtamu Fekadu Gemeda, Centre for Food Science and Nutrition, Addis Ababa University, Addis Ababa, Ethiopia
E-mail: fekadu_habtamu@yahoo.com

each kebele) were selected randomly for interview after preliminary screening. In order to obtain an insight into the processing operations, ingredient proportions, consumption patterns and undisclosed things, focus group discussions were carried out in three kebeles with 30 women who were selected based on availability. Data were collected on the preparation techniques, types and proportions of ingredients, sources of energy, types of equipment, sensory properties, shelf-life and economic importance of *cheka* as well as constraints in its production, marketing and consumption [8]. Cooking temperatures were recorded during on the spot interviews at Karat, Gato and Gidole where people use warm water for diluting *cheka*.

Result and Discussion

Raw materials utilized for the preparation of *cheka*

In the study areas, *cheka* is mainly prepared from cereals such as sorghum (*Sorghum bicolor*) and maize (*Zea mays*) and vegetables such as leaf cabbage (*Brassica spp.*), moringa, (*Moringa stenoptella*), and decne (*Leptadenia hastata*). In addition, brewers in Dirashe use the root part of taro, whereas producers in Konso rarely use the leaf part of taro. In some localities, few households also use dried edible leftovers of *injera*, *kitta* or *kurkufa*. The informants of this study also disclosed that few brewers use hop to make *cheka* taste bitter as consumers judge this sort of *cheka* as of good quality.

Actually, most *cheka* producers (48.9%) use a mixture of cereals for *cheka* preparation since they have more than one farm and therefore, can produce different cereal crops during one production season. They also believe that the quality and sensory properties of cheka is determined by the type and combination of cereals utilized. For instance, brewers in most localities reported that if maize is exclusively utilized, the prepared *cheka* become sour before expected time and on the other hand, if only red sorghum is used, the *cheka* takes much time to become mature. However, maize is frequently utilized for producing Konso *cheka* which needs short fermentation period (usually 4 days). The ingredients utilized and their possible combinations were found to vary within and between households regardless of localities and processing methods employed. The proportion of malt to unmalted ingredients varies with the processing method, climatic conditions and strength of the product of second phase of *cheka* fermentation. The proportion of the malt used during the whole phases of *cheka* fermentation varies between 20 and 25% of the total unmalted ingredients. The raw materials utilized for *cheka* preparation are selected based on availability, price, purpose of production (e.g. for home consumption, social events, etc.), processing activities involved and preferences of the brewers. Seasonal variations in the price of various cereals also affect the choice of ingredients in both localities.

Preparation of malt

Malt used for *cheka* preparation can be prepared from a single or mixture of cereals. The cereals utilized for this purpose include maize, sorghum, barley (*Hordeum vulgare*) and finger millet (*Eleusine coracana*). The latter two cereals always are not utilized alone and barley could be utilized as malted or unmalted ingredient. About 38% of the respondents reported maize was the most appropriate raw materials for malt. According to the brewers barley is utilized to make the *cheka* more alcoholic and it is often used in small quantities (10-15% of the malted ingredient). Cereals stored in silos if not damaged by pests such as weevils are appropriate for malt preparation. In some instances, grains stored underground for few weeks to a year might be used for malt.

Brewers always begin malt preparation by cleaning the grains through winnowing and the floatation method during soaking. The grains are steeped in water overnight for sorghum, barley and finger millet but it requires 24 hours for maize. In case, maize is mixed with other cereals for malt preparation, the soaking time is reduced from the usual 24 hours. This is because other cereals absorb excess water and become spoiled instead of sprouting. The steep water is poured out and/or drained off when storing in a sack. Historically, the swollen grains were allowed to germinate in a basket while covered with leafs of castor oil or ensete. The duration of germination varies with the type of cereal used and the interest of the brewer. In most areas, the germination takes 2-3 days for all cereals. However, some producers in Konso allow the grains to germinate for more than 4 days at ambient temperatures so that mold could develop which is desired for foam formation. In this case, the sack is tightly wrapped in order not to allow excess air to enter into the sack. After four days, the sack is opened for about 2 minute and again wrapped for extra 3-4 days. The germinated grains are then spread out on an animal hide, a sheet of plastic material or mat made of leaves of *Phoenix reclinata* and let to dry in the sun for 2-5 days depending on weather conditions and stored in a dry place until required.

Depending on the volume of cheka to be produced, the entire malt or its portion can be milled for immediate use. Some brewers mix the malt with unmalted barley when they need to mill it. None of the respondents utilize wet malt that was reported to be used in borde preparation [8].

Sources of fuel and equipment utilized for *cheka* preparation

Cheka producers depend on firewood and dry crop residues such as maize or sorghum stalks, straw and corncob. Locally available rudimentary equipment is used by producers for traditional preparation of *cheka*. Large clay pots (*gan or insira*), plastic containers or metal barrels (whose capacity varies from 50 litres to 200 litres), plastic buckets, plates and bowls made from woods (*Gebete*) and car tires are utilized for cheka fermentation. Baskets of varying shapes and sizes made from bamboo and sieves made from leaf of *Phoenix reclinata* (*Yezembaba kitel*) and circular flat metal mesh are used. Traditional pestle and mortar made from wood and circular flat trays made by interweaving bamboo splints are utilized when cleaning grains. Modern flour mills are available for milling purpose but grinding stones are still used for milling fermented vegetables and malt in some villages. Metal and clay pots of different size are used for cooking fermented products and boiling water during *cheka* preparation and consumption. Large gourd bottles with long necks (10-15 litres), plastic jars and clay pots (10-25 litres) are used for transporting *cheka* to farm for workers as well as in case of social events such as wedding and funeral ceremonies. Depending on age a single person may use small screw-cap plastic bottles and gourd bottles or jars (2-5 litres) when going to farm or looking after livestock. At villages and market places *cheka* is served in plastic or metal containers (cans) and gourd bottles (not long necked) whose capacity is approximately 1 litre (Table 1).

Description of the methods and steps in *cheka* preparation

The processes of *cheka* preparation are very complex and vary among households, villages and localities. The duration of *cheka* fermentation varies from 12 hours (½ day) for *menna* to months for *parshota*. Processes which involve short fermentation time are followed by those brewers producing *cheka* for sale. The variation among respondents opinion concerning proportions of raw materials utilized for *cheka* preparation makes the estimation of the amount of

Operations	Equipment
Drying of grains and malt	Plastic sheets, animal hide, mats, mosquito net, blanket
Malt preparation	Metal pots, bucket, bowl, sack, baskets
Cleaning of grains	Traditional flat trays (*sefed*), mortar and pestle
Milling	Flour mill, grinding stones
Filtering and sieving	Traditional sieve (*wonfit*)
Fermentation of leafy vegetables	Small traditional bowl, buckets, plastic plates, broken clay pots or jars
Cooking and boiling	Metal pots, barrel, insira
Crashing of dough balls	Beer bottle, cylindrical stone, pestle-like wood (*tomambyta* or *korya kabotat*)
Main fermentation and storage of cheka	Large bowl, plastic container, barrel
Serving utensils	Small metal or plastic containers, gourd bottles

Table 1: *Cheka* processing operations and equipment utilized for the purpose.

the ingredients and water used problematic. The type and proportion of ingredients depend on the volume of *cheka* to be produced, the availability of the ingredient and type of the *cheka* being produced. Three types of cheka are produced in the study districts such as *hiba* (*parshota*), *chaqa* (*fasha*) and *menna* (*poh-kedha or madhot*). Most cheka preparation methods involve three major phases that are marked by cooking. *Menna* is prepared in a similar way to konso cheka. The only difference is that in the case of *menna*, the initial fermentation lasts within 12-14 hours and leafy vegetables are not used at all. *Menna* fermentation may also involve single phase of fermentation. In this case, only malted porridge is allowed to ferment overnight which results in cheka that tastes sweet.

Phase I: In phase I, grain flour is thoroughly kneaded with water in *gebete* and allowed to ferment for 14 hours (for *menna* preparation in both localities) to over a month (in low-land rural areas of Dirashe). For home consumption and occasionally for sale, brewers in Konso use the leaves of taro to produce *cheka*. In this case, taro leaves are chopped and cooked in a metal or clay pot. The overcooked taro leaves are allowed to ferment for about 6 days in a gebete. The fourth day, the fermented product is mixed with a handful of malt and left to ferment for extra 2 days. Brewers believe that the added malt facilitate the decomposition of the leaves. After that the fermented taro is mixed with fresh flour as usual and is kneaded with water which also ferments for 36-40 hours. This fermenting material is commonly referred to as *pulota*.

In Dirashe, leaf cabbage is chopped into pieces with traditional double-bladed knife prepared only for this purpose. The chopped cabbage is put in a bowl or bath and little quantity of water is sprayed on it. Then, it is tightly covered with leaves of ensete or plastic sheet. Some producers spread small quantity of flour on the surface of cabbage. These prevent the entry of air that otherwise causes the fermenting cabbage develop bad odor which could be sensed by consumers during consumption. The cabbage is allowed to ferment for 4 to 6 days and is then blended with small quantity of flour. In the past, people used to use grain grits for the same purpose. After fermenting for additional 2-3 days, the fermented cabbage is milled with a grinding stone. The milled product is blended with excess water in a bath and is sieved through *wonfit* (traditional sieve). The filtrate is mixed with fresh flour, exhaustively kneaded and is allowed to ferment overnight. Some brewers may blend cooked and smashed taro roots kneaded with little flour and the fermented product. However, several brewers in the low land areas of Dirashe allow the chopped cabbage, leaves of moringa or decne to dry and then mill them with some grain (15-20 kilograms) and dried food leftovers, if any. Then, the flour is kneaded with water and allowed to ferment for at least 1 month while being uncovered and the fermenting product is kneaded with little amount of water with an interval of 2-3 days. Respondents reported that if it is neglected even

for about five days, larvae appears on the product because after four days insects including flies start to settle on it and lay their eggs in the cracks formed as it is dehydrating. In addition, the product may develop undesirable odor. When leafy vegetables are unavailable, only flour can be used and the fermentation time becomes relatively short. The fermented product is then blended with fresh flour one day earlier before the day it is desired to cook.

Phase II: The fermented product (*pulota*) is kneaded with little or no water and then made into dough balls called *qabot* (*gafuma*). The dough balls shouldn't be less or much moistened. If the balls are less moistened, they become uncooked at the centre and if too moistened they are too tiresome for kneading. During cooking, pieces of dried hop wood or peeled barks of some plants are placed at the bottom of the pot or barrel and excess water is added to prevent the dough balls from burning. If a lot of balls are prepared, most brewers add the dough balls thrice at an interval of 10-15 minutes. The balls are added when the water is boiled (93-95.5°C) and the barrel or pot is covered with a lid or a gourd that fits the pot. The dough balls are cooked for about 45 minutes to 1½ hours depending on the amount of balls and intensity of the fire. Cooking of the dough balls in water would be expected to gelatinize cereal starch granules and thereby increase the efficiency of starch degradation by amylase. The process of gelatinization occurs over a temperature range depending on the type and size of granules and starch to water ratio. Leaching of amylose occurs during gelatinization and thus create available carbohydrate for the proliferation of fermentation microorganisms [16]. Brewers often insert stick into the balls to check whether they are cooked well or not. When the dough balls are cooked well producers take one ball at a time and dip their hands quickly into water in a container handled by the other hand to avoid damage to them. Then, the *qabot* is smashed in *gebete* using a beer bottle or a round-headed (pestle-like) material made from wood called *tomambayt*. Once the dough balls are broken down into pieces, they are kneaded with little water and spread on a plastic sheet, large sized gebete or a bed made from wood to cool for few minutes to 7 hours. However, the time of cooling not only depends on the amount of the product, but also the thickness of the product spread on the plastic sheet or gebete. After cooling, it is mixed with adequate milled malt, thoroughly kneaded and allowed to ferment overnight in a *gebete*. However, most brewers in Dirashe allow this product to ferment for 36-40 hours to enhance the bitterness of the product. Most brewers spread a handful of malt on the surface of the kneaded product. The proportion of malt added during this phase can be as high as 25% of the unmalted ingredient. Next day early in the morning, the product is transferred into large fermentation vessel (barrel or *rotto*); water is added and is then well mixed together. This actively fermenting material is commonly referred to as *sokatet* (*difdif*). *Sokatet* can be stored for more than a week and so brewers may utilize a portion of it for preparing *cheka* for home consumption. Some consumers would like to use this product and it is usually given to respectable people such as hard-workers and close relatives.

Phase III: On the same day the *Sokatet* is transferred into large containers and mixed with water, a very thick porridge (*koldhumat or hanshalt*) is prepared by pouring boiling water (94.5-97°C) on to flour in *gebete* and thorough mixing using a material made from wood for this purpose or a flat cattle bone (Scapula). The porridge is allowed to cool to room temperature for 5-7 hours and malt is kneaded with the cooled porridge. The respondents indicated that the amount of malt added at this stage depends on the strength of the *sokatet* and amount of *cheka* being produced. If the *sokatet* tastes much bitter, small quantity of malt is added or otherwise it would increase. Then, the *koldhumat* (equivalent term in Dirashe is *hanshalt*) is added into the vessel containing the *sokatet*; sufficient water is added and is thoroughly mixed together using a thick stick with flat end. In some cases, brewers use their hands to mix the two products and also to adjust the consistency of the mixed product. The *cheka* is ready for consumption after 4-12 hours of fermentation (Table 2). As the duration of fermentation in the preparation of *hiba* (Dirashe *cheka*)is too long, the *sokatet* becomes much bitter and as a result the amount of malt added into *hanshalt* in the preparation of *fasha* (Konso *cheka*) is slightly larger than for *hiba* and also the proportion of the *sokatet* in the final product is much greater than *hanshalt* in *fasha*. The amount of malt used during *menna* preparation is smaller than amount utilized during both hiba and fasha production (Figure 1).

Sensory properties and consumption pattern of *cheka*

Cheka is produced in both rural and urban communities of Dirashe and Konso for household consumption, income generation and also for special occasions like *debo, waleta* (a group of affluent people who have good contact so that they invite one another to drink cheka together), mahiber, wedding and funeral. The way of preparing *cheka* differs as between households; ethnic groups and depends on tradition, economic situation and consumer preferences. According to the respondents, the sensory properties of cheka vary with the type of *cheka* and raw materials utilized. However, the *cheka* which is often produced for sale and at special occasions should have a bitter taste, yellowish to green foam, refreshing aroma, consistent texture, a very small residue (*atela* which is given to animals) and a fairly longer shelf-life. In addition, it shouldn't contain excess foreign materials such as chaff, dead weevils or other else even though it is consumed unfiltered. Brewers in Konso villages and Gato kebele who produce *fasha* indicated that increasing the proportion of malt during the third phase of fermentation adds to the bitterness and overall quality of *cheka* whereas producers in

Dirashe believe that addition of excess malt makes the cheka sourer within a day (Figure 2).

Unlike *borde* which must be consumed within a day [8], *cheka* has a shelf-life of 2 to 4 days. But it is usually produced on a small-scale basis to avoid loss and if it is produced following Konso's processing method, it should preferably be sold within one day because consumers usually wouldn't like to drink *cheka* on the next day once it is ready for consumption. Quality deterioration starts when the active fermentation slows down and the sparkling foam doesn't appear on the surface of *cheka*. In both districts, *cheka* is retailed only at vendors' house and is often consumed locally. At times of scarcity people also consume a very sour *cheka* either by diluting with excess water or mixing it with fine ash separated through sieve. But, some people currently use orange powder to make cheka less sour. Up on addition in both cases (ash and orange powder) the *cheka* starts to form foam and tastes less acidic. The neutralising effect of ash on sour *cheka* would be because wood ashes to some extent may contain the oxides and carbonates which serve as liming agents, raising pH and thereby helping to neutralize acidic *cheka*. Most people in Dirashe like to drink sour cheka and are often suffer from stomach ache and other health problems. Smooth and reddish lower lips and skin lesions which are common among Konso people and some Dirashe people were attributed to sour *cheka* and *arake*.

Cheka drawn from fermentation vessel is very thick and is normally diluted with cold or warm water (boiled at 65-80°C; especially in Konso and some villages in Dirashe like Gato and Gidole) during serving. Depending on the thickness of the *cheka* being served and the desire of an individual, *cheka* can be diluted by 20 to 50% water. For this reason, the amount of nutrients a given person can get per a given amount of *cheka* could greatly vary with the extent to which the *cheka* is being diluted. It is consumed daily by both adults and children as a drink and meal replacement. In the study areas, solid foods are not available during day time at most households and as a result it is *cheka* that is consumed all day long. Most people particularly adult's start drinking it early on an empty stomach and people in Konso on average drink 3-5 litres of *cheka* per day but, in Dirashe adults can drink up to 8 litres per day. Since it can be obtained for free in most villages of Dirashe district, the amount a single individual drinks per day may go beyond 8 litres.

Although most mothers do not give cheka for their under 1 year's infants due to its high alcohol contents, children die due to *cheka* as few care givers try to give it to infants even less than 8 months and usually not on demand. Based on the informants' opinion, what matters is not only giving *cheka* for infants but also the feeding practice. Care givers hold one of their hands beneath the lower lip of the child to contain the

Ingredients	Proportion (w/w or w/v)			
	Fasha	Hiba*	Hiba**	Mena
Chopped cooked taro leaves (cooked) : malt : water	0.5:0.2:0.2	-	-	-
Chopped leaf cabbage : water	-	0.6:0.2	-	-
Flour of dried moringa/ leaf cabbage + grain flour : water	-	-	1.4:1	-
Grain four: Fermented product : water	5:0.6 :4.0	5:0.8 :4	5:2.4:4	-
Pulota : kneading water : cooking water	7:0.5: 4.0	9 : 0.5:4	10.4:0.5:4	-
Cooled smashed dough balls : malt : kneading water	9:1.4:1	9:1.4:1	9 : 1.4: 1	-
Sokatet: diluting water	9:2.5	9:2.5	9 : 2.5	-
Grain flour : boiling water for porridge preparation	5:3.5	5:3.5	5:3.5	5:3.5
Sokatet : porridge : Malt: Water for mixing	9:1.4:1: 2	9:0.9:1:2	9:0.9:1:2	5:0.1:2.5

*When producing *cheka* following the method used by people in highland areas of Dirashe

**When producing *cheka* following the method used by people in lowland areas of Dirashe

Table 2: The proportion of ingredients during *cheka* fermentation.

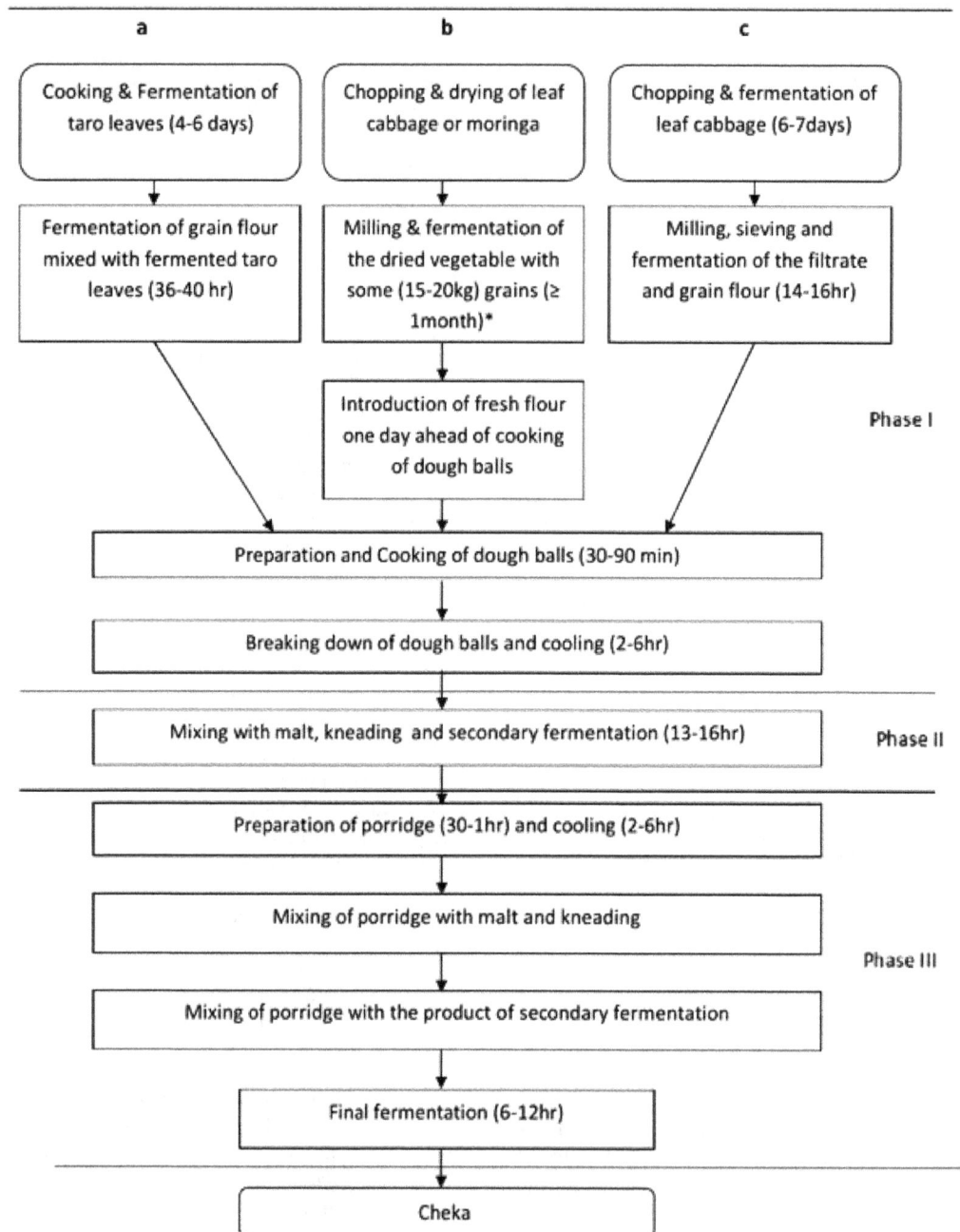

Figure 1: Flow chart for cheka preparation in Konso and Dirashe districts.

a-When producing cheka following the method used by people in Konso and in few villages of Dirashe
b-When producing cheka following the method used by people in lowland areas of Dirashe
c-When producing cheka following the method used by people in highland areas of Dirashe
*Involves frequent kneading of the fermenting product usually at an interval of 2-3 days

cheka and after adding the *cheka*, they close the nose of the child with the other hand in order to prevent the entry of *cheka* via nose. During this moment, the child is struggling to breath and this increases the likelihood of the child being choked. Even during this survey, a child of about 8 months old has lost his life due to this practice in one of the survey areas called *Shelele* kebele.

Both in Dirashe and Konso, drinking *cheka* is a common feature of social gatherings. *Cheka* is consumed in large quantities at collective work gatherings (*debo*), on market days and possibly on weekends and at special occasions. In rural communities of Dirashe, it is common that parents of a marrying man write letters to their close relatives to help them with *cheka* to be served for the invited people on the wedding and accordingly every requested household provides up to 6 jerry cans or clay pots (20-25 litres each) of *cheka*. Some people also willingly supply 1-2 jerry cans of *cheka* on other occasions like funerals. These indicate that cheka plays a vital role in building the social interaction of the society.

a- Fasha or Chaqa
b-Hiba or parshota produced in lowland areas of Dirashe
c- Hiba or parshota produced in highland areas of Dirashe
d-Menna

Figure 2: Cheka (Photos taken by the investigators).

Cheka is considered as a low-cost meal (about ETB 2 per litre) for low-income people including government employees who cannot afford factory produced beverages and restaurant foods. *Cheka* and *kurkufa* (a dish prepared by cooking leafy vegetables particularly moringa and rounded balls of maize, sorghum or wheat flour) are believed to enhance lactation and thus, lactating women are encouraged to use them. Those consumers who drink excess *cheka* and additionally use other alcoholic beverages such as *arake* do not eat other solid foods. These indicates that dependence on *cheka* alone can affect the nutritional well-being and general health of individuals. However, most *cheka* consumers eat other foods like *nufro* or *nufiti* (salted boiled maize and/or haricot bean), *kurkufa, kitta* (unleavened bread) and *kollo* (roasted maize, chickpea, sunflower or their combination). Ground chili pepper spiced with ginger, garlic, coriander, rue, basil and salt and sometimes mixed with raw tomato or cooked vegetables is served with *cheka* as appetizers and to reduce satiation. Usually people in the study areas eat the cooked dough balls and also drink the product of phase II (*sokatet*). Some people also drink a mixture of thin porridge and *cheka* literally called *hoskidha*.

Economic importance of cheka and constraints encountered by brewers

Cheka is a good source of economic opportunity in particular for the women as its preparation is not physically demanding. As the cost of entry to *cheka* vending is minimal, many women in Konso and Dirashe sell *cheka* and earn income for their family and basically for themselves. Some producers in Karat, town in Konso, sell *cheka* on a daily basis and most producers in both districts sell *cheka* twice per week or every other day. On a single day a given woman can produce up to 1000 litres of *cheka* which can generate a profit over ETB. 400. By virtue of *cheka*, she can also sell other foods such as *kollo*, tomato and cooked leafy vegetables and boiled haricot bean (separately or mixed) which help her to get extra income. Since the cost of raw materials and fuel in the study areas is fluctuating with seasons, the profit a brewer gets from cheka sale can be variable. Most informants in rural communities of Dirashe reported that they usually do not get a fair profit from cheka if their labour and cost of fuel is considered.

Atella (residue of *cheka*) is used to enhance the livestock nutrition and is believed to improve their health particularly in times of feed shortage, thereby strengthening the livelihood system. As *cheka* fermentation involves labour intensive activities such as milling

fermented vegetables and kneading, it serves as source of both direct and indirect employment for women. In Karat and Gidole (town in Dirashe), two or more women are employed per household that cover the major complex tasks of *cheka* fermentation. These women get paid ETB. 40-50 per day and besides obtain their daily meal from there. In some cases, if the brewer cannot provide solid foods that consumer take while drinking, neighbouring women may help with it and indirectly generate money for her. Moreover, *cheka* fermentation has played a huge role in gender development. In rural communities of Dirashe and Konso, women who produce quality *cheka* are given more status and have a greater say in a family and community.

Although *cheka* serves as source of cash for households, the short keeping quality of *cheka*, lack of clean water, electricity and fire wood, seasonality of *cheka* marketing (particularly in rural communities), inconsistency in the product quality and lack of encouragement from masculine family members are the major challenges which reduce the profitability of *cheka*. In most rural communities of Dirashe especially during harvesting and threshing seasons, people can get *cheka* for free. Consequently, brewers who depend on *cheka* vending cannot get enough customers during such times and become non-profitable. For this reason and others, most women in Dirashe do not engage in *cheka* marketing.

Conclusion

Cheka is a cereal and vegetable-based beverage which serves as source of nutrients for hundreds of thousands of people in Southwestern Ethiopia. Survey results showed that the preparation of *cheka* involves very complex and tedious operations such as repeated cooking and kneading of the fermentation products. Diverse methods of preparing *cheka* exist in Dirashe and Konso with differences in some ingredients utilized. In this study, it was found that the duration of *cheka* fermentation varies among localities (from a day to months); as a consequence, *cheka* with different sensory properties is produced. *Cheka* is being consumed while it is actively fermenting and has a short shelf-life of two to four days. The investigators believe that *cheka* fermentation has not received the scientific attention that it deserves. Studying the nutritional value, alcoholic content, and microbial dynamics of *cheka* as well as understanding process variables and properties of raw materials during its preparations will help in making it as a commercially viable enterprise.

Acknowledgement

The authors would like to thank the informants in both districts for providing their indigenous knowledge and reliable information which constituted the principal part of this study.

References

1. Fellows P (2000) Food processing technology: Principles and practice, (2nd edition), Baca Raton, CRC press LLC, USA.

2. Rose AH (1997) Alcoholic beverage. In: Economic Microbiology I (Rose, A.H. ed.), Academic Press, UK.

3. FAO (2012) Traditional fermented food and beverage for improved livelihoods. A Global Perspective (Agricultural Services Bulletin No. 21). Rome, Italy.

4. Fite A, Tadesse A, Urga K, Seyoum E (1991) Methanol fusel oil and ethanol contents of some Ethiopian traditional alcoholic beverages. SINET: Ethiop J Sci 14: 19-27.

5. Alemu F, Amha-Selassie T, Kelbessa U, Elias S (1991) Methanol, fuel oil and ethanol contents of some Ethiopian traditional alcoholic beverages. SINET: Ethiop J Sci 14:19-27.

6. Ghebrekidan H (1992) The effect of different chemical and physical agents on the viability of Cysticercus bovis: a preliminary report. Ethiop Med J 30: 23-31.

7. Kohajdova Z, Karovicova J (2007) Fermentation of cereals for specific purpose. J Food Nutr Res 46: 51-57.

8. Abegaz K, Beyene F, Langsrud T, Judith AN (2002a) Indigenous processing methods and raw materials of borde, a Ethiopian traditional fermented beverage. J Food Technology Africa; 7: 59-64.

9. Achi OK (2005) The potential for upgrading traditional fermented foods through biotechnology. African J Biotechnol 4: 375-380.

10. Desta B (1977) A survey of the alcohol content of traditional beverages. Ethiop Med J 15: 65-68.

11. Bahiru B, Mehari T, Ashenafi M (2006) Yeast and lactic acid flora of tej, an indigenous Ethiopian honey wine: variations within and between production units. Food Microbiol 23: 277-282.

12. Yohannes T, Fekadu M, Khalid S (2013) Preparation and physiochemical analysis of some Ethiopian traditional alcoholic beverages. African J Food Sci 7: 399-403.

13. Berza B, Wolde A (2014) Fermenter technology modification changes microbiological and physicochemical parameters, improves sensory characteristics in the fermentation of tella: An Ethiopian traditional fermented alcoholic beverage. J Food Process Technol 5: 316.

14. Abegaz K, Beyene F, Langsrud T, Judith AN (2002b) Parameters of processing and microbial changes during fermentation of borde, a traditional Ethiopian beverage. J Food Technol Africa 7: 85-92.

15. Abegaz K, Beyene F, Langsrud T, Judith AN (2004) The effect of technological modifications on the fermentation of borde, an Ethiopian traditional fermented cereal beverage. J Food Technol Africa 9: 3-12.

16. Liu H, Corke H, Ramsden L (1999) Functional properties and enzymatic digestibility of cationic and cross-linked cationic ae, wx, and normal maize starch. J Agric Food Chem 47: 2523-2528.

3

Effect of Processing and Drying Methods on the Nutritional Characteristic of the Multi-cereals and Legume Flour

Kumari PV[1]* and Sangeetha N[2]

[1]*IICPT, Thanjavur, Tamilnadu, India*
[2]*Department of Food Science and Technology, Pondicherry University, Puducherry, India*

Abstract

Cereals and legumes of today are more nutritious and healthful than ever before. Cereals and legumes processing is one of the oldest and the most essential part of all food technologies. Besides, it forms a large and indispensable component of the food production chain. The cereals and legume processing industry is as diverse as its range of products. Drying and dewatering plays an important role in food manufacturing and food processing activities worldwide often one of the last operations in the food processing. In this study three sets of ingredients were chosen they subject to different treatment and drying conditions. The results found that significant difference was observed in the nutritional composition of different treatment and different drying conditions. The developed composite mix has possessed to have good nutritional properties and it possess o have good health benefits. This mix can be used for further product development.

Keywords: Cereals; Legumes; Drying; Germination; Milling

Introduction

The consumer demand for nutritious cereal–based food products with minimal artificial additives has been met with increased research and development from the food industry. Nutritional quality is eventually important in considering processed flour as a food ingredient and its successful performance depends principally on functional characteristics imparted to the final products. The versatility of processed flour as a base for many food products emphasizes the need for a better understanding of its functional characteristics and nutritional characteristics of the processed flour [1]. Composite flours containing cereal and legumes have proven practical uses in many parts of the world to improve the nutritional and functional properties of flour. Basically, composite flour technology refers to the process of mixing wheat flour with cereals or legumes to make use of indigenous raw materials to produce high quality food products in an economical way. Being nutridense, these formulations could form sustainable strategy for combating malnutrition [2]. One of the chief challenges of nutritionists is to diminish human sufferings due to nutritional stress as Indian population suffers from food and nutritional insecurity. Hence the present study was carried out with object to study the nutrition characteristics of the multi cereal and legume mix subjected to different drying methods.

Materials and Methods

Selection of raw materials

The whole cereals and legumes with immense nutritional potentiality were selected for the formulation of processed multi-cereals and legume flour. The good quality raw materials were purchased from the wholesale shop in bulk quantity to maintain uniform quality throughout the processing. The selected raw materials are listed in Table 1.

Steps involved in the formulation of processed multi-cereals and legume flour from 3 sets of raw materials

The selected raw materials were washed to remove the unwanted dust particles. The listed sets (Table 1) were weighed and soaked in water for a period of 12 h, after which the excess water was drained and each set was divided in two groups. Group A was allowed to sprout for

Set 1 (Cereals)	Set 2 (Legumes)	Set 3 (Cereals and Legumes)	Quantity (%)
Parboiled rice (*Oryza punctata*)	Dry peas (*Pisum sativum*)	Parboiled rice (*Oryza punctata*)	20
Kodo millet (*Paspalum scrobiculatum*)	Red Gram (*Cajanus cajan*)	Kodo millet (*Paspalum scrobiculatum*)	10
Sorghum (*Sorghum bicolor*)	Rajma (*Phaseolus vulgaris*)	Sorghum (*Sorghum bicolor*)	10
Foxtail millet (*Setaria italica*)	Horse gram (*Macrotyloma uniflorum*)	Foxtail millet (*Setaria italica*)	10
Oats (*Avena sativa*)	Green gram (*Vigna radiata*)	Green gram (*Vigna radiata*)	10
Maize (*Zea mays*)	Bengal gram (*Cicer arietinum*)	Bengal gram (*Cicer arietinum*)	10
Barley (*Hordeum vulgare*)	Black gram (*Vigna mungo*)	Black gram (*Vigna mungo*)	10
Wheat (*Triticum aestivum*)	Cow pea (*Vigna unguiculata*)	Cow pea (*Vigna unguiculata*)	10
Bajra (*Pennisetum glaucum*)	Soyabean (*Glycine max*)	Soyabean (*Glycine max*)	10

Table 1: Raw materials selected for processing and product development.

a period of 24 h and the group B was not sprouted. The processed raw materials were subjected to drying methods namely sun drying (SD for 12 h), forced convection tray drying (FCTD at 60°C for 8 h) and fluidized bed drying (FBD at 60°C for 6 h). All the dried raw materials were milled in stone miller to obtain fine flour. The process involved in the formulation of multi-cereal and legume flour from 3 sets of raw materials is shown in Figure 1.

*Corresponding author: Kumari PV, IICPT, Thanjavur, Tamilnadu, India
E-mail: Vasanthi.phd@gmail.com

Chemical composition

The chemical composition of the cereal and legume mix namely carbohydrate [3] proteins [4] fats [5] energy [6] moisture [7] and ash were carried out using standard procedure. Carbohydrate by anthrone method, protein by the kjeldhal method with (NX6.25), fat by soxlet method with automated soxpluse, energy using boam calorimeter, moisture determined by hot air oven method and ash using muffle furnance.

Results and Discussion

Carbohydrate (g/100 g) content of the multi-cereals and legume flour

Carbohydrates are the macronutrient which plays a major role in human diets, comprising 40% to 75% of energy intake. Their most important nutritional property is digestibility in the small intestine. In terms of their physiological or nutritional role, they are often classified as available and unavailable carbohydrates (Table 2). The carbohydrate content of the multi-cereals and legume flour were found to be in the range of 46.66 g/100 g to 67.33 g/100 g in the non-sprouted sets subjected to sun drying, where as in the sprouted sets subjected to sun drying was found to be in the range of 36 g/100 g to 2 g/100 g. In the case of fluidized bed drying, the carbohydrate content of the non- sprouted sets was found to be in the range of 43.66 g/100 g to 65.33 g/100 g and 34 g/100 g to 51.66 g/100 g in the sprouted samples of all the three sets. As

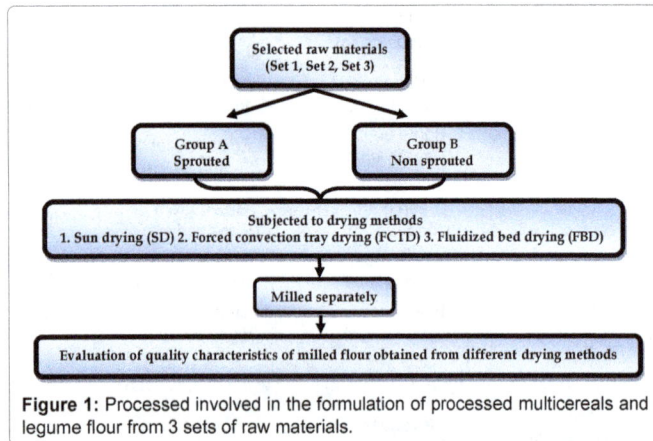

Figure 1: Processed involved in the formulation of processed multicereals and legume flour from 3 sets of raw materials.

Processing methods	Sets	SD	FBD	FCTD
NS	Set 1	67.33 ± 2.08	65.33 ± 1.52	64.66 ± 2.51
	Set 2	46.66 ± 1.52	43.66 ± 1.52	41.66 ± 2.08
	Set 3	52.66 ± 2.51	51.66 ± 1.52	48.66 ± 1.52
SP	Set 1	52.00 ± 2.00	51.66 ± 1.52	52.66 ± 1.52
	Set 2	36.00 ± 1.00	34.00 ± 2.00	34.00 ± 2.00
	Set 3	50.00 ± 1.00	45.66 ± 2.08	44.00 ± 1.00
p–value	Drying methods			
	Processing methods	P ≤ 0.05*		
	Sets			

All values are means of triplicate determinations ± Standard Deviation (S.D)
NS: Non-Sprouted; SP: Sprouted: SD: Sun Drying; FBD: Fluidized Bed Drying;
FCTD: Forced Convection Tray Drying; Set 1: Cereals; Set 2: Legumes; Set 3: Cereals and Legumes
*Significantly different (p ≤ 0.05) by ANOVA

Table 2: Carbohydrate (g/100 g) content of the multi-cereals and legume flour.

Processing methods	Sets	SD	FBD	FCTD
NS	Set 1	11.56 ± 0.32	9.70 ± 0.10	10.26 ± 0.20
	Set 2	28.50 ± 1.04	28.83 ± 0.15	28.56 ± 0.67
	Set 3	13.96 ± 0.05	14.88 ± 0.06	12.50 ± 0.40
SP	Set 1	12.40 ± 0.10	10.86 ± 0.15	11.80 ± 0.10
	Set 2	30.83 ± 0.77	26.80 ± 0.30	31.00 ± 1.00
	Set 3	12.58 ± 0.50	15.61 ± 0.12	12.60 ± 0.43
p–value	Drying methods			
	Processing methods	P ≤ 0.05*		
	Sets			

All values are means of triplicate determinations ± Standard Deviation (S.D)
NS: Non-Sprouted; SP: Sprouted: SD: Sun Drying; FBD: Fluidized Bed Drying;
FCTD: Forced Convection Tray Drying; Set 1: Cereals; Set 2: Legumes; Set 3: Cereals and Legumes
*Significantly different (p ≤ 0.05) by ANOVA

Table 3: Protein (g/100 g) content of the multi-cereals and legume flour.

far as the forced convection tray drying is concerned, the carbohydrate content of non-sprouted and sprouted sets was in the range of 41.66 g/100 g to 64.66 g/100 g and 34 g/100 g to 52.66 g/100 g respectively. The sprouted samples of all the three sets showed a significant decrease in the carbohydrate content when compared to samples which was not allowed for sprouting. The set 1 which constitute the cereals resulted in increased carbohydrate content. The concentration of carbohydrate in the samples exposed to different drying methods was almost similar and slightly higher in sun dried sample which was found to be statistically significant (p ≤ 0.05). Vidal–Valverde et al. [8] explained that during germination, carbohydrate was used as source of energy for embryonic growth which could explain the changes of carbohydrate content after germination. The findings of the carbohydrate content of the multi-cereals and legume flour was in par with results of Khetarpaul and Goyal [9] who reported that during germination, mobilization and hydrolysis of seed polysaccharides takes place. Polysaccharides can promote hydrolysis by fermenting microbes which possess both alpha and beta amylases [10]. A sharp decrease in total and reducing sugar at 48 hrs may be due to microbial utilization.

Protein (g/100 g) content of multi-cereals and legume flour

Cereal grains and legumes are a valuable source of food proteins. In comparison to cereal grains, the seeds of legumes are rich in good quality protein, providing man with a highly nutritious food resource. Similarly, in the present study, the protein content in both the processing methods adopted was found to be considerably (p ≤ 0.05) high especially in set 2 comprising of legumes. The increase in the protein content is due to presence of whole legumes. The protein content of the sets subjected to sprouting and non-sprouting which are exposed to sun drying was in the range of 12.40 g/100 g to 30.83 g/100 g and 11.56 g/100 g to 28.50 g/100 g respectively. In the case of non-sprouted and sprouted sets dried using fluidized bed drying the protein content was found to be in the range of 14.88 g/100 g to 28.83 g/100 g and 10.86 g/100 g to 26.80 g/100 g respectively (Table 3). As far as the protein content of non-sprouted and sprouted samples in all the three sets exposed to forced convection tray drying are concerned, the values were in the range of 10.26 g/100 g to 28.56 g/100 g and 11.80 g/100 g to 31.00 g/100 g respectively. Grain legumes enhance the protein content of cereal–based diets and may improve the nutritional status as well. Cereal proteins are deficient in certain essential amino acids, particularly lysine [11]. On the other hand, legumes have been reported to contain adequate amounts of lysine, but are deficient in S-containing amino acids (methionine, cystine and cysteine) [12]. Chavan and Kadam [13] stated that complex qualitative changes are reported to occur during soaking and sprouting of seeds. The conversion of storage proteins of cereal grains into albumins and globulins during sprouting may improve the quality of cereal proteins. Many studies have shown an increase in the content of the amino acid lysine upon sprouting. Increase in proteolytic activity during sprouting is desirable for nutritional improvement of cereals because it leads to hydrolysis of prolamins and the liberation of amino acids such as glutamine and proline which are converted to limiting amino acids such as lysine.

Fat (g/100 g) content of multi-cereal and legume flour

Generally, cereals and legumes are low in fat. The fat composition was notably ($p \leq 0.05$) less in all the sprouted samples, since the fat was used as source of energy during germination. The fat content in the non-sprouted and sprouted samples of all the three sets subjected to different drying methods was found to be in range of 3.36 g/100 g to 3.50 g/100 g and 1.16 g/100 g to 2.76 g/100 g in the non-sprouted and sprouted sets subjected to sun drying; 1.63 g/100 g to 3.43 g/100 g and 0.90 g/100 g to 2.66 g/100 g in the non-sprouted and sprouted sets subjected to fluidized bed drying; 1.90-3.10 g/100 g and 1.10 g/100 g to 2.40 g/100 g in the non-sprouted and sprouted sets subjected to forced convection tray drying respectively (Table 4). On germination, there existed a significant ($p \leq 0.05$) decrease of fat content in three sets

Processing methods	Sets	SD	FBD	FCTD
NS	Set 1	2.66 ± 0.15	1.63 ± 0.20	1.90 ± 0.10
	Set 2	3.50 ± 0.20	3.43 ± 0.25	3.10 ± 0.10
	Set 3	3.36 ± 0.30	3.40 ± 0.20	2.86 ± 0.15
SP	Set 1	1.16 ± 0.15	0.90 ± 0.10	1.10 ± 0.10
	Set 2	2.50 ± 0.20	2.16 ± 0.15	2.40 ± 0.30
	Set 3	2.76 ± 0.15	2.66 ± 0.25	1.53 ± 0.25
p–value	Drying methods	P ≤ 0.05*		
	Processing methods			
	Sets			

All values are means of triplicate determinations ± Standard Deviation (S.D)
NS: Non-Sprouted; SP: Sprouted: SD: Sun Drying; FBD: Fluidized Bed Drying; FCTD: Forced Convection Tray Drying; Set 1: Cereals; Set 2: Legumes; Set 3: Cereals and Legumes
*Significantly different ($p \leq 0.05$) by ANOVA

Table 4: Fat (g/100 g) content of the multicereals and legume flour.

Processing methods	Sets	SD	FBD	FCTD
NS	Set 1	1.16 ± 0.15	1.46 ± 0.25	1.66 ± 0.15
	Set 2	2.94 ± 0.03	2.52 ± 0.02	3.33 ± 0.03
	Set 3	1.91 ± 0.01	1.92 ± 0.02	2.02 ± 0.03
SP	Set 1	0.90 ± 0.10	1.40 ± 0.10	1.00 ± 0.10
	Set 2	2.93 ± 0.03	2.51 ± 0.01	3.06 ± 0.06
	Set 3	1.64 ± 0.02	1.51 ± 0.02	1.73 ± 0.02
p–value	Drying methods	P ≤ 0.05*		
	Processing methods			
	Sets			

All values are means of triplicate determinations ± Standard Deviation (S.D)
NS: Non-Sprouted; SP: Sprouted: SD: Sun Drying; FBD: Fluidized Bed Drying; FCTD: Forced Convection Tray Drying; Set 1: Cereals; Set 2: Legumes; Set 3: Cereals and Legumes
*Significantly different ($p \leq 0.05$) by ANOVA

Table 5: Ash (g/100 g) content of the multicereals and legume flour.

Processing methods	Sets	SD	FBD	FCTD
NS	Set 1	303 ± 2.51	295 ± 3.05	292 ± 2.51
	Set 2	334 ± 2.51	312 ± 2.08	325 ± 2.00
	Set 3	311± 2.00	287 ± 2.51	293 ± 3.60
SP	Set 1	283 ± 1.52	291 ± 2.08	273 ± 1.52
	Set 2	306 ± 3.05	304 ± 3.60	295 ± 2.51
	Set 3	294 ± 3.51	270 ± 1.52	271 ± 1.00
p–value	Drying methods	P ≤ 0.05*		
	Processing methods			
	Sets			

All values are means of triplicate determinations ± Standard Deviation (S.D)
NS: Non-Sprouted; SP: Sprouted: SD: Sun Drying; FBD: Fluidized Bed Drying; FCTD: Forced Convection Tray Drying; Set 1: Cereals; Set 2: Legumes; Set 3: Cereals and Legumes
*Significantly different ($p \leq 0.05$) by ANOVA

Table 6: Energy (Kcal/100 g) value of the multi-cereals and legume flour.

subjected to different drying methods, which could be due to total solid loss during soaking prior to germination [14] or use of fat as an energy source in sprouting process. The results are comparable with findings of Vanderstoep [15] for germinated green gram and lentil.

Ash (g/100 g) content of multi-cereals and legume flour

Ash is the inorganic residue remaining after the water and organic matter have been removed by heating in the presence of oxidizing agents, which provides a measure of the total amount of minerals within a food. The ash content of all the sets subjected to different drying methods did not vary much. The sprouted sets possess to have a slight decrease in the ash content which reflects on the concentration of minerals (Table 5). Increased ash content was observed among non-sprouted samples of all the three sets subjected to different drying techniques when compared to the sprouted sets exposed to different drying techniques. Leaching out of solid matter during pre-germination soaking process could be the reason for significant reduction of mineral matter on germination. The results are in par with the findings of Okrah [16] who found that ash content of germinated sorghum was decreased which varied from 0.28% to 1.70%. While, Mubarak [17] reported that germination and cooking processes caused significant decreases in ash content. Alemu [18] observed that sorghum ash content was significantly decreased after fermentation.

Energy (Kcal/100 g) value of multi-cereals and legume flour

Calories are a measure of energy and are commonly used to describe the energy content of foods. The energy values (Kcal) represent the presence of complex mixture of carbohydrates, protein and fat. In the present study, there was no considerable difference (p>0.05) in the energy levels of set 1, set 2 and set 3. However, the energy level was significantly decreased in the sprouted samples when compared to the non-sprouted samples. Total metabolizable energy content was higher during sprouting resulted in decreased energy value after sprouting. In the present study, the energy value was found to be in the range of 303-311 Kcal/100 g in the non-sprouted sets subjected to sun drying whereas after sprouting the energy value was found to in the range of 283-306 Kcal/100 g. The non-sprouted sets subjected to fluidized bed drying was in the range of 287-312 Kcal/100 g and the energy value was found to be in the range 270-304 Kcal/100 g in the sprouted sets. The energy value of all the three sets which was not allowed for sprouting and subjected to forced convection tray drying was in the range of 292-325 Kcal/100 g where as in the sprouted sets the energy value was in the range of 271-295 Kcal/100 g (Table 6).

Processing methods	Sets	SD	FBD	FCTD
NS	Set 1	3.20 ± 0.10	3.03 ± 0.15	3.26 ± 0.15
	Set 2	3.26 ± 0.15	2.73 ± 0.15	3.30 ± 0.10
	Set 3	3.36 ± 0.15	3.06 ± 0.20	3.03 ± 0.15
SP	Set 1	3.70 ± 0.20	4.20 ± 0.10	4.03 ± 0.15
	Set 2	3.70 ± 0.10	3.36 ± 0.15	4.03 ± 0.15
	Set 3	4.23 ± 0.15	3.86 ± 0.15	3.63 ± 0.15
p–value	Drying methods	P ≤ 0.05*		
	Processing methods			
	Sets			

All values are means of triplicate determinations ± Standard Deviation (S.D)
NS: Non-Sprouted; SP: Sprouted: SD: Sun Drying; FBD: Fluidized Bed Drying; FCTD: Forced Convection Tray Drying; Set 1: Cereals; Set 2: Legumes; Set 3: Cereals and Legumes
*Significantly different (p ≤ 0.05) by ANOVA

Table 7: Crude fiber (g/100 g) content of the multicereals and legume flour.

Processing methods	Sets	SD	FBD	FCTD
NS	Set 1	7.83 ± 0.15	6.30 ± 0.26	4.23 ± 0.20
	Set 2	9.46 ± 0.25	6.50 ± 0.30	4.46 ± 0.37
	Set 3	7.93 ± 0.15	6.13 ± 0.15	4.10 ± 0.10
SP	Set 1	8.20 ± 0.26	7.00 ± 0.10	4.70 ± 0.20
	Set 2	9.53 ± 0.40	6.76 ± 0.25	5.03 ± 0.15
	Set 3	8.40 ± 0.36	6.63 ± 0.30	4.96 ± 0.15
p–value	Drying methods	P ≤ 0.05*		
	Processing methods			
	Sets			

All values are means of triplicate determinations ± Standard Deviation (S.D)
NS: Non-Sprouted; SP: Sprouted: SD: Sun Drying; FBD: Fluidized Bed Drying; FCTD: Forced Convection Tray Drying; Set 1: Cereals; Set 2: Legumes; Set 3: Cereals and Legumes
*Significantly different (p ≤ 0.05) by ANOVA

Table 8: Moisture (g/100 g) content of the multicereals and legume flour.

Crude fiber (g/100 g) content of multi-cereals and legume flour

Measurements of crude fiber are the only index of food fiber content available at the present time. However, crude fiber values reflect only a portion of the fiber present in food. Crude fiber has been reported to recover on the average 20% of the hemicelluloses, 10% to 50% of the lignin and 50% to 80% of the cellulose [19]. In the present study, significant difference was observed (p ≤ 0.05) between all the sets which was sprouted and non-sprouted and subjected to three different types of drying. The results of the present study were in par with Martin Cabrejas et al. [20], who found that the fibre increased substantially during germination by about 100% in peas. This increase in fibre was reported to be mostly due to changes in the polysaccharides found in the cell wall such as cellulose, glucose and mannose, suggesting that the changes were due to an increase in the cellular structure of the plant during germination [20]. The fibre content of set 2 comprising of legumes was appreciably higher (p ≤ 0.05) than the set 1 and set 2 (Table 7).

Moisture (g/100 g) content of multi-cereals and legume flour

Moisture is the integral part of the cereals and legumes. In the present study, the moisture content of multi-cereals and legumes exposed to sun drying was greater when compared to other two drying methods. However, investigations have shown that low moisture content of food samples is a desirable phenomenon, since the microbial activity is reduced [21]. Low moisture content in food samples increased the

storage periods of the food products [22] while high moisture content in foods encourages microbial growth leading to spoilage of foods [23]. Moisture content of raw material affects gelatinization process and lower moisture content indicates better gelatinization process [24-26] and better gelatinization process results in better swelling of the extrudate (Table 8). In the present study, significant difference (p ≤ 0.05) was observed in the moisture content of the sets exposed to different drying techniques. On comparing the different drying techniques, sets subjected to forced convection tray drying possessed less moisture content when compared to other two drying techniques. The overall observation showed that the sets subjected to sun drying had higher retention of moisture content which is not acceptable for the storage of foods for a longer duration of time.

Conclusion

Nutritional quality is eventually important in considering processed flour as a food ingredient and its successful performance depends principally on functional characteristics imparted to the final products. The developed composite mix has possessed to have good nutritional properties and it possess o have good health benefits. This mix can be used for further product development.

References

1. McWatters K, Holmes M (1979) Influence of moist heat on solubility and emulsification properties of soy and peanut flours. J Food Sci 44: 774-776.

2. Gahlawat P, Sehgal S (1994) Protein quality of weaning foods based on locally available cereal and pulse combination. Plant Food Human Nutri 46: 245-253.

3. AOAC (1995) Official Methods of Analysis. (16th edn) Association of official analytical chemists, Arlington, VA, USA.

4. Raghuramulu N, Madhavan K, Kalyansundaram S (2014) Food analysis: A Manual of Laboratory.

5. Ranganna S (2005) Vitamins: Hand book of Analysis and Quality Control for fruit and vegetable products. (2nd edn) Tata McGraw Hill publishing Co. Ltd, New Delhi.

6. AOAC (1990) Official methods of analysis. (15th edn), official analytical chemists, Arlington, VA, USA, 1990.

7. AACC (2005) Approved methods of the AACC. American association of cereal chemists. (11th edn), St. Paul, MN.

8. Vidal-Valverde C, Frias J, Sierra I, Blazquez I, Lambein F, et al. (2002) New functional legume foods by germination: Effect on the nutritive value of beans, lentils and peas. Europe Food Res Technol 215: 472-477.

9. Khetarpaul N, Goyal R (1995) Effect of germination and probiotic fermentation on pH, titratable acidity, dietary fibre, β-Glucan and vitamin content of sorghum based food mixtures. J Nutri Food Sci 2: 2.

10. Bernfeld S (1962) On psychoanalytic training. Psychoanalytic Quarterly 31: 453-482.

11. Iqbal A, Khalil IA, Shah H (2003) Nutritional yield and amino acid profile of rice protein as influenced by nitrogen fertilizer. Sarhad J Agri (Pakistan) 26: 1237-1245.

12. Farzana W, Khalil IA (1999) Protein quality of tropical food legumes. J Sci Technol 23: 13-19.

13. Chavan JK, Kadamn SS (1989) Nutritional improvement of cereals by sprouting. Critical Review Food Sci Technol 28: 401-437.

14. Wang N, Lewis MJ, Brennan JG, Westby A (1997) Effect of processing methods on nutrients and anti-nutritional factors in cowpea. Food Chem 58: 59-68.

15. Vanderstoep J (1981) Effect of germination on the nutritive value of legumes. Food Technol 26: 3121-3125.

16. Okrah SG (2008) Screening of six local sorghum varieties for their malting and brewing qualities. Kwame Nkrumah University Science and Technology, Ghana.

17. Mubarak AE (2005) Nutritional composition and antinutritional factors of

mung bean seeds (*Phaseolus aureus*) as affected by some home traditional processes. Food Chem 89: 489-495.

18. Alemu MK (2009) The effect of natural fermentation on some anti-nutritional factors, minerals, proximate composition and sensory characteristics in sorghum based weaning food. University of Addis Ababa, Ethiopia.

19. Van Soest PJ, McQueen RW (1973) The chemistry and estimation of fibre. Proceeding Nutrition Society 32: 123-130.

20. Martín-Cabrejas MA, Ariza N, Esteban R, Mollá E, Waldron K, et al. (2003) Effect of germination on the carbohydrate composition of the dietary fiber of peas (*Pisum sativum* L.) J Agri Food Chem 51: 1254-1259.

21. Oyenuga VA (1968) Nigeria's foods and feeding stuffs. University Press, Nigerian University.

22. Alozie Y, Akpanabiatu MI, Eyong EU, Umoh IB, Alozie G (2009) Amino acid composition of *Dioscorea dumetorum* varieties. Pak J Nutri 8: 103-105.

23. Temple VJ, Badamosi EJ, Ladeji O, Solomon M (1996) Proximate chemical composition of three locally formulated complementary foods. West Africa J Biol Sci 5: 134-143

24. Miller RC (1985) Low moisture extrusion: Effect of cooking moisture on product characteristics. J Food Sci 50: 249-253.

25. Santosa BAS, Sudaryono S, Widowati S (2005) Technology evaluation of flour instant popcorn and quality. J Penelitian Pascapanen Pertanian 2: 66-75.

26. Foubion JM, Hoseney RC, Seib PA (1982) Functions quality of grain components in extrusion. Cereal Food World 27: 212- 216.

4

Functional and Pasting Properties of Maize 'Ogi' Supplemented with Fermented Moringa Seeds

Jude-Ojei BS[1], Lola A[2]*, Ajayi IO[3] and Ilemobayo Seun[2]

[1]Department of Nutrition and Dietetics, Rufus Giwa Polytechnic, Owo Ondo State, Nigeria
[2]Department of Food Science and Technology, Rufus Giwa Polytechnic, Owo Ondo State, Nigeria
[3]Department of Science Laboratory Technology, Rufus Giwa Polytechnic, Owo Ondo State, Nigeria

Abstract

The process of Ogi production results in remarkable nutrient loss, *Moringa* seed flour, rich in micronutrients and vitamins, could increase the micronutrient and macronutrient contents of ogi. This study aimed at evaluating the functional and pasting properties of 'ogi' supplemented with fermented *Moringa* seeds. *Moringa* seeds was de-feathered and fermented for 48 h, dried and milled into flour. Maize 'ogi' was produced following traditional methods. Maize-*Moringa* Ogi was formulated by mixing the samples in ratio 90:10, 80:20 and 70:30 while 100% maize and 100% *Moringa* flour serves as control. The functional properties, shows that the swelling capacity ranged between (0.94 ml to 0.74 ml), water absorption (18 ml to 13 ml) and bulk density (0.66 g/ml to 0.36 g/ml), and the least gelation for 10% to 30% *Moringa* seed inclusion results showed no gelation at 2%, 4%, 6% and 8%, weak gel at 10%, 12%, 14% and 16% and strong gel at 18% and 22%. In pasting properties, the result of peak viscosity of the samples ranged between (3552.67 RVA to 15.00 RVA), trough (1842.33 RVA to 8.50 RVA), breakdown (1717.33 RVA to 7.00 RVA), final viscosity (3926.67 RVA to 12.00 RVA), set back (2084.67 RVA to 4.00 RVA) and peak time (5.00 to 4.47 min). The addition of *Moringa* seed flour to maize-'ogi' reduced the functional and pasting properties.

Keywords: 'Ogi'; Moringa; Pasting properties; Functional properties

Introduction

Ogi a fermented gruel from cereal has been recognized as the most popular traditional health-sustaining fermented food in Western Nigeria. It is commonly used as weaning food, food for convalescence, young children and as a standard breakfast cereals in many homes. Ogi usually has smooth texture and is boiled into porridge called pap or cooked and turned into a stiff gel called "agidi" or "eko" prior to consumption [1].

Traditional preparation of 'ogi' involved washing, steeping, milling, sieving, fermentation and drying. During these processes, nutrients including protein and minerals are lost from the grains thereby affecting nutritional quality adversely [2,3]. Various studies have been carried out to improve the nutritional value of 'ogi' by fortifying it with either plant protein (melon, okro, cowpea, and soybean) or animal protein sources (egg and milk) [4,5]. The application of *Moringa* for this purpose is however limited to the leaves [6]. *Moringa oleifera*, is known by different names such as benzolive, drumstick tree, kelor, marango, mlonge, mulangay, nébéday, saijhan, and sajna across many regions. It is the most widely cultivated species of a monogeneric family, the *Moringaceae*, which is native to the sub-Himalayan tracts of India, Pakistan, Bangladesh and Afghanistan. All the parts of the tree have been reported to be edible and are consumed in many parts of the world [7]. *Moringa oleifera* will be one of the alternatives to most imported food supplies in the treatment of malnutrition. The diet of many rural and urban dwellers is deficient in protein and high in carbohydrate. The plant seeds contain hypotensive activity, strong antioxidant activity and chelating property against arsenic toxicity [8-11]. Seed flour from *Moringa oleifera* is widely used as a natural coagulant for water treatment in developing countries [11]. It has an impressive range of medicinal uses with high nutritional quality

When starch-based foods are heated in an aqueous environment, they undergo a series of changes known as gelatinization and pasting. These are two of the most important properties that has effect on quality and aesthetic concerns in the food industry, since they affect texture and digestibility as well as the end use of starchy foods [12]. The aim of this paper therefore is to determine how the addition of fermented *Moringa* seed will affect the functional and pasting properties of maize 'ogi'.

Materials and Methods

Materials

'Swan 1'maize used in this research was purchased from let's farm Agric input store, Akure, Ondo State, Nigeria. The grains were dry when purchased and the *Moringa* seeds were harvested from the School experimental farm.

Methods

Preparation of *Moringa* flour: Fermentation of the seeds was achieved by soaking in water and allowed to ferment for 2 days at ambient temperature. Fermented samples were dried at 50°C ± 5°C until constant moisture content was obtained. The dried samples were milled using disc milling machine. The milled samples were kept in air tight plastic container until needed.

Preparation of Maize 'ogi' flour: The method described by Akingbala [2] was used for 'ogi' manufacture. Maize grain (1 kg) was soaked in water for 48 h. The grain was milled with attrition mill at medium speed for 7 min. The slurry was passed through muslin cloth and the suspension obtained was left to stand for 48 h for the 'ogi' to sour, the supernanant was decanted to be able to collect the 'ogi'. 'Ogi'

*Corresponding author: Lola A, Department of Food Science and Technology, Rufus Giwa Polytechnic, Owo Ondo State, Nigeria
E-mail: lola_ajala2006@yahoo.co.uk

Level of substitution	Swelling capacity (g/ml)	Water absorption (ml)	Bulk density (g/ml)
100% maize 'ogi'	0.94[a]	18[a]	0.66[a]
10% FMOSD	0.89[b]	16[b]	0.57[b]
20% FMOSD	0.86[c]	15[b]	0.57[b]
30% FMOSD	0.84[c]	14[bc]	0.55[c]
100% Moringa	0.74[d]	13[c]	0.36[d]
Key FMOSD: Fermented Moringa 'Ogi' Seed			

Table 1: Functional properties of maize-'ogi' supplemented with fermented Moringa seed.

was dried at 50°C ± 5°C until constant weight was obtained and was crushed manually.

Preparation of Moringa-'Ogi' flour mixtures: Moringa-'ogi' flour was produced by mixing dry 'ogi' powder with fermented Moringa seed flour. Formulation of Moringa-'ogi' was at different mixing ratio of 0:100, 10:90, 20:80, and 30:70 of Moringa and 'ogi', respectively. Samples were kept in air tight container until when needed.

Determination of Functional Properties

Water absorption capacity

The procedure of Sathe et al. [13] was used. To 1.0 g of each sample was added 10 ml of water, the suspension was then stirred using magnetic stirrer for 5 min. and was transferred into centrifuge tubes and centrifuged at 3,500 rpm for 30 min. The supernatant that was obtained was measured using a 10 ml measuring cylinder. The density of the water was assumed to be 1 g/ml. The water absorbed was calculated as the difference between the initial volume of water used and the volume of the supernatant obtained after centrifugation. The result was expressed as a percentage of water absorbed by the samples on percentage g/g basis.

Least gelation concentration (LGC)

The LGC of the flour blends was determined using the modified method of Coffman and Garcia [14]. Sample suspensions of 2%, 4%, 6%, 8%, 12%, 14%, 16%, 18% and 20% (w/v) was prepared in 10 ml distilled water in test tubes. The tubes containing the suspensions was then heated for 1 h in a gentle boiling water bath. The tubes were cooled rapidly in water at 40°C for 2 h. Each tube was inverted one after the other. The LGC was taken as the concentration when the sample from the inverted test tube did not fall or slip.

Swelling capacity and solubility

The method described by Leach et al. [15] was used with slight modifications Sample (1 g) was weighed and transferred into a clean, dry test tube and weighed (W_1). The flour was then dispersed in 50 ml of distilled water using a magnetic stirrer. The resulting slurry was heated at desired temperatures (40°C, 50°C, 60°C and 70°C) for 30 min in a thermostatically controlled water bath. The mixture was cooled to room temperature and centrifuged at 2,200 rpm for 15 min. Aliquot of the supernatant (5 ml) was dried to a constant weight at 120°C. The residue obtained after drying represented the amount of starch solubilized in water. Solubility was calculated as per 100 g of starch on dry weight basis. The residue obtained after centrifugation with the water it retained was transferred to the clean, dried test tube and weighed (W_2).

$$Swelling\ capacity\ of\ starch = \frac{w_2 - w_1}{Weight\ of\ flour} \times 100$$

Determination of pasting properties

The pasting profile of the samples was studied using a rapid Visco-Analyzer (RVA) (Newport Scientific Pty. Ltd) with the aid of a thermocline for windows version 1.1 [16]. The RVA was connected to a PC where the pasting properties and curve were recorded. Sample suspension was then prepared by addition of the equivalent weight of 3.0g dry starch to distilled water to make a total weight of 28.0 g suspension in the RVA sample canister

Data Analysis

Data collected were statistically analyzed with the Statistical Analysis Systems (SAS) package (version 8.2 of SAS institute Inc, 1999) [17]. Statistically significant differences (p ≤ 0.05) in all data were determined by general linear model procedure (GLM) while least significant difference (LSD) was used to separate the means. Correlation coefficient between variables was obtained using Pearson correlation coefficient analysis.

Results and Discussion

The result of the functional properties of maize 'ogi' supplemented with mature Moringa seed is presented in Table 1. The results show that there was gradual decrease in the level of water absorption capacity and bulk density as the addition of Moringa seed flour increases. These compared favorably with the result obtained by substituting 'ogi' with bambara-nut flour by Theodore et al. [18], the decrease in water absorption capacity contents is in line with a conclusion made by Theodore et al. [18], which states that a weaning food should have low water absorption capacity and bulk density in order to have high energy density food which are more suitable as weaning food. According to Omueti et al. [19] lower water absorption capacity is desirable for making thinner gruels with high caloric density per unit volume. Also, according to Onuoha [20] low bulk density is an advantage because high bulk limits the caloric and nutrient intake per feed per child and infants sometimes are unable to consume enough to satisfy their energy and nutrient requirements. Swelling capacity which is the measure of the ability of starch to imbibe water and swell, ranged from 0.74 to 0.94. The 100% 'ogi' sample had the highest value (0.94), while the 100% of Moringa sample had the lowest (0.74). The mixed samples had significant different swelling capacity. The swelling index of granules reflects the extent of associative forces within the granules as reported by Sanni et al. [21]. The swelling capacity of the supplemented sample indicated that the associative forces within their granules were not strongly bonded.

Results of least gelation properties of maize 'ogi' supplemented with mature Moringa seed is presented in Table 2. Maize-'ogi' formed weak gel between 6% and 10% concentration while a strong gel was at 12% concentration and 100% Moringa formed no gel in all the concentration. Maize-Moringa 'ogi' samples at the 10%, 20%, 30% level of Moringa seed substitution formed a weak gel between 10% and 16% concentrations. A strong gel was however, formed at 18% concentration. Yadav [22] observed that the variation in the gelling properties of flours could be attributed to the relative ratio of protein, carbohydrates and lipids that makes up the flours and the interaction between the components. From this result, Food prepared from maize 'ogi' substituted with fermented Moringa seed will contain more food and therefore nutrients than food produced from 100% maize-'ogi'

The result of the pasting properties of maize 'ogi' supplemented with mature Moringa seed is presented in Table 3. From this result, the peak viscosity of the 'ogi' samples ranged between 15.00 RVA and

Samples	2%	4%	6%	8%	10%	12%	14%	16%	18%	22%
100% Maize'ogi'	X	X	Y	Y	Y	√	√	√	√	√
10% FMOSD	X	X	X	X	Y	Y	Y	Y	√	√
20% FMOSD	X	X	X	X	Y	Y	Y	Y	√	√
30% FMOSD	X	X	X	X	Y	Y	Y	Y	√	√
100% *Moringa*	X	X	X	X	X	X	X	X	X	X

KEY: X= No gel, Y=Weak gel, √ =strong gel

Table 2: Least gelation properties of maize 'ogi' supplemented with fermented *Moringa* seed.

Level of *Moringa* Substitution	Peak Viscosity (RVA)	Trough (RVA)	Break Down (RVA)	Final Viscosity (RVA)	Set Back (RVA)	Peak Time (min)	Pasting Temp (RVA)
0%	3552.67	1842.33	1717.33	3926.67	2084.67	5.00	75.02
10%	1386.00	802.00	584.00	1442.00	9640.00	5.13	77.55
20%	1185.00	686.00	499.00	1228.00	542.00	5.13	77.45
30%	944.00	567.00	377.00	963.00	396.00	5.07	79.05
100%	15.00	8.50	7.00	12.00	4.00	4.47	------

Table 3: Pasting properties of maize-ogi supplemented with fermented moringa seed.

3552.67 RVA. The 100% 'ogi' sample had the highest peak viscosity while 100% *Moringa* seed flour has the lowest. The peak viscosity in the 'ogi' samples reduced with increase in the level of substitution with fermented *Moringa* seed. This is an indication that fermented *Moringa* seed flour has no ability to gel. This trend was also observed in some of the earlier studies [4,23,24] on the substitution of 'ogi' with okra seed flour and in the study of Theodore et al. [18] on the substitution of 'ogi' with Bambara-nut. Two factors interact to determine the peak viscosity of cooked starch paste: the extent of granule swelling (swelling capacity) and solubility. Higher swelling index is indicative of higher peak viscosity while higher solubility as a result of starch degradation results in reduced paste viscosity.

The trough of the 'ogi' samples ranged between 8.50 RVA and 1842.33 RVA. 100% 'ogi' sample had the highest value of 1842.33 RVA while 100% *Moringa* had the lowest value of 8.50 RVA. The trough which shows the holding capacity of the starch granules showed that 100% maize 'ogi' had superior holding capacity due to the crystalline and strength of the starch molecules in it. This implied that 100% maize 'ogi' has ability to withstand breakdown cooling. The breakdown viscosity value is an index of the stability of starch [25]. The value in this study ranged between 7.00 RVA and 1717.33 RVA. 100% maize 'ogi' had the highest value of 1717.33 RVA while 100% *Moringa* 'ogi' had the lowest value of 7.00RVA. The breakdown viscosity values for all the samples were lower than the peak viscosity values and this is altered by nature of the material, degree of mixing, the temperature used and shear applied to the mixture [16]. The final viscosity which is the change in the viscosity after holding cooked starch at 50°C, indicates the ability of the material to form a viscous gel or paste after cooking and cooling as well as the resistance of the paste to shear force during stirring [26].

The final viscosity of the 'ogi' samples ranged between 12.00 RVA and 3926.67 RVA. 100% maize 'ogi' had the highest of 3926.67 RVA while 100% *Moringa* had the lowest value of 12.00 RVA. The final viscosity was also affected by the mixture of *Moringa* into the maize. Setback viscosity is an indication of the stability of cooked paste against retrogradation and can be used to predict the storage life of a product prepared from the flour. The setback value ranged between 2084.67 RVA and 396.00 RVA. From the result 100% maize-'ogi' had the highest value of 2084.67 while 100% *Moringa* had the lowest value of 4.00. The setback revealed the gelling ability or retrogradation tendency of the amylase present in the starch. The low setback values of the sample indicate low rate of retrogradation. This implication of this is that maize-*Moringa* 'ogi' may not retrogade fast [27].

The peak time of the 'ogi' samples, which is a measure of the cooking time, ranged between 4.47 minutes in 100% *Moringa* and 5.13 minutes in 10% and 20% samples were significantly different from one another. The pasting temperature ranged between 75.02°C and 79.05°C. 100% fermented *Moringa* seed flour has no ability to gel hence showed no pasting temperature. The gelatinization time during processing could not be confirmed in 100% *Moringa* 'ogi' sample. Pasting temperature is an index that characterized the initial change due to swelling of starch [27]. It can be deduced from this study that the mixture (ratio) reduced the inherent characteristic of 100% maize 'ogi' as seen as the loss in its pasting properties.

Conclusion

The addition of *Moringa* seed flour sample to 'ogi' increased the protein content, fibre, ash and fat content but reduced the moisture content. And the reduction of functional and pasting properties controlled the starchy swollen granules in the food sample. Therefore, it would be necessary to encourage the use *Moringa* seed flour as protein source to supplement the local/traditional cereals such as maize, millet and sorghum based which will reduce the incidence of malnutrition.

Acknowledgement

This Research was sponsored by Tertiary Education Trust Fund and we appreciate the sponsorship. I also acknowledge the effort of Dr. (Mrs.) Lola Ajala for guidance and effort for the completion of this research work.

References

1. Faber M, Jogessar VB, Benade AJ (2001) Nutritional status and dietary intakes of children aged 2-5 years and their care-givers in a rural South African community. Int J Food Sci Nutr 52: 401-411.

2. Akingbala JO, Rooney LW, Faubion JM (1981) A laboratory procedure for the preparation of ogi, a Nigerian fermented food. J Food Sci 46: 1523-1526.

3. Adeyemi IA, Beckley O (1986) Effect of period of maize fermentation and souring on chemical properties and amylograph pasting viscosity of ogi. J Cereal Sci 4: 353-360.

4. Aminigo ER, Akingbala JO (2004) Nutritive composition and sensory properties of ogi fortified with okra seed meal. J Appl Sci Environ Manag 8: 23-28.

5. Oyarekua MA (2010) Sensory evaluation, nutritional quality and anti-nutritional factors of traditionally co-fermented cereals/cowpea mixtures as infant complementary food. Agri Biol J North America 1: 950-956.

6. Olorode OO, Idowu MA, Ilori OA (2013) Effect of benoil (Moringa oleifera) leaf powder on the quality characteristics of 'Ogi'. America J Food Nutri Sci Huβ.

7. Fahey JW (2015) Moringa oleifera: A review of the medical evidence for its nutritional, therapeutic, and prophylactic properties. Tree Life J 1: 5.

8. Arabshahi DS, Devi DV, Urooj A (2007) Evaluation of antioxidant activity of some plant extracts and their heat, pH and storage stability. Food Chem 100: 1100-1105.

9. Ghasi S, Nwobodo E, Ofili JO (2000) Hypocholesterolemic effects of crude extract of leaf of Moringa oleifera lam in high-fat diet fed wistar rats. J Ethnopharmacol 69: 21-25.

10. Mehta LK, Balaraman R, Amin AH, Bafna PA, Gulati OD (2003) Effect of fruits of Moringa oleifera on the lipid profile of normal and hyper cholesterolaemic rabbits. J Ethnopharmacol 86: 191-195.

11. Santos AFS, Luciana A, Adriana CCA, Teixeira JA, Paiva PMG, et al. (2009) Isolation of a seed coagulant Moringa oleifera lectin. Process Biochem 44: 504-508.

12. Adebowale YA, Adeyemi IA, Oshodi AA (2005) Functional and physicochemical properties of flours of six Mucuna species. Afr J Biotechn 4: 1461-1468.

13. Sathe SK, Deshpande SS, Salunkhe DK (1982) Functional properties of winged bean (Psophocarpus tetragonolobus) proteins. J Food Sci 47: 544-549.

14. Coffman CW, Gracia VV (1977) Functional properties of amino acid content of a protein isolate from mung bean flour. J Food Technol 12: 473-484.

15. Leach HW, McCowen LD, Scoch TJ (1959) Structure of starch granules, swelling and solubility pattern of various starches. Cereal Chem 36: 534-544.

16. Newport Scientific (1998) Applications manual for the rapid visco analyzer using thermocline for windows. Newport Scientific Pty Ltd., Apollo Street, Warriewood NSW, Australia.

17. SAS (2002) Statistical analysis system proprietary software. Release 8.3 SAS Institute Inc., Carry, NC.

18. Theodore IM, Ikenebomeh MJ, Ezeibe S (2009) Evaluation of mineral content and functional properties of fermented maize (Generic and Specific) flour blended with Bambara Groundnut (Vigna subterranean). Afr J of Fd Sci 3: 107-112.

19. Omueti O, Otegbayo B, Jaiyeola O, Afolabi O (2009) Functional properties of complementary diets developed from soybean (Glycine max) and Groundnut 8: 563-573.

20. Onuoha OG, Chibuzo E, Badau M (2014) Studies on the potential of malted Digitaria exilis, Cyperus esculentus and Colocasia esculenta flour blends as weaning food formulation. Nigeria Food J 32: 40-47.

21. Sanni L, Maziya DB, Akanya J, Okoro CI, Alaya V, et al. (2005) Standards for cassava products and guidelines for export. IITA, Ibadan, Nigeria 93.

22. Yadav RB, Yadav BS, Dhull N (2012) Effect of incorporation of plantain and chickpea flours on the Quality Characteristics of biscuits. J Food Sci Technol 49: 207-13.

23. Akingbala JO, Akinwande BA, Uzo-Peters PI (2005) Effects of color and flavor changes on acceptability of ogi supplemented with okra seed meals. Plt Fd Hum Nutri 58: 1-9.

24. Otunola ET, Sunny-Roberts EO, Solademi AO (2007) Influence of the addition of okra seed flour on the properties of 'ogi', a Nigerian fermented maize food. University of Göttingen, Germany.

25. Fernande De Tonella, ML Berry JW (1989) Rheological properties of flour and sensory characteristics of bread made from germinated chick peas. Int J Food Sci Technol 24: 103-110.

26. Adeyemi IA, Idowu MA (1990) The evaluation of pre-gelatinized maize flour in the development of maissa, a baked product. Nigeria Food J 8: 63-73.

27. Eniola L, Delarosa LC (1981) Physiochemical characteristics of yam starches. J Food Biochemistry 5: 115-130.

5

Impression of Instinctive Cookery Methods along with Altered Processing Time on the Potential Antioxidants, Color, Texture, Vitamin C and β-Carotene of Selected Vegetables

Ali M[1]*, Khan MR[2], Rakha A[2], Khalil AA[2], Lillah K[2] and Murtaza G[2]

[1]School of Food Science and Biotechnology, Key Laboratory of Fruits and Vegetables, Zhejiang Gongshang University, Hangzhou, China
[2]National Institute of Food Science and Technology, Fruits and Vegetables Processing Laboratory, University of Agriculture Faisalabad, Pakistan

Abstract

In the current millennium, consumers are becoming more conscious about their dietary patterns with special concern to nutrient retention during cooking methods. There is a need to assess the most convenient and nutritionally better thermal cooking method which causes the least nutrient abuse. The current study investigated the consequence of three cookery methods viz. conventional boiling, steaming and microwave cooking on the physical parameters, β-carotene, vitamin C, total phenolic contents (TPC), total flavonoid contents (TFC) and antioxidant activity (DPPH%) of the particular vegetables. Results revealed that both cooking methods and length of time exerted positive and negative influence on nutritional composition of vegetables. L^*, a^* and b^* values decreased in all samples. In texture analysis, highest force N (Newton) determined in control and microwave cooked samples followed by steaming and boiled samples. Cooking of vegetable by microwaving had the maximum retention for vitamin C, TPC and DPPH% after control. While, ß-carotene contents increased in microwave cooking than control. Total flavonoid contents were tending to a decreasing trend in all cooking methods but highest contents were retained in boiling cooking. Amongst the three cookery methods adopted, microwave cooking method emerged as the most appropriate method in terms of retention of nutrients in vegetables.

Keywords: Carrot; Cabbage; Cooking; Boiling; Steaming; Microwave

Introduction

In current era, there is a mounting trend towards the assessment of beneficial phytochemicals and efficient ingredients from natural dietetic sources like fruits and vegetables [1,2]. Fresh vegetables contain nutritional constituents including phytochemicals, vitamins and minerals. However, these are extremely perishable [3]. Numerous epidemiological investigation has showed the defensive impacts of vegetable utilization against the danger of several age-related illnesses like tumor, cardiovascular diseases, cataract and muscular disintegration [4,5]. Carrot and Cabbage are usually consumed after adopting the cooking procedures such like boiling, steaming and microwave before use. These vegetables are consisting of various bioactive components like phytochemicals, carotenoids, vitamin C and minerals. In this way, these compounds prevent the people from certain diseases like hypertension, stroke and heart disorders [6]. The role of carrot carotenoids as the precursors of vitamin A and excellent antioxidants source has been generally known [7,8]. Cabbage is a cruciferous green leafy vegetable, which contains high amounts of fibre, vitamins, and minerals [9,10]. Cabbage also attains beneficial phytochemicals and carotenoid contents in significant amounts [11].

The nutritional profile of vegetables is highly dependent on handling and processing techniques. Even minor thermal treatments, like blanching exert detrimental effects on functional constituents [12]. Cookery methods initiate substantial alteration in natural composition, manipulation in chemical concentration and bio-accessibility of bioactive component in vegetables. Alternatively, impacts of both aspects, positive and negative have been accounted which are dependent upon changes in process conditions, morphological parameters and dietary properties of vegetables [13]. Defective cooking methods expressively affect the physical parameters, ß-carotene, vitamin C, TPC, TFC and DPPH% of vegetables. Such cooking methods alter the antioxidant and anti-nutrient components [14]. While, the degree of alteration largely depends upon length of time and adopted cooking methods [15]. Cookery methods can also lead to interruption of the food matrix, growing the bio-accessibility of many phytochemicals and therefore improve the nutritional quality of vegetables [16].

After processing, vegetables quality was gradually lowered as the result of nutrient loss. The result of these changes shows poor acceptability [17]. Hence, the influence of different cooking methods on qualitative and quantitative values of nutrients should be investigated. Therefore, current study was planned to inspect the values of nutritionally active components and physical characteristics of carrot and cabbage before and after boiling, steaming and microwave cooking protocols.

Material and Methods

Research was conducted in National Institute of Food Science and Technology, University of Agriculture Faisalabad.

Chemicals

All chemicals used were of analytical grade supplied by Merck and Sigma. All the measurements were done in triplicates.

Procurement of raw material

The vegetables cabbage and carrot were purchased from the local market of Faisalabad, Pakistan.

*Corresponding author: Maratab Ali, School of Food Science and Biotechnology, Key Laboratory of Fruits and Vegetables, Zhejiang Gongshang University, Hangzhou 310018, China, E-mail: maratab.786@gmail.com

Preparation of raw material

The vegetables were subjected to washing, peeling and dicing prior to subjecting for cooking process.

Cooking methods

Conventional boiling, microwave and steam cooking methods were adopted to cook the vegetable [12]. In all cooking methods, 500 g of each vegetable was taken. In conventional cooking method, vegetables were subjected to boiling in sufficient volume of water. While in steam cooking, vegetables were subjected to a domestic steamer. After cooking, vegetables were drained using a strainer. For microwave cooking, vegetables samples were subjected to a microwave oven (Panasonic 600 W power) in which no added water was used. All vegetable samples were cooked for both 10 min and 15 min separately. The cooked material was packed in polythene zip bags and stored at (-42°C) for further analysis

Treatments

Raw vegetables were taken as T_0 showing no cooking treatment employed, T_1 and T_2 showed vegetables subjected to boiling for 10 and 15 minutes, T_3 and T_4 showed vegetable subjected to steaming for 10 and 15 minutes while T_5 and T_6 showed vegetables subjected to microwave cooking for 10 and 15 minutes.

Physicochemical Analysis

Color analysis

The color measurements of cooked vegetables were determined using ColorTec-PCM™ spectrophotometer (Accuracy Microsensors, Inc. Pittsford, New York, USA) method [17]. The color was expressed in terms of L*, a* and b*.

Texture analysis

Textural analyses of all the samples were determined using the texture analyzer (TA-XT2, Stable Microsystems, Surrey, UK) method [18].

Determination of ascorbic acid

Vegetables samples were assessed for evaluating the ascorbic acid content by the method of Association of Vitamin Chemists [19]. The blue color created by the reduction of 2,6-dichlorophenyl indophenols dye by ascorbic acid was recorded calorimetrically.

Determination of ß-carotene

Beta-carotene contents were analyzed by the spectrophotometer method [20]. In this method, 500-gram sample was taken in a pestle and mortar, grinded using acetone. Extraction procedure was repeated for 2-3 times. The extract was collected and subjected to filtration process. The filtrate was shifted to separating funnel and mixed with 10-15 mL of petroleum ether. The pigments were shifted into the petroleum ether phase. The staying period was given to separate the extract thoroughly. After staying time the bottom layer was drained off while top layer extract was collected into a 250 mL conical flask. The absorbance of the extract was estimated at 452nm by spectrophotometer.

Preparation of extract

Extract was prepared by mixing the 5 grams of each sample into 50 mL 80% methanol by using ultrasonic bath for 20 minutes. An aliquot (2 mL) of the extracts was ultracentrifuged for 15 minutes at 2200 rpm at room temperature. The clear extract solution was analyzed for the estimation of TPC, TFC and antioxidant activity.

Determination of total phenolic contents

The TPC of sample extract was estimated by using Folin-Ciocalteu's reagent method [21]. In this method, 1 mL of extract was mixed with 9 mL of distilled water. Latterly, 1 mL of Folin-Ciocalteu's phenol reagent was incorporated with extract. After time interval of 5 min, 10 mL of 7% Na_2CO_3 was incorporated. Final volume was made up to 25 mL with the incorporation of 4 mL of distilled water. After 90 min of incubation period at room temperature, the absorbance was recorded at 750 nm by using spectrophotometer. The total phenolics were stated as mg of gallic acid equivalents (GAE)/g fresh matter of vegetable (mg/g sample).

Determination of total flavonoids contents

The TFC of vegetables extract was estimated according to the aluminium chloride colorimetric method [22]. In this method, 1 mL of extract was added in 4 mL of distilled water and incorporated with 0.3 mL 5% $NaNO_2$. 0.3 mL of 10% $AlCl_3$ was mixed after time interval of 5 min. This mixture was agitated thoroughly with 2 mL of 1 M NaOH for 6 min. After it final volume of mixture was made up to 10 mL by distilled water incorporation. At wavelength of 510 nm, the reaction mixture absorbance was recorded. The findings were stated as mg of quercitin equivalent (QE)/g fresh matter of vegetable.

Determination of antioxidant activity (DPPH%)

The antioxidant activity of extract was estimated by adopting the spectrophotometer method [23]. In this method, 200 μL of extracts and 0.8 mL methanol was incorporated in 2 mL of 0.1 mM DPPH methanol solution. Then, final mixture was agitated systematically and placed in the dark place for 60 min at room temperature. The control was prepared by the incorporation of 2 mL of DPPH with 1 mL of methanol. Finally, absorbance was estimated at 517 nm. Percent inhibition was calculated by using the following formula;

Reduction in absorbance (%) = [Abs control - Abs sample / Abs control] × 100

Statistical Analysis

Statistical analysis was carried out by using two factor factorials under completely randomized Design (CRD) to determine the level of significance [24].

Results and Discussion

Color

Statistical results regarding color values of vegetables are presented in Table 1. Highest L* (lightness), a* (red color) and b* (yellow color) values were noticed in T_5 after T_0 followed by T_6, T_3 and T_4. Lowest values were recorded in T_2 and T_1. The decreasing trend of L* a* and b* value in boiling cooking was observed due to the fact of degradation of chlorophyll pigments mainly in cabbage and carotenoid contents in carrot. The deteriorated effect of heat along with leach down effect of boiling cooking was the major reason of color lessening. α- and β- carotene contents collectively resolute the final color of cooked vegetables. However, such carotene compounds were identified as moderately heat sensitive. So, these compounds isomerize into numerous cis-isomers during cooking. The reduction in L*, a*, b* values observed in all treated vegetables which may be related to α- and β- carotene decrease and their isomerization [25]. Heat induces

Treatment	L*-value	a*-value	b*-value	Texture (N)
		Carrot		
T$_0$	57.50 ± 1.75[a]	33.57 ± 1.06[a]	42.41 ± 1.76[a]	151.82 ± 2.16[a]
T$_1$	42.59 ± 1.17[ef]	13.63 ± 1.11[de]	29.64 ± 1.27[c]	122.48 ± 2.02[e]
T$_2$	40.16 ± 1.47[f]	11.92 ± 1.04[e]	24.81 ± 1.51[d]	117.82 ± 2.26[f]
T$_3$	48.50 ± 1.38[cd]	17.05 ± 1.11[d]	36.41 ± 1.20[b]	132.48 ± 2.02[d]
T$_4$	45.83 ± 1.52[de]	14.67 ± 1.45[de]	32.19 ± 1.45[c]	126.15 ± 2.72[e]
T$_5$	54.50 ± 1.75[ab]	26.17 ± 1.38[b]	41.81 ± 1.01[a]	142.82 ± 2.49[c]
T$_6$	51.84 ± 1.62[bc]	22.28 ± 1.72[c]	38.55 ± 1.37[ab]	147.48 ± 2.74[b]
		Cabbage		
T$_0$	52.82 ± 1.55[a]	-6.59 ± 0.06[b]	8.69 ± 0.43[a]	105.82 ± 2.50[b]
T$_1$	35.47 ± 1.23[e]	-1.64 ± 0.60[a]	5.36 ± 0.81[de]	67.48 ± 2.74[f]
T$_2$	31.51 ± 1.11[f]	-1.53 ± 0.55[a]	4.86 ± 0.60[e]	62.41 ± 3.00[c]
T$_3$	41.83 ± 1.21[cd]	-5.93 ± 0.53[b]	6.69 ± 0.51[bcd]	77.43 ± 2.01[d]
T$_4$	39.68 ± 1.25[d]	-5.73 ± 0.57[b]	6.09 ± 0.95[cde]	69.67 ± 0.78[ab]
T$_5$	45.67 ± 1.08[b]	-6.53 ± 0.97[b]	8.23 ± 0.39[ab]	93.15 ± 2.02[f]
T$_6$	43.28 ± 1.04[bc]	-6.13 ± 0.50[b]	7.90 ± 0.62[abc]	97.27 ± 2.74[f]

Values carrying same letters are non-significantly different with each other.
T$_0$= Vegetable without any cooking technique, T1= Vegetable subjected to boiling for 10 minutes, T2 = Vegetable subjected to boiling for 15 minutes, T3 = Vegetable subjected to steaming for 10 minutes, T4 = Vegetable subjected to steaming for 15 minutes, T5 = Vegetable subjected to microwaving for 10 minutes, T6 = Vegetable subjected to microwaving for 15 minutes.

Table 1: Effects of different cooking methods on the physical characteristics of vegetables.

modifications on carotenoid pigment which results in color variation in vegetables. The results were parallel with the findings of Miglio [13], Nwanekezi [26] who noticed pronounced L*, a*, b* values in vegetables during different cooking methods.

Texture

Different thermal cooking methods influence the texture attributes of vegetables. Firmness of cooked samples significantly decreased relative to the control sample. The statistical results are presented in Table 1. Highest shear force N (Newton) was taken by T$_6$ after T$_0$ followed by T$_5$, T$_3$ and T$_4$. While the Lowest shear force was gained by T$_2$ and T$_1$. Cooking of vegetables triggered a decrease in the force required to shred the vegetable. Heat effect of cooking represents the decrease of softness and consequently softening of the vegetable internal and external structures. The findings of this study were covenant with Miglio [13], Maria [18] who studied the effect of microwaving and conventional cooking methods on nutrient profile and textural analysis of different vegetables (Table 1).

ß-Carotene content

Statistical values regarding ß-Carotene content are presented in Table 2. Highest β-carotene content was observed in T$_6$ followed by T$_5$, T$_0$, T$_1$ and T$_2$. Lowest β-carotene contents were determined in T$_4$ and T$_3$. Steaming caused major losses due to the oxidation of conjugate double bonds in β-carotene along with higher temperatures degradation. It was observed that heating triggered the both coloring pigment deprivation and an extractability upturn due to the breakdown of protein-carotenoid complexes: in the skin the first outcome prevailed as of the thinness of soft tissue, which permitted easy and fast heat transmission and results in extreme water leaching [27]. Results of beta carotenoid contents during the study of different cooking treatment in different vegetables were in similarity with the findings of some previous studies [13,20,28,29].

Vitamin C

Different cooking methods showed significant variation on vitamin C contents of vegetables. Statistical values of all treatments are presented in Table 2. The highest vitamin C contents were recorded in

T$_5$ after T$_0$ followed by T$_6$, T$_3$ and T$_4$. Unlike to that, maximum loss of vitamin C content was seen in T$_2$ and T$_1$. Consequently, microwaving did not far destroy the vitamin C as compared to boiling and steaming [30]. Boiling largely decreased the vitamin C content when vegetable was subjected for 15 minutes. In boiling cooking, decreasing trend of vitamin C occurred due to the oxidation of vitamin C in the presence of molecular oxygen initiated by inherent enzymes (vitamin C oxidase and peroxidase). The findings of this study regarding vitamin C concentrations were similar with Lee SK [31] who determined the ascorbic acid contents decreased in vegetables from 3.02 mg/100 g to 2.47 mg/100 g during different cooking methods. Lower content of vitamin C during cooking were also determined [20,28] (Table 2).

Total phenolic content

It is obvious from the results that the total phenolic contents were highly significantly affected due to the differences in cooking methods and different cooking time length. Statistical values regarding total phenolic content of vegetables are presented in Table 3. This indicated that the maximum TPC were recorded in T$_6$ after T$_0$ followed by T$_5$, T$_3$ and T$_4$. The lowest TPC were observed in T$_1$ and T$_2$ in both vegetables. The total phenolic contents were effect significantly by the cooking methods and water contents. Low moisture contact results in lowest losses of total phenolic contents. Application of thermal treatment regulates firmness and break down of cellular structures with consequential discharge of these components into the hot water. The higher degree of softness was recorded for boiled samples, which clarifies the maximum loss of phenolic compounds in boiling as compared to steamed and microwave samples. The results were in conformity with the findings of Miglio [13] who observed that the TPC decreased in vegetables during different cooking methods. Changes of TPC during different cooking methods were also in conformity with the research findings of Hunter [30] and Ismail [32] showed that TPC decreased during different cooking methods. Our results of this study were also in matching with the conclusions of Sahlin [33].

Total flavonoid content

Cooking had both positive and negative influence on TFC depending on the kind of vegetables [34,35]. Statistical values regarding

Treatment	β – Carotene (mg/100g)	Vitamin C (mg/100g)	β-Carotene (mg/100g)	Vitamin C (mg/100g)
	Carrot		Cabbage	
T_0	55.36 ± 2.88[bc]	30.56 ± 1.31[a]	1.63 ± 0.02[b]	18.11 ± 1.88[a]
T_1	53.58 ± 2.05[c]	19.14 ± 0.22[d]	1.57 ± 0.01[c]	12.02 ± 1.42[de]
T_2	51.66 ± 2.07[c]	18.18 ± 0.27[d]	1.54 ± 0.01[cd]	10.60 ± 1.87[e]
T_3	47.56 ± 2.17[d]	26.11 ± 1.27[bc]	0.52 ± 0.03[de]	14.44 ± 1.34[bc]
T_4	45.56 ± 2.78[d]	24.26 ± 1.52[c]	0.49 ± 0.01[e]	13.11 ± 1.27[cd]
T_5	57.83 ± 1.63[ab]	29.21 ± 0.33[ab]	1.67 ± 0.03[a]	17.44 ± 0.37[ab]
T_6	59.42 ± 1.06[a]	28.17 ± 1.90[b]	1.69 ± 0.01[a]	16.11 ± 0.51[ab]

Values carrying same letters are non-significantly different with each other.
T_0= Vegetable without any cooking technique, T_1= Vegetable subjected to boiling for 10 minutes, T_2 = Vegetable subjected to boiling for 15 minutes, T_3 = Vegetable subjected to steaming for 10 minutes, T_4 = Vegetable subjected to steaming for 15 minutes, T_5 = Vegetable subjected to microwaving for 10 minutes, T_6 = Vegetable subjected to microwaving for 15 minutes.

Table 2: Effects of different cooking methods on the β – Carotene and Vitamin C contents of vegetables.

Treatment	TPC (mg GAE/100 g)	TPC (mg GAE/150 g)	TFC (mg QE/100g) DPPH%
	Carrot		
T_0	13.55 ± 1.84a	5.79 ± 0.15a	73.67 ± 2.14a
T_1	8.21 ± 1.03[d]	5.40 ± 0.21[b]	62.35 ± 2.57[c]
T_2	7.96 ± 0.81[d]	5.04 ± 0.10[c]	56.42 ± 2.91[d]
T_3	10.55 ± 1.28[bc]	5.32 ± 0.10[b]	69.15 ± 0.56[b]
T_4	9.21 ± 1.06[cd]	4.94 ± 0.04[c]	64.82 ± 2.24[c]
T_5	11.88 ± 1.01[ab]	4.01 ± 0.02[d]	70.16 ± 1.07[b]
T_6	12.1 ± 1.06[ab]	3.97 ± 0.04[d]	72.01 ± 1.02[ab]
	Cabbage		
T_0	16.70 ± 1.11[a]	10.91 ± 0.49[a]	64.10 ± 1.80[a]
T_1	9.36 ± 0.52[de]	9.29 ± 0.15[b]	47.95 ± 1.92[e]
T_2	8.03 ± 0.11[e]	8.18 ± 0.11[c]	44.10 ± 1.57[f]
T_3	11.36 ± 1.04[cd]	8.42 ± 0.20[c]	55.67 ± 1.23[c]
T_4	10.41 ± 0.58[de]	7.37 ± 0.04[d]	52.89 ± 1.00[d]
T_5	13.04 ± 1.81[ab]	6.77 ± 0.18[e]	58.25 ± 1.41[c]
T_6	15.63 ± 1.14[bc]	6.34 ± 0.19[e]	61.30 ± 1.80[b]

Values carrying same letters are non-significantly different with each other.
T_0= Vegetable without any cooking technique, T_1= Vegetable subjected to boiling for 10 minutes, T_2 = Vegetable subjected to boiling for 15 minutes, T_3 = Vegetable subjected to steaming for 10 minutes, T_4 = Vegetable subjected to steaming for 15 minutes, T_5 = Vegetable subjected to microwaving for 10 minutes, T_6 = Vegetable subjected to microwaving for 15 minutes.

Table 3: Effects of different cooking methods on the TPC, TFC and DPPH% of vegetables.

total flavonoid content of both vegetables are presented in Table 3. This indicated that the highest total flavonoid contents were recorded in T_1 after T_0 followed by T_3, T_2, T_4, and T_5. While the lowest total flavonoid contents were identified in T_6. Decreasing trend of TFC in all cooking methods was due to the fact of food processing, like cutting action of the vegetable tissues and influence of employed temperatures. This can lead to cellular destruction and separation of some flavonoids compounds from cellular assemblies such as lignin and causing them to be highly extractable and freely identified [36]. The TFC was found to decrease highly in microwave cooking due to the application of dry heat, which leads to the maximum moisture reduction along with oxidation of volatile compounds in the presence of light and oxygen. Commonly, thermal actions have damaging influence on the flavonoid and phenolic compounds as they are extremely heat sensitive compounds [32]. The results were parallel with the findings of Khwairakpam B [20] who observed that the total flavonoid contents decreased in vegetables from 5.7 to 4.5mg QE/100g during different cooking methods.

Antioxidant activity (DPPH%)

Statistical values regarding antioxidant activity (DPPH%) of both vegetables are described in Table 3. This indicated that the maximum antioxidant activity (DPPH%) was observed in T_6 after T_0 followed by T_5, T_3, T_4 and T_1. While the lowest antioxidant activity (DPPH%) was observed in T_2. The decreasing trend of antioxidant activity (DPPH%) was due to the fact of reduction and extractability of total phenolic compounds in cooking methods. No association was establish between TPC and antioxidant activity (DPPH%) in the study by Kahkonen MJ [37] on some vegetable extracts having phenolic contents. While the current study showed a strong correlation between phenolic contents and DPPH radical scavenging activity. antioxidant activity (DPPH%) was found to decrease by cooking irrespective of the leach down of high extent of total phenolic contents caused by thermal damage of cellular and sub-cellular compartment walls and radical exclusion by heat and chemical reaction. Overall microwave cooking was identified as an optimum method of cooking which resulted in highest DPPH% inhibition. The results were in accordance with the findings of Khwairakpam [20] who observed that the antioxidant activity increased from 11.20% to 13.75% during microwave cooking method. Changes of antioxidant activity during different cooking treatments were also in conformity with the findings of Faller ALK [38-40] (Table 3).

Conclusion

Certain nutrients are lost during processing of vegetables. Vegetables were cooked by boiling, steaming and microwave cooking methods. Physical characteristics of cooked vegetables were highly affected by all adopted cooking methods. Vitamin C, beta carotene,

TPC, TFC and antioxidants activity (DPPH%) significantly decreased by thermal action of applied cooking methods along with longer time of cooking. Short time length of cooking significantly reduced the nutrient loss. Overall, microwave cooking method was recognized the optimum cooking method which resulted in highest retention of vegetables constituents. Consequently, it is suggested that cook the vegetables for short time up to just softening the tissue to improve their digestibility and reserve maximum nutritional profile.

Acknowledgement

I am so grateful to Dr. Moazzam Rafiq Khan for providing research material and keen guidance throughout the completion of research project.

Research

1. Iqbal S, Bhanger MI, Anwar F (2007) Antioxidant properties and components of bran extracts from selected wheat varieties commercially available in Pakistan. Food Sci Technol 40: 361-367.

2. Son Y, Kim J, Lim JC, Chung Y, Lee JC (2003) Ripe fruits of Solanum nigrum L inhibit cell growth and induces apoptosis in MCF-7 cells. Food Chem Toxic 41: 1421-1428.

3. Miah MAS, Khan MMRL, Jabin SA, Abedin N, Islam MF, et al. (2016) Nutritional quality and safety Aspects of wild vegetables consume in Bangladesh. Asian Pac Trop Biomed 6: 125-131.

4. Gosslau A, Chen KY (2004) Nutraceuticals, apoptosis and disease prevention. Nutri 20: 95-102.

5. Siddhuraju P, Becker K (2003) Antioxidant properties of various solvent extracts of total phenolic constituents from three different agroclimatic origins of drumstick tree (Moring oleifera) Lam. leaves. Agric Food Chem 51: 2144-2155.

6. Lampi AM, Kamal-Eldinand A, Piironen P (2002) Tocopherols and tocotrienols from oil and cereal grains. Func Food-Biochem. Proc Asp 55: 230-237.

7. Sharma KD, Karki S, Thakur NS, Attri S (2012) Chemical composition, functional properties and processing of carrot. Food Sci Technol. 49: 22-32.

8. Bembem K, Sadana B (2014) Effect of different cooking methods on the antioxidant components of carrot. Bioscience Discovery. 5: 112-116.

9. Oboh G, Raddatz H, Henle T (2008) Antioxidant properties of polar and non-polar extracts of some tropical green leafy vegetables. Sci Food Agric 88: 2486-2492.

10. Subhasree B, Baskar R, Keerthana RL, Susan RL, Rajasekaran P (2009) Evaluation of antioxidant potential in selected green leafy vegetables. Food Chem. 115: 1213-1220.

11. Mizgier P, Kucharska AZ, Łętowska AS, Ostek JK, Kidoń M, et al. (2016) Characterization of phenolic compounds and antioxidant and anti-inflammatory properties of red cabbage and purple carrot extracts. Funct Foods 21: 133-146.

12. Bureau SS, Mouhoubi L, Touloumet L, Garcia C, Moreau F, et al. (2015) Are folates, carotenoid s and vitamin C affected by cooking? Four domestic procedures are compared on a large diversity of frozen vegetables. LWT-Food Sci Tech 64: 735-741.

13. Miglio C, Chiavaro E, Visconti A, Fogliano V, Pellegrini N (2008) Effects of different cooking methods on nutritional and physicochemical characteristics of selected vegetables. Agric Food Chem 56: 139-147.

14. Turkmen N, Poyrazoglu ES, Sari F, Velioglu YS (2006) Effects of cooking methods on chlorophylls, pheophytins and color of selected green vegetables. Int J Food Sci Technol 41: 281-288.

15. Singh S, Singh DR, Salim KM, Nayak D, Roy SD (2015) Changes in phytochemicals, anti-nutrients and antioxidant activity in leafy vegetables by microwave boiling with normal and 5% NaCl solution. Food Chem 176: 244-253.

16. Roy MK, Juneja LR, Sobe S, Tsushida T (2009) Steam processed broccoli (Brassica oleracea) has higher antioxidant activity in chemical and cellular assay systems. Food Chem 114: 263-269.

17. Mazzeo T, Maria P, Emma C, Vincenzonad V, Tommaso G (2015) Impact of the industrial freezing process on selected vegetables-Part II. Colour and bioactive compounds. Food Res Int 75: 89-97.

18. Maria P, Tommaso G, Nicoletta P, Massimiliano R, Maria Z, et al. (2015) Impact of the industrial freezing process on selected vegetables Part I. Structure, texture and antioxidant capacity. Food Res Int 74: 329-337.

19. Schultze MO (1996) Methods of vitamin assay. Association of vitamin Chemists Inc, Intersci Publi: 306-312.

20. Khwairakpam B, Balwinder S (2014) Effect of different cooking methods on the antioxidant components of carrot. Biosci Disc 5: 112-116.

21. Singleton VL, Rossi JA (1965) Colorimetry of Total Phenolics with Phosphomolybdic-Phosphotungstic Acid Reagents. Am J Enol Vitic 16: 144-158.

22. Marinova D, Ribarova F, Atanassova M (2005) Total phenolics and total flavonoids in Bulgaria fruits and vegetables. Chem. Technol. Metallurgy 40: 255-260.

23. Akowuah GA, Ismail Z, Norhayati I, Sadikun A (2005). The effects of different extraction solvents of varying polarities of polyphenols of Orthosiphon stamineus and evaluation of the free radical-scavenging activity. Food Chem 93: 311-317.

24. Steel RGD, Torrie JH, Dickey D (1997) Principles and procedures of statistics: a biometrical approach. 3rd edn. McGraw Hill Book Co. Inc, New York.

25. Gonclaves EM, Pinheiro J, Abreu M, Brandao TRS, Silva CLM (2010) Carrot (Daucus carota) peroxidase inactivation, phenolic content and physical changes kinetics due to blanching. Food Eng 97: 574-581.

26. Nwanekezi EC, Okorie SU (2005) Effects of processing and storage on the physicochemical and sensory properties of Okra (Hibiscus esculentum). Pak J Food Sci. 15: 25-30.

27. Laura D, Boccia GL, Lucarini M (2010) Influence of heat treatments on carotenoid content of cherry tomatoes. J Foods 2: 353-363.

28. Ishiwu C, Wouno I, James O, Tochkwu E (2014) Effect of thermal processing on lycopene, beta-carotene and Vitamin C content of tomato. Food Nutri Sci 2: 87-92.

29. Michaela M, Stuparic M, Schieber A, Carle R (2003) Effects of thermal processing on trans–cis-isomerization of b-carotene in carrot juices and carotene-containing preparations. Food Chem 83: 609-617.

30. Hunter KJ, Fletcher JM (2002) The antioxidant activity and composition of fresh, frozen, jarred and canned vegetables. Innovative Food Sci. Technol 3: 399-406.

31. Lee SK, Kader AA (2000) Preharvest and postharvest factors influencing vitamin C content of horticultural crops. Postharvest Biol Tech 20: 207-220.

32. Ismail A, Marjan Z, Foong CW (2004) Total antioxidant activity and phenolic content in selected vegetables. Food Chem 87: 581-586.

33. Sahlin E, Savage GP, Lister CE (2004) Investigation of the antioxidant properties of tomatoes after processing. Food Compo Anal 17: 635-647.

34. Zhang D, Hamauzu Y (2004) Phenolics, ascorbic acid, carotenoids and antioxidant activity of broccoli and their changes during conventional and microwave cooking. Food Chem 88: 503-509.

35. Sangeeta S, Charu LM (2013) Effect of steaming, boiling and microwave cooking on the total phenolics, flavonoids and antioxidant properties of different vegetables of Assam, India. Int J Food and Nutri Sci 2: 47-53.

36. Bernhardt S, Schlich E (2005) Impact of different cooking methods on food quality: Retention of lipophilic vitamins in fresh and frozen vegetables. Food Eng 77: 327-333.

37. Kahkonen MJ, Hopia AI, Vuore HJ (1999) Antioxidant activity of plant extracts containing phenolic compounds. J. Agric. Food Chem 47: 3954-3962.

38. Faller ALK, Fialho E (2009) The antioxidant capacity and polyphenol content of organic and conventional retail vegetables after domestic cooking. Food Res Int 42: 210-215.

39. Parr AJ, Bolwell GP (2000) Phenols in the plant and in man. The potential for possible nutritional enhancement of the diet by modifying the phenols content or profile. Food Sci Agric 80: 985-1012.

40. Sultana B, Anwar F, Iqbal S (2008) Effect of different cooking methods on the antioxidant activity of some vegetables from Pakistan. Int J Food Sci 43: 560-567.

Development and Quality Evaluation of Tamarind Plum Blended Squash During Storage

Ibrahim Khan[1]*, Rehman AU[1], Khan SH[3], Qazi IM[1], Arsalan khan[2], Shah FN[2] and Rehman TU[4]

[1]The University of Agriculture Peshawar, Khyber Pakhtunkhwa, Pakistan
[2]Agricultural Research Institute ARI Tarnab Peshawar, Khyber Pakhtunkhwa, Pakistan
[3]Gomal University of D.I. Khan, Pakistan
[4]Abdul Wali Khan University, Mardan, Pakistan

Abstract

The achievement was done to study the combination of tamarind plum blended squash for 90 days' interval at room temperature. Tamarind and plum was added at a combination of 750: 0, 650: 100, 550: 200, 450: 300, 350:400, 250:500, 150:600 and 50:700 represent each treatment. The prepared tamarind plum blended squash was analyzed physio-chemically for TSS, Ascorbic acid, acidity, sugar acid ration, pH, reducing and non-reducing sugar, organoleptically for taste, color, texture and overall acceptability for a total period of 90 days. The result of the statistical analysis showed that treatment and storage interval shows a significant (P<0.05) effect both physio-chemical and organoleptic evaluation. Results also revealed that the decrease occurred in ascorbic acid content from (39.49 mg/100 gm to 27.40 mg/100 gm), titratable acidity (1.09% to 0.98%),non-reducing sugar (44.36% to 21.97%), and sensory evaluation included taste (6.85 to 5.83), color (6.33 to 5.36), flavor (7.54 to 5.75) and overall acceptability (8.03 to 6.14) while increased was found in total soluble solid (48.98°brix to 49.61°brix), sugar-acid ratio (44.94 to 50.79), pH (2.77 to 2.84), reducing sugar (17.21% to 31.23%) during storage. The maximum mean values were observed for TSS is TPS_7 (51.64°brix), ascorbic acid TPS_7 (37.87 mg/ 100 gm), titratable acidityTPS_1 (2.31%), sugar acid ratio TPS_0 (50.55), pH TPS_7 (2.93), reducing sugar TPS_0 (25.32%), non-reducing sugar TPS_4 (37.64%), color TPS_5 (6.70), flavourTPS_5 (7.54), taste TPS_5 (7.00) and overall acceptability TPS_5 (7.76). Among all the treatment TPS_5 was found to be the best. The result revealed that significant (P<0.05) decreased was found in physio-chemical and organoleptic parameter of treatment TPS_5.

Keywords: Plum; Tamarind; Vitamin C; Sodium benzoate; Acidity

Introduction

Tamarind (*Tamarind indica L.*) belongs to *Caesalpiniaceae* family. It is mostly grow in tropical Africa but has become naturalized in North and South America from Florida to Brazil, also grown in subtropical China, India, Pakistan, China, Thailand, Philippines, Indonesia and Spain. Tamarind fruit can be used for many purposes such is digestive, carminative, laxative, expectorant and tonic blood [1]. Tamarind pulp has medicinal purposes also and continues to be used by many people in Africa, Asia and America [2]. Tamarind juice have certain disadvantages such as unappetizing color, loss of fresh taste and spoiled easily [3] and hypoglycemic activity [4]. Tamarind pulp is mainly used for souring food products like chutneys, sambar, curries and sauces. Tamarind pulp is also used in preparations of jams, jellies, ice-creams, wine like beverages, canned tamarind juice and syrup. It is also enjoyed in the form of refreshing drinks and beverages. Fruit are commonly processed into juices, nectars, fruit punch, concentrates, glazed and crystallized fruit. The pulp can be used with original flavor after thermal processing [2]. Tamarind fruit contain low water content and is difficult to extract pulp from the fruit. With the advancing of technologies pulp of tamarind can be extracted by conventional processing techniques like soaking, maceration and straining. With the use of such techniques we can easily extract pulp [5]. The pulp of tamarind contains tartaric acid, reducing sugars, pectin, proteins, fiber, and cellulosic materials. The acid and sugar contents differ from sample to sample; for example, tartaric acid: 8%-18%, reducing sugars 25% to 45%, pectin 2% to 3.5% and proteins 2% to 3% [6]. Tamarind pulp has rich aroma and pleasant acidic taste which is widely used as a chief souring agent for curries, sauces, and certain beverages. The pulp also used as a raw material for the preparation of wine like beverages [7].

Plum (*Prunus domestica* L.) is highly perishable climacteric stone fruit and has short shelf-life at optimal temperatures. Decay of plum fruit may be due to mold growth and rapid ripening during storage.

Shelf life of plum can be extend through proper handling, transportation and marketing chain and also to kept in low temperature storage to extent postharvest quality of the fruit [8]. Plum also called as stones fruits consist of a solid covering with seed enclosed. The enclosed seed of plum is richest in proteins, lipids thus, they maybe a cheap source of different substances that could be useful for food, cosmetic, and pharmaceutical industries. The lipid content of plum seeds has already been explored. Plums contain red flesh and peel and are very exciting fruit due to their high content on bioactive compounds, such as the anthocyanins and other polyphenolic compounds with a high antioxidant capacity [9]. These natural substances found in plum acts to prevent diseases such as diabetes and cancer [10]. Concentrated soft drinks are used for refreshing purpose and are very popular drink contains certain proportion of juice. The summer season of Pakistan is long there for mostly people uses such type of beverages. Such type of activities like production, preservation and sale of these beverages provide commercial importance to our country [11]. Fruit beverages are a combination of products containing pulp, juice and water as well as sweetener, coloring, flavoring, and preservatives. Although fruit ingredient present in beverages has a dominant role of providing flavor and overall character, such types of products differ from fruit

*Corresponding author: Ibrahim Khan, Food Science and Technology, Rahat Abad, House No 41, Peshawar, Khyber Pakhtoon Khwa-25000, Pakistan
E-mail: ibrahimfst339@gmail.com

Treatments	Tamarind juice (ml)	Plum juice (ml)	CMC (g)	Sugar (kg)	Water (ml)	potassium meta-bi-sulphite (%)
TPS_0	750	–	2	1	250	0.1
TPS_1	650	100	2	1	250	0.1
TPS_2	550	200	2	1	250	0.1
TPS_3	450	300	2	1	250	0.1
TPS_4	350	400	2	1	250	0.1
TPS_5	250	500	2	1	250	0.1
TPS_6	150	600	2	1	250	0.1
TPS_7	50	700	2	1	250	0.1

Table 1: Proposed plan of study for research.

juices and are labeled accordingly [12]. Keeping in view the importance of tamarind plum fruit; the plum and tamarind blended squash is developed.

Objectives

a. To produce value added beverage from blends of tamarind plum.

b. To develop suitable combination of tamarind plum blended squash.

c. To analyze tamarind plum blended squash for physicochemical and sensory characteristic during storage.

Materials and Methods

Selection of fruits

Tamarind and Plum fruit at optimum maturity were purchased from the local market of Peshawar and was brought to the laboratory of Food Technology section, ARI Tarnab, Peshawar, for preparation of tamarind plum blended squash.

Pretreatment of blended squash

Tamarind and Plum fruit were carefully sorted to discard diseased, damaged, bruised and immature fruits. Then sorted fruits were thoroughly washed with tap water and the water was drained off. The unwanted portion was removed by trimming. The pulp was extracted by using pulping machine (Model.35027, Rochdale England).

Preparation of blended Squash

Tamarind and plum fruit blended squash were prepared following the method of Archana and Laxman [13], showed in (Table 1). The materials were added following the ratio 4:3:1 of sugar, pulp and water respectively.

Packaging and storage

The prepared squash was packed in PET bottles and was stored at room temperature for 3 months and was study for phsico-chemical characteristics and sensory attributes at 15 days of intervals.

Physicochemical analysis

The prepared squash was examining for pH, TSS, Titratable acidity, Vit C, reducing and non-reducing sugar, sugar acid ratio was calculated from the data of TSS and titratable acidity and was measured by method of AOAC [14].

Total soluble solids

TSS (°brix) were find out by the standard method of AOAC [14], method no, 932.14 and 932.12. TSS (°brix) of the blended squash was finding out using hand refractometer. The instrument was calibrated and takes the reading accurately by putting a minute quantity of tamarind plum blended squash.

Titratable acidity

Preparations of standard solution 0.1 N NaOH: Take 6.30 g of oxalic acid and 4.5 g of NaOH in a volumetric flask and add distilled water in it to make a volume of 1 liter separately. Take 10 ml of 0.1 N solution of NaOH and titrate against 0.1 N solution of oxalic acid. Add 3 drops of phenolphthalein (indicator). Repeat the experiment for three times. Taken the reading till pink color is appears.

Titration of sample: Take 10 ml of squash sample, dissolved in distilled water to make a volume of 100 ml. Then take 10 ml of sample solution and add two drops of phenolphthalein and titrate along 0.1N NaOH solution. Repeat the experiment for 3 times to reduce error. Take the reading when pink color is appears.

$$Acidity(\%) = \frac{CF \times N \times T \times D \times 100}{V \times S}$$

Where:

C.F = Correction Factor for acidity.

N = Normality of sodium hydroxide used.

T = ml of sodium hydroxide used.

D = Dilution Factor for sample.

V = Sample taken for dilution.

S = Sample taken for titration.

Sugar acid ratio

Sugar acid ratio for tamarind plum blended squash was calculated using the formula.

$$Sugar\ acid\ ratio = \frac{Total\ Soluble\ Solids\,(TSS)}{Titratble\ acidity\,(\%)}$$

pH

The pH;is hydrogen ion concentration and it ranges from 1 to 14 that shows acidity and alkalinity of the sample, while the pH with 7 is neutral that is pure water indication. To find out pH of the sample, proper method of AOAC [14], 2005.02 was applied. Switch on the pH meter and standardized with the buffer solution of pH 4 and pH 7, respectively. Take10 ml of tamarind plum sample in a beaker and put the electrode in it and note the result.

Reducing sugar

To analyze/ reducing sugar of tamarind plum blended squash standard method' of AOAC [14], 920.183 was' applied.

Reagents

Fehling A: Dissolved. 34.65g of $CuSO_4.5H_2O$ in 500 ml of distilled, water.

Fehling B: Take 173 g of potassium/ titrate and 50gg of NaOH in beaker, dissolve it in 10 ml of water. The prepared solution was; taken and put into 500 ml conical flask and volume was prepared up to the mark by means of distillation water.

Methylene blue: Methylene blue is an indicator. Take 0.2 g of methylene; blue in 100 ml of volumetric flask and dissolve it in 150 ml of distilled water and the level was]made up to the spot, through further addition distilled water.

Procedure: Take 10 ml of tamarind plum blended squash sample and add distilled water to make the exact volume of 100 ml. Then take 5 ml of Fehling A and 5 ml of Fehling B, with 10 ml of distilled water was taken in conical flask. Heat was given to the flask till boiling start. Add the solution from the burette drop by drop till color becomes bricks red. 2 drops of methylene blue was added in a boiling solution. If color changes from red to blue the reaction needed to add extra tamarind plum solution till brick red color persists.

Calculation: Amount of Fehling A is 5 ml + % ml of Fehling B = X ml of the 10% of sample solution is equal of 0.05 g of reducing sugar × 100 ml of 10 % sample solution will contain.

$$100 \text{ ml of } 10\% \text{ solution will contain} = \frac{0.05 \times 100}{X \text{ ml}} = Y \text{ g of reducing sugar}$$

$$\text{Reducing sugar}(\%) = \frac{Y \times 100}{10}$$

Non-reducing sugars

To investigate non-reducing sugar of tamarind plum blended squash standard method of AOAC [14], 920.184 was applied.

Procedure: 10 ml of sample was taken in volumetric flask and volume was made 100 ml with distill water. 20 ml of solution was taken and dilute with 10 ml of 1 N HCl. Mixture was heated till boiling, 10 ml of 1 N of NaOH was added after cooling and volume was made 250 ml. Take 5 ml Fehling A and B solution and dilute with 10 ml distilled water. Heat the solution to boiling and add tamarind plum blended diluted solution drop by drop till red brick color appears. Add 2 drops of methylene blue to check either the reaction is completed or not. For determination of non-reducing sugar the following formula was applied.

Calculations: Solution is equal to X ml = 0.05 g of reducing sugars

250 ml of sample contains = 259 × 0.05 / ml = Y g of reducing sugars

This 250 ml of sample solution was prepared from 20 ml of 10%.

Sample solution contains Y × 100 / 20 = P g reducing sugar.

10 ml of sample solution contain = P g of reducing sugar.

100 ml of sample solution contain = P × 100/10 = Q g of total reducing sugar.

Q g of reducing sugar = inverted sugar + free reducing sugar.

Formula for non-reducing sugar is = total reducing sugar − free reducing sugar.

Ascorbic acid

Preparation of standard solutions: 42 mg of sodium bicarbonates (NaHCO₃) and 50 mg of 2,6 dichlorophenol indophenols dye to make the volume of 250 ml with distilled water. To prepare standard solution of Vitamin C take 50 mg of ascorbic acid and poured in 50 ml 0.4%

of oxalic acid solution. Keep the solution for 24 hours. Take 5 ml of ascorbic acid solution and titrate along dye till pink color appears and persists for one minute. Formula used to find out dye factor.

$$\text{Dye factor}(F) = \frac{\text{vitamin C solution taken in ml}}{\text{Volume of used dye}}$$

Titration of sample: Take 10 ml of tamarind plum blended squash and make a volume of 100 ml with 0.4% oxalic acid solution. 10 ml of sample solution were taken in a flask and titrate along dye to appear pink color and persist for 15 sec. Formula for Vitamin C content is:

$$\text{Ascorbic acid (mg/100 g)} = \frac{F \times T \times 100}{S \times D}$$

Where,

F = Standardization factor = ml of ascorbic acid / ml of pigment used.

T = ml of pigment used for sample.

S = ml of diluted sample taken for titration.

D = ml of sample taken for dilution.

Sensory evaluation

The samples of tamarind plum blended squash were sensory evaluated for color, texture, flavor and overall acceptability by 10 trained judge's panel. Organoleptic study was carried out for about 3 month. The evaluations were done using 9 points hedonic scale of Larmand [15].

Statistical analysis

All the data concerning treatments and storage interval were statistically analyzed by means of complete Randomized Design (CRD) 2 Factorial as recommended by Hicks [12] and the means were find out using least significant difference (LSD) Test at 5% possibility level.

Result and Discussion

Total soluble solids (°brix)

According to Table 2 the sample of tamarind plum blended squash were studied for TSS (°brix). The data of the samples shows significant ($P<0.05$) increase during keeping time of storage. The sample of the tamarind plum blended squash were in the range of 47.45 (TPS_2) to 51.35 (TPS_7). The TSS of all the samples was gradually rises from 48.29 (TPS_0) to 51.94 (TPS_7) during 90 days of storage. Table 2 also showed that minimum mean TSS value was recorded for TPS_2 (47.78) while TPS_7 had maximum mean value of (51.64). Similarly, this is also observed from the data that maximum percent increase in TSS was found in TPS_2 (1.76), while a sample TPS_5 (1.05) had a minimum percent increase. The data showed a significant change in blended squash of tamarind plum during storage. Similar observations were recorded by Kotecha and Kadam [16] in tamarind syrup and Nath et al. [17] in ginger blended with mandarin squash that because of hydrolysis of polysaccharides like starch and pectic substances into simpler substances during processing increases in TSS. Gillani [18] investigated increase in TSS in different mango cultivar. With the use of chemical preservative TSS of apple pulp increases Kinh et al. [19]. It is concluded that TSS of tamarind plum blended squash increased with storage and treatment.

Ascorbic acid (Vitamin C)

The data from Table 3 shows significant ($p < 0.05$) effect on

Treatment	Storage interval							% Inc	Means
	Initial day	15	30	45	60	75	90		
TPS$_0$	47.55	47.69	47.78	47.88	47.92	48.12	48.29	1.53	47.89g
TPS$_1$	47.65	47.74	47.84	47.96	48.11	48.23	48.33	1.41	47.98f
TPS$_2$	47.45	47.54	47.65	47.74	47.84	47.93	48.03	1.76	47.78h
TPS$_3$	48.35	48.44	48.56	48.63	48.72	48.81	48.92	1.17	48.63e
TPS$_4$	49.25	49.34	49.43	49.52	49.62	49.73	49.82	1.14	49.53d
TPS$_5$	49.95	50.04	50.11	50.2	50.29	50.39	50.48	1.05	50.21c
TPS$_6$	50.25	50.34	50.43	50.54	50.62	50.73	50.81	1.10	50.53b
TPS$_7$	51.35	51.45	51.54	51.63	51.74	51.85	51.94	1.14	51.64a
Mean	48.98g	49.07f	49.17e	49.26d	49.36c	49.47b	49.61a	--	--

a-g Values of different alphabetic letter shows significant (P<0.05) difference from each other.

Table 2: TSS (°brix) of squash prepared from blending of tamarind and plum juice at different levels.

Treatments	Storage interval							% Dec	Means
	Initial day	15	30	45	60	75	90		
TPS$_0$	35.79	33.35	31.09	29.78	27.87	25.35	23.76	33.61	29.69g
TPS$_1$	36.23	35.98	33.98	31.09	26.98	25.18	23.93	33.95	30.60f
TPS$_2$	37.98	36.01	34.98	31.35	28.64	26.98	25.75	32.20	31.68e
TPS$_3$	38.09	37.98	33.09	32.29	30.01	28.87	26.75	31.23	32.68d
TPS$_4$	39.91	37.45	35.67	33.91	31.25	29.65	27.85	30.22	33.67c
TPS$_5$	41.25	39.28	38.01	36.61	34.44	32.11	30.01	27.03	35.98b
TPS$_6$	41.99	39.24	37.35	36.12	33.42	31.81	29.19	30.48	35.59b
TPS$_7$	43.86	41.23	39.09	37.54	36.15	34.54	31.87	27.34	37.87a
Mean	39.49a	37.58b	35.72c	33.69d	31.11e	29.31f	27.40g	--	--

a-g Values of different alphabetic letter shows significant (P<0.05) difference from each other.

Table 3: Ascorbic acid of squash prepared from blending of tamarind and plum juice at different levels.

Treatments	Storage interval							% Dec	Mean
	Initial day	15	30	45	60	75	90		
TPS$_0$	1.00	0.98	0.96	0.95	0.93	0.92	0.90	10.00	2.08h
TPS$_1$	1.03	1.01	0.99	0.97	0.95	0.93	0.91	11.65	2.31g
TPS$_2$	1.05	1.03	1.01	0.99	0.97	0.95	0.93	11.43	2.29f
TPS$_3$	1.08	1.06	1.04	1.03	1.01	0.99	0.97	10.19	2.17e
TPS$_4$	1.11	1.09	1.07	1.05	1.03	1.02	1.00	9.91	2.16d
TPS$_5$	1.15	1.13	1.11	1.09	1.08	1.06	1.04	9.57	2.15b
TPS$_6$	1.13	1.11	1.09	1.07	1.05	1.04	1.02	9.73	2.16c
TPS$_7$	1.18	1.16	1.15	1.13	1.11	1.08	1.06	10.17	2.25a
Mean	1.09a	1.07b	1.05c	1.04d	1.02e	1.00f	0.98g	--	--

a-g Values of different alphabetic letter shows significant (P<0.05) difference from each other.

Table 4: Titratable acidity of squash prepared from blending of tamarind and plum juice at different levels.

storage and treatment of blended squash of tamarind plum. There shows a significant (p < 0.05) decrease in vitamin C. The ascorbic acid of tamarind plum squash was in the zero days from 35.79 (TPS$_0$) to 43.86 (TPS$_7$) which is then gradually decrease from 23.76 (TPS$_0$) to 31.87 (TPS$_0$) during storage period of 90 days. Mean value of ascorbic acid was recorded 39.49 at zero-day interval, while 27.40 at for the period of 90 days. According to Table 3, TPS$_7$ had a highest mean value (37.87), while TPS$_0$ had minimum (29.69) mean value. Sample TPS$_1$ (33.95) shows highest percent decrease, while sample TPS$_5$ (27.03) has lowest. The above results are in agreement with Kinh et al. [19], studied lower percent of ascorbic acid found in apple pulp affected by both temperature and light. Saleem et al. [20] studied that time interval also decreases ascorbic acid value. Bezman et al. [21] also concluded that ascorbic acid of grape juice also decreased during time of storage in room temperature. Storage interval, oxygen, light and heat treatment decrease the effect of ascorbic acid by both enzymatic and non-enzymatic catalyst [22]. In most liable nutrients, Vitamin C is very important because its degradation is used as an indicator of quality.

Titratable acidity

In Table 4 samples of tamarind plum squash shows significant (P<0.05) difference during time period of storage. The % acidity of the squash samples was in the range of 1 (TPS$_0$) to 1.18 (TPS$_7$) at initial day, while showed a decreasing trend of 0.9 (TPS$_0$) to 1.06 (TPS$_7$) correspondingly during 90 days of interval. Mean value at initial day was 1.09, decreases to 0.98 at 90 days intervals. The sample TPS$_1$ (2.31) shows high value of mean while sample TPS$_5$ (2.15) shows minimum TPS$_1$ (11.65) had a maximum % decrease in acidity, while TPS$_5$ (9.57) showed the minimum decrease in percent acidity. Increase of acidity is because of storage condition and pectic substance break down [23]. Hye et al. [24] found increasing trend in acidity, while pH decrease of fruit juices during processing and storage time. Analogous result was reported by Gajanana [25] that hydrolysis of polysaccharides and non-reducing sugars reduces acid of amla juice, where the acid is converting to hexose sugars or complexes in the presence of metal ions. Lakshmi et al. [26] and Nidhi et al. [27] also observed reduction in acidity during the storage period of the tamarind RTS and RTS bael-guava beverages

Treatment	Storage interval							% Inc	Mean
	Initial day	15	30	45	60	75	90		
TPS_0	47.55	48.66	49.77	50.40	51.53	52.30	53.66	11.38	50.55a
TPS_1	46.26	47.27	48.32	49.44	50.64	51.86	53.11	12.89	49.56b
TPS_2	45.19	46.16	47.18	48.22	49.32	50.45	51.94	12.99	48.35c
TPS_3	44.77	45.70	46.69	47.21	48.24	49.30	50.43	11.23	47.48d
TPS_4	44.37	45.27	46.20	47.16	48.17	48.75	49.82	10.94	47.11e
TPS_5	43.70	44.55	45.43	46.37	46.87	47.86	48.86	10.56	46.23f
TPS_6	44.20	45.08	45.97	46.92	47.90	48.45	49.49	10.68	46.86e
TPS_7	43.52	44.35	44.82	45.69	46.61	48.01	49.00	11.19	46.00f
Mean	44.94g	45.88f	46.80e	47.68d	48.66c	49.62b	50.79a	--	--

a-g Values of different alphabetic letter shows significant (P<0.05) difference from each other.

Table 5: Sugar acid ratio of squash prepared from blending of tamarind and plum juice at different levels.

Treatment	Storage interval							% Dec	Means
	Initial day	15	30	45	60	75	90		
TPS_0	2.68	2.69	2.71	2.72	2.73	2.74	2.76	2.90	2.72h
TPS_1	2.69	2.07	2.72	2.73	2.75	2.76	2.77	2.89	2.73g
TPS_2	2.72	2.73	2.74	2.75	2.76	2.78	2.79	2.51	2.75f
TPS_3	2.72	2.74	2.75	2.76	2.78	2.79	2.08	2.86	2.76e
TPS_4	2.75	2.76	2.77	2.79	2.81	2.82	2.83	2.83	2.79d
TPS_5	2.86	2.88	2.89	2.91	2.92	2.92	2.93	2.39	2.90b
TPS_6	2.81	2.83	2.84	2.85	2.86	2.87	2.88	2.43	2.85c
TPS_7	2.89	2.9	2.92	2.93	2.94	2.95	2.97	2.69	2.93a
Mean	2.77g	2.78f	2.79e	2.81d	2.82c	2.83b	2.84a	--	--

a-g Values of different alphabetic letter shows significant (P<0.05) difference from each other.

Table 6: pH of squash prepared from blending of tamarind and plum juice at different levels.

respectively. It's released from the data that the titratable acidity decreases with storage and treatment.

Sugar acid ratio

Table 5 shows a significant (P<0.05) effect on both the treatment effect and storage effect on blended squash of tamarind plum. The ratio sugar acid of squash samples was in the range of 43.52 (TPS_7) to 47.55 (TPS_0) at initial day, while showed an increasing trend of 48.86 (TPS_6) to 53.66 (TPS_1) correspondingly during 90 days storage interval. Initial day storage mean was 44.94, which increase to 50.79. Sample TPS_0 (50.55) show high mean value while the sample TPS_7 (46.00) with lowest value of mean. Sample TPS_0 showed % increase of maximum (11.38), while TPS_5 (10.56) showed the minimum increase in percent sugar acid ratio. According to Chyau et al. [28] substances like pectin, reducing sugar, total sugar and acidity of guava fruit decreases at ripe stage while the sugar/acid ratio of the fruit guava increased. It is concluded from the data that sugar acid ratio increased with time by storage and treatment.

pH

The data of tamarind plum blended squash shows a decreasing trend during the period of storage intervals. Tamarind plum squash pH was in between 2.68 (TPS_0) to 2.89 (TPS_7) at zero days of interval which gradually increases from 2.76 (TPS_0) to 2.97 (TPS_7) during 90 days of storage time. Mean of the data at 1st day was 2.77, and then decreased to 3.73 during keeping time of storage. Sample TPS_7 of tamarind plum squash shows high mean 2.93, while sample TPS_0 of tamarind plum has a lowest mean 2.72. Sample TPS_0 has high percent decrease of (2.90) in case of pH. However, squash sample TPS_5 (2.39) of the minimum pH with percent decrease found. There found a significant (P<0.05) effect of tamarind plum blended squash in case of time and treatment. Nath et al. [17] investigate same results for kinnow (mandarin) ginger

squash. According to Jitareerat et al. [29] pH of fruits and vegetables changes because of heat treatment on biochemical substances, decrease of respiration and metabolic process. Cecilia and Maia [30] studied a decreasing trend in pH of apple juice during keeping time. With the increase of acidity and pectin hydrolysis pH of the juice decline [31]. Thus, concluded that pH increases with treatment and storage effects on tamarind plum blended squash.

Reducing sugar

Table 6 shows effect of time interval and treatment on blended squash of tamarind plum. Reducing sugars of tamarind plum squash was in between 17.10 (TPS_7) to 17.32 (TPS_1) at initial day. There shows an increasing trend of 28.10 (TPS_0) to 33.23 (TPS_3) during 90 days of storage time period. Initial day mean of tamarind plum was 17.21, which shows gradual increase of 31.23 during the storage time period. Tamarind plum sample TPS_0 (25.32) showed the maximum mean value, however sample TPS_5 (23.08) had minimum mean value. Squash sample TPS_3 (48.18) found with maximum percent increase, while the sample TPS_5 (39.07) with lowest percent increase in reducing sugar. There found a significant (P<0.05) effect on tamarind plum blended squash during treatments and storage intervals of times. The above results show similarity with the report of Kotecha and Kadam [16] and Sahu et al. [32] on tamarind syrup and mango lemongrass beverage respectively reported an increase trend in total and reducing sugars. Both acidity and temperature has caused positive effect on reducing sugar (convert sucrose to glucose and fructose) [33]. Reducing sugar of fruits increases because of sucrose reduction. It is concluded that the reducing sugars of the treatment increases with time interval.

Non-reducing sugar

In the Table 7 tamarind plum blended squash data are significantly (P<0.05) reduced during storage and treatment intervals. The data of squash samples was in the range of 40.20 (TPS$_0$) to 47.1 (TPS$_4$) at initial day. While during storage period the non-reducing sugar content decrease gradually from 19.35 (TPS$_0$) to 25.76 (TPS$_5$) at 90 days of interval. Initial mean data was 44.36, which shows a reducing trend of 21.97. Tamarind plum blended sample TPS$_4$ (37.64) with maximum mean, while squash sample TPS$_7$ (30.87) with minimum mean. High percent decrease (54.95), for sample TPS$_2$. However, TPS$_5$ (44.54) showed the minimum % decrease. Kotecha and Kadam [17] and Sahu et al. [32] reported same results of increasing total sugar as well as reducing sugar, while decreasing of non-reducing sugar for tamarind syrup and mango lemongrass beverage respectively, during storage. Main cause of reducing sugar conversion to non-reducing sugar is glycogenesis, also change of vitamins, sugar and organic acid change during storage intervals in carrot pulp. Thus, concluded that the non-reducing sugar decreases with treatment and storage condition.

Taste

According to the data of Table 8, statistically shows a reducing trend significantly (P<0.05) of tamarind plum blended squash during

treatment and storage condition. The sensory score for taste of tamarind plum blended squash were in the range of 6.6 (TPS$_0$, TPS$_7$) to 7.4 (TPS$_5$) at zero days of interval, there found a gradual decrease of 5.5 (TPS$_1$, TPS$_7$) to 6.5 (TPS$_5$) during the storage period of 90 days. Mean data for initial day was 6.85, which gradually down to 5.83. The squash of tamarind plum sample TPS$_5$ (7.00) shows highest mean, while with lowest score of sample TPS$_7$ (6.04). Sample (TPS$_1$) with maximum decrease of 17.91%, while minimum decrease of 12.16% was observed by TPS$_5$. The data above had a significant effect on taste of tamarind plum blended squash during storage and treatment time intervals. During RTS beverages light effects acids and ascorbic acid (Vitamin C) present in orange squashes [34]. The RTS of tamarind shows same results according to Kotecha and Kadam [17]. The depletion of taste is effected by acid, pH fluctuation [35].

Color

There shows a decreasing effect significantly (P<0.05) on color of tamarind plum squash during period of time interval. At zero day interval, the sensory score for color of tamarind plum squash samples

Treatment	Storage interval							% Inc	Mean
	Initial day	15	30	45	60	75	90		
TPS$_0$	17.25	20.25	23.76	25.48	28.15	30.35	31.98	46.06	25.32a
TPS$_1$	17.32	19.01	21.24	25.31	28.54	30.35	32.31	46.39	24.88ab
TPS$_2$	17.28	18.09	21.29	24.32	26.45	28.54	31.46	45.07	24.03bc
TPS$_3$	17.22	20.02	22.25	25.32	28.45	30.21	33.23	48.18	25.27a
TPS$_4$	17.24	19.25	22.87	24.54	26.93	28.54	30.65	43.75	24.29bc
TPS$_5$	17.12	20.21	22.23	23.35	24.43	26.15	28.01	39.07	23.08d
TPS$_6$	17.15	19.01	22.98	24.84	26.89	29.45	31.98	46.37	24.63abc
TPS$_7$	17.01	19.19	21.09	23.65	26.98	28.09	30.15	43.28	23.98c
Mean	17.21g	19.53f	22.32e	24.60d	27.10c	29.06b	31.23a	--	--

a-g Values of different alphabetic letter shows significant (P<0.05) difference from each other.

Table 7: Reducing sugar of squash prepared from blending of tamarind and plum juice at different levels.

Treatment	Storage interval							% Dec	Mean
	Initial day	15	30	45	60	75	90		
TPS$_0$	40.02	39.56	35.43	31.87	27.85	22.96	19.35	51.87	31.03d
TPS$_1$	42.35	40.12	35.24	31.01	27.43	24.45	20.25	52.18	31.55d
TPS$_2$	44.95	42.01	38.65	34.21	30.24	23.46	20.25	54.95	33.41c
TPS$_3$	45.98	43.25	40.13	35.65	31.35	25.78	21.98	52.20	34.87b
TPS$_4$	47.01	45.24	41.21	39.19	34.99	30.01	25.65	45.54	37.64a
TPS$_5$	46.45	43.87	40.24	35.87	30.87	28.31	25.76	44.54	35.91b
TPS$_6$	46.01	45.01	40.95	37.45	30.14	25.09	22.45	51.30	35.44b
TPS$_7$	41.76	39.35	34.76	30.12	27.09	22.12	20.01	51.87	30.87d
Mean	44.36a	42.32b	38.33c	34.42d	30.10e	25.39f	21.97g	--	--

a-g Values of different alphabetic letter shows significant (P<0.05) difference from each other.

Table 8: Non-reducing sugar of squash prepared from blending of tamarind and plum juice at different levels.

Treatment	Storage interval							% Dec	Mean
	Initial day	15	30	45	60	75	90		
TPS$_0$	6.6	6.4	6.3	6.2	6.0	5.8	5.7	13.64	6.14de
TPS$_1$	6.7	6.6	6.4	6.1	6.0	5.8	5.5	17.91	6.16cd
TPS$_2$	6.8	6.6	6.5	6.2	6.0	5.9	5.7	16.18	6.24cd
TPS$_3$	6.9	6.7	6.5	6.4	6.3	6.2	6.0	13.04	6.43b
TPS$_4$	7.0	6.9	6.7	6.5	6.3	6.1	6.0	14.29	6.50b
TPS$_5$	7.4	7.3	7.2	7.0	6.9	6.7	6.5	12.16	7.00a
TPS$_6$	6.8	6.6	6.5	6.2	6.1	5.9	5.7	16.18	6.26c
TPS$_7$	6.6	6.3	6.2	6.0	5.9	5.8	5.5	16.67	6.04e
Mean	6.85a	6.68b	6.54c	6.33d	6.19e	6.03f	5.83g	--	--

a-g Values of different alphabetic letter shows significant (P<0.05) difference from each other.

Table 9: Taste of squash prepared from blending of tamarind and plum juice at different levels.

Treatment	Storage interval							% Dec	Mean
	0	15	30	45	60	75	90		
TPS$_0$	6.5	6.4	6.2	6.1	5.9	5.6	5.5	15.38	6.03b
TPS$_1$	6.3	6.2	6.0	5.9	5.6	5.5	5.3	15.87	5.83c
TPS$_2$	6.1	6.0	5.8	5.6	5.5	5.3	5.1	16.39	5.63e
TPS$_3$	6.0	5.9	5.7	5.4	5.3	5.2	5.0	16.67	5.50f
TPS$_4$	6.2	6.0	5.9	5.8	5.7	5.4	5.3	14.52	5.76d
TPS$_5$	7.1	7.0	6.9	6.7	6.6	6.4	6.2	12.68	6.70a
TPS$_6$	6.2	6.0	5.9	5.6	5.4	5.3	5.2	16.13	5.66e
TPS$_7$	6.2	6.0	5.9	5.7	5.6	5.5	5.3	14.52	5.74d
Mean	6.33a	6.19b	6.04c	5.85d	5.70e	5.53f	5.36g	--	--

a-g Values of different alphabetic letter shows significant (P<0.05) difference from each other.

Table 10: Color of squash prepared from blending of tamarind and plum juice at different levels.

Treatment	Storage interval							% Dec	Mean
	Initial day	15	30	45	60	75	90		
TPS$_0$	7.1	6.6	6.0	5.4	4.4	3.3	2.3	67.61	5.01c
TPS$_1$	7.8	7.5	7.1	6.7	6.2	5.6	4.9	37.18	6.54b
TPS$_2$	7.6	7.5	7.3	7.2	7.0	6.6	6.3	17.11	7.07ab
TPS$_3$	7.5	7.4	7.2	7.0	6.9	6.8	6.7	10.67	7.07ab
TPS$_4$	7.7	7.4	7.2	6.8	6.6	6.3	6.0	22.08	6.86b
TPS$_5$	7.9	7.8	7.7	7.6	7.4	7.3	7.1	10.13	7.54a
TPS$_6$	7.4	7.3	7.1	7.0	6.8	6.7	6.3	14.86	6.94b
TPS$_7$	7.3	7.2	7.1	7.0	6.9	6.8	6.4	12.33	6.96b
Mean	7.54a	7.34ab	7.09ab	6.84bc	6.53cd	6.18de	5.75e	--	--

a-g Values of different alphabetic letter shows significant (P<0.05) difference from each other.

Table 11: Flavor of squash prepared from blending of tamarind and plum juice at different levels.

Treatment	Storage interval							% Dec	Mean
	Initial day	15	30	45	60	75	90		
TPS$_0$	7.8	7.3	6.8	6.1	5.3	4.1	3.6	53.85	5.86d
TPS$_1$	8.0	7.7	7.4	7.0	6.4	5.8	5.0	37.5	6.76c
TPS$_2$	8.2	7.9	7.5	7.1	6.6	6.0	5.4	34.15	6.96bc
TPS$_3$	8.0	7.8	7.6	7.5	7.4	7.1	6.9	13.75	7.47ab
TPS$_4$	8.1	8.0	7.8	7.5	7.4	7.1	7.0	13.58	7.56a
TPS$_5$	8.2	8.0	7.9	7.8	7.7	7.4	7.3	10.98	7.76a
TPS$_6$	8.0	7.9	7.5	7.4	7.3	7.1	6.9	13.75	7.44ab
TPS$_7$	7.9	7.7	7.6	7.5	7.3	7.1	7.0	11.39	7.44ab
Mean	8.03a	7.79ab	7.51bc	7.24cd	6.93de	6.46ef	6.14f	--	--

a-g Values of different alphabetic letter shows significant (P<0.05) difference from each other.

Table 12: Overall acceptability of squash prepared from blending of tamarind and plum juice at different levels.

from 6 (TPS$_3$) to 7.1 (TPS$_5$) which decreased gradually from 5 (TPS$_3$) to 6.3 (TPS$_5$) through 90 days of intervals. Initial day mean was 6.33, which decreases to 5.36. The sample TPS$_5$ with maximum mean of 6.70 were found, while there found lowest mean of 5.50 for sample TPS$_3$. Decrease of 16.67 % was observed at sample TPS$_3$ while the minimum % decrease was noted at TPS$_5$ (12.68). There found a significant (P<0.05) effect on color of tamarind plum blended squash during storage interval. The result was in favor of Jain et al. [36], reported a decreasing trend in color during 90 days storage of squash. Color of the beverages decreases because of presence of 2 Methyl 3 furanthiol and methanol gives rotten flavors in stored orange juices [21]. Brennder et al. [37] studied that presence of SO$_2$ decreases fruits and vegetables browning.

Flavor

Table 9 shows data of tamarind plum blended squash. The mean sensory scores for flavor of squash decreased significantly (P<0.05) on both treatments and storage time intervals. The judges panel scores for flavor of tamarind plum blended squash from 7.1 (TPS$_0$) to 7.9 (TPS$_5$) during zero days of intervals. However, during storage interval of 90 days' flavor of the squash samples decreased gradually from 2.3 (TPS$_0$) to 7.1 (TPS$_5$). Mean flavor was found 7.54, which decreased to 5.75 throughout the storage period of time intervals. TPS$_5$ was found to be high mean (7.54), while the low score mean (5.01) was obtained for TPS$_0$. The maximum percent decrease in flavor of the squash was recorded in TPS$_0$ (67.61), while minimum decrease of 10.13% was observed at TPS$_5$. The tamarind plum squash was significantly (P<0.05) differ in case of treatment and time interval. Results of physiochemical, sensory properties of orange drink shows similarity were reported by Jain et al. [35]. According to Martin [38] results on pasteurized orange juice shows depletion of organoleptic quality kept in glass bottles. Similar with these results of Paracha [39], that loss of flavor of guava squashes during storage of 3 months of storage interval. A slight difference in flavor may be due to storage conditions and storage time.

Overall acceptability

Table 10 shows the effect of both the treatments and storage interval on overall quality of tamarind plum blended squash. The acceptability of overall quality of the blended squash reduces considerably (P<0.05) on both treatments and storage time interval. The overall acceptance score of tamarind plum squash at initial days ranges from 7.8 (TPS$_0$) to 8.2 (TPS$_5$, TPS$_2$), which fall gradually from 3.6 (TPS$_0$) to 7.3 (TPS$_5$) during the 90 days of storage period of time. Mean value for over-all acceptance was 8.03, which decrease down to 6.14 during the storage period. The highest score of mean (7.76) was observed at TPS$_5$, while minimum score of mean (5.86) was observed at TPS$_0$. The highest percent decrease of 53.85 was recorded at TPS$_0$, while minimum percent decrease of 10.98 was observed at TPS$_5$. The overall acceptability of tamarind plum blended squash is significantly (P<0.05) influenced by treatments and storage interval (Tables 11 and 12). Rosario [34] observed that with the increasing of days' storage overall quality of acceptance decreases. Loss of overall quality were affected by processing like, temperature and storage time [24].

Conclusion and Recommendations

Conclusion

Present work of tamarind plum blended squash was carried out with different proportions. Chemical preservatives were used to inhibit the growth of microbial activity in tamarind plum blended squash. Prepared squash was packed in plastic bottles and stored at room temperature for 90 days of storage. Prepared squash was then evaluated for physicochemical and sensory properties during 90 days of storage. Some physicochemical and sensory analysis was examined to be changed but not affected overall quality of the squash. On the basis of above results it was concluded that sample TPS$_5$ show best in keeping quality during storage time intervals. Hence, the results of sample TPS$_5$ of tamarind plum blended squash is more recommended in terms of commercial use and for large scale industrial production. Squash prepared from tamarind and plum are more acceptable to consumers because of sour test, need commercialization.

Recommendations

1. Different proportion of tamarind pulp can also be used with other fruit pulp.

2. It is suggested to study the influence of storage condition and packaging materials on tamarind plum blended squash.

3. This is recommended to carry a research on non-caloric tamarind plum blended squash.

References

1. Komutarin T, Azadi S, Butterworth L, Keil D, Chitsomboon B, et al. (2004) Extract of the seed coat of *Tamarindus indica* inhibits nitric oxide production by murine macrophages *in vitro* and *in vivo*. Food Chem Toxicol 42: 649-658.

2. Siddig KE, Gunasena HP, Prasad BA, Pushpakumar DK, Ramana KV, et al. (2006) Tamarind monograph. Southampton centre for underutilized crops, Southampton, UK pp: 1-198.

3. Martinello F, Soares SM, Franco JJ, Santos AJ, Sugohara A, et al. (2006) Hypolipemic and antioxidant activities from *Tamarindus indica* L. Pulp fruit extract in hypercholesterolemic hamsters. Food Chem Toxicol 44: 810-818.

4. Maiti R, Das UK, Ghosh D (2005) Attenuation of hyperglycemia and hyperlipidemia in streptozotocin-induced diabetic rats by aqueous extract of seed of *Tamarindus indica*. Bio Pharm Bulletin 28: 1172-1176.

5. Joshi AA, Kshirsagar RB, Sawate AR (2012) Studies on standardization of enzyme concentration and process for extraction of tamarind pulp, variety Ajanta. J Food Process Technol 3: 1-3.

6. Shankaracharya NB (1998) Tamarind chemistry, technology and uses: A critical appraisal. J Food Sci Technol 35: 193-208.

7. Sanchez PC (1985) Tropical fruit wines. A lucrative business Research at Los Banos 3: 10-13.

8. Wang CY (1993) Approaches to reduce chilling injury of fruits and vegetables. Hort Rev 15: 63-95.

9. Diaz MHM, Zapata PJ, Guillen F, Romero DM, Castillo S (2009) Changes in hydrophilic and lipophilic antioxidant activity and related bioactive compounds during postharvest storage of yellow and purple plum cultivars. Postharvest Bio Technol 51: 354-363.

10. Seeram NP, Adams LS, Zhang YJ, Lee R, Sand D (2006) Blackberry, black raspberry, blueberry, cranberry, red raspberry, and strawberry extracts inhibit growth and stimulate apoptosis of human cancer cells *in vitro*. J Agri Food Chem 54: 9329-9339.

11. Ismail S, Rehman S (1995) Beverages: Science and Technology.

12. Hicks D (1990) Production and packaging of non-carbonated fruit juices and fruit beverages. Van Nostrland Reinhold, New York.

13. Archana P, Laxman K (2014) Studies on preparation and storage of tamarind squash. J Spice Aromatic Crop 24: 254-261.

14. AOAC (2012) Official method of analysis. (19th edn.) The Association for Official Analysis in Chemistry, Rockwille, USA.

15. Larmand E (1977) Laboratory method for sensory evaluation of food. Pub. Canada Department of Agriculture, Ottawa.

16. Kotecha PM, Kadam SS (2003) Preparation of ready to serve beverage, syrup and concentrate from tamarind. J Food Sci Tech 40: 76-79.

17. Nath A, Yadav DS, Sarma P, Dey B (2005) Standardization of ginger-kinnow squash and its storage. J Food Sci Tech 42: 520-522.

18. Gillani SSN (2002) Development of mango squash from four different cultivars of mango. Department of Food Science and Technology, NWFP Agricultural University, Peshawar.

19. Kinh AE, Shearer CP, Dunne S, Hoover DG (2001) Preparation and preservation of apple pulp with chemical preservatives and mild heat. J Food Process 28: 111-114.

20. Saleem N, Kamran M, Shaikh SA, Tarar OM, Jamil K (2011) Studies on processing and preparation of peach squash. Pak J Biochem Mol Biol 44: 12-17.

21. Bezman Y, Russell L, Rouseff D, Naim M (2001) 2-Methyl-3-furanthiol and methional are possible off-flavors in stored orange juice. J Agric Food Chem 49: 425-432.

22. Mapson LW (1970) Vitamins in fruits. In: Hulme AC (Ed.) The Biochemistry of Fruits and their Products. Academic Press, London pp: 369-384.

23. Hashmi MS, Alam S, Riaz A, Shah AS (2007) Studies on microbial and sensory quality of mango pulp storage with chemical preservatives. Pak J Nutr 6: 85-88.

24. Hye WY, Streaker CB, Zhang QH, Min DB (2000) Effect of pasteurized electric field on the quality of orange juice and comparison with heat pasteurization. J Agri Fd Chem 48: 4597-4605.

25. Gajanana K (2002) Processing of aonla (*Emblica officinalisGaertn.*) fruits. University of Agriculture Sciences, Dharwad, India.

26. Lakshmi K, Kumar KAKV, Rao LJ, Naidu MM (2005) Quality evaluation of flavored RTS beverage and beverage concentrate. J Food Sci Tech 42: 411-414.

27. Nidhi R, Gehlot R, Singh, Rana MK (2008) Changes in chemical components of RTS bael-guava blended beverages during storage. J Food Sci Tech 45: 378-380.

28. Chayu CC, Wu SY, Chen CM (1992) Differences of volatile and nonvolatile constituents between mature and ripe guava fruit. J Agric and Food Chem 40: 846-849.

29. Jitareerat P, Paumchai S, Kanlayanarat S (2007) Effect of chitosan on ripening enzymatic activity and disease development in papaya (*Carica papaya*) fruit. New Zealand J Crop Hort Sci 35: 211-218.

30. Cecilia E, Maia GA (2002) Storage stability of cashew apple juice preserved by hot fill and aseptic process. University of Ceara, Brazil.

31. Imran A, Rafiullah K, Muhammad A (2000) Effect of added sugar at various concentration on storage stability of guava pulp. Sarhad J of Agric 7: 35-39.

32. Sahu C, Choudhary PL, Patel L, Sahu R (2006) Physico-chemical and sensory characteristics of whey based mango herbal (lemon grass) beverage. Ind Food Packer 60: 127-132.

33. Singh S, Shivhare US, Ahmed J, Raghavan GSV (1999) Osmotic concentration kinetics and quality of carrot preserve. J Food Res Int 32 : 509-514.

34. Muhammad R, Ahmed M, Chaudhry MA, Hussain B, Khan I (1987) Ascorbic acid quality retention in orange squashes as related to exposure to light and container type. J Pak Sci Ind Res 30: 480-483.

35. Rosario MJG (1996) Formulation of ready to drink blends from fruits and vegetables juices. J Philippines 9: 201-209.

36. Jain S, Sankhla APK, Dashora A, Sankhla AK (2003) Physiochemical and sensory properties of orange drink. J Food Sci Tech Ind 40: 656-659.

37. Brenndor K, Oswin CO, Trim DS, Mrema GC, Werek GC (1985) Solar driers and their role in post-harvest processing. Common wealth Sci council 2: 78-83.

38. Martin JJ, Solances E, Bota E, Sancho J (1995) Chemical and organoleptic changes in pasteurized orange juice. Alimentaria 216: 59-63.

39. Paracha GM (2004) Development and storage stability of low caloric guava squash. Agricultural University, Peshawar, Pakistan.

7

Mass Transfer Kinetics of Osmotic Dehydration of Pineapple

Insha Zahoor and Khan MA*

Department of Post-Harvest Engineering and Technology, Faculty of Agricultural Sciences, Aligarh Muslim University, Aligarh, India

Abstract

Current study deals with the kinetics and mathematical modelling of osmotic dehydration of pineapple. Pineapple (*Ananas comosus*), 10 mm thick slices weighing 50 g each, was studied for the osmotic dehydration using hypertonic solutions of sucrose and fructose. The osmotic dehydration process was performed using three levels of temperature 40°C, 50°C and 60°C, three levels of osmotic solution concentration (40%, 50% and 60%) with sample to solution ratio maintained at 1:4, 1:5 and 1:6 respectively. After each interval of time, moisture loss and solid gain was recorded. It was found that moisture loss and solid gain increased with increase in osmotic temperature and osmotic solution concentration. The highest mass transfer was observed at concentration of 60% and temperature of 60°C. Three models (Handerson and Pabis model, Logarithmic model and Lewis model) were used to analyze osmotic dehydration data. Among the three models, Logarithmic model showed a best fit to the osmotic dehydration data with higher value of coefficient of determination (R^2).

Keywords: Osmotic dehydration; Mass transfer kinetics; Pineapple; Solid gain; Moisture loss

Introduction

Pineapple, also known as Queen of fruits is one of the important commercial fruit crops in the world [1]. The fruit is known for its exceptional juiciness, excellent flavor, taste and numerous health benefits. The fruit is highly perishable containing about 14% of sugar, good amount of vitamin A and B, citric acid, malic acid and bromelin [2]. The bromelin, a protein digesting enzyme, aids in the digestion of proteins when taken with meals [3]. Various food items like squash, syrup, jelly are produced from pineapple. Vinegar, alcohol, citric acid, calcium citrate etc. are also produced from pineapple. Pineapple is also recommended as medical diet for certain diseased persons [4]. Physically, the fruit is hard on the outside and soft on the inside and can be eaten raw or added to desserts and fruit salads. In addition to this, squash, syrup and jelly like food items are also made from pineapple. Thailand, Philippines, Brazil and china are the main pineapple producers in the world supplying nearly about 50% of the total output [5]. The commercial cultivation of pineapple in India is believed to be only four decades old and is largely grown in states like Assam, Meghalaya, Tripura, Sikkim, Mizoram, West Bengal, Kerala, Karnataka and Goa. Osmotic dehydration is basically a water removal process in which materials such as fruits are placed into a concentrated solution of soluble solutes. By doing this, a major part of water is removed from substance and time required for relatively high temperature air drying is reduced. Conventional air drying is energy intensive and cost intensive because it is simultaneous heat and mass transfer process accompanied by phase change [6]. Even though the pineapple is available round the year but there is some peak harvest season at which harvest is so abundant that some of the fruit has to be left in the field or sold at a very low price. One way to increase the value of this crop is by drying it. Conventional air drying may result in browning or caramelization of sugar due prolonged exposure to the heat. Osmotically pre drying pineapple would reduce this problem. The effects of sucrose concentration, processing time, temperature, slice thickness, fruit to syrup ratio on weight reduction and total soluble solids were studied by Singh et al. [7]. It was observed that percent weight reduction and total soluble solids increases with increase in sucrose concentration and temperature. It has been found that 60% sucrose solution at 50°C, 1:4 fruit to syrup ratio and 10 mm thickness give best results [7]. A significant amount of weight loss (47.40) within 4 hours of osmosis was showed by mango slices when osmosed in 67.4°

brix of osmotic solution at 40°C having sample to solution ratio of 1:3.34 [8]. About 50% of water was removed from the 5 mm of banana slices when 63°brix sugar solutions was maintained at 75°C within one hour of osmosis and 57.9% of water was removed when slices were osmosed for 2.5 hours [9]. The specific objective of this work was to study the effect of osmotic solution concentration, sample to solution ratio and temperature on mass transfer of the osmotic dehydration of pineapple and to determine the best mathematical model that can describe the kinetics of osmotic dehydration process.

Materials and Methods

Raw material preparation

The experiments were conducted on fresh, ripe and good quality pineapples. The fully ripen pineapples were peeled manually, cored and then sliced into 10 mm thick slices and further divided into four pie wedge shaped pieces. To inactivate enzymes, pineapple slices were blanched at 80°C for a min [10]. Moisture content was determined by placing the samples in an oven at 100°C for 16 to 18 hours or till constant weight was achieved [11]. The samples were then subjected to osmotic dehydration treatment. Figure 1 shows the procedure involved in the osmotic dehydration treatment of the pineapple slices.

Osmotic dehydration treatment

The sucrose and fructose solution made with concentration levels of 40%, 50%, 60% with sample to solution ratio of 1:4, 1:5 and 1:6 respectively were used for each experiment. The samples weighing 50 g were used for each experiment and then immersed in osmotic solutions for 10, 20, 30, 40, 50, 60, 90, 120, 150, 180 and 240 min at a temperature of 40°C, 50°C, and 60°C. The temperature was controlled with hot

***Corresponding author:** Khan MA, Department of Post-Harvest Engineering and Technology, Faculty of Agricultural Sciences, Aligarh Muslim University, Aligarh-202002, India, E-mail: makamu4@yahoo.co.in

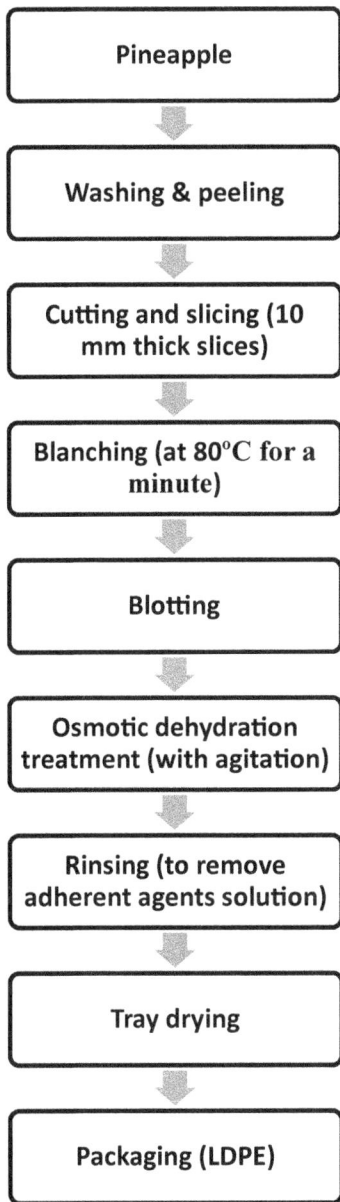

Figure 1: Procedure involved in osmotic dehydration of pineapple slices.

water bath. The sample/solution ratio was kept as 1:4, 1:5, and 1:6 [12]. After each interval of time moisture loss and solid gain were recorded. The osmotically dehydrated samples were then blotted dry with tissue paper and then weighed in weighing balance. The samples were then dried in tray dryer.

Determination of process parameters

The moisture content of the sample was found by using following equation [11]:

$$MC \ (\% \ w.b) \ = \frac{(initial \ weight - final \ weight)}{initial \ weight} \times \ 100 \qquad (1)$$

The moisture loss during osmotic dehydration treatment was determined using the following equation [13,14]:

$$ML \ (\%) \ = \frac{(Wt \ of \ initial \ moisture \, (g) - Wt \ of \ final \ moisture \, (g))}{initial \ weight \ of \ sample \ in \ gms} \times 100 \qquad (2)$$

The solid gain during osmotic dehydration treatment was determined by using following equation [13,14]:

$$SG \ (\%) \ = \frac{(Wt. \ of \ final \ solid \, (g) - Wt. \ of \ initial \ solid \, (g))}{initial \ wt \ of \ the \ sample \ in \ gms} \times \ 100 \qquad (3)$$

Mass transfer kinetics

For determination of moisture and solid change during osmotic dehydration under different treatment, as a function of dehydration time, the rate of change of a quality factor C can be represented by:

$$\frac{dc}{dt} = -kC^n \qquad (4)$$

Where C is the concentration of a quality factor at time t, k is the kinetic rate constant and n is the order of the reaction. For the majority of foods, the time–dependence relations appear to be described by zero order [15,16] or first order kinetic models [15,16], by integrating eq. (4), zero order eq.(5) and first order kinetic models eq. (6) can be derived as:

$$C \ = \ C_0 \pm \ kt \qquad (5)$$

$$C \ = \ C_0 \ exp \ (\pm kt) \qquad (6)$$

Where C_0 is the initial value of mass transfer parameter and C is the mass transfer value at a specific time. In the equation, ($\pm$) indicates gain and loss of any mass transfer parameter.

Statistical analysis

The XLSTAT software package (XLSTAT evaluation version 2016) was used for regression analysis. The correlation coefficient (R^2) and RMSE were considered as the criteria for selecting the best equation. The higher the value of R^2 and the lower the value of RMSE, the better the model was taken to fit. By equation:

$$R^2 = \frac{\sum\limits_{i=1}^{N} (C_{pre,i} - C_{pre,avg})^2}{\sum\limits_{i=1}^{N} (C_{evp,i} - C_{evp,avg})^2} \qquad (7)$$

$$RMSE = \sqrt{\frac{\sum\limits_{i=1}^{N} (C_{pre,i} - C_{evp,i})}{N}} \qquad (8)$$

Where, $C_{exp,i}$ is the ith experimental value, $C_{exp,i}$ is the ith predicted value and N is the total number of observations in particular model.

Mathematical modeling

Following formulae were used for calculation of moisture ratio (MR) during osmotic dehydration experiment.

$$MR = \frac{[Mt - Me]}{[Mi - Me]} \qquad (9)$$

Where,

MR is the moisture ratio.

M_t = Moisture content at any time, t

M_i = initial moisture content.

M_e = Equilibrium moisture content (at the end of drying).

The following drying models were used for osmotic dehydration data:

a. Handerson and Pabis model:

$$MR = exp(-kt) \qquad (10)$$

b. Logarithmic model:

$$MR = a.exp(-kt) + c \qquad (11)$$

c. Lewis model:

$$MR = exp(-kt) \qquad (12)$$

Results and Discussion

The specific objective of this research was to examine the effect of osmotic temperature, osmotic solution concentration and sample to solution ratio on osmotic dehydration behavior of pineapple slices. Three models (Handerson and Pabis model, Logarithmic Model, Lewis model) were used for osmotic dehydration data. The results are presented below.

Mass transfer kinetics during osmotic dehydration of pineapple

The experiments were carried out at osmotic solution concentration of 40%, 50% and 60%, osmotic temperature of 40°C, 50°C and 60°C and sample to solution ratio of 1:4, 1:5 and 1:6, respectively. To study the osmotic kinetics at each experimental condition, the osmotic dehydration treatment was carried out from 10 to 240 min with varying time interval.

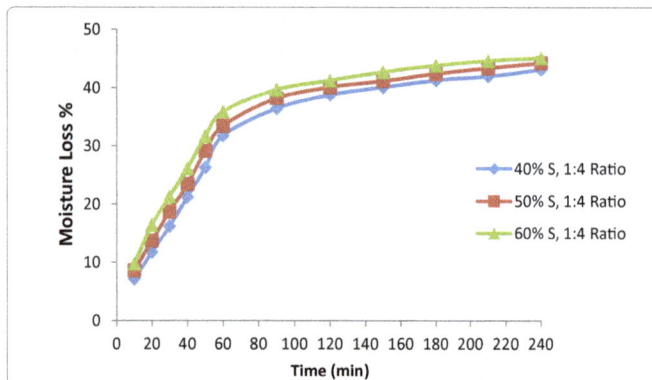

Figure 2: Effect of osmotic concentrations of sucrose on moisture loss of pineapple at 40°C and 1:4 sample to solution ratio.

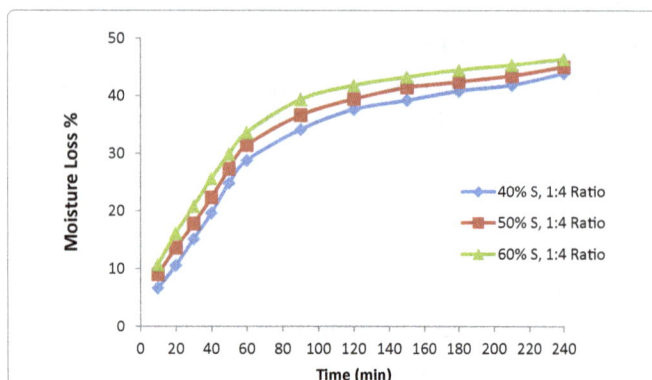

Figure 3: Effect of osmotic concentrations of sucrose on moisture loss of pineapple at 60°C and 1:4 sample to solution ratio.

Model	Sucrose		Fructose	
	R^2	RMSE	R^2	RMSE
Zero order: $C = C_0 \pm kt$	0.838	5.203	0.854	6.865
First order: $C = C_0\,exp\,(\pm kt)$	0.743	5.857	0.762	7.891

Table 1: R^2 and RMSE values of zero and first order kinetic models for moisture loss.

Effect of osmotic solution concentration, osmotic temperature and sample to solution ratio on moisture loss of pineapple slices during osmotic dehydration treatment

Plots for moisture loss verses dehydration time as shown in Figures 2 and 3 shows the effects of osmotic solution concentration and osmotic temperature on moisture loss of pineapple samples. The plots showed that the moisture loss increased in a nonlinear mannered with time at all concentration of sucrose at different temperatures. Similar results have been obtained for experimental data for fructose. It was found that moisture loss increased with increase in temperature from 40°C to 60°C. Similar results were reported by Rahman and Lamb [17]. It was also observed that moisture loss was faster during initial period of osmotic dehydration and then rate decreased. During osmotic dehydration treatment moisture loss after 240 min in sucrose solution was found to be 39.079% to 46.32% of initial weight of pineapple samples. However, moisture loss after 240 min of osmotic dehydration in fructose solution was found to be 39.079% to 45.385% of initial weight of pineapple samples.

Kinetic models for moisture loss

For the mass transfer kinetics, zero order and first order kinetic models were used during osmotic dehydration of pineapple in sucrose. From the Table 1 it can be seen that the data for the moisture loss fitted to zero order kinetic model compared to the first order kinetic model with high values of coefficient of determination (R^2) and low values of root mean square error (RMSE) as shown by mass transfer kinetic studies.

Effect of osmotic solution concentration, osmotic temperature and sample to solution ratio on solid gain of pineapple slices

Solid gain of osmotic dehydrated pineapple slices were calculated in order to determine the amount of solute penetrated during osmotic dehydration process from the osmotic solution. The effect of osmotic solution concentration of sucrose on solid gain of pineapple samples at 40°C and 60°C temperature having sample to solution ratio of 1:6 with respect to time are shown in Figures 4 and 5. From the graphs, it can be observed that solid gain increased nonlinearly with time. Moreover, it was also found that solid gain increased in the initial period of osmotic dehydration treatment and then rate decreased. It was also found that the solid gain increased with increase in temperature from 40°C to 60°C.

Kinetic models for solid gain

For the mass transfer kinetics during osmotic dehydration of pineapple slices, zero order and first order kinetics models were also used. It was found that the data for solid gain was fitted to zero order kinetic model compared to the first order kinetic model with high values of coefficients of determination and low value of root mean square error (RMSE) as shown in Table 2.

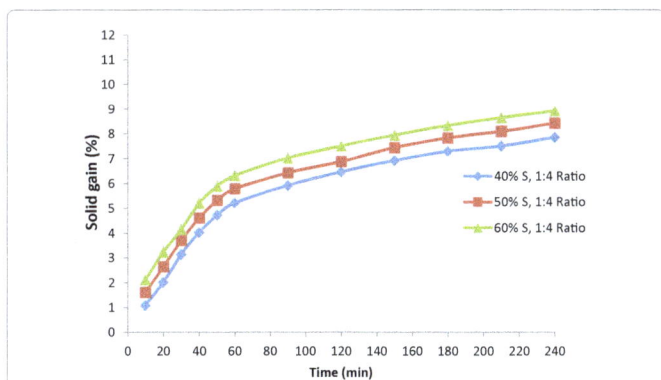

Figure 4: Effect of osmotic concentrations of sucrose on solid gain of pineapple at 40°C and 1:4 sample to solution ratio.

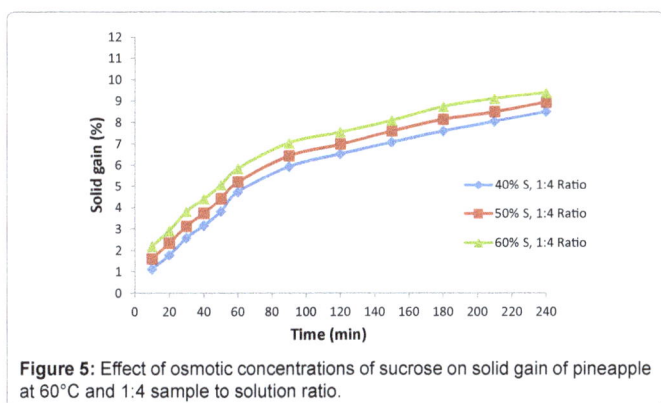

Figure 5: Effect of osmotic concentrations of sucrose on solid gain of pineapple at 60°C and 1:4 sample to solution ratio.

Model	Sucrose		Fructose	
	R²	RMSE	R²	RMSE
Zero order: $C = C_0 \pm kt$	1.00	0.00	0.852	0.901
First order: $C = C_0 \, exp \, (\pm kt)$	0.854	0.907	0.770	1.081

Table 2: R² and RMSE values of zero and first order kinetic models for solid gain.

Osmotic agent	Models					
	Handerson and Pabis model $MR = a.exp(-kt)+c$		Logarithmic model $MR = a.exp(-kt)+c$		Lewis model $MR = exp(-kt)$	
	R²	RMSE	R²	RMSE	R²	RMSE
Sucrose	0.995	0.052	0.998	0.050	0.993	0.079
Fructose	0.997	0.046	0.999	0.047	0.997	0.067

Table 3: R² and RMSE values of different models.

Validity of models

Three models i.e. Handerson and Pabis model [18], Logarithmic model [19], and Lewis model [20], were tested to select the best model. In the proposed drying models, the moisture ratio (MR) is a nonlinear function of time. Nonlinear regression modelling of experimental data was carried out to obtain the values of constants of these models. The correlation coefficient (R²) and RMSE were considered as the criteria for selecting the best equation. The models were fitted to the experimental data using XLSTAT. The R² and RMSE values for each of the tested models are given in Table 3.

From the Table 3 it can be seen that, for the Handerson and Pabis model, the R² was found to be 0.995 using sucrose as an osmotic agent and 0.997 for fructose while as RMSE was found to be 0.052 for sucrose and 0.046 for fructose which indicates good fit of the Handerson and Pabis model. For logarithmic model, the R² was found to be 0.998 using sucrose as an osmotic agent and 0.999 for fructose while as RMSE was found to be 0.050 for sucrose and 0.047 for fructose which indicates the best fit of the logarithmic model as shown in Table 3. Similarly for the Lewis model , the R² was found to be 0.997 using sucrose as an osmotic agent and 0.993 for fructose while as RMSE was found to be 0.079 for sucrose and 0.067 for fructose which indicates a good fit of the Lewis model (Table 3) Thus, Logarithmic model is the most acceptable one and fits best to the given set of experimental data for osmotic dehydration of pineapple.

Conclusion

On the basis of the finding of the present study it can be concluded that there was remarkable effect of osmotic solution concentration, osmotic temperature and sample to solution ratio on osmotic dehydration of pineapple slices. It was found that moisture loss and solid gain increased with increase in osmotic solution concentration from 40% to 60%. Temperature was found to have proportional effect on moisture loss and solid gain. Both the moisture loss and solid gain were higher during initial period of osmotic dehydration treatment than in the later period. Both the data for moisture loss and solid gain were fitted to zero order kinetic model compared to a first order kinetic model with high value of coefficient of determination. Among the three models that were used Logarithmic model showed a best fit to the experimental data of osmotic dehydration with higher value of coefficient of determination and low values of RMSE.

References

1. Baruwa OI (2013) Profitability and constraints of pineapple production in Osun State. Nigeria J Horti Res 21: 59-64.

2. Joy PP (2010) Benefits and uses of pineapple. Pineapple research station, Kerala Agricultural University, Kerala, India.

3. Debnath P, Dey P, Chanda A, Bhakta T (2012) A survey on pineapple and its medicinal value. Scholar Academic J Pharma.

4. Moniruzzaman FM (1988) Fruit cultivation in Bangladesh. (2ndedn), Bangla Academy, Dhaka.

5. FAO (2004) Food and Agriculture Organization of United Nations.

6. Barbanti D, Mastrocola D, Severine C (1994) Drying of plums a comparison among twelve cultivars. Sciences des Aliments 14: 61-73.

7. Singh AK, Nath N, Mittal BK (1998) Studies on osmo–air drying of apricot grown in Kumanon hills of UP. Proceedings of 4th International Food Convention, CFTRI, Mysore, India.

8. Samsher A, Khan K, Srivastava PK, Alam SS (1998) Suitability of packaging materials for storage of dehydrated mango slices under ambient condition. Proceedings of 4th International Food convention, CFTRI, Mysore, India.

9. Thangavel K, Jhon KZ, Kailappan R, Amaladas RH (1998) Osmotic dehydration of red banana. Proceedings of 4th International Food Convention CFTRI Mysore India.

10. Saputra C (2001) Osmotic dehydration of pineapple. Dry Technol 19: 415-425.

11. Ranganna S (2001) Handbook of analysis and quality control of fruit and vegetables product. (3rdedn), Tata McGraw-Hill Publ.co, New Delhi, India.

12. Kar A, Gupta DK (2003) Air drying of osmosed button mushrooms. J Food Sci Technol 40: 23-27.

13. Lenart A, Flink JM (1984) Osmotic dehydration of potato: Criteria for the end point of the osmosis process. J Food Technol 19: 65-89.

14. Hawkes J, Flink JM (1978) Osmotic concentration of fruit slices prior to freeze dehydration. J Food Process Preserv 2: 265-284.

15. Maskan M (2000) Microwave/air and microwave finish drying of banana. J Food Eng 44: 71 -78.

16. Chen C, Ramaswamy HS (2002) Colour and texture change kinetics in ripening bananas. Food Sci Technol 35: 415- 419.

17. Rahman MS, Lamb L (1990) Osmotic dehydration of pineapple. J Food Sci Technol 27: 50-152.

18. Handerson SM, Pabis S (1961) Grain drying theory I: Temperature effect on drying coefficient. J Agri Eng Res 6: 169-174

19. Togrul IT, Pehlivan D (2002) Mathematical modeling of solar drying of apricots in thin layers. J Food Engineering 55: 209-216.

20. Bruce DM (1985) Exposed-layer barley drying, three models fitted to new data up to 150°C. J Agricultural Engineering Research 32: 337-347.

8

Hygienic Assessment of Tools Used for Juice Extraction by Street Juice Vendors in Close Vicinity of Sam Higginbottom Institute of Agriculture, Technology and Sciences, Allahabad, Uttar Pradesh, India

Anane MA and Immanuel G*

Department of Food Processing Engineering, Sam Higginbottom Institute of Agriculture, Technology and Sciences (SHIATS), Allahabad, UP, India

Abstract

This study was conducted to assess the hygiene of tools used for juice extraction by street vendors in close vicinity of Sam Higginbottom Institute of Agriculture, Technology and Sciences (India) and the quality evaluation of juice prepared. Structured questionnaires and observational checklist was used in the data collection. A total of 18 juice vendors were interviewed followed by swab test of tools used by the vendors such as; knives, extractors, collectors and sieve used during extraction. General hygiene of the juice vendors that is the sanitary practices in the extraction of juice and washing of tools and utensils were observed to be very poor. The total coliform counts on knives, collectors, sieve and extractors were above the recommended level of TPC. The mean coliform count on extractor was 1.56×10^3 cfu/ cm^2. The mean coliform count on utensils (collectors) was 1.02×10^3 cfu/ cm^2. The mean coliform count on sieve was 1.08×10^3 cfu/ cm^2 and the mean coliform count on knives was 0.19×10^3 cfu/ cm^2. The maximum coliform count was found on extractors and the minimum was found on knives. Most of the tools had total coliform count above the recommended maximum levels which is 1.0×10^3 cfu/cm^2.

Keywords: Swab test; Microbial quality; Microorganisms; Fruit juices; Tools

Introduction

In developing countries, fruit juices sold by street vendors are widely consumed by millions of people. These juices provide a source of available and affordable source of nutrients to many sectors of the population, especially to the poor. From the survey of the population, it is evident that unpasteurized fruit juices are preferred because of the fresh flavor it has, hence its increase in demand. Fresh blended juices are preferred to the pasteurized by most consumers because of the fear of added preservative to the pasteurized fruit juices. They are simply prepared by extracting the liquid and pulp out of ripped fruit. Despite the potential benefits offered by street fruit juices, concerns over their safety and quality have been raised. Preparation of unpasteurized fresh fruit juices can easily get contaminated due to its exposure to poor environmental conditions and without any quality control checks, careful hygienic measures such as proper washing and keeping of fruits and extraction equipment's must be done and kept during the preparation of such. Hygiene is the least considered factor by the fruit vendors due to their low illiteracy level. They have no idea about the importance of hygiene with respect to the vending of juices. Street fruit vending doesn't have any specificity as to where to be locate. Vendors could easily be located; sometimes more than a single vendor could be seen at a particular location rendering the same services. Getting a place to sell is a vendor's priority than the hygienic condition of the location. Hence, the compromised consideration of selling fruits at the roadsides. Heavy traffic causes lot of dust in the environment and the dust ends up accumulating on the surfaces of the tools used for the preparation of juices which cause cross contamination of the fresh juice prepared, most of the places where the juices are being sold are close to refuse dumps and sewages which increases the contamination fidelity of pathogens making consumption of such juices harmful. It has been observed that most of the fruit vendors do not wash their tools thoroughly after preparation of the juices, they rather use water to rinse and dangerously exposes tools like knives, sieve, and utensils (collectors) to atmospheric conditions, such as dust and pathogens. Juice extraction equipment ranges from hand operated crushers to tones/hour mechanical extractors and the most commonly used by the street vendors in rural areas are the hand operated crushers. The construction of the extractor should be made to reduce the interactions between the fruits and the extractor; to reduce the contaminations which may be caused by wear and tear of the extraction machine, to reduce fruit-extractor interactions, corners and edges of the extractor during manufacturing should be smooth in lieu of rough and sharp edges which could trap some of the fruit particles and if not washed properly, could cause growth of microbes within. This, therefore brings about the fact that equipment should be easy to wash and clean. The surface of the equipment should not react with detergents used for washing the equipment.

Salmonella, a commonly infections associated with animal-derived foods, such as meat, seafood, dairy, and egg products was also associated with fresh juice, occurred as far back as 1922 in France [1,2]. Early outbreaks resulting in typhoid fever were associated with poor hygiene by asymptomatic *S. typhi* shedding food handlers. In USA, more recent outbreaks of non- typhoidal salmonellosis in fresh juice have been attributed to fecal-associated contamination of fruit or poor processing practices [1,2]. In 2005, 152 cases of *S. typhimurium* infection associated with commercially distributed unpasteurized orange juice were reported in USA. Upon inspection by Food and Drug Administration (FDA), it was found that the production facility did not comply with the HACCP plan and that noncompliance likely contributed to this outbreak [3,4]. Hygienic assessment on the equipment and tools used in the extraction of the juice is equally

*Corresponding author: Immanuel G, Department of Food Processing Engineering, Sam Higginbottom Institute of Agriculture, Technology and Sciences (SHIATS, Allahabad, UP, India
E-mail: genithaimmanuel@yahoo.co.in

important as assessment on the juice prepared but many times, apathy has been observed towards the quality of street fruit juices. This work seeks to check the sanitary practices on the tools used in extraction of street juices produced in close vicinity of Sam Higginbottom Institute of Agriculture, Technology and Sciences mainly in Mahewa, Dandi and Naini.

Objectives

1. To assess the sanitary practices on tools used for street fruit juice extraction.

2. To evaluate the microbiological quality of the tools.

Materials and Methods

Materials required

Chemicals and reagents: Nutrient Agar, Red Bile Agar and Salmonella-Shigella Agar were used for microbial testing. Distilled water was used for preparation of the agar and buffered peptone water was used as a pre-enrichment medium for recovery of injured Salmonella species.

Methodology

Description of study area: The study was conducted from 3 different locations (Mahewa, Dandi and Naini) which are in close vicinity of SHIATS in Allahabad district, in the state of Uttar Pradesh, India. The study was conducted from February to June, 2016.

Collection of samples: Swab samples of the tools were collected from 2 street vendors from each location that is Mahewa, Naini and Dandi. It was taken from juice vendors who were available at the time of collection and also those who were willing to participate in the study and ready to give required information. Swap samples were taken from the extractors, knives, utensils and sieve. The swabs was put in sterile containers and then taken to the laboratory for microbial testing. a total of 18 respondents were involved in the study. The experiment was repeated twice for each tool.

Data collection

Questionnaires: Structured questionnaires were used to collect information from the juice vendors. Most questionnaires were made with closed and open-ended questions. The questionnaires were used to collect sociological information from the respondents as regards to demographic characteristics, as well as serving tools and vending sites.

Observational checklists: The observational checklist was made with some of the Codex recommended general principles (CAC/RCP 1-1969, Rev.4-2003) of food hygiene for food preparation settings, washing processes, general hygiene of the vendors and premises, waste management and general sanitary practices on juice making. On the other the checklist was provided with a YES and NO sections for each parameter for observation [5].

Swab testing procedure for tools

Swab test was done to determine the compliance with the requirements given in individual specification. It is the counting of total number of aerobic bacteria, yeast and molds on any surface. This test was used to check the hygienic condition of tools and other accessories used during the extraction of the juices. A cotton swab was moistened with normal saline (0.9% NaCl) and was placed in a suitable test tube. The mouth was closed with a cotton plug and rapped with aluminum foil. The test tube containing the swab was then sterilized by autoclaving at 121°C, for 15 mins. The swab was then taken from the test tube carefully, wearing hand gloves and the surface of the equipment and other accessories was swab. The swab was then carefully put in the test tube and closed and taken to the laboratory for microbial analysis.

Microbiological testing

10 ml of Ringer's solution was added to the test tube containing the swab. It was well shaken and 1 ml of the solution was transferred to Petri dishes and the media was added. Nutrient Agar, Red Bile Agar and Salmonella- Shigella Agar were used. The cultures were incubated for about 24 hrs at a temperature of 37°C.

Counting of colonies: After the incubation, the numbers of colony forming units were counted

$$Number\ of\ colonies = \frac{(colonies\ present\ on\ agar \times 10\ mL\ neutralizer\ in\ swab)}{(area\ swabbed)\,25\,cm^2} \qquad (1)$$

Statistical analysis

All experiments were repeated three times with duplicate samples and data were analyzed by analysis of variance using the ANOVA procedure with replication. The null hypothesis states that the mean coliform count value of the tools from the 3 different locations is equal. When the p-value is greater than the tabulated value, the null hypothesis is rejected which indicates that there are no significant differences of the mean coliform count from the three locations.

Results and Discussion

Analysis of questionnaire

Equipment handling practices: Observational study was first done concerning this objective. It was observed that most of the vendors do not wash their equipment and utensils after use. It was observed that the sieve, collector and knives were just lying on the table with houseflies all around them. Some were just covered with a red cloth which is also not clean and could cause cross contamination of microbes. The vendors were also asked and most of them admitted not washing their equipment after use. Out of 18 vendors interviewed, 9 of them admitted not washing their equipment at all and the rest of them wash it once only at the end of the day. Lakshmanan and Schaffner [6] in the study of orange squeezing machines found that, some of the machines had scraps of oranges in internal tubing which were then reflected in the formation of bacteria biofilms. Tambekar et al. [7] found that street fruit juices could often prove to be a public health threat due to their quick methods of cleaning utensils, handling and extraction.

Microbial assessment of tools: Very few guidelines have been published on the acceptable level of microorganisms on surfaces. The US Public Health Service recommends that cleaned and disinfected food service equipment should not exceed 10 viable microorganisms per cm^2. The Public Health Laboratory Service (PHLS) in the UK recommends guidelines for cleaned surfaces ready for use: less than 80 cfu/ cm^2 is satisfactory, 80- 10^3 cfu/cm^2 is borderline and over 10^3 cfu/ cm^2 is unsatisfactory. The microbial analysis of fruit juice contact surfaces on extractors is shown in Table 1. In Table 1, the maximum mean value recorded for the total coliform count on extractors was 1.75 × 10^3 cfu/ cm^2 from Mahewa and the minimum mean value obtained was from Naini which had a mean coliform count of 1.45 × 10^3 cfu/ cm^2. with respect to locations, the extractors from Dandi recorded a minimum total coliform count of 1.3 × 10^3 cfu/ cm^2 and a maximum total coliform count of 1.7 × 10^3 cfu/ cm^2. The extractor from Mahewa recorded a minimum total coliform count of 1.4 × 10^3 cfu/ cm^2 and a maximum total coliform count of 2.1×10^3 cfu/ cm². Finally, the extractors from Naini recorded a minimum total coliform count of

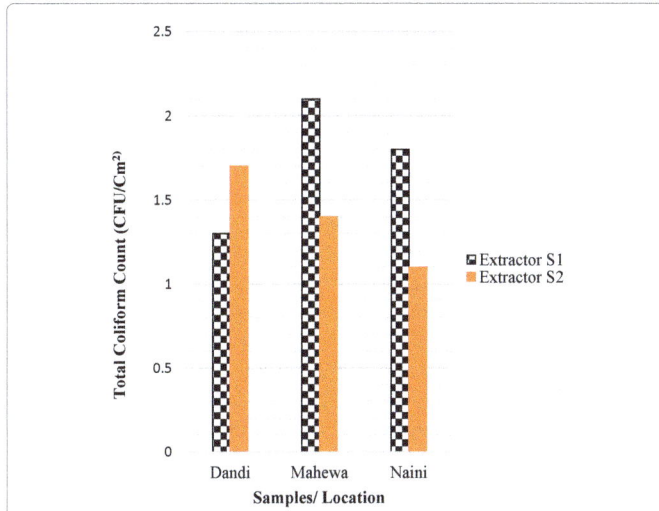

Figure 1: Total coliform count of the fruit juice contact surfaces on extractor.

Location	Total Coliform Count × 10³ (cfu/ cm²)		
	Extractor		Mean
	S1	S2	SM
Dandi	1.3	1.7	1.5
Mahewa	2.1	1.4	1.75
Naini	1.8	1.1	1.45
Critical Difference			0.501

Table 1: Total coliform count of the fruit juice contact surfaces on extractors.

Location	Total Coliform Count × 10³ (cfu/ cm²)		
	Collector		Mean
	S1	S2	SM
Dandi	1.0	0.8	0.9
Mahewa	0.9	1.1	1.0
Naini	1.4	0.9	1.15
Critical Difference Value			0.5

Table 2: Total coliform count the fruit juice contact surfaces on utensils (Collectors).

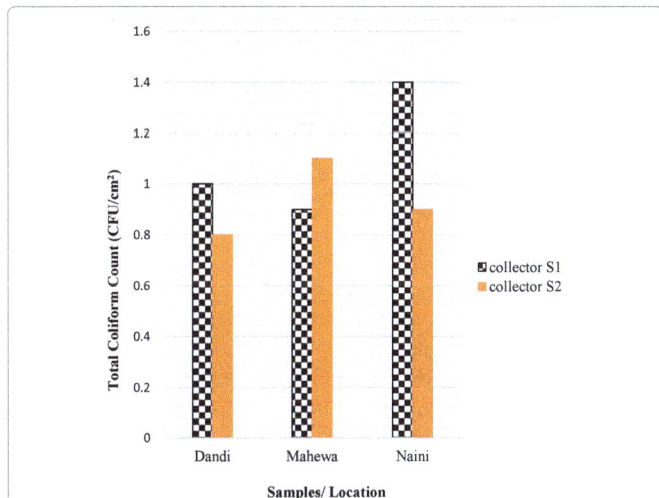

Figure 2: Total coliform count of the fruit juice contact surfaces on utensils (Collector).

Location	Total Coliform Count × 10³ (cfu/ cm²)		
	Sieve		Mean
	S1	S2	SM
Dandi	1.2	1.0	1.1
Mahewa	0.9	1.9	1.4
Naini	0.4	1.1	0.75

Table 3: Total coliform count of the fruit juice contact surfaces on sieve.

1.1×10^3 cfu/ cm² and a maximum total coliform count of 1.8×10^3 cfu/ cm². Comparing these results to the recommended guidelines for cleans surfaces with respect to food contact, it can be seen that it is above the standard required. This indicates that, the sanitary practices on extractor is very poor, thus causes cross contamination of the fruit juices prepared. This means the juice is unhygienic and not safe for human consumption and could cause harm to the consumer (Figure 1). In these results, the null hypothesis states that the mean coliform count value of the utensils (collectors) from the 3 different locations are equal using ANOVA. From Table 1, with respect to location, the p-value is 0.459 which is less than the calculated value 0.826. Thus the null hypothesis is accepted which indicates that there are no significant differences of the mean coliform count value in the extractors from these three locations. With respect to the replication of samples, the p-value is greater than the calculated value (0.796>0.256), thus the null hypothesis is accepted. Which means that there are significant differences in each replication of sample thus critical value for each sample was calculated. The critical value calculated was 0.501. Comparing the tested mean of each samples which were 0.25, 0.30 and 0.05 to the critical value 0.501, they were lesser than the critical value thus differs significantly among each other. The microbial analysis of fruit juice contact surfaces on utensils (collectors) is shown in Table 2.

In Table 2, the maximum mean value recorded for the total coliform count on collectors was 1.15×10^3 cfu/ cm² from Naini and the minimum mean value obtained was from Dandi which had a mean coliform count of 0.9×10^3 cfu/ cm². With respect to locations, the collectors from Dandi recorded a minimum total coliform count of 0.8×10^3 cfu/ cm²and a maximum total coliform count of 1.0×10^3 cfu/ cm². The collectors from Mahewa recorded a minimum total coliform count of 0.9×10^3 cfu/ cm² and a maximum total coliform count of 1.1×10^3 cfu/ cm². Finally, the collectors from Naini recorded a minimum total coliform count of 0.9×10^3 cfu/ cm²and a maximum total coliform count of 1.4×10^3 cfu/ cm². Comparing these results to the recommended guidelines for cleans surfaces with respect to food contact, it can be seen that it is above the standard required. This indicates that, the sanitary practices on collector is very poor, thus causes cross contamination of the fruit juices prepared. This means the juice is unhygienic and not safe for human consumption and could cause harm to the consumer (Figure 2).

In these results, the null hypothesis states that the mean coliform count value of the utensils (collectors) from the 3 different locations are equal using ANOVA. From the Table 2, with respect to location, the p-value is 0.497481 which is less than the tabulated value 0.675676. this means that the null hypothesis is rejected which indicates that there are no significant differences of the mean coliform count value on the utensils (collectors) from these three locations. With respect to the replication of samples, the p-value is greater than the tabulated value (0.6607>0.5134), thus the null hypothesis is accepted. Which means that there are significant differences in each replication of sample thus critical value for each sample was calculated. The critical difference value calculated was 0.5. the tested mean of each samples were 0.1, 0.25 and 0.15 which were all less than the critical value 0.5 thus not

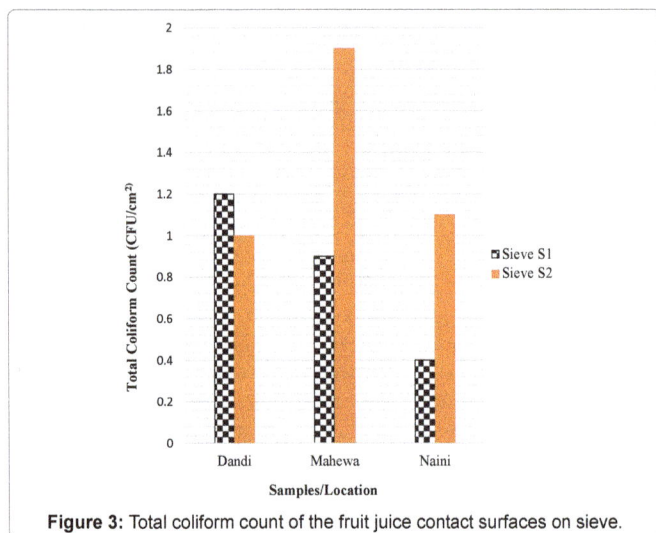

Figure 3: Total coliform count of the fruit juice contact surfaces on sieve.

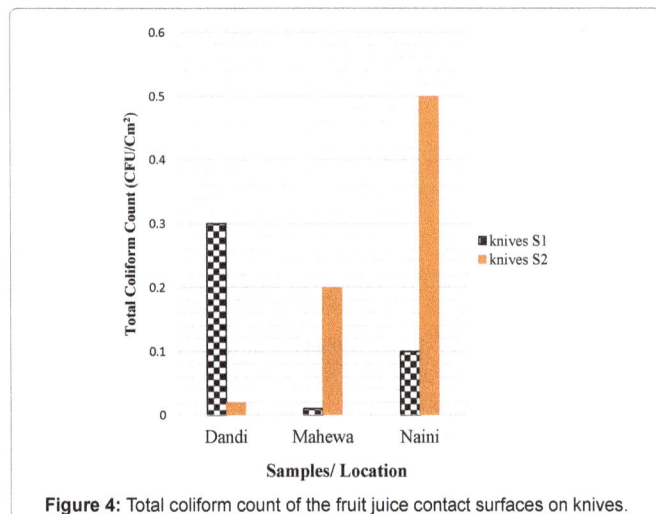

Figure 4: Total coliform count of the fruit juice contact surfaces on knives.

Location	Total Coliform Count × 10³ (CFU/ cm²)		
	Knives		Mean
	S1	S2	SM
Dandi	0.3	0.02	0.16
Mahewa	0.01	0.2	0.11
Naini	0.1	0.5	0.3
Critical difference			0.496

Table 4: Total coliform count of the fruit juice contact surfaces on knives.

significantly different from each other. The microbial analysis of fruit juice contact surfaces on sieve is shown in Table 3.

In Table 3, the maximum mean value recorded for the total coliform count on the sieves was 1.4×10^3 cfu/ cm² from Mahewa and the minimum mean value obtained was from Naini which had a mean coliform count of 0.75×10^3 cfu/cm². With respect to locations, the sieve from Dandi recorded a minimum total coliform count of 1.0×10^3 cfu/ cm²and a maximum total coliform count of 1.2×10^3 cfu/ cm². The sieves from Mahewa recorded a minimum total coliform count of 0.9×10^3 cfu/ cm² and a maximum total coliform count of 1.9×10^3 cfu/ cm². Finally, the sieve from Naini recorded a minimum total coliform count of 0.4×10^3 cfu/cm² and a maximum total coliform count of 1.1×10^3 cfu/ cm². Comparing these results to the recommended guidelines

for cleans surfaces with respect to food contact, it can be seen that it is above the standard required. Few of the sieves had total coliform count on the borderlines while most had unsatisfactory counts. This indicates that, the sanitary practices on sieve is very poor, thus causes cross contamination of the fruit juices prepared. This makes the juice unhygienic and safe for consumption and could cause harm to the consumer (Figure 3).

In these results, the null hypothesis states that the mean coliform count value of the sieve from the 3 different locations are equal using ANOVA. From the Table 3, with respect to location, the p-value is 0.299 which is less than the tabulated value 1.923. This means that the null hypothesis is rejected which indicates that there are no significant differences of the mean coliform count value on the sieves from these three locations. With respect to the replication of samples, the p-value is also less than the tabulated value (0.479>1.085), thus the null hypothesis is also rejected. Which means that there were also no significant differences in each replication of samples. The microbial analysis of fruit juice contact surfaces on knives is shown in Table 4. In Table 4, the maximum mean value recorded for the total coliform count on the knives was 0.3×10^3 cfu/ cm² from Naini and the minimum mean value obtained was from Mahewa which had a mean coliform count of 0.11×10^3 cfu/ cm². With respect to locations, the knives from Dandi recorded a minimum total coliform count of 0.02×10^3 cfu/ cm²and a maximum total coliform count of 0.3×10^3 cfu/ cm². The knives from Mahewa recorded a minimum total coliform count of 0.01×10^3 cfu/ cm² and a maximum total coliform count of 0.2×10^3 cfu/ cm². Finally, the knives from Naini recorded a minimum total coliform count of 0.1×10^3 cfu/ cm² and a maximum total coliform count of 0.5×10^3 cfu/cm². Comparing these results to the recommended guidelines for cleans surfaces with respect to food contact, it can be seen that most if the knives had total coliform count on the borderlines while some had satisfactory counts. This indicates that, the sanitary practices on knives is very poor, thus causes cross contamination of the fruit juices prepared (Figure 4).

In these results, the null hypothesis states that the mean coliform count value of the sieve from the 3 different locations are equal using ANOVA. From the Table 4, with respect to location, the p-value is 0.658 which is greater than the tabulated value 0.264. This means that the null hypothesis is accepted which indicates that there are significant differences of the mean coliform count value on the knives from these three locations. With respect to the replication of samples, the p-value is greater than the tabulated value (0.749>0.020), thus the null hypothesis is accepted. Which means that there are significant differences in each replication of sample thus critical value for each sample was calculated. The critical value calculated was 0.496. Comparing the tested mean of each samples to the critical value which were 0.19, 0.14 and 0.05 which are all less than the critical value, hence does not differ significantly from each other.

Conclusion

Even though the dangers of these unpasteurized fruits could be as a results from contamination due to exposure of tools and even the fruits to poor atmospheric conditions such as dust, flies and other airborne pathogens. The business of unpasteurized juice production by street vendors has increased dramatically due to the following factors (a) it is inexpensive to establish (b) no special skill is required (c) no certificate is required from any governmental quality control agency to commence such business and lastly (d) it is the preferential business avenue for the less privileged especially in the rural communities in

India. The findings, as recorded in diagrams above, it will be noticed that it is evident that the tools used by these street fruit juice vendors are easily contaminated by dust, flies and other air-borne pathogens and the juice prepared, even though are fresh, could be harmful for human consumption. It can be concluded that the handling practices toward juice preparation, extraction methods and washing of equipment were observed to be very poor. General hygiene of vendors and premises were poor. Poor waste management within preparation and vending sites encouraged contamination of the juices. It is therefore recommended that (a) sanitation sensitization task force be formed by the local government to educate these fruit juice vendors about the harmful effects of exposure of their extraction tools to unfavourable atmospheric condition which could cause contaminations and sickness to patrons of these fruit juices (b) because the fruit juice vending is the livelihood of the poor population, tougher sanction could collapse their business, hence the need for hygienic sensitization.

References

1. Bevilacqua A, Corbo MR, Campaniello D, D'Amato D, Gallo M, et al. (2004) Bacteriological profile of street foods in Mangalore. Indian J Med Microbiol 22: 197-199.

2. Danyluk MD, Goodrich-Schneider RM, Schneider KR, Harris LJ, Worobo RW (2012) Outbreaks of food-borne disease associated with fruit and vegetable juices, 1922–2010. Food Science and Human Nutrition Department, Florida Cooperative Extension Service, Institute of Food and Agricultural Sciences, USA.

3. FDA (2008) Internalization of microorganisms into fruits and vegetables. Literature review of common food safety, Food and Drugs Administration.

4. Keller SE, Miller AJ (2006) Microbiological safety of fresh citrus and apple juices. In: Sapers GM, Gorny JR, Yousef AE (eds.) Microbiology of fruit and vegetables. CRC Press Taylor and Francis Group, Boca Raton, FL pp: 211-224.

5. FAO (2003) Recommended international code of practice general principles of food hygiene. (CAC/RCP 1-1969, Rev.4- 2003). Food and Agriculture Organization, Geneva, Switzerland.

6. Lakshmanan C, Schaffner DW (2005) Understanding and controlling microbiological contamination of beverage dispensers in university foodservice operations. Food Protect Trend 26: 27-31.

7. Tambekar DH, Jaiswal VJ, Dhanorkar DV, Gulhane PB, Dudhane MN (2009) Microbial quality and safety of street vended fruit juices: A case study of Amravati city. Internet J Food Safety 10: 72-76.

9

Microbial Evaluation and Control of Microbes in Commercially Available Date (Phoenix dactylifera Lynn.) Fruits

Ragava SC*, Loganathan M, Vidhyalakshmi R and Vimalin HJ

Department of Microbiology, Hindusthan College of Arts and Sciences, Coimbatore, Tamil Nadu, India

Abstract

Commercially available date fruit samples in different forms were collected and tested for the initial microbial load. Various types of bacterial and fungal species were observed from the test results. Food borne pathogens like *S. aureus* was identified in collected samples. Pet Jar Seedless and Polythene Seeded packs contained more amounts of microbes. So samples from these types are taken and subjected to physical control parameters like refrigeration, deep freezing, heat treatment and irradiation. The radiation and hot air oven treatments contain different time intervals and packed aseptically and stored in room temperature. Samples also kept in refrigerator and in deep freezer for cold treatments. Then the samples kept as undisturbed for 30 days. Based on 15th and 30th day microbial counts from samples, it is observed that deep freezing and refrigeration has more positive effect to control the level of microbes and it is concluded that refrigeration is the best way to control microbes in date (*Phoenix. dactylifera. Lynn.*) fruits.

Keywords: Dates; Microbes; Control; Radiation; Freezing; Hot air oven

Introduction

Date fruits (*Phoenix dactylifera Lynn.*) is one of the oldest fruits in the world. It has been used from 6000 Years ago [1]. It is grown in many countries worldwide, majorly Algeria, China, Egypt, Iran, Iraq, Pakistan, Saudi Arabia, Sudan and United Arab Emirates (UAE) [2]. The fruit weights from 4.60 g to 11.62 g. 100 g of date flesh provide 73.5 g of carbohydrates, 2.3 g of proteins, 1.5 g of ash and 0.2 g of fat. It also contains minerals like calcium, iron, magnesium, phosphorous, potassium, sodium, zinc, copper manganese and selenium [3]. Dates fruits are generally resistant to microbial contamination due to their high sugar contents but they are affected by various fungal species, insect infestations and bacterial contaminants [4]. It reduces the texture, taste and shelf life of dates. Majorly lactic acid bacteria and acetic acid bacteria colonize in higher number. A numerous fungal species also involved in the contamination of date fruits. Fungal contamination is a direct relationship with both the physical initial dates and environmental conditions of the premises including the storage temperature and humidity can alter the organoleptic parameters of dates, and consequently decrease the market value. So control parameters like sun drying, chemical fumigation, modified atmospheric packaging, radiation, deep freezing, refrigeration, Water washing, oil coating were followed to prevent the entry of microbes (Table 1).

Materials and Methods

Sample collection

A total of seven different varieties of dates samples were collected which are commercially available. They are vacuum packed, date's syrup, loosely available, both seeded and seedless in polythene packed, pet jar packed. They were covered and kept in aseptic conditions. In laminar chamber, it was opened in sterile condition for processing.

Total microbial analysis

Each was measured 10 grams and mixed with 90 ml of distilled sterile water and mixed well. Then the samples are serially diluted up to 10^{-7} dilutions in sterile distilled water. For bacterial enumeration, 10^{-4} and 10^{-6} dilutions were taken and for fungal enumerations, 10^{-3} and 10^{-5} dilutions were taken. For bacterial count Plate count agar is used and for fungal count Rose Bengal agar supplemented with chloramphenicol is used. Pour plate technique performed for inoculating the samples into media. The plate count agar plates were incubated at 37°C for 24 hours and fungal plates were incubated at 27°C for 5 days. For each dilution three replicas were made and after proper incubation, the colonies counted on colony counter and statistically analyzed [5].

Possible pathogen detection

Some of the colonies grown in plate count agar doubted as pathogens were studied microscopically and biochemically. Based on the confirmative tests they confirmed as pathogens. So samples from all the collected date fruits were analysed for the presence of common food borne pathogens.

Pathogens analysis

To detect the common food pathogens like *Staphylococcus aureus*, *Bacillus cereus*, *Salmonella*, *Shigella*, *E. coli* and other coliforms contamination samples from dilution 10^{-2} is taken and inoculated into specific selective media and incubated. For *Staphylococcus*

Kingdom	Plantae
Order	Arecales
Family	Arecaeceae
Genus	*Phoenix*
Species	*dactylifera*
Common name	Date palm fruit, Dates

Table 1: Classification of date plant.

*Corresponding author: Ragava SC, Department of Microbiology, Hindusthan College of Arts and Sciences, Coimbatore, Tamil Nadu, India
E-mail: ragavasanthosh5394@gmail.com

aureus Mannitol Salt Agar, *Bacillus cereus* Bacillus HiVeg Medium, *Salmonella* and *Shigella* SS agar, *E. coli* Eosin Methylene Blue agar and for Coliforms Violet Bile Red Agar is used. After incubation based on growth, the samples analyzed for the presence of pathogens.

Control parameters

To control the microbial load, physical parameters like deep freezing, refrigeration, hot air oven and radiation was followed in different time intervals. The physical parameters followed in different time intervals to check the effectiveness and to find out the optimum contact time of particular parameter. For heat treatment in hot air oven, 10 g of sample kept in 100°C for 1 min, 2 mins and 3 mins. For each time interval, three replicas were made and packed aseptically. Then they are serially diluted and plated for initial microbial load. After 15th and 30th day, the samples was plated and checked for total microbial load by that day. In radiation, the samples kept at 240 nm radiations of UV rays from 15 mins, 30 mins and 45 mins. In each time interval, the samples taken and aseptically packed and analyzed immediately for initial load and in 15th and 30th day. Likewise samples kept in 4°C and -19°C also analyzed in 15th and 30th day.

Statistical analysis

The colonies were counted by using digital colony counter and the counting's was applied on the following formula to get a neutral value from the replicas [6].

$$\text{Number of colonies} = \frac{\Sigma C}{(N_1 + 0.1N_2)\,D}$$

Where,

$\Sigma C =$ Total number of colonies,

$N_1 =$ Number of dishes obtained in first dilution,

$N_2 =$ Number of dishes obtained in second dilution,

$D =$ Dilution factor corresponding to first dilution.

Results and Discussion

Total microbial load

The total number of colonies was represented in Table 2. The load of both bacteria and fungi are higher than the normal permitted level of not more than 10,000 bacterial count in 1 gram of food and yeast and mould should be absent in 0.1 g of food as per Prevention of food adulteration rules, India, 1956. To the maximum, in seedless samples in pet jar packed, 8.75×10^5 cfu/g bacterial loads observed and in fungal load, the maximum number of 16.36×10^3 cfu/g was observed in locally packed dates. The observed results were matched with previous researches from Salah et al. [7] and Raimi [8]. In plating, colonies with various types of morphology were observed. In fungi, most of the colonies resembled *Aspergillus* species. They were observed under light microscope and confirmed. The high number of microbial load may occurred due to poor post-harvest handling, storage temperature variance, damages in the skin of fruit, mycorhyzzal association relationship etc. [9].

Possible pathogen detection

Some colonies were taken and observed under light microscope. Their structure resembled alike *Staphylococcus* species and *Bacillus* species. So biochemical tests were performed like IMViC, catalase and oxidase. Based on the test results, the colonies were identified as *Staphylococcus aureus* and *Bacillus cereus*.

Enteric pathogen detection

Because of the presence of pathogens detected in some of the samples, tests performed to check out the detection of enteric pathogens in all the food samples. The results were represented in Table 3. Presence of *Staphylococcus aureus* observed in polythene seedless, pet jar seeded and in locally packed samples. *Staphylococcus* is a serious food borne pathogen. It is presented in human skin as normal flora. So through improper post-harvest handling, it gains enter into the food. It produces heat resistant enterotoxin. It produces symptoms like nausea, vomiting, abdominal cramps (which are usually quite severe), diarrhoea, sweating, headache, prostration, and sometimes a fall in body temperature generally lasting from 24 to 48 hours, and the mortality rate is very low or nil. The usual treatment for healthy persons consists of bed rest and maintenance of fluid balance [10]. *Bacillus* species was observed in pet jar seedless and locally packed samples. Members of the *B. cereus* group are ubiquitously distributed in the environment, mainly because of their spore-forming capabilities. Thus *B. cereus* can easily contaminate various types of foods, especially products of plant origin. It produces diarrhoea, gastroenteritis, emetic syndrome, abdominal pain etc., Its mortality rate is very low but deaths also observed due to bodily complications which occurred due to the toxicity of *Bacillus* [11]. However, no other common food pathogens like *Salmonella*, *Shigella*, Coliforms or *E. coli* found in samples.

Control parameters

Control parameters like Cold Storage, Heat treatment and radiation was followed and the effectiveness was checked on initially, 15th day and 30th day. They are represented in Figures 1a and 1b.

Cold storage: Chilling and freezing are successful methods for food preserving because chemical reactions and microbial are heat dependent. So, by keeping the food sample under low temperature conditions ceases these reactions so that the shelf life of food prolonged. Though psychrotrophs can grow in chilled foods they do so only relatively slowly so that the onset of spoilage is delayed. In this respect temperature changes within the chill temperature range can have pronounced effects [12]. Figures 1c and 1d show that there were

S.No.	Type	Bacteria (× 105 cfu/g)	Fungi (× 103 cfu/g)
1	Polythene seeded	6.42	7.27
2	Polythene seedless	5.48	7.88
3	Pet jar seeded	4.54	12.42
4	Pet jar seedless	8.75	8.79
5	Vacuum	1.64	11.21
6	Local packed	6.21	16.36
7	Syrup	1.67	2.12

Table 2: Stock load.

S.no.	Type of Sample	MSA (×10^{-2} cfu/g)	EMB (×10^{-2} cfu/g)	VBRA (×10^{-2} cfu/g)	SSA (×10^{-2} cfu/g)	Bacillus Hi Veg Medium (×10^{-2} cfu/g)
1	Polythene seeded	0	0	0	0	0
2	Polythene seedless	2	0	0	0	0
3	Petjar seeded	0	0	0	0	0
4	Petjar seedless	1	0	0	0	3
5	Vacuum	0	0	0	0	0
6	Local packed	2	0	0	0	2
7	Syrup	0	0	0	0	0

Table 3: Pathogen testing on stock sample.

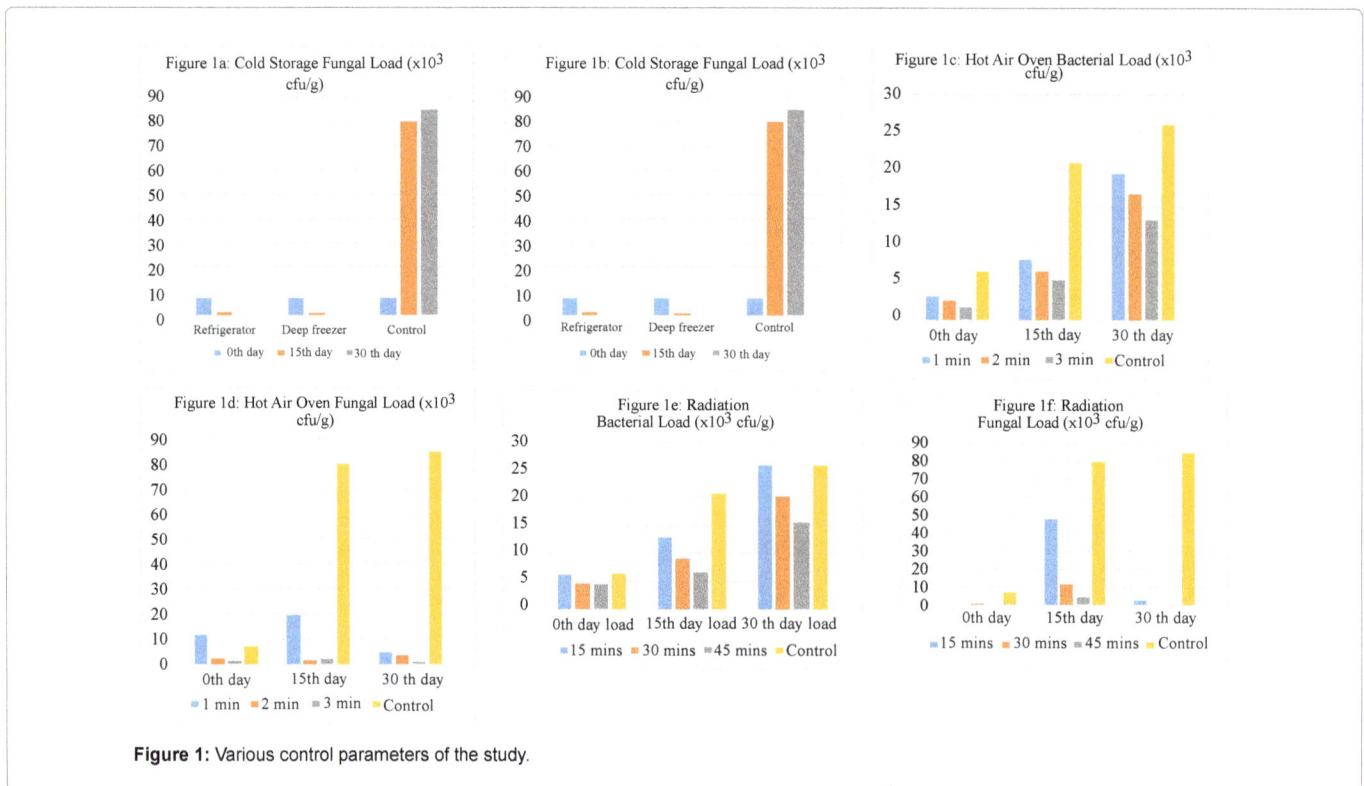

Figure 1: Various control parameters of the study.

very low growth of microbes, which are in negligible level observed. The results observed were matched with Al-Sahib [3]. After 30th day, pathogens also killed from stored samples.

Heat treatment: Most microorganisms can be killed at 100°C. Fungal spores are slightly resistant than the normal vegetative cells so that they can be killed at below 100°C [12]. From the results, it was observed that most of the microbial load was killed by heating but due to high moisture content still remained at the centre of the fruit; the endospores of bacteria gained withstanding and got growth when subjected to suitable environment again. But maximum fungal spores which were presented on the surface of fruit got lysed due to high temperature. No sensory changes observed till the end of storage time and the fruit's taste remained same.

Radiation: Microorganisms are more susceptible of UV rays. They kill the microbes by forming thymine dimers at 200 nm to 280 nm. Since most of the date contaminants present in the surface UV rays is used for preservation. Irradiation, like heat, kills microbial cells and destroys their spores at a predictable rate that is basically dependent on dose level, exposure time, and microbial type [13]. Its effectiveness on microbial control in food has been studied well since the beginning. Figures 1e and 1f showed that the radiation is more effective on fungal sterilization because of its surface presence. Because of the presence of bacteria in the endocorpic region, the UV rays are unable to penetrate inside the fruit so that the bacteria gains growth [14,15]. But the microbial growth was merely controlled than the fruit samples kept as controls.

Conclusion

Since radiation and heat treatment are more effective, due to less penetration power of UV and high moisture content of the fruit, these methods becomes ineffective. By storing the date fruits in such a very

low temperature of 4°C or -19°C the metabolic activities of microbes become arrested so that the fruits remain same for a long time. During the research, the pathogens also identified from the samples so that the public are advised to be more cautious while buying the fruit. The future work may carry for detecting the survived organisms in such harsh conditions where the physical parameters used.

References

1. Kwaasi AA (2003) Date palm and sandstorm-borne allergens. Clini Exp Allerg 33: 419-426.

2. Ibrahim AM, Khalif MNH (1998) The date palm: Its cultivation, care and production in the Arab world. Almaarif Public Company, Alexandria, Egypt.

3. Al-Shahib W, Marshal RM (2003) The fruit of the date palm: Its possible uses as the best food for the future. Int J Food Sci Nutri 54: 247-259.

4. Siddig HH (2012) The microbial quality of processed date fruits collected from a factory in Al-Hofuf City, Kingdom of Saudi Arabia. Emir J Food Agri 24: 105-112.

5. Umar ZD, Bilkisu A, Bashir A (2014) Bacteriological analysis of date palm fruits sold in katsina metropolis. Int J Environ 3: 83-86.

6. FSSAI (2012) Manual of methods of analysis of foods, microbiological testing. Lab Manual 14. Food Safety and Standards Authority of India.

7. Salah MA, Bakri H, Salah AA, Safar H, Sobhy MI, et al. (2014) Microbial loads and physicochemical characteristics of fruits from four Saudi date palm tree cultivars: Conformity with applicable date standards. J Food Nutri Sci 5: 316-327

8. Raimi OR (2013) Microbiological assessment of date fruits purchased from owode market, in Offa, Kwara State Nigeria. J Environ Sci Toxic Food Technol 4: 23-26.

9. Al-Sheikh H (2009) Date palm fruit spoilage and seed borne fungi of Saudi Arabia. Res J Microbiol 4: 208-213.

10. James JJ, Martin J, Loessner, David AG (2005) Modern food microbiology (7thedn). Springer Publications.

11. Riemann H, Oliver D (2006) Foodborne Infections and intoxications (3rdedn). Elsevier Publications.

12. Adams Martin R, Moss Maurice O (2008) Food microbiology (3rdedn). RSC Publishing.

13. Ray B (2005) Fundamental food microbiology (5thedn). CRC Press LLC

14. Mohammed SJ (2010) Effect of storage temperatures on microbial load of some dates palm fruit sold in Saudi Arabia Market. Afri J Food Sci 4: 359-363.

15. Muhammad N, Salim R, Faqir MA, Ijaz AB (2011) Quality evaluation of some Pakistani date varieties. Pak J Agri Sci 48: 305-313.

10

Effect of Roselle Calyces Concentrate with Other Ingredients on the Physiochemical and Sensory Properties of Cupcakes

Abdel-Moemin AR*

Department of Nutrition and Food Science, Faculty of Home Economics, Helwan University, Bolak, Cairo, Egypt

Abstract

Purpose: Roselle calyces are a major crop for export and used to make a common drink in Egypt. The objective of this research was to determine the physiochemical and the sensory properties of cupcakes formulated with roselle calyces concentrate incubated with 11 different food grade ingredients (FGI) prior to addition to the cupcake batters.

Methodology: Anthocyanins, fibre, moisture, colour and sensory evaluations were done along with batter and baking quality.

Findings: Roselle calyces cupcakes incubated with molasses and orange zest had the highest sensory scores ($P<0.05$). The parameter a* was significantly redder when roselle calyx concentrates were incubated with vinegar, lemon or orange juice. One hundred g of roselle cupcakes with lemon juice provided 420 mg/100 g anthocyanins and 10% of total dietary fibre.

Practical implications: The FGI is available and inexpensive. Roselle calyces cupcake with the FGI can be made at home and is less sour than the roselle calyces drink. These cupcakes would have a "clean" label.

Originality: This is one of the first studies to use FGI to treat roselle calyces concentrate. The FGI are sources of acids such as juices or vinegar, natural sweeteners such as honey and molasses that enhance the stability of anthocyanins and which may themselves have numerous phytochemicals.

Keywords: Roselle calyces; *Hibiscus sabdariffa* L.; Anthocyanins; Cupcakes; Fibre; Cost effective product; Needs assessment

Introduction

Roselle calyces (*Hibiscus sabdariffa* L.) is a tropical plant in the *Malvaceae* family and is known in Egypt as *Karkadah*. It is probably a native of West Africa and is now widely cultivated throughout the tropics and subtropics, e.g., Sudan, China, Thailand, Egypt, Mexico, and the West India [1]. Roselle calyces are one of the major Egyptian crops and are used in food, drinks, and cosmetics.

Roselle plant is cultivated mainly in Upper Egypt [2]. There are two types of roselle calyces: light and dark red colours. The dark red roselle calyx, (*H. sabdariffa* var. *sabdariffa*) is shorter and bushier, and planted for its edible calyces. The light red roselle (*H. sabdariffa* var. *altissimoe wester*) is an erect, sparsely branched annual growing to 4.8 m high, which is cultivated for its jute-like fibre.

Roselle calyx anthocyanins might be used as a natural food colourant [3], safer than most synthetic dyes that contain azo functional groups and aromatic rings, they may have negative effects on health including allergic and asthmatic reactions [4], DNA damage [5], and hyperactivity [6]. Some synthetic dyes are even considered to be potentially carcinogenic and mutagenic to humans [7]. However, using roselle calyces into food products would provide a "clean" label and would also add additional phytochemicals that might be beneficial to health while protecting the roselle calyces anthocyanins.

The pigments roselle calyces anthocyanin are inherently unstable, but when acylated becoming more stable. Stability is being improved in solutions and in baking [8,9], or by encapsulation [10] using relatively pure citric acid, acetic acid, citrus pectin, tartaric acid, and adipinic acid.

The colour stability of anthocyanins depends on a combination of factors including: structures of the anthocyanins, pH, temperature, oxygen, light, and water activity [11]. Enzymatic degradation and interactions with other food components such as ascorbic acid, sugars, metal ions, sulfur dioxide, co-pigments and food matrices are also important [9,11]. The bioactive compounds in roselle calyces, such as some flavonoids and anthocyanins, are unstable and may be degraded during beverage preparation to colourless or brown-coloured products, which represent a loss of their beneficial health properties [12]. Cupcakes are a convenient bakery product for consumers to buy or make at home. Therefore, the current study aimed to enhance roselle calyx anthocyanins pigment stability for both improved colour and health in cupcakes to evaluate 11 food grade ingredients (FGI): honey, molasses, vinegar, lemon rind, lemon juice, lemon zest, Nescafe, Arabica coffee, orange rind, orange juice, and orange zest.

Materials and Methods

Cupcake ingredients

Cupcake ingredients included all purpose wheat flour (72% extraction), sun-dried loose dark red roselle calyces, caster sugar, Nescafe, honey, unsalted butter, liquid skimmed milk, baking powder and iodized salt, eggs, pure vanilla powder, oranges, Arabica coffee, molasses, and sugar cane vinegar (5%) were all purchased from local markets in Cairo, Egypt. Vinegar, Arabica coffee, Nescafe, honey, and molasses were added to the batters as purchased. However, orange and lemon rinds, juices, and zests were obtained by peeling, squeezing or by grating as needed.

Chemicals

Cyanidin-3-glucoside was purchased from Sigma–Aldrich

---

*Corresponding author: Abdel-Moemin AR, Department of Nutrition and Food Science, Faculty of Home Economics, Helwan University, Bolak, Cairo 11221, Egypt, E-mail: aly.moemin@heco.helwan.edu.eg

(Darmstadt, Germany). Heat stable α-amylase (*Bacillus licheniformis*, solution, A3306), protease (*Bacillus licheniformis*, lyophilized powder, P3910), and amyloglucosidase (*Aspergillus niger*, solution, A9913) were purchased from Sigma-Aldrich (St. Louis, MO, USA). All other chemicals and solvents were Analar grade.

Preparation of roselle calyces concentrate

A 100 g of dried loose roselle calyces were cleaned by removing visually observed non-calyces matter, then dried in a vacuum dryer at 28°C for 3 hr, cooled at room temperature (23°C), weighed and ground using a coffee machine. Ground roselle calyces were soaked overnight in 200 mL distilled water (DW). In the morning, the suspension was heated at 80°C for 1 hr after adding 450 mL (DW) in a 2 L Erlenmeyer flask. The suspension was strained, 10 g unprocessed sample of each FGI was incubated with 40 g of roselle calyces concentrate for 2 hours at room temperature, and each of these mixtures was added to the other cupcake ingredients to test their ability to keep the roselle calyces red.

Batter preparation

The basic formulation, modified from Gisslen [13], was used for the preparation of 250 g of each cupcake batter and is shown in Table 1. The dry ingredients were sieved together and then the melted butter, eggs, vanilla, and skimmed milk or roselle calyces concentrate (replacing the milk) or roselle calyces concentrate with a FGI were mixed together for 5 min. Cupcake papers were fitted into each of 12 wells cupcake tray (34 cm × 26 cm). Four cupcake papers were filled with each batter on a balance and 60 g batter added, and then baked in a preheated gas oven and baked at 175°C for ~20 min. After baking, cupcakes were left to cool and then packed in polyethylene bags and stored for 12 hr (overnight) in a dry place before sensory testing began. Both moisture determination and sensory testing were begun that morning.

Analytical methods

Moisture was determined using an Infrared Moisture Determination Balance (FD-610-Kett Electric Laboratory, Tokyo, Japan) by weighing 5 g of each cupcake crumb and measured at 80°C for 60 min.

The water loss was calculated according to the following equation:

$$\% \ ML \ (Moisture \ loss) = \frac{W_1 - W_2}{W_1} \times 100$$

Where,

W_1 is the weight of cupcake batter actually transferred into each cupcake paper (~60 g) and W_2 is the weight of the baked cupcake 12 hrs after baking [14]. Total dietary fibre of cupcake samples was determined according to AOAC method 960.52 [15]. Cupcakes samples were lyophilized (Snijders Scientific, Tilburg, Holland, capacity 3 kg ice). After lyophilization cupcakes were weighed again, ground, sieved through a 40-mesh sieve, and stored at -20°C for up to 15 days until analysis. Total dietary fibre was the weight of the residue less the weight of the protein and ash. Protein was determined by Kjeldahl using AOAC official method 991.20 [16]. Hydrolysis was done using a Tecator Digestion System 20, 1015 Digestor (Tecator, Höganäs, Sweden). Ash was determined gravimetrically according AOAC official method 930.30 [17] using a muffle furnace (Nabertherm, D2804, Lilenthial-Bremen, Germany) at 550°C for at least 6 hr. Calculation of total dietary fibre was done according to the following equation:

$$\% \ TDF = \left[\frac{\left(R_{Sample} - P_{Sample} - A_{Sample} - B \right)}{SW} \right] \times 100$$

Where,

Ingredient (g)	Cupcake batter formulation		
	Control	Roselle calyces control	Cupcake samples
Egg	40.0	40.0	40.0
Pure vanilla powder	1.0	1.0	1.0
Caster sugar	55.0	55.0	55.0
Wheat flour	80.0	80.0	80.0
Salt	0.25	0.25	0.25
Milk	50.0	0.0	0.0
Butter	21.0	21.0	21.0
Baking powder	2.75	2.75	2.75
Roselle calyces concentrate	0.0	50.0	0.0
*Treated roselle calyces	0.0	0.0	50.0
Total	250.0	250.0	250.0

*Eleven food grade ingredients were incubated with roselle calyces concentrate: honey, molasses, vinegar, lemon rind, lemon juice, lemon zest, Nescafe, Arabica coffee, orange rind, orange juice, and orange zest.

Table 1: The ingredients used in the cupcake batter formulations.

TDF = Total dietary fibre; R = Average residue weight (mg); P = Average protein weight (mg); A = Average ash weight (mg); SW = Average sample weight (mg); The Residue Weight = W_2-W_1; Ash Weight = W_3-W_1; B = R_{Blank} - P_{Blank} - A_{Blank}; W_1= Celite + crucible weight; W_2 = Residue + celite + crucible weight; W_3 = Ash + celite + crucible weight

Measurement of pH

The pH of roselle calyces concentrate and cupcake samples were measured according to the method of Von Elbe et al. [18] with slight modification. A 0.5 g sample of ground cupcake was mixed with 20 mL of DIW and vortexed for 3 min. The mixture was held at room temperature for 1 hr to separate the solids and liquid. After centrifugation of the liquid for 3 min at 3,050 $x g$, the pH of supernatants was measured.

Determination of anthocyanins

Anthocyanins were determined according to the method of Lee et al. [19]. The method is based on the monomeric anthocyanin pigments reversibly changing colour with a change in pH; the coloured oxonium form exists at pH 1.0, and the colourless hemiketal form predominates at pH 4.5. The difference in the absorbance at 520 nm at these two pH is proportional to the pigment concentration.

The absorption of anthocyanins was measured at both pH at 520 nm and 700 nm using an E-Chrom Tech Spectrophotometer (CT-2200, Taipei, Taiwan) with a DIW blank. The absorbances were measured within 20-50 min of preparation. Calibration curves were prepared using Cyd-3-Glu concentrations of 0, 150, 300, 600, and 1200 mg/L in DIW based on powder weight.

The concentration of anthocyanins were calculated in the samples and expressed as Cyd-3-Glu equivalents (mg/100 g), as follows:

$$Concentration \ of \ anthocyanins = \frac{A \times MW \times DF \times 10^3}{\varepsilon \times 1}$$

Where A = ($A_{520 \, nm}$ – $A_{700 \, nm}$) pH 1.0 – ($A_{520 \, nm}$ – $A_{700 \, nm}$) pH 4.5; MW (molecular weight) = 449.2 g/mol for Cyd-3-Glu; DF = dilution factor = 1.5; l = path length in cm; ε = 26,900 molar extinction coefficient of Cyd-3-Glu [19].

Known amounts of Cyd-3-Glu were also added to the samples

before testing. The recovery ranged from 88-106% for the samples and 95% for the controls. This recovery was within the acceptable range.

Batter viscosity

A simple method to compare different batter viscosities was done using the method described by Ebeler et al. [20]. A funnel with a top inside diameter of 10 cm and a bottom inside diameter of 1.6 cm was used. The funnel was filled to the top with batter and then the batter was allowed to flow for 15 s, to stop the batter flow a palette knife was used to block the outlet. The amount of batter was weighed, divided by 15 to give flow rate in g/s, i.e., higher values indicate lower viscosities.

Batter specific gravity

The specific gravity (density) of the batters was estimated by dividing the weight of a certain volume of batter by the weight of an equal volume of water using the following equation:

$$\text{Specific gravity } (g/cm^3) = \frac{(\text{Weight of batter} - \text{filled container} - \text{Weight of container})}{(\text{Weight of water} - \text{filled container} - \text{Weight of container})}$$

A handmade plastic container having an internal volume of ~100 cm³ was used. The batter was put in using a rubber spatula to carefully fill the dry clean container with minimal air pockets and carefully levelled off with the spatula.

Volume index

The volume index of baked cupcakes was measured one day after baking according to AACC methods 10-91 [21]. Cupcakes were cut vertically through their centre and the heights of the samples were measured at three points (B, C, D; B and D are 3/5 away from the centre (C) along the cross-sectioned cupcakes using the index template. These heights were used to calculate the volume index, contour, and symmetry as described in the official method:

The volume index $= B + C + D$

Contour $= (2C - B - D)$

Symmetry $= |B - D|$

Cupcake density

Since the baked cupcake density and specific volume are reciprocal only the cupcake density was measured using the seed displacement method (AACC, method 74-09) [22].

$$W_{seeds} = W_{total} - W_{cupcake} - W_{container}$$

$$V_{seeds} = \frac{W_{seeds}}{\rho_{seeds}}$$

$$V_{cupcake} = V_{container} - V_{seeds}$$

$$\text{Cupcake density} = \frac{W_{cupcake}}{V_{cupcake}}$$

Where,

W represents weight (g), V represents volume (cm³), and ρ represents density (g/cm³). Measurement of seed's density was done by measuring the internal dimensions of a cylindrical container in cm using a ruler to obtain the volume of the container (V). The container was weighed empty (W_1), the container was filled with rapeseeds (obtained from the Egyptian Ministry of Agriculture, Cairo, Egypt), levelled off and reweighed (W_2).

$$\text{The density of the seeds } (g/cm^3) = \frac{(\text{Weight of container full of seeds} - \text{Weight of container empty})}{\text{Volume of container}}$$

Colour measurement

Colour measurements (CIE L*a*b*) of baked cupcakes were done using an image analysis technique [23]. A colour image was obtained using a digital camera (Canon, Power Shot A470, 3.4X optical zoom, 7.1 megapixels, Shanghai, China) under controlled and defined illumination conditions using 2 Philips Natural Daylight 18 W fluorescent lamps with a colour temperature of 6500 K according to the manufacturer. The colour of the surface and the crumb of the cupcakes were measured. Sample photos were taken 30.5 cm above the cupcakes at an angle of 45° using a tripod. The pictures were downloaded to a personal computer using a USB digital film reader. Once the colour images of the cupcake samples were captured, the colour was analyzed quantitatively using Photoshop [24]. The captured photos were viewed in the Adobe Photoshop window and from the Info Palette and Histogram Window (Figure 1). The L*, a*, b*, values were calculated (Figure 1), using the following equations according to Yam and Papadakis [23] and Chakraborty et al. [25].

$$L* = \frac{Lightness}{255} \times 100$$

$$a* = \frac{240a}{255} - 120$$

$$b* = \frac{240b}{255} - 120$$

Sensory evaluation

Sensory evaluation was used to assess the sensory acceptability of the cupcakes using an acceptance test. Cupcakes were subjected to sensory evaluation by untrained panellists. Prospective panellists were screened using the following criteria: 1) female >20 yr ($n = 50$), 2) employees who have eaten cake or cupcakes at least once a week for the last three months, and (3) have no food allergies ($n = 15$). They were neither trained nor given prior information about any of the ingredients

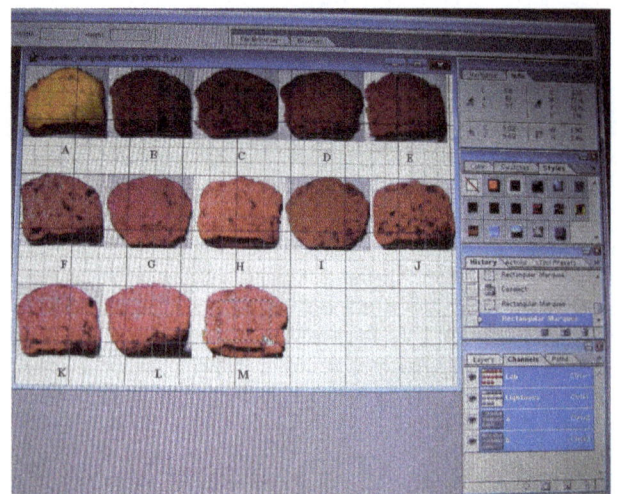

Figure 1: The Photoshop Window Screen of cupcake crumb samples appearing the determined area in M sample that will be measured, show the transversal cut of cupcakes, A= Control; B = Control roselle calyces concentrate; C = Honey; D = Molasses; E = Vinegar; F = Lemon rind; G = Lemon juice; H = Lemon zest; I = Nescafe; J = Arabica coffee; K = Orange rind; L = Orange Juice; M = Orange zest.

use. The room temperature cupcakes were presented to the panellists in random order [26,27]. The panellists were instructed to score their liking for each of the attributes being studied. A 5-point hedonic scale with 1 = dislike very much, 2 = dislike moderately, 3 = neutral, 4 = like moderately to 5 = like very much was used to evaluate the colour, appearance, texture (visual and eating), taste, volume, lightness, aroma and overall liking of the cupcakes.

Statistical analysis

Moisture content, pH, anthocyanins, batter viscosity, batter specific gravity, volume index, colour measurements, cupcake density and sensory evaluations were done in triplicate, and mean values and standard deviations were calculated using Excel for Microsoft Windows Operating System. One-way ANOVA was done and the differences of the means were evaluated using Tukey's HSD test (P<0.05) using SPSS 16.0 for Windows (SPSS Inc., Chicago, IL, USA).

Results and Discussion

The results of anthocyanin, fibre, moisture, water loss and pH of baked cupcake samples are summarized in Table 2.

Anthocyanin content

Monomeric anthocyanins in the roselle calyces concentrate were 520 mg/100 g Cyd-3-Glu. Anthocyanins increased statistically (P<0.05) for all cupcakes compared to the control; the most effective additives in retaining anthocyanins were vinegar, lemon and orange juice (Table 2). Incubating roselle calyces concentrate with honey, molasses, Arabica coffee and Nescafe significantly reduced anthocyanins (P< 0.05) and cupcakes became dark brown. The red flavylium cation may be degraded due to the high temperatures used during baking and converted into colourless carbinol pseudo base, which subsequently undergoes an opening of the pyrilum ring, leading to the formation of brown-coloured chalcone [8].

Žilić et al. [9] recently showed that adding citric acid to pigmented corn dough increased the phenolic compounds in corn cookies

Sample	Anthocyanins mg/100g	Fibre %	Moisture %	Water loss	pH
Roselle calyces C.	520 ± 0.7	15 ± 0.8	63 ± 2	NA	5.5 ± 1.3
Control	0	10 ± 1e	25 ± 2d	8 ± 0.2b	7.9 ± 2.1a
Roselle calyces	400 ± 0.1a	14 ± 1a	23 ± 2f	5.6 ± 0.3c	7.5 ± 1.5a
Honey	250 ± 0.3c	9 ± 0.3f	27 ± 2b	5 ± 0.4d	5.7 ± 0.6d
Molasses	220 ± 0.1c	9 ± 0.6f	26 ± 2c	5 ± 0.2d	7.1 ± 0.5b
Vinegar	410 ± 0.1a	8 ± 0.3g	27 ± 3b	4 ± 0.5e	4.5 ± 0.3e
Lemon r.	270 ± 0.3c	11 ± 0.2d	27 ± 1.7b	3 ± 0.4f	7.7 ± 0.4a
Orange r.	300 ± 0.2b	12 ± 1c	27 ± 1b	2.8 ± 1.2f	6.4 ± 0.7c
Lemon j.	420 ± 0.3a	10 ± 1e	26 ± 3c	5 ± 1d	4.6 ± 0.9e
Orange j.	380 ± 0.2b	9 ± 0.4f	26 ± 3c	4.4 ± 1.5e	5.2 ± 0.6e
Lemon z.	390 ± 0.1b	11 ± 0.1d	27 ± 2b	4.7 ± 0.9d	5.5 ± 0.1d
Orange z.	260 ± 0.2c	13 ± 0.5b	28 ± 3a	2.7 ± 0.1f	5 ± 0.5e
Arabica c.	230 ± 0.2d	11 ± 1d	24 ± 2e	10 ± 3a	5.8 ± 0.8d
Nescafe	250 ± 0.2c	9 ± 0.2f	27 ± 3b	4.7 ± 0.3d	7.1 ± 0.4b

Means that do not share the same letter in a column are significantly different according to Tukey HSD test (P□0.05). (n = 3 cupcakes/sample). C = control, r = rind, j = juice, z = zest, c = coffee.

Table 2: Average anthocyanin, fibre, moisture, water loss, and pH of baked cupcake samples.

stabilizing the anthocyanins. The citric acid helped retain anthocyanins by lowering pH and by acylation of their sugar residues or flavylium cation.

Selim et al. [10] showed that low pH values (1.5 to 3) optimized anthocyanin stability. They showed that roselle calyces extracts heated for 30 min at temperatures of 60, 70, 80, 90, and 100°C, retained 99.9, 99.2, 94, 86, and 79% of their anthocyanins contents, respectively. In this study the heating above 50°C occurred at 2 stages: heating ground roselle calyces concentrate with water to extract the pigment at 80°C for 1 hr, and baking cupcakes at 175°C for ~20 min.

No studies were found that evaluated the effect of honey and molasses on anthocyanin stability during baking. Honey and molasses significantly reduced (P<0.05) anthocyanins in roselle calyces cupcakes. The reduction of anthocyanins in baked cupcakes ranged between 19-58%. Fructose in honey can interact with amino acids to give brown Maillard reaction products [28]. The brown colour may also be due to brown pigment in coffee and molasses along with some caramelization during heating.

Fibre content

Roselle calyces concentrate had 15% dietary fibres while the dietary fibre in cupcakes ranged from 8 to 14%. Duke and Atchely [29] found 2.3% fibre in roselle calyces while Gabb [30] found 12% fibre. These differences may be due to different cultivars or to differences in the methodologies for fibre determinations. The orange zest contained 13% dietary fibre and the lemon and orange rinds increased the dietary fibre in cupcakes to 11-12%, which may also reflect the high pectin (~30%) in orange rind [31]. Although the addition of fibre to baked products is associated with a decrease in volume due to partial dilution of gluten [32] yet the volume of the roselle calyces control samples gave a similar volume as the control. In another study, incorporation of 10% red raspberry juice decreased the volume of muffins [33]. They suggested that the rapid release of carbon dioxide (CO_2) due to the reduction in pH led to the decreased volume. The acidification also led to air pockets (pores) inside the muffins, which was also observed in the cupcakes treated with acids.

Moisture content and pH

Moisture content ranged between ~23-28%. Incorporation of 4% FGI significantly (P<0.05) increased the moisture content of cupcakes with a maximum of 28% with lemon and orange rind and orange zest. This could be attributed to the hygroscopic nature of lemon and orange. Incorporation of roselle calyces concentrate at 20% decreased the moisture (23.3%) compared to the control cupcakes. The roselle calyces concentrate's pH was 5.5 while it ranged in cupcakes from 4.5 to 7.7 in cupcakes with vinegar and lemon rind, respectively.

Batter viscosity

Batter viscosity (Table 3) ranged from 1.1 g/s to 1.9 g/s for vinegar (highest viscosity) and orange rind, respectively. The higher batter viscosity for molasses and honey may reflect their capability to absorb water and slow down the rate of gas diffusion during the early stages of baking [34]. The increase in the water absorption capacity of ingredients such as honey and molasses reduces the amount of free water available to facilitate the movement of particles in batters and consequently gives high viscosity to cupcake batters [35].

During baking, the wheat gluten forms a network that traps gas in the batter and provides a smooth texture. However, increased amount of CO_2 could produce peaks and tunnels in the crumb, which is not desired [36]. These were observed in the honey, lemon juice and Arabica coffee cupcakes. Wong [37] also reported that decreasing the pH would dissociate the micro fibrils within the gliadin protein fraction into its monomers, thereby increasing the formation of air pockets.

Batter density

Low batter density is desired in cupcake batter because it indicates that more air is incorporated into the batters, which distinguish the consistency of cakes [38]. Batter density ranged from 1.03 g/cm³ to 1.2 g/cm³ for orange juice, Nescafe and lemon juice, and Arabica coffee, respectively (Table 3). A low viscosity batter can not hold the air bubbles sufficiently. Batter density decreases during cake baking because of the loss of water and the increase of cupcake volume because of gas expansion. However, the final cupcake volume is not only dependent on initial air incorporated into the batter but also its capacity to retain air during baking [39].

Cupcake density

The cupcake density ranged from 0.391 to 0.5 g/cm³ for molasses and Arabica coffee, respectively (Table 3). Excessive batter viscosity may cause disruption of air bubble production during baking. Thus, there is an optimum cupcake batter viscosity to achieve cupcakes with high volume; if the viscosity of the batter is too low or too high, the batter can not hold the air bubbles sufficiently and the cupcakes collapse in the oven [35].

The cupcake volume index ranged from 105 mm to 115 mm for lemon juice and orange rind, respectively (Table 3). The volume index indicates the amount of air entrapped in the cupcake crumb. Although a high volume index does not always indicate to a desirable cupcake, low volumes generally indicate a heavy and less desirable crumb [40]. The lemon and orange rind and Arabica coffee gave the cupcakes a small structure. It was also noticed that the crust of these cupcakes were cracked in the middle of the crust suggesting that these ingredients increased the gas production during baking that may have led to structure collapse.

Cupcake contour and symmetry

Cupcake contour ranged from 6 mm to 8.5 mm for Nescafe and orange rind, respectively (Table 3). The cupcakes with higher volumes showed higher central height. Generally, a peaked cupcake would have a higher contour value and a flat cake would have a lower value. All treatments resulted in cupcakes with intermediate contour values,

reflecting cupcakes with appropriate rounded surfaces. Cupcake symmetry ranged from 1.2 mm to 2.5 mm for vinegar and orange rind, respectively (Table 3). A low symmetry value is more desirable.

Crust and crumb colour of cupcakes

Results in Table 4 showed that the crust colour was dependent on the FGI. It seems that FGI and baking temperature were the main factors that influenced the colour of the crust. The crumb of the control cupcake was lighter than the crust as the Maillard reactions occur mainly on the surface. The a* value increased significantly ($P<0.05$) with the acids (lemon and orange juice, and vinegar). The b* was significantly ($P<0.05$) reduced by addition of roselle calyces and all FGI. Some of these additives turned the cupcakes purple, red and brown (Figure 1). The pinkish colour increased significantly ($P<0.05$) when vinegar, and lemon or orange juice were used. However, when baking with honey or molasses a dark brown colour was obtained (Table 4, Figure 1).

A recent study evaluated the effect of various organic acids on colour retention during fruit juice storage, and found that acetic acid improved the colour stability in both elderberry and black currant juices, whilst the citric and tartaric acids only improved colour stability in elderberry juice [8]. It could be concluded that adding acids to the roselle calyces concentrate help to stabilize the red colour of roselle calyces probably by lowering the pH. Bronnum-Hansen et al. [41] noted that the efficiency of extracting solvents for anthocyanins increased with increasing concentration of citric acid. The lemon and orange juices currently used also improved the colour [8,9].

Sensory evaluation

The results are shown in Table 5. Cupcakes with FGI had high acceptance scores for texture (2.5-4.5) and aroma (2-5). However, cupcakes with molasses, Arabica coffee and Nescafe showed a significant ($P<0.05$) decrease in redness due to colour (dark brown). There was a significant difference ($P<0.05$) with respect to the acceptance of colour of the treated cupcakes compared to control. Cupcakes with honey and Arabic coffee had a lower acceptance for all attributes (3-3.3) although this was still above the value of 2.8 that was set as the acceptable point for overall liking. Smith and Johnson [42] noted when honey replaced sucrose in different types of cakes the quality was poor due to dense structure, low volume, dark crumb, and an undesirable flavour. However the volume of the honey cupcakes was not affected with low amount (4%) used. The panellists gave lower scores for taste and aroma of the honey samples. Most panellists liked the appearance and

Sample	Batter viscosity (g/s)	Batter density (g/cm³)	Cupcake density (g/cm³)	Volume index (mm)	Contour (mm)	Symmetry (mm)
Control	1.6 ± 0.1[b]	1.0 ± 0.03[c]	0.4 ± 0.04[b]	106 ± 3[d]	5.0 ± 0.3[f]	1.5 ± 0.3[d]
Roselle calyces	1.5 ± 0.03[c]	1.0 ± 0.02[c]	0.4 ± 0.03[b]	107 ± 6[c]	5.6 ± 0.4[e]	1.6 ± 0.1[d]
Honey	1.3 ± 0.05[c]	1.0 ± 0.05[c]	0.5 ± 0.01[a]	110 ± 2[b]	6.2 ± 0.5[e]	2.2 ± 0.2[b]
Molasses	1.4 ± 0.01[c]	1.0 ± 0.03[c]	0.4 ± 0.01[b]	108 ± 3[c]	5.8 ± 0.1[e]	2.0 ± 0.2[c]
Vinegar	1.9 ± 0.05[a]	1.1 ± 0.1[b]	0.4 ± 0.03[b]	107 ± 4[c]	6.3 ± 1.5[e]	1.6 ± 0.3[d]
Lemon r.	1.4 ± 0.02[c]	1.1 ± 0.1[b]	0.5 ± 0.02[a]	114 ± 5[a]	7.0 ± 0.2[c]	2.4 ± 0.1[a]
Orange r.	1.4 ± 0.04[c]	1.1 ± 0.07[b]	0.5 ± 0.01[a]	115 ± 2[a]	8.5 ± 1[a]	2.5 ± 0.4[a]
Lemon j.	1.6 ± 0.08[b]	1.0 ± 0.02[c]	0.5 ± 0.04[a]	105 ± 5[d]	7.5 ± 0.4[b]	2.0 ± 0.3[c]
Orange j.	1.6 ± 0.03[b]	1.0 ± 0.01[c]	0.4 ± 0.03[b]	109 ± 2[c]	6.7 ± 0.4[d]	1.8 ± 0.1[c]
Lemon z.	1.4 ± 0.05[c]	1.1 ± 0.05[b]	0.4 ± 0.02[b]	109 ± 4[c]	5.7 ± 0.2[e]	2.0 ± 0.2[c]
Orange z.	1.4 ± 0.04[c]	1.1 ± 0.1[b]	0.5 ± 0.01[a]	110 ± 5[b]	6.5 ± 0.1[d]	2.3 ± 0.1[b]
Arabica c.	1.0 ± 0.06[d]	1.2 ± 0.06[a]	0.5 ± 0.01[a]	112 ± 2[b]	6.5 ± 0.5[d]	2.3 ± 0.4[b]
Nescafe	1.5 ± 0.05[c]	1.0 ± 0.1[c]	0.4 ± 0.01[b]	109 ± 2[c]	6.0 ± 0.4[e]	1.9 ± 0.2[c]

Means that do not share the same letter in a column are significantly different according to Tukey HSD test (P□0.05). (n = 3 cupcakes/sample). r = rind, j = juice, z = zest, c = coffee.

Table 3: The effect of different roselle calyces concentrates treatments on batter and cupcake quality parameters.

texture of cupcakes with Nescafe, molasses and orange zest compared to the control cupcake. The Arabica coffee cupcakes were hard and dry. Overall the FGI treated samples had significantly (P<0.05) improved sensory attributes for various reasons. Panellist comments suggested that the molasses had a distinguished brown chocolate colour and spongy crumbs while both the Nescafe and molasses had an aroma and crumb texture was highly liked.

Molasses, Nescafe, lemon and orange zest cupcakes were the most liked (liking scores of 4.5-4.6). Low density, high volume, desirable lightness and, appearance and pleasant aroma, and appropriate colour seem to be the important parameters that resulted in the highest liking score. Least like were the honey, lemon and orange rind, and vinegar (Table 5). Cupcakes containing orange zest and molasses showed the highest desirability for almost all sensory parameters.

The FGI lowered the acceptability of the volume probably because these additives may have inhibited the swelling of starch granules and also by forming a film around the granules [43,44] thereby increasing the gelatinization temperature [45].

Needs assessment

Cupcake is popular in Egypt and hundreds of bakery and specialised shops produce cupcakes. The market of cupcakes in Egypt is large, the market is targeting people through cupcake shops, TV cooking programmes and specialised cooking channels. However, our target in this study is children as they like coloured cupcakes and the most sensitive group for over consumption of synthetic food colourants as these shops use them apparently without official control.

Our cupcakes are red, purple and brown with clean label. Also our target is older people who looking for healthy and soft products, that can be made commercially to be labelled as functional cupcakes. Roselle calyces cupcake can provide food safety for children and protective agents for older people. Beneficial health effects associated with anthocyanin consumption include lowering the risk of cardiovascular disease (CVD) was reported [46]. The Egyptian National Hypertension Project found an adjusted overall prevalence of coronary heart disease of 8.3% and a high prevalence of hypertension (26%) that was an important driver of adverse cardiovascular outcomes [47].

Cost-effective product

The current study focused on using roselle calyces and FGI to produce a new cupcake-like product (CLP); as a cost effective (Roselle calyces and FGI are available and cheap) and functional cupcake food with high nutritional value (specifically bioactive compounds). The cost effective in the new product is reasonably cheap compared to the control cupcake costs. One of the most important factors that influence the cost effectiveness of producing cupcake is the choice and the prices of ingredients such as flour. The price of flour based on a retailer price in Egypt is between L.E 7-8 (1 kg), while the commercial cupcakes are L.E 10. The real calculation of each cupcake in this study as only ingredients for roeslle cupcakes is L.E 2.60 compared to control (using milk) L.E 3.20. With the assumption of reducing the cost to ~ 19% as (~ L.E 0.6) when roselle calyces and FGI added, in addition to its higher nutritional value. As we've found in the literature, total phenols, anthocyanin, and fibre content exert their beneficial effects in reducing risks of coronary heart disease [26,48-51]. Additionally, its use could contribute to reducing the cost of national heath services in the treatment of heart diseases [52].

Sample	Crust			Crumb		
	L*	a*	b*	L*	a*	b*
Control	71 ± 3ᵃ	10 ± 0.2ᶠ	61 ± 3ᵃ	89 ± 3ᵃ	-0.5 ± 0.1ᵉ	65 ± 3ᵃ
Roselle calyces	16 ± 1ᵍ	15 ± 1ᵈ	8 ± 0.3ᶜ	54 ± 1ᶜ	40 ± 3ᶜ	25 ± 1ᶠ
Honey	24 ± 1ᵉ	19 ± 1ᶜ	10 ± 0.2ᵇ	51 ± 1ᵈ	42 ± 2ᶜ	25 ± 0.4ᶠ
Molasses	16 ± 1ᵍ	9 ± 0.4ᶠ	8 ± 0.4ᶜ	56 ± 2ᶜ	41 ± 1ᶜ	40 ± 2ᶜ
Vinegar	27 ± 1ᵈ	25 ± 1ᵇ	8 ± 0.2ᶜ	56 ± 3ᶜ	48 ± 3ᵃ	28 ± 2ᵉ
Lemon r.	22 ± 1ᵉ	11 ± 1ᶠ	3 ± 0.2ᵈ	61 ± 2ᵇ	29 ± 1ᵈ	25 ± 2ᶠ
Orange r.	23 ± 1ᵉ	17 ± 0.4ᵉ	5 ± 0.4ᵈ	54 ± 3ᶜ	45 ± 3ᵇ	30 ± 5ᵉ
Lemon j.	39 ± 1ᵇ	40 ± 2ᵃ	9 ± 0.4ᶜ	53 ± 2ᶜ	48 ± 2ᵃ	21 ± 3ᵍ
Orange j.	36 ± 1ᶜ	26 ± 1ᵇ	11 ± 3ᵇ	60 ± 4ᵇ	44 ± 2ᵇ	28 ± 8ᵉ
Lemon z.	27 ± 1ᵈ	20 ± 1ᶜ	7 ± 4ᶜ	60 ± 2ᵇ	45 ± 2ᵇ	34 ± 5ᵈ
Orange z.	29 ± 6ᵈ	19 ± 1ᶜ	7 ± 5ᶜ	61 ± 3ᵇ	41 ± 2ᶜ	32 ± 6ᵈ
Arabica c.	20 ± 7ᶠ	15 ± 1ᵈ	7 ± 1ᶜ	55 ± 2ᶜ	43 ± 2ᶜ	41 ± 3ᶜ
Nescafe	29 ± 2ᵈ	20 ± 1ᶜ	11 ± 1ᵇ	53 ± 3ᶜ	44 ± 2ᵇ	49 ± 2ᵇ

Means that do not share the same letter in a column are significantly different according to Tukey HSD test (P☐0.05). (n =3 cupcakes/sample). The surface and crumb colour of cupcakes are reported as average L* (lightness), a* (redness), b* (yellowness) values. r = rind, j = juice, z = zest, c = coffee.

Table 4: Effect of roselle calyces cupcake on the colour of the crust and crumb of cupcakes.

Sample	Colour	Appearance	Texture	Taste	Volume	Lightness	Aroma	Overall liking
Control	5.0 ± 0.1ᵃ	4.8 ± 0.2ᵃ	4.9 ± 0.3ᵃ	4.8 ± 0.3ᵃ	5.0 ± 0.1ᵃ	5.0 ± 0.2ᵃ	4.8 ± 0.3ᵃ	4.9 ± 0.3ᵃ
Roselle calyces	3.9 ± 0.2ᵇ	4.2 ± 0.2ᵇ	4.3 ± 0.4ᵇ	3.7 ± 0.2ᶜ	4.5 ± 0.2ᵃ	4.4 ± 0.2ᵇ	4.3 ± 0.4ᵇ	4.2 ± 0.2ᵇ
Honey	3.4 ± 0.3ᵇ	3.6 ± 0.3ᶜ	3.7 ± 0.3ᶜ	2.2 ± 0.1ᵉ	3.3 ± 0.3ᶜ	3.5 ± 0.3ᶜ	2.0 ± 0.3ᶜ	3.0 ± 0.1ᵈ
Molasses	4.5 ± 0.1ᵃ	4.8 ± 0.1ᵃ	4.5 ± 0.3ᵃ	4.6 ± 0.2ᵃ	4.6 ± 0.1ᵃ	4.8 ± 0.1ᵃ	4.5 ± 0.3ᵃ	4.6 ± 0.2ᵃ
Vinegar	3.8 ± 0.1ᵇ	3.7 ± 0.1ᶜ	3.4 ± 0.3ᶜ	3.5 ± 0.2ᶜ	3.4 ± 0.1ᵇ	3.8 ± 0.1ᶜ	3.4 ± 0.3ᶜ	3.5 ± 0.2ᶜ
Lemon r.	3.7 ± 0.3ᵇ	4.2 ± 0.4ᵇ	3.0 ± 0.3ᵈ	3.2 ± 0.3ᶜ	3.4 ± 0.3ᵇ	3.3 ± 0.4ᵈ	3.5 ± 0.3ᶜ	3.0 ± 0.3ᵈ
Orange r.	3.0 ± 0.2ᶜ	2.8 ± 0.2ᵈ	2.3 ± 0.2ᵉ	3.0 ± 0.3ᵈ	2.7 ± 0.2ᵈ	2.8 ± 0.2ᵃ	3.8 ± 0.2ᶜ	2.9 ± 0.3ᵈ
Lemon j.	5.0 ± 0.3ᵃ	4.5 ± 0.6ᵃ	4.0 ± 0.4ᵇ	4.0 ± 0.1ᵇ	4.5 ± 0.3ᵃ	3.9 ± 0.6ᶜ	4.0 ± 0.4ᵇ	4.0 ± 0.1ᵇ
Orange j.	4.4 ± 0.2ᵇ	3.8 ± 0.2ᶜ	3.5 ± 0.2ᶜ	3.7 ± 0.3ᶜ	3.4 ± 0.2ᵇ	3.8 ± 0.2ᶜ	3.8 ± 0.2ᶜ	3.8 ± 0.3ᶜ
Lemon z.	4.5 ± 0.1ᵃ	4.2 ± 0.2ᵇ	4.3 ± 0.2ᵇ	4.1 ± 0.1ᵇ	3.9 ± 0.1ᵇ	4.2 ± 0.2ᵇ	4.5 ± 0.2ᵃ	4.5 ± 0.1ᵃ
Orange z.	4.6 ± 0.2ᵃ	4.4 ± 0.2ᵃ	4.8 ± 0.2ᵃ	5.0 ± 0.3ᵃ	5.0 ± 0.2ᵃ	4.8 ± 0.2ᵃ	5.0 ± 0.2ᵃ	4.6 ± 0.3ᵃ
Arabica c.	2.3 ± 0.2ᵈ	3.5 ± 0.2ᶜ	2.5 ± 0.2ᵈ	2.0 ± 0.3ᵉ	2.2 ± 0.2ᵈ	2.1 ± 0.2ᵈ	3.5 ± 0.2ᶜ	3.3 ± 0.3ᶜ
Nescafe	4.6 ± 0.2ᵃ	4.8 ± 0.2ᵃ	4.7 ± 0.2ᵃ	4.9 ± 0.3ᵃ	4.3 ± 0.2ᵇ	4.5 ± 0.2ᵃ	4.9 ± 0.2ᵃ	4.5 ± 0.3ᵃ

Means that do not share the same letter in a row are significantly different according to Tukey HSD test (P☐0.05). A 5-point hedonic scale ranging from 1 = Dislike very much; 3 = Neither like or dislike; 5 = Like very much was used to evaluate the attributes in the table. r = rind, j = juice, z = zest, c = coffee. (n = 13 samples, n = 15 panellists ± S.D.; P☐0.05), one cupcake each. The total n for all values is 102 (50 panellists and 13 samples x 4 tables).

Table 5: Sensory evaluation scores for cupcakes.

Conclusion

Roselle calyx concentrates were incubated with 11 food grade ingredients (FGI), and added to cupcake batters. The sensory panellists gave the best score (statistically significant at $P<0.05$) to the roselle calyx cupcakes with molasses and orange zest. The a* value was increased significantly while the b* value was significantly reduced. The consumption of 100 g of the roselle calyx cupcakes with lemon juice provided 420 mg/100 g anthocyanins and 10% total dietary fibre.

Acknowledgments

The author would like to send special thanks to J.M. Regenstein, Cornell University, Ithaca, NY, USA, for reviewing the manuscript and also for his guiding to improve the shape of the manuscript. Also special thanks to Prof Mona S Halaby and Prof. Abdel-Rahman Attia Nutrition and Food Science, Helwan University, for using their unpublished guides in food science and technology. I would like to acknowledge the cooperation of the members in the Gezira Youth Centre for voluntary participation in the sensory evaluation test.

References

1. Plotto A (2004) Roselle calyces: Post-production management for improved market access.

2. EMALR (2016) The Egyptian Ministry of Agriculture and Land Reclamation, Agricultural Research Center, Information Unit-The Central Administration Agricultural Extension.

3. Mercadante AZ, Bobbio FO (2008) Anthocyanins in foods: occurrence and physicochemical properties. In: Socaciu C (ed.) Food colorants: Chemical and functional properties. CRC Press Inc., Boca Raton, FL, USA pp: 241-276.

4. Dipalma JR (1990) Tartrazine sensitivity. American Family Physician 42: 1347-1350.

5. Sasaki YF, Kawaguchi S, Kamaya A, Ohshita M, Kabasawa K, et al. (2002) The comet assay with 8 mouse organs: results with 39 currently used food additives. Mutation Res 519: 103-119.

6. McCann D, Barrett A, Cooper A, Crumpler D, Dalen L, et al. (2007) Food additives and hyperactive behaviour in 3-year-old and 8/9-year-old children in the community: a randomised, double-blinded, placebo-controlled trial. Lancet 370: 1560-1567.

7. EFSA (2005) Review the toxicology of a number of dyes illegally present in food. Europe Food Safe Author J 263: 1-71.

8. Hubbermann EM, Steffen-Heins A, Stöckmann H, Schwarz K (2006) Influence of acids, salt, sugars and hydrocolloids on the colour stability of anthocyanin rich black currant and elderberry concentrates. Europe Food Res Technol 223: 83-90.

9. Žilić S, Kocadağl T, Vančetović J, Gökmen V (2016) Effects of baking conditions and dough formulations on phenolic compound stability, antioxidant capacity and color of cookies made from anthocyanin-rich corn flour. LWT-Food Sci Technol 65: 597-603.

10. Selim KA, Khalil KE, Abdel-Bary MS, Abdel-Azeim NM (2008) Extraction, encapsulation and utilization of red pigments from Roselle (*Roselle calyces sabdariffa* L.) as natural food colorants. Alex J Food Sci Technol.

11. Jackman RL, Smith JL (1996) Anthocyanins and betalains. In: Hendry GAF, Houghton JD (eds.) Natural Food Colorants. Blackie and Son Ltd, London, UK.

12. Domínguez-López A, Remondetto GE, Salvador G (2008) Thermal kinetic degradation of anthocyanins in a roselle (*Roselle calyces sabdariffa* L 'Criollo') infusion. Int J Food Sci Technol 43: 322-325.

13. Gisslen W (2004) Professional baking: For flour extraction, for cupcake preparation. John Wiley & Sons, Hoboken, NJ, USA.

14. Rahmati NF, Tehrani MM (2014) Influence of different emulsifiers on characteristics of eggless cake containing soy milk: Modelling of physical and sensory properties by mixture experimental design. J Food Sci Technol 51: 1697-1710.

15. AOAC (1997) Methods 960.52. The Association of Official Analytical Chemists. Gaithersburg, MD, USA.

16. AOAC (2000) Methods 991.20. The Association of Official Analytical Chemists. Gaithersburg, MA, USA.

17. AOAC (2000) Methods 930.30. The Association of Official Analytical Chemists. Gaithersburg, MA, USA

18. Von Elbe JH, Maing IY, Amundson CH (1974) Colour stability of betanin. J Food Sci 39: 334-337.

19. Lee J, Durst R, Wrolstad R (2005) Determination of total monomeric anthocyanin pigment content of fruit Juices, beverages, natural colorants, and wines by the pH differential method. J Asso Off Analy Chem Int 88: 1269-1278.

20. Ebeler SE, Breyer LM, Walker CE (1986) White layer cake batter emulsion characteristics: effects of sucrose ester emulsifiers. J Food Sci 51: 1276-1278.

21. AACC (2000) Methods 10-91. Approved methods of the AACC, American Association of Cereal Chemists, St. Pauls, MN, USA.

22. AACC (1988) Method 74-09. Approved methods of the AACC, American Association of Cereal Chemists, St. Paul, MN, USA.

23. Yam KL, Papadakis SE (2004) A simple digital imaging method for measuring and analyzing color of food surfaces. J Food Eng 61: 137-142.

24. Adobe Systems (2002) Adobe PhotoShop 7.0 User Guide. Adobe Systems Inc, San Jose, CA, USA.

25. Chakraborty SK, Singh DS, Kumbhar BK (2014) Influence of extrusion conditions on the colour of millet-legume extrudates using digital imagery. Irish J Agri Food Res 53: 65-74.

26. Abdel-Moemin AR (2015) Healthy cookies from cooked fish bones. Food Biosci 12: 114-121.

27. Abdel-Moemin AR (2016) Analysis of phenolic acids and anthocyanins of pasta-like product enriched with date kernels (*Phoenix dactylifera* L.) and purple carrots (*Daucus carota* L. sp. *sativus var. atrorubens*). Food Measur Char.

28. Demetriades K, Guffey C, Khalil MH (1995) Evaluating the role of honey in fat free potato chips. Food Technol 49: 66-67.

29. Duke JA, Atchley AA (1984) Proximate analysis. In: Christie BR (ed.) The Handbook of Plant Science in Agriculture. CRC Press Inc, Boca Raton, FL, USA pp: 427-434.

30. Gabb S (1997) Sudanese Karkadeh, A Brief Introduction, Economics, File No 12. The Sudan Foundation, London, UK.

31. Benkebalia N (2014) Polysaccharides: Natural fibres in food and nutrition. CRC Press, Boca Raton, FL, USA.

32. Pomeranz Y, Shogren MD, Finney KF, Bechtel DB (1977) Fibre in bread making - effects on functional properties. Cereal Chemistry 54: 25-41.

33. Rosales-Soto MU, Powers JR, Alldredge JR (2012) Effect of mixing time, freeze-drying and baking on phenolics, anthocyanins and antioxidant capacity of raspberry juice during processing of muffins. J Sci Food Agri 92: 1511-1518.

34. Gomez M, Ronda F, Caballero PA, Blanco CA, Rosell CM, et al. (2007) Functionality of different hydrocolloids on the quality and shelf life of yellow layer cakes. Food Hydrocoll 21: 167-173.

35. Ronda F, Gomez M, Blanco CA, Caballero PA (2011) Effects of polyols and non digestible oligosaccharides on the quality of sugar-free sponge cakes. Food Chem 90: 549-555.

36. Griswold RM (1962) The experimental study of foods. Houghton Mifflin, Boston, MA, USA.

37. Wong DWS (1989) Mechanism and theory in food chemistry. AVI, New York, NY, USA.

38. Turabi E, Sumnu G, Sahin S (2008) Rheological properties and quality of rice cakes formulated with different gums and an emulsifier blend. Food Hydrocoll 22: 305-312.

39. Frye AM, Setser CS (1991) Optimizing texture of reduced calorie sponge cakes. Cereal Chem 69: 338-343.

40. Zhou J, Faubion JM, Walker CA (2011) Evaluation of different types of fats for use in high-ratio layer cakes. LWT - Food Sci Technol 44: 1802-1808.

41. Bronnum-Hansen K, Jacobsen F, Flink MJ (1985) Anthocyanin colourants from elderberry (*Sambucus nigra* L.): Process considerations for production of liquid extract. J Food Technol 20: 703-711.

42. Smith LB, Johnson JA (1952) The use of honey in cake and sweet doughs. Bak Digest 26: 113-118.

43. Siswoyo TA, Morita N (2001) Influence of acyl chain lengths in mono and diacyl-sn-glycerophosphatidylcholine on gelatinization and retrogradation of starch. J Agri Food Chem 49: 4688-4693.

44. Richardson G, Langton M, Faldt P, Hermansson AM (2002) Microstructure of α-crystalline emulsifiers and their influence on air incorporation in cake batter. Cereal Chem 79: 546-552.

45. O'Brien RD (2003) Fats and oils: In formulating and processing for applications. CRC Press Inc., Boca Raton. FL, USA.

46. Robert L, Agnès N, Edmond R, Christian, D, Andrzej M, et al. (2006) Entire potato consumption improves lipid metabolism and antioxidant status in cholesterol-fed rat. Europe J Nutri 45: 267-274.

47. Almahmeed W, Arnaout MS, Chettaoui R, Ibrahim M, Kurdi MI, et al. (2012) Coronary artery disease in Africa and the Middle East. Thera Clinil Risk Manag 8: 65-72.

48. Gaithersburg MA (2003) The Association of Official Analytical Chemists. St. Paul, MN, USA.

49. Assous MTM, Abdel-Hady MM, Medany GM (2014) Evaluation of red pigment extracted from purple carrots and its utilization as antioxidant and natural food colorants. Annal Agri Sci 59: 1-7.

50. Bennion EB, Bent AJ, Bamford GST (1997) The technology of cake making. Blackie Academic and Professional, London, UK.

51. Kopjar M, Jaksic K, Pilizota V (2012) Influence of sugars and chlorogenic acid addition on anthocyanin content, antioxidant activity and colour of blackberry juice during storage. J Food Process Preserv 36: 545-552.

52. Tee PL, Yusof S, Mohamed S, Umar NA, Mustapha NM, et al. (2002) Effect of roselle (Hibiscus sabdariffa L.) on serum lipids of Sprague Dawley rats. Nutri Food Sci 32: 190 -196.

11

Development and Quality Evaluation of Jamun Seed Powder Fortified Biscuit Using Finger Millet

Kalse SB*, Swami SB, Sawant AA and Thakor NJ

Department of Agricultural Process Engineering, College of Agricultural Engineering and Technology, Dr. BS Konkan Krishi Vidyapeeth, Ratnagir, India

Abstract

Jamun seed are popular among alternative medicine systems to control different ailments such as diabetes, cardio-vascular and gastro-intestinal disorders. Owing to such attributes, the most important aspect of this study to develop jamun seed powder fortified biscuits have been commercialized to meet these purposes. Efforts were made to prepare biscuits having different combinations of Maida (M), finger millet (FM) and jamun seed powder (JSP) were prepared by mixing them in different proportions viz., T_1 - 87% + 10% + 3%, T_2- 84% + 10% +6%, T_3-81% + 10% + 9%, T_4-78% +10% + 12%. The biscuits were baked in a thermally controlled oven at temperature of 170°C for 20 min. The prepared biscuits were subjected to textural analysis and compared with the control biscuit containing 100% maida flour.

The physical and textural properties of biscuits made by various blending were determined. The qualities of the product were determined with the help sensory evaluation. In sensory analysis treatment T_3 (81% maida + 9% jamun seed powder + 10% finger millet flour) secured maximum score for colour, taste, flavour and acceptability. Therefore, treatment T_3 was more acceptable so it was optimised treatment than others.

Keywords: Jamun seed power; Biscuit; Texture analysis; Diabetes

Introduction

Biscuit is a term used for a variety of baked, commonly flour-based food products. Indian Biscuits Industry is the largest among all the food industries and has a turnover of around Rs. 3000 crores. India is known to be the second largest manufacturer of biscuits, the first being USA. Biscuits were assumed as sick-man's diet in earlier days. Now, it has become one of the most loved fast food product for every age group. Biscuits are easy to carry, tasty to eat, cholesterol free and reasonable at cost.

Jamun seed powder has been used for centuries as a natural form for balancing the healthy blood sugar level. It is a very delicious, detoxifying herb which has properties that helps to maintain natural urination and sweating. It also acts as liver stimulant, digestive, coolant and a blood Purifier. Jamun seeds contain a glycoside, named *Jamboline* which helps in the maintenance of glucose levels as in the normal limits.

Finger millet (*Eleusine coracana* L.) is one of the important millet grown extensively in various regions of India and Africa. Regarding protein (6% to 8%) and fat (1% to 2%) it is comparable to rice and with respect to mineral and micronutrient contents it is superior to rice and wheat. Nutritionally; it has high content of calcium (344 mg/100g), dietary fiber (15% to 20%) and phenolic compounds (0.3% to 3%). This minor millet contains important amino acids viz isoleucine, leucine, methionine and phenyl alanine which are deficient in other starchy meals. It is also known for several health benefits such as anti-diabetic, anti-tumerogenic, atherosclerogenic effects, antioxidant, which are mainly attributed due to its polyphenol and dietary fibre contents. Being indigenous minor millet it is used in the preparation of various foods both in natural and malted forms. Grains of this millet are converted into flours for preparation of products like porridge, puddings, pancakes, biscuits, roti, bread, noodles, and other snacks. Besides this it is also used as a nourishing food for infants when malted and is regarded as wholesome food for diabetic's patients. Finger millet being staple food in different parts of India and abroad is promoted as an extremely healthy food [1].

Jamun seed powder-containing biscuit has been developed and incorporated into the diabetic diet. It has been found to be effective in reducing the postprandial rise in the blood glucose level and in improving glycaemic control [2]. These biscuits can be used for dealing with the symptoms of indigestion. These biscuits can also stimulate the liver functions [3].

Finger millet has the highest calcium content among all the food grains, but it is not highly assimilable. The protein content in millet is very close to that of wheat; both provide about 11% protein by weight, on a dry matter basis. Ayurvedic text suggests that 1-3 g of jamun seed powder per day is an average dose for the treatment of diabetes [4]. The direct consumption of jamun seed powder is uneasy. Therefore, this work has been undertaken to develop the biscuit so that diabetic people will consume it easily and get recommended dose of jamun seed powder.

Materials and Methods

Processing of raw material

The pulp and seed of jamun fruit was separated by pulper. Then the seed washed in water and dried in tray dryer at 60°C for 48 hours still complete drying and ground the seed in pulveriser to fine powder of average particle size 0.58 mm. Milling of wheat and finger millet was done to obtain fine flour with help of attrition mill. The proposed research was carried out in the bakery training centre, Department of

*Corresponding author: Kalse SB, Department of Agricultural Process Engineering, College of Agricultural Engineering and Technology, Dr. BS Konkan Krishi Vidyapeeth, Dapoli, Ratnagir, India
E-mail: sandeep.kalse@gmail.com

Agricultural Process Engineering, College of Agricultural Engineering and Technology, Dr. Balasaheb Sawant Konkan Krishi Vidhyapeth, Dapoli.

Treatment details

a. Control – 100% (M).

b. T_1 - 87% (M) + 10% (FM) + 3% (JSP).

c. T_2 - 84% (M) + 10% (FM) +6% (JSP).

d. T_3 - 81% (M) + 10% (FM) + 9% (JSP).

e. T_4 - 78% (M) +10% (FM) + 12% (JSP).

Process for preparation of biscuit

The biscuits were prepared by mixing of ingredients like maida, finger millet (10%) and jamun seed powder (3%, 6%, 9%, and 12%), salt, sugar etc. were put together for dough formation (Table 1). The dough was kept for resting for 10 minutes. The sheet of appropriate thickness was prepared with the help of wooden roller (bellon). The prepared sheet was cut by using mould (rhomboidal shape). The cutting part of sheet was kept in convective oven as shown in Figure 1 at 170°C for 20 min. for baking. After completion of baking the biscuits were allowed to cool at room temperature (26 ± 2°C) for 1 hour.

The ingredients required for the preparation of Jamun seed powder fortified biscuit by using finger millet for different compositions are as follow.

Results and Discussion

The biscuits were prepared by making different proportion of jamun seed powder of average particle size 0.58 mm. The standard procedure was used for preparation of biscuits as described in Figure 1. The maida, finger millet and jamun seed powder and other ingredients were taken as mentioned in Table 1. Then the Physical properties, calorific value and Textural properties was measured. The results obtained are as follows.

Physical properties of biscuit

The biscuits were prepared with help of mould which is having rhomboidal shape. All the biscuits were prepared with the help of same mould hence the shape of the all the biscuits were same i.e. rhomboidal shape. The physical properties measured for all treatments are shown in Table 2.

All the biscuits were prepared with the help of same mould so there was minute difference for the length and breadth for all the treatments. The length of T_1, T_2 and T_4 was 7.3 ± 0.05 and for T3 it was 7.2 ± 0.05 cm. The breadth for treatment T_1 was 3.6 ± 0.057 cm and that for treatment T_2, T_3 and T_4 was 3.5 ± 0.057 cm. Thickness for the treatment T_1, T_2, T_3 and T_4 were 0.70 ± 0.009, 0.75 ± 0.008, 0.77 ± 0.008, 0.76 ± 0.007 mm respectively. The unit weight for the treatment T_1 T_2, T_3 and T_4 were 9.325 ± 0.09, 9.356 ± 0.07, 9.414 ± 0.06, 9.420 ± 0.057 cm respectively. The density for the treatment T_1, T_2, T_3 and T_4 were 0.345 ± 0.0071, 0.345 ± 0.0069, 0.347 ± 0.0068, 0.346 ± 0.007 cm respectively.

Textural analysis

The texture of biscuit was measure with QTS Texture Analyser made by M/s. Brookfield Engineering Labs, Inc., USA. The experiment was repeated for three times for its replication and average peak force (g) be reported. In the test, the probe was allowed to penetrate in the specimen up to 3 mm at a constant speed of 0.5 mm/sec. Figure 2

shows the mean values of hardness of biscuit samples having different proportions of jamun seed powder as calculated in various experiments. From Figure 3, it was found that maximum hardness was obtained in treatment T_2 whereas minimum hardness was obtained in treatment T_1. The hardness values obtained for the biscuits of various blends were in the range of 468.33 to 4535 g.

Calorific value

The Calorific value of the developed biscuits was determined by using the Bomb Calorimeter (ASTM D271-70) [5]. Table 3 shows the results of the calorific value of the developed biscuits. The calorific value of all type of biscuits (experimental samples) varied between 402.23 to 482.68 kcal/100g. The calorific value of the treatment T_3 was found more i.e. 482.68 kcal/100 g followed by the treatment T_2, i.e. 453.426 kcal/100g.

Sensory analysis

Sensory analysis has been carried out in NAIP laboratory of Department. of Agricultural Process Engineering and Technology, CAET, Dapoli. Product of different treatments was analyzed by different subjects in our college faculty and students. They were

Ingredients (g)	T_1	T_2	T_3	T_4
Maida	870	840	810	780
Jamun seed powder	30	60	90	120
Finger millet	100	100	100	100
Baking powder	10	10	10	10
Milk powder	20	20	20	20
Salt	5	5	5	5
Sugar	200	200	200	200
Dalda	250	250	250	250

Table 1: Ingredients required for the preparation of jamun seed powder biscuits (per 1 kg flour basis).

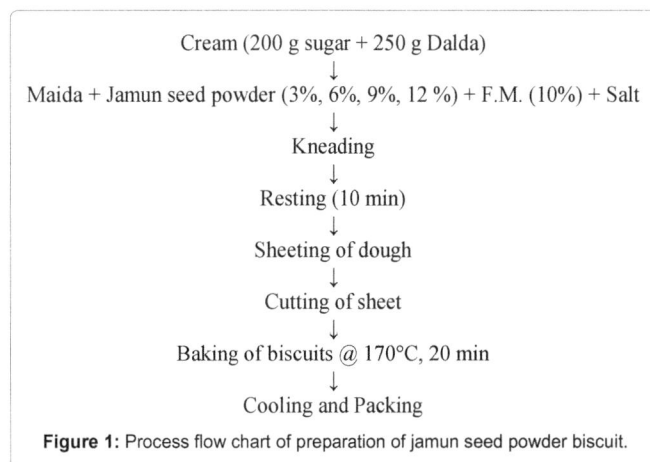

Cream (200 g sugar + 250 g Dalda)
↓
Maida + Jamun seed powder (3%, 6%, 9%, 12 %) + F.M. (10%) + Salt
↓
Kneading
↓
Resting (10 min)
↓
Sheeting of dough
↓
Cutting of sheet
↓
Baking of biscuits @ 170°C, 20 min
↓
Cooling and Packing

Figure 1: Process flow chart of preparation of jamun seed powder biscuit.

S.no.	Properties	T_1 (3% JSP)	T_2 (6% JSP)	T_3 (9% JSP)	T_4 (12% JSP)
1.	Length (cm)	7.3 ± 0.05	7.3 ± 0.05	7.2 ± 0.05	7.3 ± 0.05
2.	Breadth (cm)	3.6 ± 0.057	3.5 ± 0.057	3.5 ± 0.057	3.5 ± 0.057
3.	Thickness (cm)	0.70 ± 0.009	0.75 ± 0.008	0.77 ± 0.008	0.76 ± 0.007
4.	Unit weight (g)	9.325 ± 0.09	9.356 ± 0.07	9.414 ± 0.06	9.420 ± 0.057
5.	Density (g/cc)	0.345 ± 0.0071	0.345 ± 0.0069	0.347 ± 0.0068	0.346 ± 0.007

Table 2: Physical properties of biscuit for different treatment.

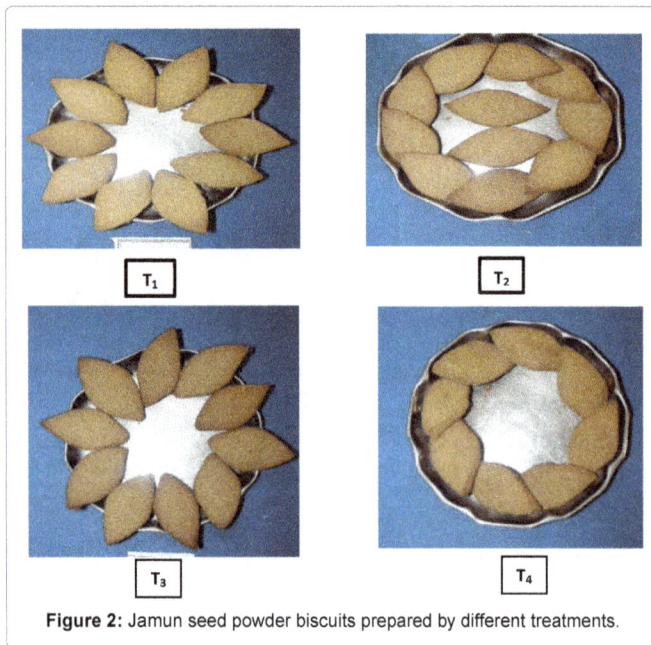

Figure 2: Jamun seed powder biscuits prepared by different treatments.

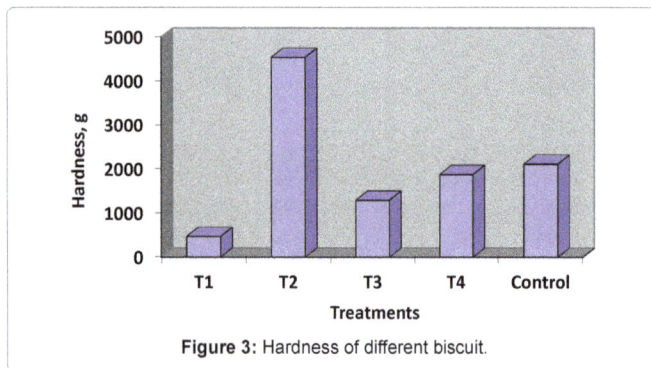

Figure 3: Hardness of different biscuit.

Treatment	Calorific value (kcal/100g)
T_1	402.23
T_2	453.426
T_3	482.68
T_4	446.18
Control	409.546

Table 3: Calorific value of developed biscuits.

Sample Code	Sensory Parameters Score (Out of 9)				
	Colour (9)	Taste (9)	Flavour (9)	Texture (9)	Acceptability (9)
T_1	7.5	8.5	7.5	7.0	7.5
T_2	7.0	7.2	7.0	6.8	7.2
T_3	8.0	7.9	8.1	7.9	7.8
T_4	4.7	6.0	6.8	5.2	7.6
Control	7.8	8.1	8.0	8.0	7.6

Table 4: Results of sensory analysis.

provided with standard evaluation sheets based on nine-point hedonic scale by 20 members' consumer test panel for colour, texture, flavour, taste, appearance and overall acceptability for Jamun seed powder based biscuits. From collected data following results were derived as shown in Table 4.

The sensory analysis of the developed biscuits was carried out by the 20 panel of judges from the faculty and students of the college. The treatment T_3 (81% M + 10% FM + 9% JSP) secured maximum score for colour, taste flavour and acceptability i.e. 8.0, 7.9, 8.1, 7.9 and 7.8 respectively.

Summary and Conclusion

Jamun seed powder in combination with finger millet and maida were used to prepare biscuit by using traditional creamy method. The results pertaining to standardization of composite flour for biscuit preparation revealed that sensorial quality characteristics of biscuits could be improved with incorporation of maida, finger millet flour and jamun seed powder. The various physical properties i.e. Length, width, thickness, unit weight, density, and calorific value, Textural properties of biscuits were determined.

The following conclusions were drawn from the analysis.

I. The density of Jamun seed powder biscuits was found to be in range of 0.345 to 0.347 g/cc.

II. The calorific value of all type of biscuits varied between 402.23 to 482.68 kcal/100g. The calorific value of treatment T_3 was found more i.e. 482.68 kcal/100g compared to others.

III. In textural analysis, the hardness values of biscuit T_1, T_2, T_3 and T_4 were 468.33, 4535, 1291.66 and 1873.33 g respectively. The treatment T_2 has more hardness i.e. 4535 g and moderate hardness was found in treatment T_3 (81% M + 9% JSP + 10% FM).

IV. In sensory analysis treatment T_3 (81% M + 9% JSP + 10% FM) secured maximum score for colour, taste, flavour and acceptability. Therefore, treatment T_3 was more acceptable so it was optimised treatment than others.

References

1. Amir G, Jan R, Nayik GA, Prasand K, Kumar P (2014) Significance of finger millet in nutrition, health and value added products: A review. J Environ Sci Comp Sci Eng Technol 3: 1601-1608.

2. Bhargava S (1991) Efficiency of bitter gourd and jamun fruit seed in the treatment of diebetis mellitus. Department of Food and Nutrition, College of Home Science, Udaipur, RAJASTHAN (INDIA).

3. Shorti DS, Kelkar M, Deshmukh VK, Aiman R (1962) Investigation of hypoglycaemic properties of *Vinca rosea* and *Eugenia jambolina*. Indian Med 3: 51-62.

4. Swami SB, Nayansingh Thakur J, Patil M, Haladankar P (2012) Jamun (*Syzygium cumini* L): A review of its food and medicinal uses. Food Nutri Sci 1102

5. ASTM (2006) Laboratory sampling and analysis of coal and coke. ATSM Method D271-70.

12

Optimisation of Process for Development of Nutritionally Enriched Multigrain Bread

Hafiya Malik[1], Gulzar Ahmad Nayik[2]* and Dar BN[1]

[1]*Department of Food Technology, Islamic University of Science and Technology, Awantipora Pulwama, Jammu & Kasmir, India*
[2]*Department of Food Engineering and Technology, Sant Longowal Institute of Engineering and Technology, Longowal, Punjab, India*

Abstract

The main aim for the development of multigrain breads was to meet the increasing demand of healthy diet with reference to economy. The multigrain breads were developed by replacing wheat flour by 5.10, 15, 20 and 30% of oat, barley, maize and rice flours and 1% flax seeds were incorporated in bread making to increase its pharmaceutical value. A prominent change was observed in case of protein content by altering the substitution levels. Similarly fat, fibre and ash also vary by varying the flour ratios. The colour analysis showed certain change in L˙, a˙ and b˙ values. More the fibre content was introduced in the samples, more the brown colour appeared. The texture profile analysis (hardness, springiness, chewiness & cohesiveness) increased by increasing the percentage of composite flours in the blends. Physical characteristics (bread volume, dough expansion and specific volume) increased by decreasing the percentage of blends in the bread samples and vice versa.

Keywords: Multigrain; Composite flours; Proximate analysis; Colour analysis; Bread optimization

Introduction

India is a developing country with a large segment of population depending upon wheat, rice and maize as staple food which provide calories and proteins. Traditionally only wheat has been used as a whole wheat meal *(atta)* in production of chapattis, paratha and poori whereas refined flour *(maida)* finds great application in manufacture of bakery foods like bread and cookies [1]. 75 per cent wheat is produced as whole wheat flour and only 25 per cent is used in preparation of bakery goods [2]. It has been proved that regular consumption of single items affect health directly e.g. regular consumption of wheat causes lysine deficiency while gluten protein may cause allergic reactions in some people. Diet should be balanced besides being it should be wholesome, appetizing, palatable and satisfying. It has been proved that right food can cure several diet related disorders. With increasing consumer awareness, improved educational status and standard of living, knowledge about natural foods, change in food habits and increased cost of medicines, there is an increased trend in consumption of healthy foods and hence alternate wheat flour and meal serves as excellent source to provide functional ingredients from other natural sources in our diet. The multigrain products feature a combination of grains such as wheat, oat, barley, maize, rice, flax etc. and provide opportunity for snack manufacturers to develop products within an imaginative appearance, featuring new texture and colour with a beneficial nutritional profile [2]. Multigrain products must be of course whole grain to offer maximum nutritional benefits. The use multi-grains are well established in other food sectors particularly bakery and breakfast cereals [3]. They make a positive contribution to the taste and texture of products and consumer readily accept the health benefits. Multigrain products can contribute to a healthy digestive system, help in weight control, reduce the risk of diabetes reduce the risk of cardiac failures and prevent the chances of bowel cancer. The flax seeds are commonly consumed in one of the three forms, whole seed, ground or powder form and flax seed oil. Most of the benefits reported from flax seeds are believed due to the presence of alpha linolenic acids (ALA), lignans and fibre [4]. Flax seeds are reported to have lot of health benefits e.g. flax seeds are most commonly used as laxative, flax seed oil is used for various conditions like arthritis, both flax seed and flax oil have been used to prevent high cholesterol levels and reduce the

risk of cancer [5]. Bread is an ideal functional product, since it is an important part of our daily diet. Bread is consumed in large quantity in world in different types and forms depending upon cultural habits [6]. Bread is usually made from wheat flour dough that is cultured with yeast, allowed to rise, and finally baked in an oven [7]. Multigrain breads are reported to have lot of health benefits. Multigrain breads introduce more fibre in the diet than other types of breads. Multigrain breads also provide required quantity of thiamine, phosphorous, potassium, riboflavin, pantothenic acid, calcium, iron, zinc and copper [8]. The vitamin B in multigrain breads helps to convert in energy. There was a need to quantify the different levels of various grains for development of multigrain breads. Such information will increase the understanding of the functionality of multigrain bread in the diet to harness the potential benefits of various grains. Therefore the present investigation was planned to optimise the different levels of various grains for development of nutritionally enriched multigrain bread.

Materials and methods

The work was carried out in the Department of Food Technology, Islamic University of Science and Technology Awantipora during the year 2011-2012. Wheat, maize, oat, barley and rice flour were purchased from local market of Srinagar, J&K. The flours were separately stored in air tight plastic containers at refrigerated temperatures until used. Flax seeds, shortening, compressed yeast; salt and sugar were purchased from local market of Srinagar. The formulation for development of multigrain breads enriched with flax seeds were according to Table 1. Breads were prepared from blended flours (wheat, oat, barley, maize,

*Corresponding author: Gulzar Ahmad Nayik, Department of Food Engineering and Technology, Sant Longowal Institute of Engineering and Technology, Longowal, Punjab, India, Email: gulzarnaik@gmail.com

Parameter	Treatments									
	T1	T2	T3	T4	T5	T6	T7	T8	T9	Control
Protein (%)	7.2[b]	8.03[e]	8.00[e]	7.30[bcd]	8.00[e]	8.4[gh]	8.07[ef]	8.00[e]	6.00[a]	7.00[b]
Fat (%)	3.60[e]	4.60[f]	3.00[d]	3.60[e]	2.60[c]	3.00[d]	2.30[b]	3.00[d]	2.00[a]	2.60[c]
Fibre (%)	15.00[e]	13.00[d]	16.00[f]	13.00[d]	13.00[d]	11.00[c]	14.00[e]	10.00[a]	11.00[c]	10.30[ab]
Ash (%)	0.50[a]	1.00[b]	1.50[c]	2.50[e]	1.50[c]	2.00[d]	1.50[c]	1.00[bc]	2.00[d]	1.00[b]
Carbohydrate(%)	35.64[c]	47.07[f]	42.98[e]	60.00[f]	36.60[cd]	47.66[fg]	29.33[a]	32.98[b]	50.36[h]	54.10[i]

Mean value in a row with same superscript do not differ significantly (p<0.05)
T: Treatment

Table 1: Proximate composition of bread samples.

rice) along with flax seeds. The ingredients were weighed accurately and the yeast was activated in warm water (55°C). All the ingredients were mixed in a vessel and yeast was added while taking into account the amount of water. The dough was then placed in an incubator at 37°C for fermentation. Dough was taken out after 2.5 hours and then knocked back to remove the excess gases. The dough was again placed in incubator for fermentation and removed after 30 min, moulded into pans and then allowed to ferment for another 35 min. The pans were then placed in baking oven (model SM 601 T) at 225°C for 30-35 min. The breads were taken out, cooled and then sliced. The breads were stored at room temperature and packaged in LDPE material.

Proximate analysis

Moisture content

Moisture (%) in the bread was determined by Gravemetric method [9]. 1gm sample was pre weighed (W1) in a petriplates and placed in a hot air oven (model NSW 144) at 105°C for 24 hrs. The sample was removed from oven, cooled in desiccators and re weighed (W2). Moisture percentage was calculated according to formula:

$$\text{Moisture (\%)} = \frac{(W_1 - W_2)}{W_1} \times 100$$

Total ash

Total ash content was determined as total inorganic matter by incineration of sample at 600°C [9]. Sample 1g was weighed into pre weighed porcelain crucible and incinerated overnight in a muffle furnace (model NSW-101) at 600°C. The crucible was removed from muffle furnace, cooled in desiccators and weighed. Ash content was calculated according to formula:

$$\text{Total Ash (\%)} = \frac{\text{Ash weight}}{\text{Sample weight}} \times 100$$

Crude protein

Crude protein was determined by Kjeldhal's method [9]. 700g of defatted and dried sample was placed in Kjeldhal's digestion flask. 5 g K_2SO_4 + 0.5 g $CuSO_4$ and 25 ml concentrated sulphuric acid was added to sample. The sample was digested for 1h. 20 ml deionised water was added to sample after allowing it to cool and transferred to a 50 ml volumetric flask. The volume was made upto mark with the distilled water. 10 ml of aliquot was taken and transferred in a distillation assembly followed by addition of 10 ml of 40% NaOH. On distillation ammonia liberated was collected in a 250 ml conical flask containing 10 ml of 0.01N HCl to which methyl red indicator (2-3 drops) was priorly added. It was titrated with 0.01N HCl. A blank was prepared and treated in the same manner except that the tube was free of sample. Protein percentage was calculated according to formula:

$$\text{Crude protein (\%)} = \frac{(\text{Sample titre - Blank titre}) \times 14 \times 6.25 \times 100}{\text{Sample weight}}$$

Where, 14 is molecular weight of nitrogen and 6.25 is nitrogen factor.

Crude fat

Crude fat was determined by employing solvent extraction using a Soxhlet unit [9]. Sample 1g was weighed in an extraction thimble and covered with absorbent cotton. 50 ml solvent (petroleum ether) was added to a pre weighed cup. Both thimble and cup were attached to extraction unit. The sample was subjected to extraction with solvent for 30 mins followed by rinsing for 1.5hrs. The solvent was evaporated from cup to the condensing column. Extracted fat in the cup was placed in an oven at 110°C for 1h and after cooling, the crude fat was calculated according to formula:

$$\text{Crude fat (\%)} = \frac{\text{Extracted fat}}{\text{sample weight}} \times 100$$

Crude fibre

Crude fibre in a sample was determined by method described by AOAC [9]. Defatted sample 1g was placed in a glass crucible and attached to extraction unit in Kjeldhal's unit. 150 ml boiling 1.25 per cent sulphuric acid was added. The sample was digested for 30mins and then the acid was drained out and the sample was washed with boiling distilled water. After this, 1.25% sodium hydroxide solution (150 ml) was added. The sample was digested for 30mins, there after the alkali was drained out and the sample was washed with boiling distilled water. Finally the crucible was removed from extraction unit and oven dried at 110°C overnight. The sample was allowed to cool in a dessicator and weighed (W1). The sample was then ashed in a muffle furnace at 500°C (model NSW-101) for 2 hrs, cooled in a dessicator and re weighed (W2). Extracted fibre was expressed as percentage of original undefatted sample and calculated according to formula:

$$\text{Crude fibre (\%)} = \frac{\text{Digested sample}(W_1) - \text{Ashed sample }(W_2)}{\text{sample weight}} \times 100$$

Texture profile analysis

Texture analyzer (TA HD Plus, stable micro Systems, Godalming, Surrey, UK) was used to measure the hardness, springiness, cohesiveness and chewiness [10]. For firmness the sample was removed from its place of storage and was placed centrally over the supports just prior to testing. The texture profile analysis was done at pre-test speed of 1.0 mm/s, test speed of 1.7 mm/s using a 5 kg load cell.

Colour analysis

Colour analysis of multigrain breads was done by using Hunter Lab colorimeter (model SM-3001476 micro sensors New York). The instrument was calibrated with user supplied black plate calibration standard that was used for zero setting, white calibration plates were used for white calibration settings. The instrument was placed at three different exposures at different places were conducted. Readings were displayed as L*, a* and b* colour parameters according to CIELAB system of colour measurement. The value of a* ranged from -100 (redness) to +100 (greenness), the b* values ranged from -100 (blueness) to +100 (yellowness) while as L˙ value indicating the measure of lightness, ranged from 0 (black) to 100 (white)

Specific volume

Loaf volume was measured after baking by rapeseed displacement method. Specific volume was calculated as loaf volume (cm³)/loaf weight (g)

Organoleptic characteristics

A panel of 10 judges evaluated the organoleptic characteristics of prepared breads. They assessed crust colour, appearance, flavour, texture, taste and overall acceptability, using 9-point Hedonic rating scale (9-Like extremely, 8-Like very much, 7-Like moderately, 6-Like slightly, 5-Neither Like nor dislike, 4-Dislike slightly, 3-Dislike moderately, 2-Dislike very much, 1-Dislike extremely).

Statistical analysis

The data was statistically analysed on a computer using design factorial in Completely Randomized Design (CRD) as suggested by Snedecor and Cochran [11].

Results and discussion

The different ingredients used in preparation of bread were flour (300 g), yeast (27 g), sugar (10 g), salt (5 g), shortening (12 g) and flax seeds (3 g). Figure 1 shows the flow chart for bread making process. It was found that flax seeds contain a high amount of fat content (37.5%), carbohydrate (29%). Dietary fibre (26%), protein (21.5%) and less percentage of moisture content (7%) and ash content (3.5%). Our results were in alignment with Halligudi [4]. The proximate composition of different flours used in the development of multigrain breads is shown in Table 2. The carbohydrate content of different flours varied from 62% (oats) to 76% (rice). However the carbohydrate content of wheat and barley were significantly same but vary from rest of the flours used in the preparation of bread samples. It is also revealed from the data in Table 2 that the protein content varied from 6.77 in rice to 11.65% in barley. The fat content varied from 0.51 in wheat to 4.58 percent in maize, although the percentage of fat is significantly same in oats and maize but differ from rest of the flours. The fibre content was found highest in barley (6.75%) and lowest in rice (0.62%) while ash content varied from 0.66% (wheat) to 2.2% (barley). Similar results were reported by Dingra [12]. The formation of various multigrain composite flours is displayed in Table 3. Nine (T1-T9) different combinations were prepared using different ranges of flours. Wheat flour percentage ranged from 50% to 60%, oat 10% to 25%, barley 5% to 20% while maize and rice 5% to 10%. The proximate composition of various multigrain breads (T1-T9) is depicted in Table 4. It is evident from the Table 1 that the protein content varied from 6.00% (T9) to 8.40% (T6) in various multigrain bread samples. Similar results were reported by Sanful and Darko [13]. The data in the Table 1 showed that minimum value of fat content (2%) was observed in T9 while as maximum value was observed in T2 (4.6%). The significant difference was observed in various multigrain bread samples. Same results were observed by Malomo, et al. [14]. The highest amount of fibre was found in in T3 (16%) while the least amount in T8 (10%). The fibre content of T6 and T9 were significantly same but differ from rest of the bread samples. The difference in the fibre content could be due to the presence of high amount of oat and barley present in different multigrain bread samples. These results are comparable to Olaoye [15]. The ash content ranged from 2.5% in T4 to 0.5% in T1 which implies that T4 possessed high amount of mineral content. The results could be due to the difference in supplementation

in different multigrain bread samples. Table 1 portrays that the carbohydrate value of different multigrain bread samples ranged from 29.33 to 60%. The highest value was observed in T4 and least value in T7, T1 and T6 were significantly same but differs from rest of the samples. These results were consistent with Malomo, et al. [14]. The sensory scores of multigrain breads are shown in Table 5. The value of colour ranged from 7.0 to 7.5. As the amount of oat is increased, the colour of the crust changed from creamy white to dull brown. These results are alluding to Gupta, et al. [16]. The data in Table 5 depicted that the crumb appearance scores for different samples are significantly similar. These results are parallel to the findings of Gupta, et al. [16]. The texture of the bread is related to external hardness or softness of bread. The texture is the quality of bread that can be decided by touch, the degree to which it is rough or smooth, hard or soft. The panellists preferred sample T1, T4, T6, T8 equally as compared to control sample. However the mean indicated that there is non-significant difference between the colours of the samples. These results are similar to the one reported by Sanful and Darko [13]. The flavour of the bread refers to its palatability. Table 6 reflects the score for flavour ranged from 6.8 (lowest) to 7.2 (highest). It is evident from data that the difference in

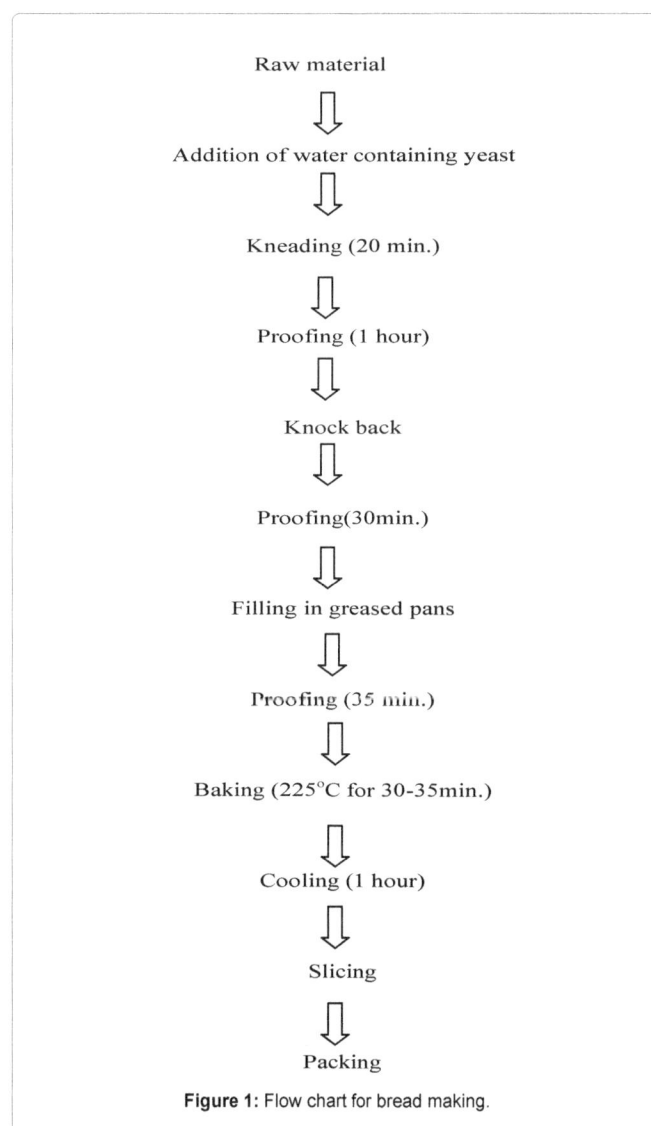

Figure 1: Flow chart for bread making.

Flours	Ash (%)	Protein (%)	Fibre (%)	Fat (%)	Carbohydrate (%)
Wheat	0.66[a]	9.55[b]	1.29[a]	0.51[a]	73.94[c]
Barley	2.20[e]	11.65[e]	6.75[e]	2.31[c]	75.00[cd]
Oat	1.70[c]	9.60[b]	5.13[d]	4.50[d]	62.00[a]
Maize	2.00[d]	9.78[bcd]	1.38[abc]	4.58[de]	70.32[b]
Rice	0.76[b]	6.77[a]	0.62[a]	0.94[b]	76.91[e]

Mean value in a column with same superscript do not differ significantly (p<0.05)

Table 2: Proximate composition of flours.

Ingredients	Combination								
	T1	T2	T3	T4	T5	T6	T7	T8	T9
Wheat (%)	50	50	50	60	60	60	70	70	70
Oat (%)	25	20	10	10	20	15	15	10	10
Barley (%)	15	20	20	20	10	15	5	10	10
Maize (%)	5	5	10	5	5	5	5	5	5
Rice (%)	5	5	10	5	5	5	5	5	5

1% flax seeds were incorporated in each preparation
T: Treatment

Table 3: Formation of multigrain composite flours.

Combination	L*	a*	B*	Hue
T1	58.09[c]	2.29[i]	29.75[h]	85.59[bc]
T2	56.92[ab]	2.33[i]	29.67[h]	85.50[a]
T3	60.05[fgh]	1.38[d]	27.44[a]	87.12[fgh]
T4	61.26[i]	2.13[h]	27.65[b]	85.59[bc]
T5	59.43[d]	1.76[fg]	28.95[g]	86.51[d]
T6	59.17[d]	2.33[ij]	29.94[ijk]	85.54[b]
T7	58.91[c]	0.90[c]	27.87[b]	88.15[i]
T8	59.50[d]	0.68[ab]	28.18[cdef]	88.61[ij]
T9	59.73[de]	1.59[e]	27.74[b]	86.71[e]
Control	71.29[j]	1.68[f]	29.19[g]	86.70[e]

Mean value in a column with same superscript do not differ significantly (p<0.05)

Table 4: Colour analysis of multigrain breads.

Parameter	Treatments									
	T1	T2	T3	T4	T5	T6	T7	T8	T9	Control
Crust colour	7.0[a]	7.0[a]	7.0[a]	7.5[d]	7.0[a]	7.4[c]	7.0[a]	7.0[a]	7.1[b]	8.0[e]
Crumb appearance	7.1[a]	7.1[a]	7.0[a]	7.2[abc]	7.0[a]	7.0[a]	7.0[a]	7.1[a]	7.1[a]	8.0[d]
Texture	7.2[c]	7.0[a]	7.1[b]	7.2[c]	7.1[b]	7.2[c]	7.1[b]	7.2[c]	7.0[a]	7.6[d]
Flavour	6.8[ab]	6.7[a]	7.1[c]	6.8[ab]	7.2[cd]	6.8[ab]	6.7[a]	6.8[ab]	8.0[e]	0.11
Overall acceptability	7.0[a]	7.1[a]	7.4[d]	7.1[a]	7.2[bc]	7.1[a]	7.1[a]	7.1[a]	8.0[e]	0.12

Mean value in a row with same superscript do not differ significantly (p<0.05)

Table 5: Sensory mean scores of multigrain breads.

Type of bread	Hardness	Springiness	Chewiness	Cohesiveness	Moisture of crumb (%)
T1	10.87[b]	0.96[a]	0.76[a]	0.43[a]	38.00[g]
T2	11.43[b]	1.20[a]	0.80[a]	0.47[a]	27.30[c]
T3	12.44[c]	1.50[b]	0.86[a]	0.48[a]	29.00[de]
T4	14.97[d]	1.96[c]	0.90[a]	0.50[a]	24.60[a]
T5	15.30[d]	2.50[d]	0.96[abcd]	0.56[bc]	38.30[gh]
T6	16.43[e]	2.90[e]	1.50[e]	0.60[d]	28.30[c]
T7	17.40[f]	3.30[f]	1.96[f]	0.66[e]	45.30[i]
T8	17.90[f]	3.50[g]	2.00[fg]	0.70[f]	45.30[i]
T9	18.50[f]	3.96[h]	2.60[h]	0.77[ghi]	35.60[f]
Control	7.87[a]	4.00[i]	3.00[i]	1.50[j]	25.00[ab]

Mean value in a column with same superscript do not differ significantly (p<0)

Table 6: Moisture content and TPA (texture profile analysis) of bread samples.

flavour scores could be due to the incorporation of maize flour in the multigrain bread samples Gupta, et al. [16]. Higher the amount of oat and barley added in the samples more is the undesirable flavour. The data in the Table 5 outlined that the majority of panellists accept the bread made out of 100 per cent wheat flour (control sample) which has score of as 8.0 on 9 point Hedonic scale. It is evident from the data that that the mean sensory scores increased upto 15% level of substitution and beyond that the trend reversed. The preference of the panellists for sensory attributes of whole wheat bread flour may be due to the familiarisation of consumers to the normal whole wheat flour. The colour analysis of all multigrain breads is shown in Table 4. There is no significant difference between the L˙ values of different multigrain bread samples. However the a˙ values showed little difference among different bread samples with increase or decrease in the amount of oat and barley flour in the samples. Similarly b˙ values also varied among the different multigrain bread samples by varying the proportion of different flours [12]. Table 6 displayed that the hardness increased in the multigrain bread samples with the increase in the fibre content while as the control sample showed least hardness. The Table 6 points out that the springiness (elasticity) of the bread samples decreased with the increase in the fibre content, less the amount of composite flours, more desirable is the chewiness. Cohesiveness also varied by varying the composition in different multigrain bread samples. The moisture content of crumb increased by addition of fibre as shown in Table 6 which could be due to the incorporation of high amount of substitution levels in the multigrain bread samples. Similar results were reported by Goesaert, et al. [17]. Figures 2-4 depicts that bread volume, dough expansion and specific volume increased by increasing the amount of whole wheat flour and decreasing the fibre content in each sample. The increase in bread volume, dough expansion and specific volume is because of the high gluten network present in wheat flour that helps to trap carbon dioxide and hence increases the volume. These results are in alignment to the findings of Ndife, et al. [18] and Aissa, et al. [19].

Conclusion

In conclusion, breads prepared from multigrain mix with flaxseed substitutions were found to be nutritionally superior to ordinary bread. Multigrain bread would serve as functional food because of its high fibre content. It can also be concluded that use of multigrain mix up to the level of 30% can be considered for the production of bread with perceptible taste of multi-grains. However, further research work should be focused on the phytochemical analysis. There is also the need

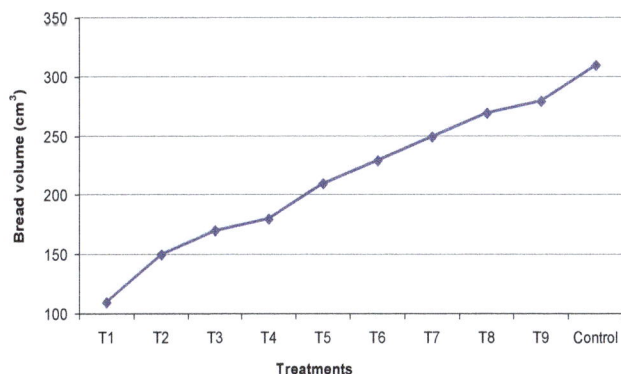

Figure 3: Graphical representation of dough expansion (cm) analysis of multi grain breads.

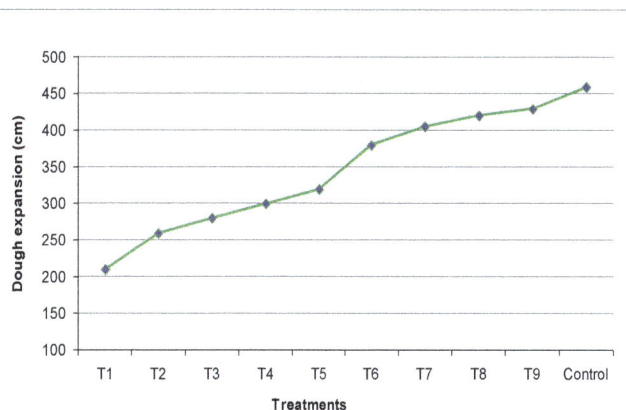

Figure 4: Graphical representation of specific volume (cm³/g) analysis of multi grain breads.

to adjust the mixing ingredients and baking techniques in order to improve the multigrain bread qualities.

References

1. Nigham V, Nambiar VS, Tuteja S, Desai R, Chakravorty B, et al. (2013) Effect of wheat ARF treatment on the baking quality of whole wheat flours of the selected varieties of wheat. Journal of Applied Pharmaceutical Science 3: 139-145

2. Indrani D, Soumya C, Rajiv J, Venkateswarao G (2010) Multigrain bread-its dough rheology, microstructure, quality and nutritional characteristics Journal of Texture Studies 41: 312-309.

3. Mandge HM, Sharma S, Dar BN (2014) Instant multigrain porridge: effect of cooking treatment on physicochemical and functional properties. Journal of Food Science and Technology 51: 97-103.

4. Halligudi N (2012) Pharmacological properties of flax seeds. Hygeia: Journal of Drugs and Medicines 4: 70-77.

5. Donaldson M (2004) A review of the evidence for an anti cancer diet. Nutritional Journal.

6. Cayot N (2007) Sensory quality of traditional foods. Food Chemistry 101: 154-162.

7. Dewettnick K, Van Bockstaele F, Kühne B, VandeWalle D, Courtens TM, et al. (2008) Nutritional value of bread: influence of processing, food interaction and consumer perception. Journal of Cereal Science 48: 243-247.

8. Angioloni A, Collar AC (2009) Multigrain breads as reinvented healthy and convenient value added goods. International Association for Cereal Sciences and Technology.

9. AOAC (2000) Official methods of analysis. Association of Official Analytical

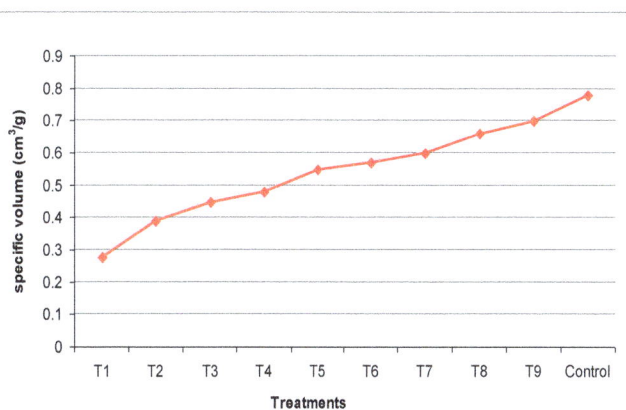

Figure 2: Graphical representation of bread volume (cm³) analysis of multi grain breads.

Chemists International. Maryland, USA.

10. AACC (2004) Approved Methods of the AACC. Methods 10-05, 74-09. American Association of Cereal Chemists, St. Paul, MN.

11. Snedecor GW and Cochran WG (1967) Statistical Methods. The Lowa State University Press, USA.

12. Dhingra S, Jood S (2004) Effect of Flour Blending on Functional, Baking and Organoleptic Characteristics of Bread. International Journal of Food Science and Technology 39: 213-222.

13. Sanful, RE, Darko S (2010) Production of Cocoyam, Cassava and Wheat Flour Composite Rock Cake. Pakistan Journal of Nutrition 9: 810-814.

14. Malolma SA, Eleyinmi AF, Fashankin JB (2011) Chemical composition, rheological properties and bread making potential of composite flours from bread fruit, bread nut and wheat. African Journal of Food Sciences 5: 400-410.

15. Olaoye OA, Onilude AA, Idowu OA (2006) Quality characteristics of bread produced from compositeflours of wheat, plantain and soybeans. African Journal Biotechnology 11: 1102-1106.

16. Gupta M, Bawa AS, Semwal AD (2011) Effect of Barley Flour Blending On Functional, Baking and Organoleptic Characteristics of High-Fiber Rusks. Journal of Food Processing and Preservation 35: 46-63.

17. Goesaert H, Leman P, Bijttebier A, Delcour JA (2009) Antifirming effects of starch degrading enzymes in bread crumb. Journal of Agricultural and Food Chemistry 57: 2346-2355.

18. Ndife J, Abdulraheem LO, Zakari UM (2011). Evaluation of the nutritional and sensory quality of functional breads produced from whole wheat and soy bean flour blends. African Journal of Food Sciences 5: 466-472.

19. Aissa MFB, Monteau JY, Perronnet A, Roelens G, Bail AL, et al. (2010) Volume Change of Bread and Bread Crumb During Cooling, Chilling and Freezing and the Impact of Baking. Journal of Cereal Science 51: 115-119.

13

Effects of Conventional and Microwave Heating Pasteurization on Physiochemical Properties of Pomelo (*Citrus maxima*) Juice

Kumar S[1]*, Khadka M[1], Mishra R[2], Kohli D[1] and Upadhaya S[1]

[1]*Department of Food Technology, Uttaranchal University, Dehradun, India*
[2]*Department of Agricultural science and Engineering, IFTM University, Moradabad, India*

Abstract

Effects of conventional and microwave heating pasteurization on physiochemical properties of pomelo (*Citrus maxima*) juice was evaluated. Microwave heating pasteurization shows less effect on pH, reducing sugar, ascorbic acid (Vitamin C) content, and total phenolic contents in compare to conventional pasteurization. Microwave heating pasteurization reduces tannin and naringin content more in compare to conventional pasteurization.

Keywords: Pomelo; Pasteurization; Microwave heating

Introduction

Fruit juices are consumed worldwide because they are good source of vitamins, minerals and fibers [1]. Citrus juices are complex mixtures of aromatic volatiles and non-volatile components. Aromatic volatiles include esters, aldehydes, ketones and alcohols while non-volatile components include organic acids and sugars [2]. *Citrus maxima* (or *Citrus grandis*), commonly known as pomelo, pomello, pummelo, pommelo, pamplemousse, jabong (Hawaii), shaddick is a citrus fruit, with the appearance of a big grapefruit, native to South and Southeast Asia. In India, it is also known as Chakotra. There are two varities of pomelo commonly classified as common which has white flesh or pigmented which has pink flesh [3]. They are generally used eaten as fruit. It tastes sweet, and is slightly acidic with a hint of bitterness. It also has appetizing, cardiac stimulant and antitoxic property [4]. Pummelo juice is a good source of ascorbic acid [5]. *Citrus maxima* have low commercial value due to the bitterness of its juice. Pomelo is a rich source of naringin, a bitter flavoured, flavanone glycoside with reported antioxidant [6].

Juice manufacturing industries aim to preserve the juices without losing organoleptic quality by the method of pasteurization. There are several methods for pasteurization. Thermal treatments are commonly applied in juice manufacturing industries for preservation of juices. However, heat-sensitive antioxidants and some of the essential compounds in the juice may deteriorate due to heat during thermal processing [7]. Microwave heating generate volumetric heating within the food material due to this microwave heating provide short heating time and produce high quality self-stable food products. Organoleptic quality characteristics of food can be protected during processing by electrical method such as microwave heating that provides inactivation of enzymes and microorganisms quickly rather than the traditional heating methods [8] and minimize the quality loses [9].

Material and Methods

Raw materials

Pomelo (*Citrus maxima*): The pomelo sample for present study was procured from the local market of Dehradun, Uttarakhand. The fruit used are of sound and good quality.

Preparation of pomelo juice

Mature, ripe pomelo fruit were washed and peeled by hand. The segments were separated and the segment membrane was separated carefully by hand and juice was prepared by squeezing.

Treatments

Conventional pasteurization (CP): Freshly prepared juice was filtered and poured into sterilized bottles and then heated at 90°C by using thermostatic water bath and maintain the temperature of juice 90°C for 15 seconds. Pasteurized juice was cooled to room temperature in a cooling water bath and was stored in a refrigerator at 4 ± 1°C.

Microwave heating pasteurization (MHP): Juice was heated at 90°C by using microwave and cooled to room temperature in a cooling water bath and was stored in a refrigerator at 4 ± 1°C.

Physicochemical analysis of Pomelo (*Citrus maxima*) juice before and after conventional and microwave heating pasteurization: Various physico-chemical properties like pH, Moisture Content, Ash content, TSS, titrable acidity, reducing sugar, Ascorbic acid content tannin content, Naringine Content and Total phenolic contents were analyzed. pH was analyzed by handy pH meter, Moisture Content, Ash content and titrable acidity was measured by the standard method of Rangana [10]. TSS were measured in °Brix by handy refractometer (ERMA), Reducing sugars were determined by the method of Lane and Eynon [11]. Ascorbic acid in raw as well as RTS were measured by standard method of Sawhney and Singh [12]. Tannin content was determined by the protein precipitation method [13] naringin content in the fruit juice was estimated by the method of Davis [14] and total phenolics contents was measured by the method of Makkar et al. [15].

Results and Discussion

The physicochemical properties of Pomelo (*Citrus maxima*) Juice before and after conventional and microwave heating pasteurization are shown in Table 1.

*Corresponding author: Sanjay Kumar, Department of Food Technology, Uttaranchal University, Dehradun, India, E-mail: mr.sanju4u@gmail.com

Parameters	Fresh Juice	After Treatment	
		Conventional Pasteurization	Microwave Heating Pasteurization
pH	3.30	3.2	3.2
Moisture Content (% wb)	94	92.53	92.03
Ash Content (%)	1	2.46	0.49
TSS (°Brix)	7.5	9.1	8.8
Titrable acidity (%)	1.248	1.4	1.68
Reducing sugar (%)	3.47	3.59	3.50
Ascorbic acid content (mg/100 ml)	67.71	52.42	54.29
Tannin content (mg/ml)	0.42	0.41	0.39
Naringin Content (µg /ml)	600	590	594
Total phenolic contents (mg GAE/L)	710	690.5	705.3

Table 1: Effect of conventional and microwave heating pasteurization on various physicochemical properties of pomelo (*Citrus maxima*) juice.

Effect of conventional and microwave heating pasteurization on pH

The pH of fresh pomelo juice was observed 3.30 as shown in Table 1. pH of pomelo juice after pasteurization and microwave heating was observed 3.2. Figure 1 shows the effect of pasteurization and microwave heating on pH. The decrease in pH may be because of lactic acid production and due to the hydrolysis of sucrose. These results are in agreement with those of Mohamed et al. [16] who also reported decrease in pH of physalis juice.

Effect of conventional and microwave heating pasteurization on moisture content

The Moisture Content of fresh pomelo juice was observed 94% as shown in Table 1. Moisture content after pasteurization and microwave heating was 92.53% and 92.03% respectively. Figure 2 shows the effect of pasteurization and microwave heating on moisture content. The decrease in moisture content may be because of the evaporation of water which causes concentration of juice to some extent by heat processing. These results are in agreement with those of Dar et al. [17].

Effect of conventional and microwave heating pasteurization on ash content

The ash content of fresh pomelo juice was observed 1% as shown in Table 1. Ash content after pasteurization and microwave heating was 2.46% and 0.49% respectively. Figure 3 shows the effect of pasteurization and microwave heating on moisture content. In our finding, we found that ash content increased in case of pasteurization but in case of microwave heating ash content was reduced.

Effect of conventional and microwave heating pasteurization on TSS

The TSS of fresh pomelo juice was observed 7.5°Brix and after treatment it was 9.1°Brix and 8.8°Brix in case of conventional and microwave heating pasteurization respectively as shown in Table 1. The increase in TSS may be because of the evaporation of water which causes concentration of juice to some extent by heat processing, or may be due to citric acid increases the TSS. These results are in agreement with those of Rivasa et al. [18] who also reported the decrease in TSS of orange and carrot juice blends. Figure 4 shows the effect of conventional and microwave heating pasteurization on TSS.

Effect of conventional and microwave heating pasteurization on titrable acidity

The titrable acidity of fresh pomelo juice was observed 1.248% and after treatment it was 1.4% and 1.68% in case of conventional and

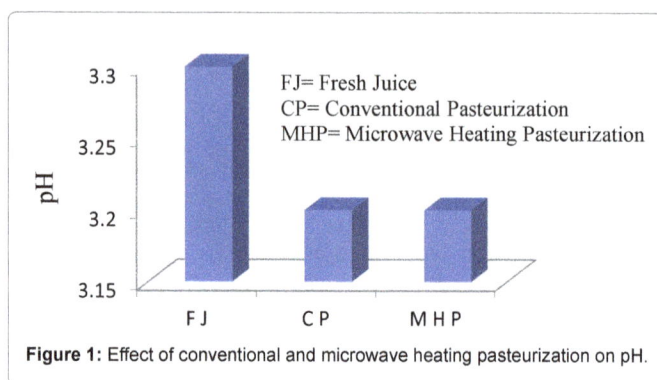

Figure 1: Effect of conventional and microwave heating pasteurization on pH.

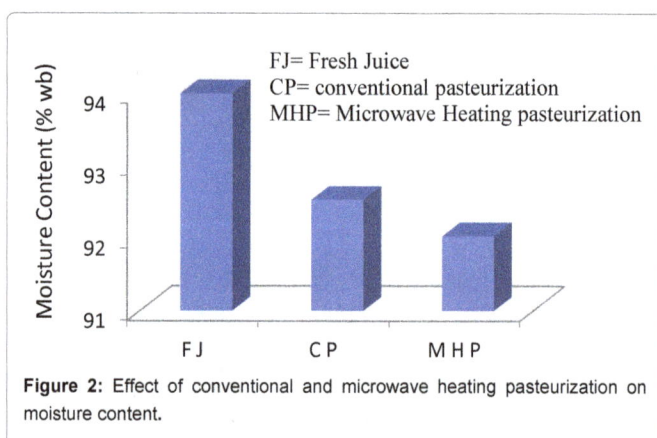

Figure 2: Effect of conventional and microwave heating pasteurization on moisture content.

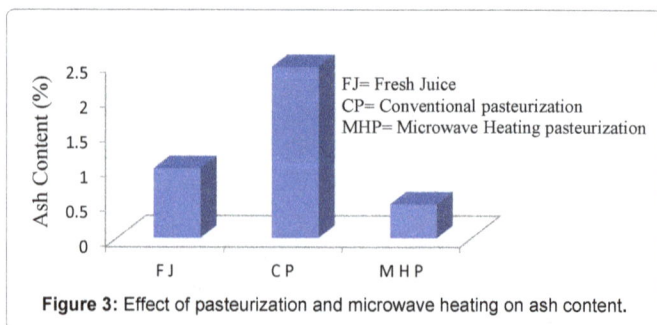

Figure 3: Effect of pasteurization and microwave heating on ash content.

microwave heating pasteurization respectively as shown in Table 1. The increase in titrable acidity may be because of Oxidation of reducing sugars can also contribute to increase in the acidity of fruits. These results are in agreement with those of Rivasa et al. [18]. Figure 5 shows the effect of conventional and microwave heating pasteurization on titrable acidity.

Effect of conventional and microwave heating pasteurization on reducing sugar

The Reducing sugar of fresh pomelo juice was observed 3.47% and after treatment it was 3.59% and 3.50% in case of conventional and microwave heating pasteurization respectively as shown in Table 1. Increase in reducing sugar may be due to the conversion of non-reducing sugars into reducing sugars. Figure 6 shows the effect of conventional and microwave heating pasteurization on reducing sugar. Our finding favors the findings of Zahid Mehmood et al. [19] who also reported the increase in reducing sugar in case of apple juice.

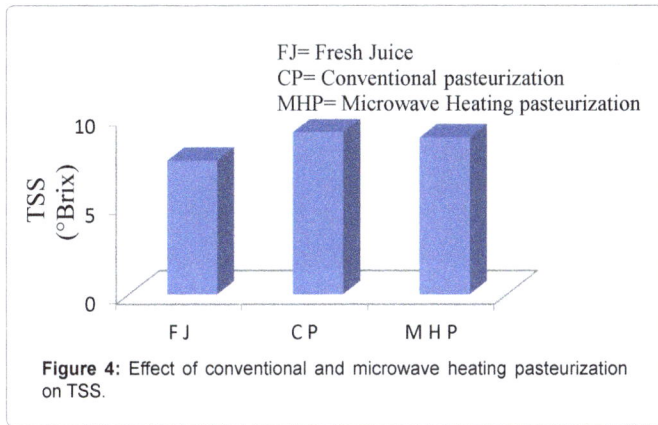

Figure 4: Effect of conventional and microwave heating pasteurization on TSS.

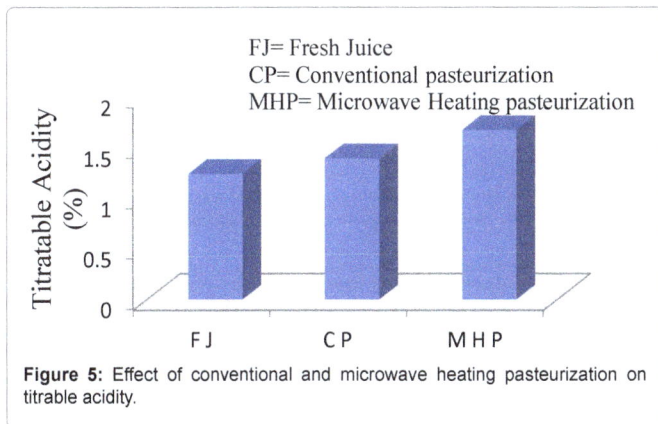

Figure 5: Effect of conventional and microwave heating pasteurization on titrable acidity.

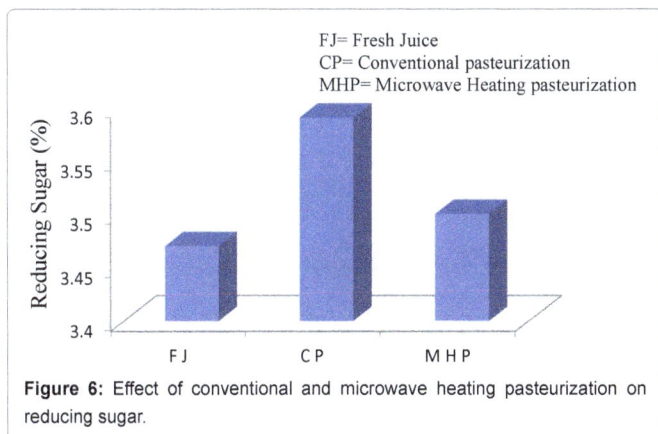

Figure 6: Effect of conventional and microwave heating pasteurization on reducing sugar.

Effect of conventional and microwave heating pasteurization on ascorbic acid (Vitamin C)

The ascorbic acid (Vitamin C) of fresh pomelo juice was observed 67.71 mg/100 ml and after treatment it was 52.42 mg/100 ml and 54.29 mg/100 ml in case of conventional and microwave heating pasteurization respectively as shown in Table 1. The decrease in ascorbic acid (vitamin C) may be because of due to degradation of vitamin C because is heat labile nutrient. Our finding favors the findings of Cinquanta et al. [20] who also reported the decrease in ascorbic acid (Vitamin C) during microwave pasteurization of orange juice. Figure 7 shows the effect of conventional and microwave heating pasteurization on Ascorbic Acid (Vitamin C).

Effect of conventional and microwave heating pasteurization on tannin content

The tannin content of fresh pomelo juice was observed 0.42 mg/ml and after treatment it was 0.41 mg/ml and 0.39 mg/ml in case of conventional and microwave heating pasteurization respectively as shown in Table 1. Figure 8 shows the effect of conventional and microwave heating pasteurization on tannin content.

Effect of conventional and microwave heating pasteurization on naringin content

The naringin content of fresh pomelo juice was observed 600 µg/ml and after treatment it was 590 µg/ml and 594 µg/ml in case of conventional and microwave heating pasteurization respectively

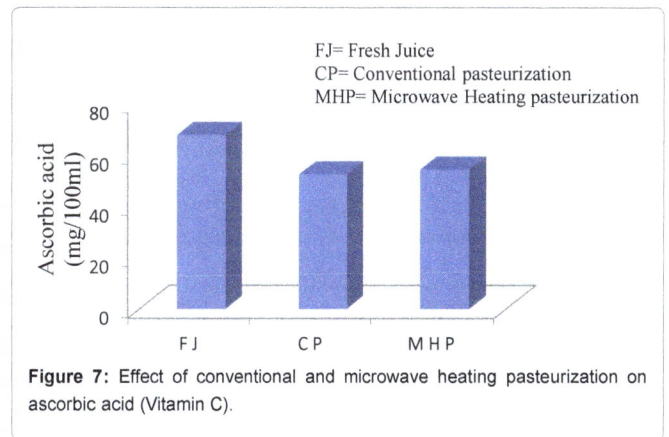

Figure 7: Effect of conventional and microwave heating pasteurization on ascorbic acid (Vitamin C).

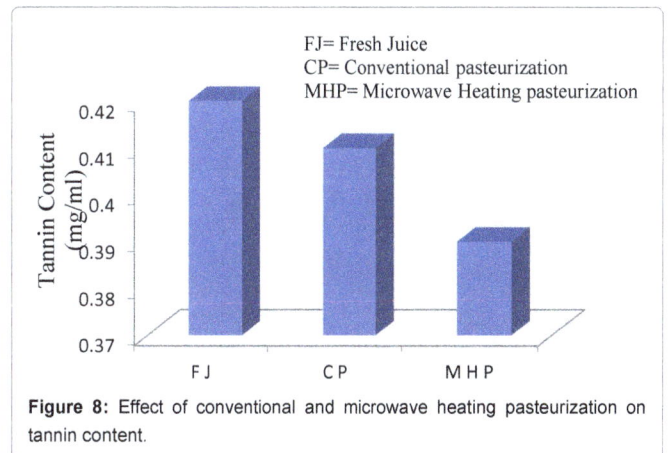

Figure 8: Effect of conventional and microwave heating pasteurization on tannin content.

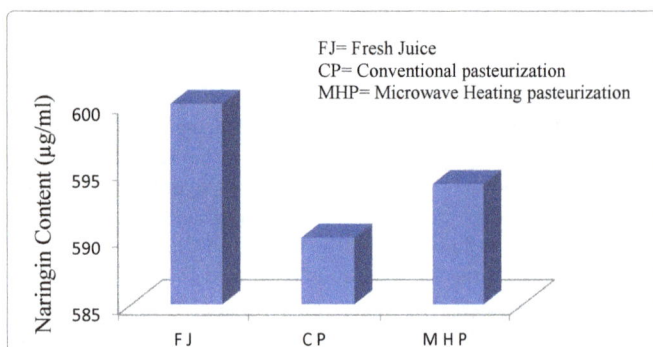

Figure 9: Effect of conventional and microwave heating pasteurization on naringin content.

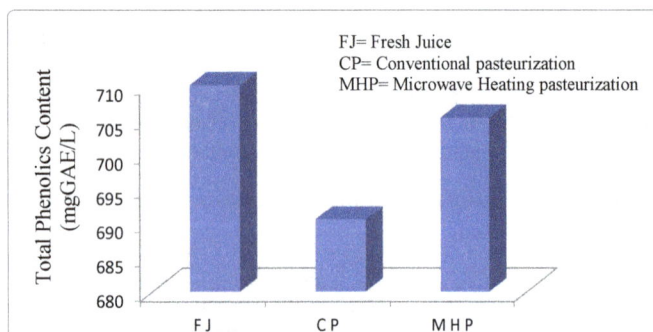

Figure 10: Effect of conventional and microwave heating pasteurization on total phenolic content.

as shown in Table 1. Figure 9 shows the effect of conventional and microwave heating pasteurization on naringin content. These results are in agreement with those of Igual et al. [21] also reported the decrease in naringin content for grapefruit juices.

Effect of conventional and microwave heating pasteurization on total phenolic contents

The total phenolic content of fresh pomelo juice was observed 710 mg GAE/L and after treatment it was 690.5 mg GAE/L and 705.3 mg GAE/L in case of conventional and microwave heating pasteurization respectively as shown in Table 1. Figures 9 and 10 show the effect of conventional and microwave heating pasteurization on total phenolic contents. Our finding favors the findings of Pala and Toklucu [22] and Mohamed et al. [22] who also reported the decrease in total phenolic contents in thermally and UV treated pomegranate juice and physalis juice respectively.

Conclusion

Present investigation has been conducted to analyze the effect of conventional and microwave heating pasteurization on various physiochemical properties of pomelo Juice. The results were analyzed for each physiochemical properties like pH, Moisture content, TSS, Titrable acidity, Reducing sugar, Ascorbic acid (Vitamin C), Tannin content, Naringin content and Total phenolic contents. On the basis of result analysis, we conclude that physiochemical properties of pomelo juice were less affected by microwave heating pasteurization in compare to conventional pasteurization.

Acknowledgement

This research was carried out in the department of food technology, Uttaranchal University, Dehradun we are sincerely thankful to HOD and Principal of UCALS for providing necessary requirements for smooth conducting of the research work.

References

1. Righetto AM, Beleia A, Ferreira SHP (1999) Physicochemical stability of natural or pre-sweetened frozen passion fruit juice. Brazil Archiv Biol Technol 42: 393-396.

2. Barboni T, Luro F, Chiaramonti N, Desjobert JM, Muselli A, et al. (2009) Volatile composition of hybrids citrus juices by headspace solid-phase micro extraction/gas chromatography/mass spectrometry. Food Chem 116: 382-390.

3. Morton J (1987) Pummelo *Citrus maxima*: Fruits of warm climates. Pummelo p. 147–151.

4. Arias BA, Ramon LL (2005) Pharmacological properties of citrus and their ancient and medicinal uses in the Mediterrean region. J Ethopharmacol 97: 89-95.

5. Pichaiyongvongdee S, Haruenkit R (2009a) Comparative studies of limonin and naringin distribution in different parts of pummelo (*Citrus grandis* (L.) Osbeck) cultivars grown in Thailand. Kasetsart J (Natur Sci) 43: 28-36.

6. Burda S, Oleszek W (2001) Antioxidant and antiradical activities of flavonoids. J Agri Food Chem 49: 2774- 2779.

7. Goh SG, Noranizan M, Leong CM, Sew CC, Sobhi B (2012) Effect of thermal and ultraviolet treatments on the stability of antioxidant compounds in single strength pineapple juice throughout refrigerated storage. Int Food Res J 19: 1131-1136.

8. Math RG, Nagender A, Nayani S, Satyanarayana A (2014) Continuous microwave processing and preservation of acidic and non-acidic juice blends. Int J Agri Food Sci Technol 5: 81-90.

9. Rayman A, Baysal T (2011)Yield and quality effects of electroplasmolysis and microwave applications on carrot juice production and storage. J Food Sci 76: 598-605.

10. Rangana S (2010) Analysis and quality control for fruit and vegetable products. Tata McGraw Hill Education Pvt Ltd, New Delhi, India.

11. Eynon L (1923) Determination of reducing sugars by Fehling solution with methylene blue as an indicator. J Soc Chem Ind London 42: 32-37.

12. Sawhney SK, Singh R (2015) Estimation of ascorbic acid in lemon juice. Introductory practical Biochemistry, Narosa Publishing House, New Delhi, India pp: 104-105.

13. Hagerman AE, Butler LG (1978) Protein precipitation method for the quantitative determination of tannins. J Agric Food Chem 26: 809-812.

14. Davis WB (1947) Determination of flavanones in citrus industry. Anal Chem 19: 476-478.

15. Makkar HPS, Bluemmel M, Borowy NK, Becker K (1993) Gravimetric determination of tannins and their correlations with chemical and protein precipitation methods. J Sci Food Agric 61: 161-165.

16. Mohamed AR, Soliman ZA, Diaconeasa SZ, Constantin B (2014) Effect of pasteurization and shelf life on the physicochemical properties of physalis (*Physalis peruviana* l.) Juice. J Food Process Preserv 39: 1051-1060.

17. Dar GH, Zargar Y, Shah GH (1992) Effect of processing operations and heat treatment on physico-chemical characteristics and microbiological load of apple juice concentrate. Indian Food Pack 46: 45-50.

18. Rivasa A, Rodrigoa D, Martı´neza A, Barbosa-Ca´ Novasb GV, Rodrigoa M (2006) Effect of PEF and heat pasteurization on the physical-chemical characteristics of blended orange and carrot juice. Swiss Society Food Sci Technol 39: 1163-1170.

19. Mehmood Z, Zeb A, Ayub M, Bibi N, Badshah A, et al. (2008) Effect of pasteurization and chemical preservatives on the quality and shelf stability of apple juice. America J Food Technol 3: 147-153.

20. Cinquanta L, Albanese D, Cuccurullo G, Dimatteo M (2010) Effect on orange juice of batch pasteurization in an improved pilot-scale microwave oven. J Food Sci 75: 46-50.

21. Igual M, García-Martínez E, Camacho MM, Martínez-Navarrete N (2011) Changes in flavonoid content of grapefruit juice caused by thermal treatment and storage. Innov Food Sci Emerg Technol 12: 153-162.

22. Pala ÇU, Toklucu AK (2011) Effect of UV-C on anthocuanin content and other quality parameters of pomegranate juice. J Food Composit Anal 24: 790-795.

14

Effect of Grain Teff, Sorghum and Soybean Blending Ratio and Processing Condition on Weaning Food Quality

Menure Heiru*

Department of Chemical Engineering, Dire Dawa University Institute of Technology, Dire Dawa, Ethiopia

Abstract

This study was conducted to evaluate the effect of grain teff, sorghum and soybean blending ratio and processing condition on weaning food quality with three specific objectives. Therefore, this study was initiated to address the protein energy malnutrition and sensory quality of weaning food. The proximate composition and sensory quality of blended samples were analyzed using standard methods. Processing condition had significant effect on nutritional and sensory properties of weaning food products. Moisture, ash, and crude fiber were significantly ($p<0.05$) higher (7.76%, 3.21%, 2.34%) respectivily. A significant high ash (3.85%) crude protein (17.50%) and crude fat (16.33%) contents were observed in weaning food blend processed via fermentation. So, the proximate analysis results obtained from feremented blend B_1 showed significantly higher ($p<0.05$) crude protien, ash and crude fat contents (16.62%, 3.47%, 11.35%) respectively and lower fiber (1.2%) content. The mean values of moisture, protein, fat, fiber, ash and carbohydrate were 4.19%, 17.17%, 14.33%, 1.26%, 3.11% and 59.91% respectively in fermented weaning foods. Sensory analysis revealed that highly acceptable product was obtained from fermented blends of teff, sorghum and soybean flour. The color, flavor, taste and overall acceptability scores of fermented blends were 5.72, 5.83, 5.77 and 5.77 (on 7- point hedonic scale), respectively. Among the treatments, fermented weaning food was found to produce acceptable weaning food gruel enriched by protein, ash and carbohydrate contents. So, fermented weaning food was enriched by proximate composition and acceptable sensory quality as compared to other processing conditions.

Keywords: Blending ratio; Soybean; Sorghum; Teff; Processing condition; Weaning food compositiont

Introduction

The growth and survival of infants after the recommended period of exclusive breast feeding for up to six months depend on the nutritional quality of the weaning food [1]. Breast milk is a sole and sufficient source of nutrition during the first six month of infant life. It contains all the nutrients and immunological factors infant require to maintain optimal health and growth. Towards the middle of the first year, breast milk is insufficient to support the growing infant. Therefore, nutritious complementary foods are needed to be introduced from six to twenty-four months of age [2]. These complementary foods are traditionally composed of staple cereals and legumes prepared either individually or as composite gruels [3], and they are supposed to serve as the main source of energy and nutrients for babies at weaning [1]. Teff (*Eragrostisis tef* (Zucc) Trotter) is one of the major and indigenous cereal crops in Ethiopia, where it is believed to have originated and has the largest share of area under cereal crop production [4]. It provides over two-thirds of the human nutrition in the country [5]. This cereal is considered high in nutritional quality, but limited information is available about its usefulness in weaning blends [6]. Teff flour is primarily used to make a fermented, sourdough type, flat bread called *Injera* [7]. Soybean (*Glycine max*) is a source of high quality cheap protein and polyunsaturated fatty acids that are often used to improve protein quantity and quality of most cereals and starch based foods. It is rich in iron, calcium and some B-vitamins though low in sulfur containing amino acids, methionine and cysteine [8]. Sorghum (*Sorghum bicolor*) *(L.)* Moench) is a critically important crop in sub-Saharan Africa on account of its drought tolerance. The main foods prepared with sorghum are: thin porridge (in Africa and Asia), stiff porridge (in West Africa), injera and bread (in Ethiopia), traditional beers (in Africa), baked products (in USA, Japan, and Africa), etc. In Africa, the majority of cereal-based foods is consumed in the form of porridge and naturally fermented products. Cereal-based thin porridge is prepared for fasting, sick or convalescent people, nursing mothers, and weaned infants [9].

Commercially made weaning foods are not available and if available most of them are priced beyond the reach of the majority of the population in less-developed countries. These foods are mostly manufactured using high technology and are sold in sophisticated packaging [10]. Such weaning foods may not be feasible in developing countries like Ethiopia due to limited income and inaccessibility. Therefore, there is a need for low-cost weaning foods which can be prepared easily at home and community kitchens from locally available raw materials such as sorghum, soybean and teff using simple technology like germination and fermentation. And, the most important nutritional problems in weaning foods consumed by the children in many parts of developing nations are protein energy malnutrition and deficient in essential macronutrients and micronutrients [11,12]. The high cost and inadequacy in production of protein-rich foods have resulted in increased protein energy malnutrition among children and other vulnerable groups in the developing world [13]. Therefore, this work was initiated to evaluate the effect of grain teff, sorghum and soybean blending ratio and processing condition on weaning food quality with the following specific objectives:

i. To determine the nutritional composition of weaning food blends of teff, sorghum and soybean.

ii. To determine the processing condition with the best potential for improving the nutritional quality of weaning food.

iii. To evaluate the sensory characteristics of weaning blends processed.

---

*Corresponding author: Menure Heiru, Department of Chemical Engineering, Dire Dawa University Institute of Technology, PO Box-1352, Dire Dawa, Ethiopia, E-mail: heirumenure782@gmail.com

Ingredients	Moisture	Ash	Crude fiber	Crude fat	Protein	Carbohydrate
Teff	8.08 ± 0.51[a]	1.84 ± 0.49[b]	1.03 ± 0.18[b]	3.19 ± 0.01[b]	9.10 ± 0.35[b]	76.74 ± 0.95[a]
Sorghum	7.80 ± 0.71[a]	1.41 ± 0.13[b]	1.77 ± 0.12[a]	2.83 ± 0.09[b]	7.87 ± 0.17[c]	78.29 ± 1.06[a]
Soybean	5.39 ± 1.24[b]	4.75 ± 0.83[a]	1.98 ± 0.18[a]	22.88 ± 0.37[a]	27.00 ± 0.89[a]	37.98 ± 1.34[b]

All values are mean ± Std. Dev on dry basis except moisture (wet basis)

Table 1: Proximate composition of grain teff, sorghum and soybean used in processing of weaning food (%).

Materials and Methods

Experimental materials

Ingredients of the composite blends were acquired from the following sources: Teff and sorghum were obtained from MARC (Melkassa Agricultural Research Center) and soybean was obtained from AARC (Awassa Agricultural Research Center) that grown 2013/2014 crop years. All grains were stored at room temperature until analyzed.

Weaning blend formulation

Weaning blends were formulated in 60% cereals to 40% legume ratios, which yield the highest projected amino acid scores based on infant lysine requirements FAO/WHO/UNU [14]. Ingredients were weighed and formulated in proportions as follows: B_1 (20% teff + 40% sorghum + 40% soybean), and B_2 (30% teff + 30% sorghum + 40% soybean) and B_3 (40% teff + 20% sorghum + 40% soybean)

Processing methods

Unprocessed control: All the test samples were cleaned, free from abnormal odors, broken seeds, dust and other foreign materials including living or dead insects before ground to flour. Sorghum and teff were milled in cyclone mill to a fine powder that able to pass through ≤ 250 µm sieve size. Then the powder obtained was placed in plastic bag and stored at room temperature prior to blend. Soybean seed was grind to flour using grinding mill.

Natural fermentation: Fermentation was performed using the microorganisms naturally present on the grain. Slurries of the three composite blends (1:4 w/v) were made from unprocessed control ingredients by mixing 200 g of flour with 800 mL of distilled water in a sterile beaker. Slurries were fermented in a temperature-controlled incubator at 30°C for 72 hrs [15]. After 72 hrs fermentation period, the slurries were transferred into aluminum pans, and then oven-dried at 55°C for 48 hrs. Fermented dry blends were further milled in to fine flour using a home coffee grinder.

Germination/sprouting: Germination was performed in a dark room following the modified method Griffith et al. [16]. Sorghum and soybean seeds were rinsed and soaked in distilled water (1:3 w/v) for 9 hrs. at ambient temperature (23°C to 25°C). Seeds were dried and placed on perforated aluminum pans lined with filter paper, then placed in a dark, temperature controlled cabinet at 30°C for 12 hrs, 24 hrs and 36 hrs germination. Germinating seeds were rinsed twice daily with distilled water to reduce microbial growth and to maintain adequate hydration. Sprouted seed was dried in forced air oven at 50°C for 20 hrs. Dried sprout sorghum and soybean were dehulled using mortar and pestle, and milled to flour by grinding mill.

Proximate composition analysis

Proximate composition of initial ingredients and blended samples of weaning food flour were conducted using standard methods. Moisture content, ash, and fiber content of ingredients and weaning blends were determined according to AOAC [17]. Protein was determined by

Micro-Kjeldal method AOAC [17]. Crude fat content was determined according to the method of AOAC [17] using soxlet apparatus and carbohydrate content was calculated as the percentage difference of proximate compositions.

Sensory evaluation

Sensory evaluation of the ready-to-eat formulated complementary foods (semi-liquid) were carried out on the taste, flavor, color and overall acceptability by 30 staff members, mothers and students using seven-point hedonic scale with score ranging from 'like extremely (7)' to 'dislike extremely (1)'in Food Technology and Process Engineering Laboratory of Haramaya University.

Statistical analysis

Analysis of variance (ANOVA) was used to test for significant differences between means of three replicate of blends and processing methods using the statistical analysis system.

Results and Discussion

Proximate composition of grain teff, sorghum and soybean used in the weaning food

The proximate composition of weaning food ingredients used in this experiment is shown in Table 1. The moisture contents were 8.08%, 7.80% and 5.39% for teff, sorghum and soybean respectively. The protein content of teff was 9.10%, which was appreciably high compared to common cereals like maize, rice and sorghum. This value was lower than 10.7% reported by Laike [18]. Compared to other cereals, teff has higher protein content than maize (8.3%), sorghum (7.1%), barley (9.0%), millet (7.2%) and almost equivalent to wheat (10.3%) [19]. The fat content of teff appeared to be lower than maize (4.6%) but higher than wheat, barley and millet and equivalent to sorghum (2.8%). Whereas the ash content was lower than millet and higher than others. Apparently, the ash content of teff observed in *Gemechis* variety was higher than sorghum (*Teshale* variety). Carbohydrate contents were 76.74%, 78.24% and 37.98% for teff, sorghum and soybean respectively. The carbohydrate content of *Gemechies* teff variety was in close agreement with National Research Council [20] content 72%.

The proximate composition of sorghum used in the weaning food were 7.80% moisture, 1.14% ash, 1.77% crude fiber, 7.87% crude protein, 2.83% crude fat and 78.29% carbohydrate respectively. Carbohydrate content was high as compared to other common cereals and legumes. So, the nutritional content of sorghum grain in close agreement with other varieties of sorghum, i.e., it contains a reasonable amount of protein (7.5% to 10.8%), ash (1.2% to 1.8%), oil (3.4% to 3.5%), fiber (2.3% to 2.7%) and carbohydrate (71.2% to 80.7%) with a dry matter ranged from 89.2% to 95.3% depending on the type of cultivars [21]. Such variations may be contributed by genotype, water availability, soil fertility, temperature and environmental condition during grain development Serna-Saldivar and Rooney [22]. The proximate composition of soybean was 5.39% moisture, 4.75% ash, and 1.98% crude fiber, and 27.00% crude protein, 22.88% crude fat and 37.98% total carbohydrate (Table 1). Famurewa and Raji [8] reported

B	Moisture	Ash	Crude fiber	Crude fat	Crude protein	Carbohydrate
Control						
B1	7.69 ± 0.45[abc]	3.47 ± 0.15[ab]	1.83 ± 0.30[cd]	10.02 ± 0.34[def]	15.98 ± 0.40[d]	60.98 ± 0.63[abc]
B2	7.39 ± 0.36[cde]	3.18 ± 0.17 [b]	0.95 ± 0.14[g]	9.63 ± 0.11[ef]	16.14 ± 0.35[cd]	62.68 ± 1.12[ab]
B3	6.80 ± 0.43[e]	3.16 ± 0.16[eb]	1.39 ± 0.14[ef]	8.83 ± 0.38[f]	16.55 ± 0.21[bcd]	63.25 ± 0.68[a]
Fermented						
B1	4.85 ± 0.49[f]	3.85 ± 0.52 [a]	1.20 ± 0.02[fg]	14.14 ± 0.85[b]	16.91 ± 0.53[abc]	59.02 ± 0.82[cd]
B2	3.42 ± 0.18[f]	2.99 ± 0.00[bc]	1.26 ± 0.13[fg]	16.33 ± 1.93[a]	17.50 ± 0.35 [a]	58.47 ± 2.17[d]
B3	4.31 ± 0.34[f]	2.50 ± 0.17[d]	1.31 ± 0.03[fg]	12.51 ± 2.3[c]	17.10 ± 0.44[ab]	62.24 ± 2.82[ab]
12 hrs. Germinated blend						
B1	7.04 ± 0.23[ed]	2.65 ± 0.00[cd]	1.61 ± 0.05[def]	11.10 ± 0.61[cde]	16.92 ± 0.21[ab]	60.68 ± 0.58[bcd]
B2	7.16 ± 0.16[cde]	3.31 ± 0.00[b]	1.91 ± 0.21[cd]	9.90 ± 0.01[def]	16.48 ± 0.25[bcd]	61.21 ± 0.17[abc]
B3	7.29 ± 0.63[cde]	3.48 ± 0.17[ab]	1.90 ± 0.15[cd]	11.04 ± 0.29[cde]	16.90 ± 0.54[ab]	59.37 ± 1.01[cd]
24 hrs. Germinated blend						
B1	7.47 ± 0.17[bcd]	3.49 ± 0.50[ab]	2.09 ± 0.14[bc]	10.96 ± 0.21[cde]	16.63 ± 0.17[abc]	59.30 ± 0.48[cd]
B2	7.69 ± 0.18[abc]	3.17 ± 0.14[b]	2.43 ± 0.34[ab]	9.77 ± 0.53[ef]	16.55 ± 0.21[bcd]	60.37 ± 0.33[bcd]
B3	8.12 ± 0.15[a]	2.98 ± 0.00[bc]	2.49 ± 0.30[ab]	10.27 ± 0.13[def]	16.78 ± 0.64[abc]	59.34 ± 0.56[cd]
36 hrs. Germinated blend						
B1	7.30 ± 0.33[cde]	3.90 ± 0.34[a]	1.75 ± 0.38[cde]	10.51 ± 0.19[de]	16.90 ± 0.14[abc]	59.62 ± 0.78[cd]
B2	7.39 ± 0.15[cde]	3.18 ± 0.16[b]	2.36 ± 0.43[ab]	11.59 ± 0.40[cd]	17.05 ± 0.99[abc]	58.41 ± 1.98[d]
B3	8.03 ± 0.20[ab]	3.19 ± 0.50[b]	2.66 ± 0.05[a]	10.06 ± 0.58[def]	16.91 ± 0.53[abc]	59.11 ± 0.75[cd]
CV	4.9	8.31	12.72	7.85	2.84	2.04

All values are expressed as Mean ± STDV of % dry basis except moisture (% wet basis),
Values in a column with the same letter are not significantly different (p<0.05).
Note: B: Blending; CV: Coefficient of variation in (%); B$_1$: 20% teff + 40% sorghum + 40% soybean; B$_2$: 30% teff + 30% sorghum + 40% soybean; B$_3$: 40% teff + 20% sorghum + 40% soybean.

Table 2: Effect of blending ratio and processing condition on proximate composition of weaning food (%).

that the proximate composition of soybean seed contains 40% protein, 21% oil 34% carbohydrate and 5% ash. The crude fat and ash content of soybean were in close agreement with Famurewa and Raji [8] but it varies in crude protein content. The variation in crude protein content might be the varieties, geographical location and soil fertility.

Proximate composition of blended weaning food

Effect of blending ratio and processing condition interaction on proximate composition of weaning food: The proximate composition of each weaning blends made by five processing method were summarized in Table 2. The moisture content of weaning blend varied significantly (p<0.05) among processing and blending methods. Moisture content of 24 hrs germinated blend had highest (8.12%) as compared to other processing methods. The interaction effect of blend was highest at 24 hrs germinated blend B$_3$ (8.12%) and lowest (3.42%) was observed in B$_2$ of fermented weaning food. This was most probably due to dry matter losses. Furthermore, the values obtained for the moisture content and the associated dry matter of the weaning blend were suitable for an increase shelf life of the food that was formulated from cereal and legumes. High moisture content aid microbial growth and reduce shelf-life of food products. Thus, the reduced moisture content of the weaning food especially the significant drop in the moisture content serve as a positive processing step that will improve the quality of the product [23]. These also reduce the cost of preservation and processing of the grain for both industrial and domestic uses.

The interaction effect of blending ratio and processing condition on ash content was highest in fermented and 36 hrs. germinated weaning food blend (B3) (3.81%) and B1 (3.90) respectively and lowest (2.50%) was in B3 of fermented weaning food flour. Lorenz [24] reported that ash content increases during germination to be apparent rather than true increases and resulted from the losses of dry matter. The interaction of blending ratio and processing condition had significant (p<0.05) effect on crude fiber content of weaning food flour (Table 2). The highest

interaction effect on crude fiber content (2.66%) was in B$_3$ of 36 hrs germinated weaning food blends and lowest (1.20%) and (1.26%) was in B$_1$ and B$_2$ of fermented weaning food flour. The crude fiber content of germinated soybean blended weaning food significantly increased as compared to unprocessed control weaning food flour. This is might be the decomposition of starch during germination to simple sugar. The expected decrease in crude fiber content during fermentation could be attributed to the partial solubilization of cellulose and hemicelluloses type of material by microbial enzymes and partly also by leaching. A previous study has reported a significant decrease of crude fiber contents after four days of maize fermentation [25]. The crude fiber content of infant food is expected to be low as foods with higher fiber content tend to cause indigestion in babies [26]. Hence sample with lower fiber content were rated good as potential weaning food. Fermentation as a process is promising to meet crude fiber stands in the preparation of weaning foods from locally available cereals. Crude fat content for all blends varied significantly (p<0.05) resulting from differences among individual ingredients and processing conditions. The highest crude fat content (16.33%) was recorded in fermented blend and lowest (8.83%) was for unprocessed control blends. Interaction effects of blend on crude fat content was highest (16.33%) in fermented weaning food blend (B$_2$) and lowest (8.83%) was in controlled weaning food flour at B$_3$. Weaning blend formulation of soybean increases fat provided more concentrated calorie source rich in the essential fatty acid, linoleic acid. In reality desirable and more expensive oils are often consumed by household members other than the targeted child [3].

The interaction of blending and processing condition had significant (p<0.05) effect on protein contents (Table 2). The highest protein content (17.50%) was observed in fermented weaning food blend (B$_2$) and lowest (15.98%) was in unprocessed control blend (B$_1$). The protein content of weaning blends was increased probably due to a reduction of phytic acid which might have contributed to the improved digestibility observed in germinated and fermented blends. Khetarpaul

Blend	Flavor/aroma	Taste	Color	Overall acceptability
	Unprocessed control			
B₁	5.50 ± 0.90[bc]	5.00 ± 0.90[c]	5.36 ± 0.49[cde]	5.16 ± 0.74[ef]
B₂	5.00 ± 1.08[cd]	5.50 ± 0.90[ab]	5.10 ± 0.54[e]	5.16 ± 0.74[ef]
B₃	5.03 ± 0.80[cd]	5.63 ± 0.96[ab]	5.20 ± 0.71[de]	5.56 ± 0.62[bcd]
	Fermented			
B₁	5.66 ± 0.75[b]	5.83 ± 0.69[a]	5.50 ± 0.50[cd]	6.00 ± 0.00[a]
B₂	6.16 ± 0.69[a]	5.83 ± 0.37[a]	5.33 ± 0.47[cde]	5.66 ± 0.47[abc]
B₃	5.66 ± 0.75[b]	5.66 ± 0.95[a]	6.33 ± 0.47[a]	5.66 ± 0.47[abc]
	12 hrs. Germinated blend			
B₁	4.80 ± 1.18[d]	5.80 ± 0.61[a]	5.33 ± 0.75[cde]	5.36 ± 0.55[cdef]
B₂	5.23 ± 0.67[bcd]	5.20 ± 0.61[bc]	5.36 ± 0.66[cde]	5.26 ± 0.63[def]
B₃	5.23 ± 0.85[bcd]	5.56 ± 0.72[ab]	5.10 ± 0.75[e]	5.33 ± 0.66[cdef]
	24 hrs. Germinated blend			
B₁	5.33 ± 0.71[bc]	5.43 ± 0.67[ab]	5.36 ± 0.49[cde]	5.46 ± 0.86[bcdef]
B₂	5.23 ± 1.13[bcd]	5.40 ± 0.72[ab]	5.66 ± 0.54[bc]	5.56 ± 0.81[bcd]
B₃	5.23 ± 0.77[bcd]	4.83 ± 0.69[c]	5.26 ± 0.52[de]	5.10 ± 0.54[f]
	36 hrs. Germinated blend			
B₁	5.10 ± 0.92[cd]	5.56 ± 0.81[ab]	5.40 ± 0.56[cde]	5.50 ± 0.68[bcde]
B₂	5.23 ± 0.50[bcd]	5.46 ± 0.57[ab]	5.66 ± 0.47[bc]	5.53 ± 0.57[bcde]
B₃	5.70 ± 0.59[b]	5.63 ± 0.66[ab]	5.86 ± 0.81[b]	5.80 ± 0.61[ab]
Mean	5.34	5.49	5.45	5.47
CV (%)	15.83	13.55	10.98	11.54

Values followed by different letters within a column indicate significant difference (p<0.05) using DMRT.
*: Mean ± SD; CV: coefficient of variation; B₁: 20% teff + 40% sorghum + 40% soybean; B₂: 30% teff + 30% sorghum + 40% soybean; B₃: 40% teff + 20% sorghum + 40% soybean.

Table 3: Effect of blending ratio and processing condition interaction on sensory quality of weaning food.

and Chauhan [27] reported improved *in-vitro* protein digestibility during germination and while improvements in *in-vitro* protein digestibility with fermentation were associated with proteolytic enzyme production by micro-organisms. Abdelhaleem et al. [28] reported that the observed increment in protein content after fermentation was probably due to shift in dry matter content through depletion during fermentation by action of the fermenting microorganisms. However, cells of the fermenting microorganisms could have contributed to the protein, therefore, fermentation of weaning blend results in an observable increase in crude protein content. In most human diets, the protein is more limiting than others. Therefore, application of fermentation process that appears to increase the protein content even at the expense of other nutrients may be advantageous nutritionally [28]. Improvements in protein quality have also been documented after fermenting blended mixtures of plant-based complementary foods based on maize and legumes, groundnut and millet and cereal and soybean blends [29]. The improvement in protein digestibility after germination, dry heating could be attributed to the reduction of anti-nutrients such as phytic acid, tannins and polyphenols, which are known to interact with proteins to form complexes [30]. Carbohydrate content was determined by difference. The interaction of blending ratio and processing condition had significant (p<0.05) effect on carbohydrate content (Table 2). The highest (63.25%) was recorded for controlled weaning food blend (B₃) and lowest (58.41%) was observed in 36 hrs germinated blends of weaning food (B₂). The processing method significantly decreases carbohydrate content of weaning foods. Moreover, fermentation and germination treatments decreased significantly the carbohydrate contents. The decrease in total carbohydrate content of weaning food ingredient may be the starch and simple sugars are the principal substrates for fermenting microorganisms; therefore, degradation and subsequent decrease in starch content are expected [25]. The decrease in carbohydrate content might be the degradation of sugar by processing conditions.

Effect of blending ratio and processing condition interaction on sensory quality of weaning food: The interaction effect of blending ratio and processing condition had significant (p<0.05) effect in the color of weaning food (gruel). The color of the thin porridge made from fermented weaning blended flour was most preferred (like very much) by the panelists, while the thin porridge prepared from 12 hrs. germinated blend and control flour were least preferred for color (like slightly). The highest weaning food gruel color (6.33) was observed in B₃ of fermented weaning food blend (like very much) and least (5.10) were obtained in B₂ (like slightly) of control weaning food gruel. The interaction of processing condition and blending ratio were significant (p<0.05) on weaning food flavor (Table 3). The highest value of gruel flavor (6.16) was recorded in fermented weaning food gruel of B₂ (like very much) and lowest 5.00 was recorded in B₂ of control weaning food gruel (like slightly). Also, interaction effect of processing condition and blending ratio had significant effect on taste of weaning food gruels. The highest value (5.83) was in fermented blend (B2 and B3) and lowest (5.00) was in B2 of control weaning food gruel. During fermentation the taste of weaning food gruel was preferred very much as compared to other processing conditions. The panelists however, noted that color, taste and overall acceptability of the gruel prepared were highly acceptable.

Overall acceptability of weaning food (gruel) were significantly (p<0.05) affected by processing condition and blending ratio (Table 3). The highest processing mean 6.00 (like very much) was recorded in fermented and lowest 5.10 (like slightly) was obtained in 24 hrs. germinated weaning food gruels. The highest (6.00) overall acceptability of weaning food was observed in fermented sample blended at B₁ and the lowest (5.10) was observed in the 24 hrs. germinated sample of blended at B₃. Generally, sugar is by far the most important addition to complementary foods and is commonly added to improve the flavor and to encourage infants to eat while fat acts as flavor retainer and increases the mouth feel of foods [31]. Oil also improves the taste/flavor of the product and reduces bulkiness of starchy food in the mixture [31]. Germination also improves the consistency, mouth feel and taste of the product [32]. Inyang and Zakari [33] reported that sensory panelists are highly rated for formulations from germinated grains for all the sensory parameters investigated.

Conclusion

1. An infant weaning food of high nutrient density could be formulated and prepared from a combination of teff, sorghum and soybean. Blend formulation showed the strongest impact on nutritional quality and should receive attentions in the design and development of an infant weaning food.

2. The present study showed that blending ratio and processing condition significantly influenced the proximate composition, mineral content and sensory characteristics blended weaning food flour.

3. Processing conditions (fermentation and sprouting) were improved the proximate composition, mineral content and sensory quality weaning food (gruel).

4. Generally, the present result suggests that blending ratio and processing condition significantly improved the nutrient density and sensory quality of weaning foods.

References

1. Ogbeide ON, Ogbeide O (2000) Mineral content of some complementary foods in Edo state, Nigeria. West Afri J Food and Nutri 2: 26-30.

2. Mamiro SP, Kolstreren P, Roberfroid D, Tatala S, Opsomer AS, et al. (2005) Feeding practices and factors contributing to wasting, stunting and iron deficiency anemia among 3-23 months old children in Kilosa District, Rural Tanzania. J Health Popul Nutri 23: 222-230.

3. Huffman SL, Martin LH (1994) First feedings: Optimal feeding of infants and toddlers. Nutri Res 14: 127-159.

4. Bultosa G (2007) Physico-chemical characterization of grain and flour in 13 tef [Eragrostis tef (ZUCC.) Trotter] grain varieties. J Appli Sci Res 3: 2042-2051.

5. Uraga K, Narasimha HV (1997) Effect of natural fermentation on the HCL-extractability of minerals from tef (Eragrostis tef). Bulletein of Chemical Society of Ethiopia 11: 3-10.

6. Cheverton MR, Chapman GP (1989) Ethiopian tef: A cereal confined to its centre of variability. In: Wickens GE, Haq N, Day P (eds.) New Crops for Food and Industry. Chapman and Hall, New York, pp: 235-238.

7. Davison J, McKnight C (2004) Teff demonstration plantings for 2003. FS 04-51. University of Nevada Reno Cooperative Extension, Reno, NV.

8. Famurewa JAV, Raji AO (2005) Parameters affecting milling qualities of undefeated soybeans. Inter J Food Eng 1: 6.

9. Dicko MH, Gruppen H, Traore AS, Voragen AGJ, Van Berkel WJH (2006) Sorghum grain as human food in Africa: relevance of content of starch and amylase activities. Afri J Biotechnol 5: 384-395.

10. Kulkarni KD, Kulkarni DN, Ingle UM (1991) Sorghum malt-based weaning food formulations: Preparation, functional properties, and nutritive value. Food Nutri Bullet.

11. Brabin BJ, Coulter JBS (2003) Nutrition associated disease. In: Cook GC, Zumla AI (eds.) Manson's Tropical Diseases. Saunders, London, pp: 561-580.

12. Millward DJ, Jackson AA (2004) Protein/energy ratios of current diets in developed and developing countries compared with a safe protein/energy ratio: Implications for recommended protein and amino acid intakes. Public Health Nutri 7: 387-405.

13. Otegbayo BO, Sobande FO, Aina JO (2002) Nutritional quality of soybean-plantain extruded snacks. Ecology Food Nutri 41: 463-474.

14. FAO/WHO/UNU (1985) Energy and protein requirements: Report of a Joint FAO/WHO/UNU Expert Consultation. Tech. Rep. Ser. No. 724. World Health Organization, Geneva, Switzerland.

15. Chavan JK, Kadam SS (1989) Nutritional improvement of cereals by fermentation. Critical Rev Food Sci Nutri 28: 349-400.

16. Griffith LD, Castell-Perez ME, Griffith ME (1998) Effects of blend and processing method on the nutritional quality of weaning food made from select cereals and legumes. J Cereal Chem 75: 105-112.

17. AOAC (1990) Official methods of analysis. (16th edn), Association of Official Analytical Chemists, Arlington, Virginia, USA.

18. Kebede L (2006) Effect of extrusion operating conditions on the physicochemical and sensory properties of grain teff puffed products. School of Graduate Studies, Haramaya University.

19. Asrat W, Frew T (2001) Utilization of Teff in Ethiopian diet. Narrowing the rift: Teff Research and Development, Proceedings of the International Workshop on Teff Genetics and Improvement, Debre Zeit, Ethiopia, pp: 239-243.

20. National Research Council (1996) Lost crops of Africa (Volume I: Grains). National Academy Press, Nutrition Society, Washington DC 56: 105-119.

21. Idris WH, Abdel Rahaman SM, El Maki HB, Babiker EE, Tinay AH (2007) Effect of malt pretreatment on HCl extractability of calcium, phosphorus and iron of sorghum cultivars. Inter J Food Sci Technol 42: 194-199.

22. Serna-Saldivar S, Rooney LW (1995) Structure and chemistry of sorghum and millets. In: Dendy DAV (ed.) Sorghum and millets chemistry and technology. American Association of Cereal Chemists, St. Paul, MN pp: 69-124.

23. Kikafunda JKL, Abenakyo, Lukwago FB (2006) Nutritional and sensory properties of high energy/nutrient dense composite flour porridges from germinated maize and roasted beans for child-weaning in developing countries: A case for Uganda. Ecology Food Nutri 45: 279-294.

24. Lorenz K (1980) Cereals sprout: Composition, nutritive value, food applications. CRC Critical Review of Food Sci Nutri 28: 353-385.

25. Ejiqui J, Savoie L, Desrosies TM (2005) Beneficial changes and drawbacks of traditional fermentation process on chemical composition and anti-nutritional factors of yellow maize (Zea mays). J biol Sci 5: 590-596.

26. Olorunfemi OB, Akinyosoye FA, Adetuyi FC (2006) Microbial and nutritional evaluation of infant weaning food from mixture of fermented food substrates. Res J Biol Sci 1: 20-23.

27. Khetarpaul N, Chauhan BM (1990) Effect of germination and fermentation on in vitro starch and protein digestibility of pearl millet. J Food Sci 55: 883-884.

28. Abdelhaleem WH, El-Tinay AH, Mustafa AI, Babiker EE (2008) Effect of fermentation, malt-pretreatment and cooking on antinutritional factors and protein digestibility of sorghum cultivars. Pak J Nutri 7: 335-341.

29. Gibson RSL, Perlas-Hotz C (2006) Improving the bioavailability of nutrients in plant foods at the household level. Proceedings of the Nutrition Society 65: 160-168.

30. Abbey TK, Alhassan A, Ameyibor K, Essiah JW, Fometu E (2001) Integrated science for senior secondary schools. Unimax Maxmillan Ltd, Accra North 75: 376-451.

31. Walker AF, Pavitt S (2007) Energy density of third world weaning foods. Nutri Bulletin 14: 88-101.

32. Helland MH, Wicklund T, Narvhus JA (2002) Effect of germination time on alpha-amylase production and viscosity of maize porridge. Food Res Inter 35: 315-321.

33. Inyang CU, Zakari UM (2008) Effect of germination and fermentation of pearl millet on proximate, chemical and sensory properties of instant "fura": A Nigerian Cereal Food. Pak J Nutri 7: 9-12.

15

Improvement of the Nutritional Value of Cereal Fermented Milk: 1- Soft Kishk Like

Nassar KS[1]*, Shamsia SM[1] and Attia IA[2]

[1]Department of Food, Dairy Science and Technology, University of Damanhour, Egypt
[2]Department of Dairy Science and Technology, Alexandria University, Egypt

Abstract

Soft Kishk like products were produced from whole wheat, barley and freek burghul with reconstituted skim milk (15% T. S) and addition of different Starter cultures. Physico-chemical, bacteriological and organoleptic properties of soft kishk like samples have been evaluated during 14 days of storage at 5 ± 1°C. The main effect on Soft Kishk like products characteristics was due to the used cereal type more than the started culture. Containing wheat burghul showed the highest pH values, Crude protein content were almost similar between all treatments as a result of the similarity of protein content in the selected seeds. Freek burghul treatment-showed higher total solids and carbohydrates whereas, lower content of ash, crude fiber, fat and crude protein contents compared to the other treatments. The c.f.u on MRS medium (mainly *Bifidobacteria*) sensitive than other starter microorganisms (mainly lactic acid bacteria) for storage. Cereal fermented dairy products containing Freek; gained the highest scores in the organoleptic properties followed by wheat. While the addition of whole barley burghul had the lowest total score at the end of storage. Therefore, a functional and nutritional Kishk like products have been successfully produced using different cereals and probiotic starter cultures.

Keywords: Fermentation; Cereal; Probiotic bacteria; Kishk; Milk

Introduction

There is a currently growing interest in certain strains of lactic acid bacteria that have been suggested or shown to provide specific health benefits when consumed as food. The history of those observations had started in the early 1900's, when Élie Metchnikoff noted the longevity of people living in the Balkans, and attributed this to their high consumption of fermented milk products [1].

Following those early observations, scientists reported that yogurt and other fermented milk products contain lactic acid bacteria that are capable of establishing and colonizing the gut, this group of bacteria referred to as probiotic bacteria [2]. Probiotics are live microbial food supplements, which beneficially affect the host animal by improving its intestinal microbial balance [3]. Recently, some foods, as called functional foods that have positive health promoting effects are already on the global market, and especially in the markets of Japan, Europe, and United states. Functional foods are defined broadly as foods that provide more than simple nutrition; they supply additional physiological benefits to the consumer. Yoghurt and other fermented milks containing probiotics may be considered the first functional foods [2-4].

Cereals are providing dietary fibers, proteins, energy, minerals and vitamins required for human health. The possible applications of cereal constituents in functional food formulations could be summarized: (a) as dietary fiber promoting several beneficial health or physiological effects; (b) as prebiotics due to their content of specific non digestible carbohydrates; and (c) as encapsulation materials for probiotics in order to enhance their stability [5]. However, the nutritional quality of cereals and the sensorial properties of their products are sometimes poor or inferior in comparison with milk and milk products. Milk proteins have high nutritional value compared to other proteins because of their relatively high content of essential amino acids and good digestibility [6].

During fermentation of dairy products like cheese and yoghurt release bioactive peptides upon enzymatic hydrolysis of milk proteins [7]. Fermentation may be the most simple and economical way of improving nutritional value, sensory properties and functional qualities of the dairy products [8]. Lactic acid fermentation of different cereals has been found effectively to reduce the amount of Phytic acid, tannins and improve protein availability [9]. Increased amounts of riboflavin, thiamine, niacin and lysine due to the action of LAB in fermented blends of cereals where also reported [10]. The traditional foods manufactured from grains usually lack flavor and aroma. Fermentation improves the sensorial value, which is very much dependent on the amounts of lactic acid, acetic acid and several aromatic volatiles such as higher alcohols, aldehydes, ethyl acetate and di-acetyl, produced via the homo-fermentative or hetero-fermentative metabolic pathways [11]. This work aimed to produce probiotic cereals fermented milk supplemented. The cereals were as wheat, barley and Fereek (green wheat), in order to introduce a diet with nutritionally balanced that needed for sensitive or elderly persons.

Materials and Methods

Cereals

Three different cereals; namely: Dry Wheat (Triticum spp.), Barley (Hordeum spp.) and Green Wheat or Fereek (Triticum spp.), were purchased from local market (Alexandria, Egypt). Dried skim milk was obtained from Alexandria market.

Starters

Three commercial freeze-dried DVS were used. They were a Yoghurt starter (YC-X11) (*Streptococcus thermophilus* + *Lactobacillus delbruckii* subsp. *bulgaricus*); Bio-yoghurt starter (ABT–2) (CHR HANSEN) + Yoghurt starter (1:2) and *Lactobacillus plantarum* + Yoghurt starter (1:2) with potential probiotic properties (From Chr.

*****Corresponding author:** Nassar KS, Department of Food, Dairy Science and Technology, University of Damanhour, Egypt

E-mail: Khalid.nassar@agr.dmu.edu.eg

Hansen laboratory, Denmark). Freeze -dried bacterial starters were propagated separately as mother cultures in autoclaved (121°C/20 min) skim milk. The cultures were incubated at 37°C for Bio-yoghurt starter, 32°C for *Lactobacillus plantarum* and 40°C for yoghurt starter, until curding of milk. Cultures were freshly prepared before using.

Preparation of whole wheat burghul, whole barley burghul and fereek burghul

The polished (wheat, Barley and Fereek) (0.5 kg) was cleaned manually by removing foreign grains and other impurities. Then, it was placed in a stainless steel sieve and washed under a strong stream of tap water with continuous stirring for two minutes followed by rinsing with distilled water. Cleaned and washed polished wheat was cooked in a Stainless-steel pot containing (1.5 L) of water. Cooking time was between 25-30 minutes from the beginning of boiling until the completion of cooking and absorption of the water. The cooked grains were left to cool then broken up a blender.

Preparation of reconstituted skim milk (15 w/w)

150 gm of the dried skim milk were dispersing by stirring in 600 gm distilled water at 40°C, then complete to 1000 gm with distilled water vigorously stirring until completely dissolving

Preparation of cereal fermented dairy products or soft kishk like

Each type of Burghul was mixed with reconstituted skim milk in a ratio of 1:4 (w/w) in addition to, suitable amount of vanillin. The final total solids of mixture were 26%. Mixture was heated to 95°C for 10 seconds, and then rapidly cooled to 45°C, addition 3% of each Starter. The ingredients (reconstituted skim Milk + Cereal + starter culture)

were mixed thoroughly before adding. The resultant paste was filled in polystyrene cups and covered then incubated at (43°C for W1, B1 and F1) and (37°C for W2, W3, B2, B3, F2 and F3) to 6 hours. After that, the fermented paste was stored at 5 ± 1°C (refrigerator temperature) (Figure 1).

Chemical Analysis

Dairy base analysis

Dried skim milk is analyzed for total solids (TS), Protein, Ash, titratable acidity, pH and Total carbohydrate content according to the Association of Official Analytical Chemists [12] (Table 1).

Cereals analysis

Cereals were analyzed for total solids (TS), Fat, Protein, Ash, Total carbohydrate and crude fiber content according to AACC [13] (Table 2).

Kishk like analysis

The samples were analyzed in triplicates for pH, titratable acidity as lactic acid, and Total solids in the fresh products and after 7 and 14 days of storage at 5 ± 1°C, whereas, total protein, carbohydrate, ash, crude fibers and the fat content were determined only in the fresh products.

Total solids content, total protein, fat, ash and crude fibers were determined as described by AOAC [12]. pH values were measured using a pH-meter model HANNA HI9321 microprocessor with a standard, combination glass electrode. Titratable acidity was estimated as percentage of lactic acid according to [14]. The carbohydrate content was calculated in the product according to the following equation:

Carbohydrate % = 100 − (Protein % + Ash % + Fat % + Fiber %).

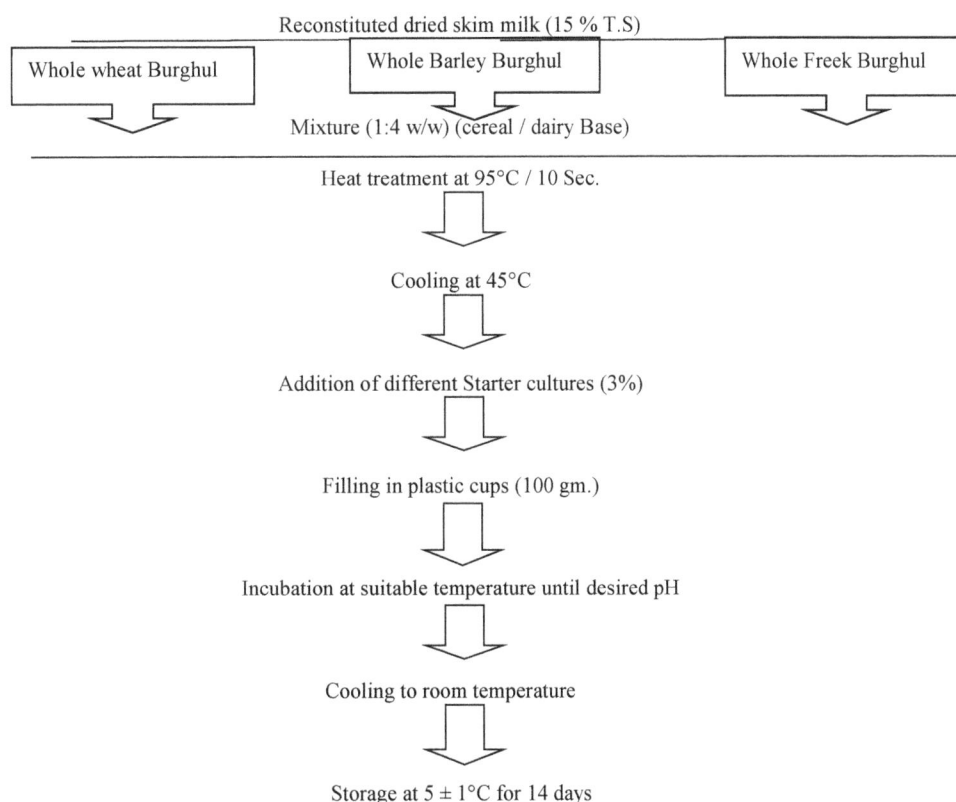

Figure 1: The manufacturing of Kishk like.

Treatments	Cereals			Dairy base	Vanillin
	Whole Wheat Burghul (W)	Whole Barley Burghul (B)	Freek Burghul (F)	Re-constituted Skim milk (15%)	
1	√		-	√	√
	-	√	-	√	√
	-	-	√	√	√
2	√		-	√	√
	-	√	-	√	√
	-	-	√	√	√
3	√	-	-	√	√
	-	√	-	√	√
	-	-	√	√	√

1: (3% Yoghurt starter)
2: (2% Yoghurt starter + 1% Bio-yoghurt starter)
3: (2% Yoghurt starter + 1% *Lactobacillus plantarum*)

Table 1: Experimental treatments.

Composition	Raw material			
	Skim Milk Powder	Wheat whole Burghul	Whole Barley Burghul	Freek Burghul
% Total protein	35.8[1] ± 0.20	4.95[2] ± 0.10	3.81[3] ± 0.20	3.62[4] ± 0.40
% fat	0.81 ± 0.20	.07.. ± 0.50	.050 ± 0.15	.05.0 ± 0.20
% carbohydrates[5]	52.08 ± 0.55	35.99 ± 0.90	31.6 ± 0.50	27.91 ± 0.55
% Ash	7075 ± 0.25	0.67 ± 0.20	0.62 ± 0.30	0.48 ± 0.20
%moisture	5005 ± 0.45	55.71 ± 0.55	60.87 ± 0.50	000.5 ± 0.20
% Crud fiber	ND[6]	1.96 ± 0.15	2.51 ± 0.2	1.44 ± 0.25

[1]Total Protein % = N × 6.38; [2,4] Total Protein % = N × 5.33
[3]Total Protein % = N × 5.36; [5]Calculated by the difference.
[6]Not determined.

Table 2: Characteristics of the raw materials used in manufacture of Kishk like.

Sensory evaluation

Organoleptic evaluation was carried out according to the scheme of Clark et al. [15]. The samples were subjected to organoleptic analysis by well-trained members of the Dairy Science and Technology Department (Fac. Agric. Alexandria Univ., Egypt). The sensory attributes evaluated were: The Flavor (1-10 points), Body and Texture (1-5 points) and appearance and Colour (1-5 points).

Microbiological analysis

Preparation of the samples: Eleven grams of each sample were weighted and transferred thoroughly under condition to sterilized flasks, contained 99 ml 2% sodium citrate solution. The necessary serial dilutions using sterilized distilled water were carried out. The samples were mixed by an electric blender for about 2 min. The following microbial were enumerated.

Lactic acid bacterial count (LAB): Using de Man Rogosa Sharpe Agar medium (MRS) as described by APHA [16]. The plates were incubated at 37°C for 48 hours.

Coliform count: Coliform bacteria were enumerated using Violet Red Bile Agar (VRBA) medium according to Difco [17]. The plates were incubated at 37°C for 24 hours.

Yeasts and moulds count: Sabouraud dextrose agar medium (Oxoid) was used for enumerating yeasts and moulds according to APHA [16]. The plates were incubated at room temperature (20°C-25°C) for 5-7 days.

Lactobacillus acidophilus count: *Lactobacillus acidophilus* was counted on MRS ager (Oxoid) supplemented with L-cystein according to Lapierra et al. [18].

Proteolytic bacteria: Proteolytic bacteria count was enumerated

on nutrient agar (NA) according to APHA [16]. The plates were incubated at 32°C for 38-48 hrs.

Spore forming bacterial count: Representative sample was heated at 80°C for 20 min in water bath and then cooled at the room temperature, the same technique as previously mentioned was followed for enumeration the spore formers but using manitol salt agar medium (MSA) Incubation was done at 32°C for 24 hrs [16].

Statistical analysis

Statistical analysis was performed by applying three ways ANOVA and multiple comparisons of means of each treatment (cereals, starter cultures and storage time) using the Least Significant Difference (LSD) test at the confidence level of 95%. Analysis of data was carried out with SAS [19].

Results and Discussion

Chemical composition

Chemical properties of cereal fermented dairy products are presented in Tables 3 and 4. The results (Table 3) revealed that the effect of cereal type on the chemical composition of the resultant cereal fermented dairy products was more pronounced (P ≤ 0.05) than that of type of starter culture used. There are significant differences (P ≤ 0.05) in pH values and acidity percentages between different cereals fermented dairy products, depending on the type of cereal or starter culture. Data in Table 3 show that the pH values of all samples were decreased gradually till 14 days of the storage, whereas the titratable acidity values were increased at the same period of cold storage. These expected results due to the starter activity similar results were obtained by Hussein [20]. The cereal fermented dairy products containing wheat (W1, W3 and W2, Respectively) were characterized by higher pH as compared with their containing of Barely (B1, B3 and B2) and Freek (F1, F3 and F2, Respectively). However, samples of cereal

Samples	Storage Period (days)	Acidity as lactic acid	pH
S.W.1	1	0.630[NM]	4.87[A]
	7	0.843[FG]	4.64[FE]
	14	0.943[BDC]	4.52[HIJ]
S.W.2	1	0.676[LM]	4.80[AB]
	7	0.883[EF]	4.46[LKMJ]
	14	0.966[ABC]	4.38[ON]
S.W.3	1	0.656[LM]	4.83[AB]
	7	0.863[EFG]	4.53[HI]
	14	0.950[ABC]	4.40[OMN]
S.B.1	1	0.690[LK]	4.80[BC]
	7	0.850[EFG]	4.55[HG]
	14	0.936[DC]	4.48[LKIJ]
S.B.2	1	0.760[IJ]	4.73[CD]
	7	0.896[ED]	4.51[HKIJ]
	14	0.986[AB]	4.44[LKMN]
S.B.3	1	0.750[IJ]	4.76[BCD]
	7	0.886[EF]	4.53[HI]
	14	0.993[A]	4.42[LMN]
S.F.1	1	0.550[O]	4.72[D]
	7	0.673[LM]	4.65[FE]
	14	0.786[IH]	4.52[HIJ]
S.F.2	1	0.603[N]	4.70[ED]
	7	0.776[IHJ]	4.52[HIJ]
	14	0.850[EFG]	4.35[O]
S.F.3	1	0.583[NO]	4.72[D]
	7	0.730[KJ]	4.62[FG]
	14	0.816[HG]	4.51[HKIJ]
SED		0.008	0.012
R-Square		0.990	0.985
Coeff. Var.		1.884	0.467

S.W. = Soft kishk like manufactured from fermented Whole Wheat Burghul Skim milk.
S.B. = Soft kishk like manufactured from fermented Whole Barley Burghul Skim milk.
S.F. = Soft kishk like manufactured from fermented Freek Burghul Skim milk.
Starter = 1: (3% Yoghurt starter), 2: (2% Yoghurt starter + 1% Bio- yoghurt starter) and 3: (2% Yoghurt starter + 1% *Lactobacillus plantarum*).
SED: Standard Error of Difference.

Table 3: Changes in pH values and acidity % of different cereal fermented dairy products during storage at 5 ± 1°C for 14 days.

fermented dairy products that fermented with only Yoghurt starter culture (1) were characterized with higher pH than that fermented with mixed cultures contain Yoghurt and Bio-Yoghurt starter (2) and mixed cultures of Yoghurt starter culture and *Lactobacillus plantarum* (3), respectively. During values of the storage at 5 ± 1°C for 14 days, significant differences (P ≤ 0.05) were recorded in pH of different cereal fermented dairy products. Moreover, a gradual decrease in pH could be observed in all samples of cereal fermented dairy products, with extending the cold storage period. The decrease in pH could be attributed to a limited growth of different bacterial starter cultures and the slow fermentation of residual lactose [20,21].

The fermentation activity can be observed by acidity and/or pH measurements [22]. During fermentation of cereal fermented dairy products, the amount of organic acids other than lactic acid production were small and neglected in acidity calculations [23]. Unlike pH, it is normal expected that the higher acidity rates would lead to lower pH values. The rates of acidity were higher in Barley samples than other samples of Wheat and Fereek [24-27]. The higher acidity of cereal fermented dairy products made with mix of Yoghurt starter culture and *Lactobacillus plantarum* culture can be attributed to the high activity of yogurt starter for splitting lactose into glucose and galactose as the first step of fermentation [28].

Table 4 illustrates the percentages of total solids, carbohydrates, ash, crude fiber, fat and crude protein of cereal fermented dairy products. There were no significant differences (P < 0.05) in the crude protein for all samples, but significant (P ≤ 0.05) in the carbohydrates, ash, crude fiber and fat contents between samples depending on type of cereals usage. The cereal fermented dairy products containing Freek were characterized with higher total solids and carbohydrates whereas, lower content of ash, crude fiber, fat and crude protein contents, compared with the other treatments. Also, depending on the type of starter culture, the cereal fermented dairy products fermented with only Yogurt starter showed very slight increase in the crude protein and carbohydrate contents, but decrease in moisture content than that fermented with mixed starter cultures except moisture content at (S. B. 1) treatment. As expected the crude fibers were higher in barley products than other samples because of its higher percent of crude fiber which reached 2.51%. Also, it ranged from 0.702% to 0.709% in whole barley burghul fermented with reconstituted skim milk, but unlike contain whole wheat burghul and Freek products that ranged between

Sample	Total Solids %	Carbohydrates %	Ash %	Crude Fiber %	Fat content %	Crude protein %
S.W.1	26.66[AB]	19.809[AB]	0.99[A]	0.388[C]	0.240[A]	5.243[A]
S.W.2	26.40[BC]	19.687[B]	0.96[AB]	0.383[C]	0.234[A]	5.143[A]
S.W.3	26.27[C]	19.603[B]	0.95[B]	0.38[C]	0.223[B]	5.128[A]
S.B.1	25.76[D]	18.930[C]	0.88[C]	0.708[A]	0.198[C]	5.18[A]
S.B.2	25.83[D]	18.831[C]	0.87[C]	0.705[AB]	0.196[C]	5.13[A]
S.B.3	25.67[E]	18.815[C]	0.84[C]	0.702[B]	0.190[CD]	5.11[A]
S.F.1	26.69[A]	20.203[A]	0.76[D]	0.285[D]	0.183[DE]	5.08[A]
S.F.2	26.37[BC]	20.165[A]	0.74[D]	0.283[D]	0.183[DE]	5.00[A]
S.F.3	26.38[BC]	20.164[A]	0.75[D]	0.282[D]	0.180[E]	5.00[A]
SED	0.053	0.088	0.007	0.001	0.001	0.057
R-Square	0.955	0.950	0.985	0.999	0.980	0.517
Coeff. Var.	0.353	0.783	1.50	0.441	1.502	1.940

S.W. = Soft kishk like manufactured from fermented Whole Wheat Burghul Skim milk.
S.B. = Soft kishk like manufactured from fermented Whole Barley Burghul Skim milk.
S.F. = Soft kishk like manufactured from fermented Freek Burghul Skim milk.
Starter = 1: (3% Yoghurt starter), 2: (2% Yoghurt starter + 1% Bio-yoghurt starter) and 3: (2% Yoghurt starter + 1% *Lactobacillus plantarum*)
SED: standard Error of Difference.

Table 4: Chemical properties of cereal fermented dairy products

0.381- 0.388 and from 0.282% to 0.285%, respectively. These differences of the chemical properties of cereal fermented dairy products, due to its chemical composition of cereal burghul and blends. It was reported that the type of starter culture used in the fermentation did not affect in the Total solids, fat and crude protein of yogurt, bio-yogurt [20-29]. Also, Salama [30] noticed that the stirred fermented milk made by ABT-4 culture (*Lb. acidophilus, Str. thermophilus* and *B. bifidum*) observed slight decrease in crude protein than made by used YO-FLEX yogurt culture (*Lb. delbruckii spp. lactis, Lb. delbruckii* spp. *bulgaricus and, Str. thermophilus*); this may be due to the limited proteolysis of milk protein by lactic acid bacteria.

Suitability of heat treatments, good hygiene and microbiological analysis of characteristics of different blends

Table 5 illustrates the changes in the viable cell counts appeared on MRS medium for fresh and 14 days stored of cereal fermented dairy products at $5 \pm 1°C$. Slight decrease were observed after 14 days of storage. Generally, the decreasing values were about 0.45 (c.f.u $\times 10^3$ g^{-1}). On the other hand, it can be noticed that the decline on MRS+L-Cysteine medium were less than that obtained in the case of MRS medium. The decreasing in values was 0.32, 0.43 and 0.44 (c.f.u $\times 10^3 g^{-1}$) for S.W. 2, S.B. 2 and S.F. 2 treatments, respectively. All the samples did not contain any growth in 0.1 gm on SDA, VRBA, NA and MSA media in either fresh or stored products through the storage period. These results are revealed the good hygiene sanitation during manufacture different products. From the above results, it could be concluded that the c.f.u on MRS medium (mainly *Bifidobacteria*) sensitive than other starter microorganisms (mainly lactic acid bacteria) for storage.

Medina and Jordano [31] studied the survival of constitutive microflora in one batch of fermented milk containing *Bifidobacteria* during storage at 7°C. Levels of *Streptococcus thermophilus, Lactobacillus bulgaricus* and *Bifidobacterium* spp. in initial population were 2.6 $\times$ 10^8, 5.1 $\times$ 10^7 and 7.4 $\times$ 10^6 c.f.u/ml respectively. *Streptococcus thermophilus* slightly increased after 10 days and then decreased. Numbers of *Bifidobacterium* and *Lactobacillus bulgaricus* decreased faster during storage. Also Shah et al. [32] studied the survival of *Lactobacillus acidophilus* and *Bifidobacterium bifidum* in commercial yoghurt during five weeks period under refrigerated storage. Viable cells of *Lactobacillus acidophilus* were 10^7 to 10^8/g in three of the five products, whereas the other two products contained less than 10^5/g. Initial count of *Bifidobacterium bifidum* was 10^6 to 10^7/g in two of five products, whereas the viable numbers were less than 10^3 /g in other three products. All the products showed a similar decline in the viable count of *Lactobacillus acidophilus* and *Bifidobacterium bifidum* during storage. Krasackoopt et al. [33] investigated the survival of the micro-capsulated probiotics, *Lactobacillus acidophilus* 547, *Bifidobacteria bifidum* ATCC1994 and *Lactobacillus casei* 01 in stirred yoghurt from UHT and conventionally treated milk during low temperature storage at 4°C for 4 weeks. They found that the survival of encapsulated probiotic bacteria was higher than free cells bacteria were mentioned above. Also, they pointed that the viability of probiotic bacteria in yoghurt from both treatments was not significantly different.

The cereals can be used as fermentable substrates for the growth of probiotic microorganisms. Also, *Lactobacillus acidophilus* exhibited the poorest growth in malt, barley and wheat media, probably due to substrate deficiency in specific nutrients [25-34]. The optimum final

Samples	Storage Period (Days)	Media					
		MRS	MRS+L- Cysteine	NA	SDA	VRBA	MSA
S.W.1	1	2.80	----				
	7	2.68	----				
	14	2.40	----				
S.W.2	1	2.80	2.56				
	7	2.72	2.46				
	14	2.44	2.24				
S.W.3	1	2.84	----				
	7	2.72	----				
	14	2.44	----				
S.B.1	1	2.92	----				
	7	2.72	----				
	14	2.44	----				
S.B.2	1	2.88	2.64				
	7	2.72	2.44		N.D		
	14	2.48	2.21				
S.B.3	1	2.80	----				
	7	2.68	----				
	14	2.48	----				
S.F.1	1	2.72	----				
	7	2.52	----				
	14	2.28	----				
S.F.2	1	2.84	2.56				
	7	2.68	2.42				
	14	2.48	2.12				
S.F.3	1	2.76	----				
	7	2.64	----				
	14	2.40	----				

S.W. = Soft kishk like manufactured from fermented Whole Wheat Burghul Skim milk.
S.B. = Soft kishk like manufactured from fermented Whole Barley Burghul Skim milk.
S.F. = Soft kishk like manufactured from fermented Freek Burghul Skim milk.
Starter: 1: (3% Yoghurt starter), 2: (2% Yoghurt starter + 1% Bio-yoghurt starter) and 3: (2% Yoghurt starter + 1% *Lactobacillus plantarum*)
ND: Not Detected in 0.1 gm; (--): Not determined; VRBA: Violet Red Bile Agar; NA: Nutrient Agar; MSA: Manitol Salt Agar; SDA: Sabouraud Dextrose Agar; MRS: Man Rogosa Sharpe Agar

Table 5: Changes in viable microbial counts (c.f.u $\times 10^{-3}$/ g) in cereal fermented dairy products during storage at 5 ± 1°C.

Sample	Storage Period (days)	Flavor (10)	Body/texture (5)	Appearance and color (5)	Total (20)
S.W.1	1	7.83[AB]	4.00[A]	4.33[A]	16.16[A]
	7	7.66[AB]	3.83[A]	3.83[ABC]	15.32[AB]
	14	6.00[BCDEF]	3.66[A]	3.66[ABC]	13.32[ABCD]
S.W.2	1	7.66[AB]	4.50[A]	4.33[A]	16.49[A]
	7	7.00[ABCD]	4.00[A]	4.00[AB]	15[AB]
	14	6.16[ABCDEF]	3.83[A]	3.5[ABC]	13.49[ABCD]
S.W.3	1	7.33A[BC]	4.33[A]	4.33[A]	15.99[AB]
	7	7.00[ABCD]	3.83[A]	4.16[AB]	14.99[AB]
	14	5.83[BCDEF]	3.66[A]	3.83 [ABC]	13.32[ABCD]
S.B.1	1	6.00[BCDEF]	3.66[A]	3.5[A]	13.16[ABCD]
	7	5.00[DEF]	3.66[A]	2.5[ABC]	11.16[BCD]
	14	4.33[EF]	3.00[A]	2.33[BC]	9.66[CD]
S.B.2	1	5.66[BCDEF]	3.66[A]	3.16[ABC]	12.48[ABCD]
	7	5.00[DEF]	3.66[A]	2.5[ABC]	11.16[BCD]
	14	4.00[F]	2.66[A]	2.33[BC]	9.33[CD]
S.B.3	1	6.33[ABCDE]	4.00[A]	3.5[ABC]	13.83[ABCD]
	7	5.00[DEF]	3.66[A]	2.5[ABC]	11.16[BCD]
	14	4.00[F]	3.00[A]	2.00[C]	9.00[D]
S.F.1	1	8.33[A]	4.00A	4.33[A]	16.66[A]
	7	6.33[ABCDE]	4.00A	4.33[A]	14.66[AB]
	14	5.66[BCDEF]	3.83A	3.5[ABC]	12.99[ABCD]
S.F.2	1	7.83[AB]	4.00[A]	4.33[A]	16.16A
	7	6.33[ABCDE]	3.66[A]	4.00[AB]	13.99[AB]
	14	5.16[CDEF]	3.00[A]	3.66[ABC]	11.82[ABCD]
S.F.3	1	7.83[AB]	4.00[A]	4.33[A]	16.16[A]
	7	7.00[ABCD]	4.00[A]	3.66[ABC]	14.66[ABCD]
	14	6.00[BCDEF]	3.33[A]	3.5[ABC]	12.83A[BCD]
SED		0.420	0.340	0.343	0.883
R-Square		0.805	0.383	0.679	0.754
Coeff. Var.		11.686	15.81	16.93	11.34

S.W. = Soft kishk like manufactured from fermented Whole Wheat Burghul Skim milk.
S.B. = Soft kishk like manufactured from fermented Whole Barley Burghul Skim milk.
S.F. = Soft kishk like manufactured from fermented Freek Burghul Skim milk.
Starter = 1: (3% Yoghurt starter), 2: (2% Yoghurt starter + 1% Bio-yoghurt starter) and 3: (2% Yoghurt starter + 1% *Lactobacillus plantarum*).
SED: Standard Error of Difference.

Table 6: Organoleptic properties of cereal-based fermented dairy products, during storage at 5 ± 1°C for 14 days.

pH and the concentration of lactic and acetic acid in fermented cereal product in relation to the properties of each specific probiotics strain have to be investigated in order to maximise the viability during storage for practical application; a pH value of the final product must be maintained above 4.6 to prevent the decline *Bifidobacteria* population [35].

The sensory evaluation

The sensory evaluation is the main factor affecting the consumption of the products and its acceptance. It was realized that the sensory evaluation could contribute pertinent, valuable information related to marketing consequences and simultaneously provide direct actionable information [15]. The scores for organoleptic properties of cereal fermented dairy products prepared using whole wheat burghul or whole barley burghul or freek burghul with different types of starter cultures, during storage at 5 ± 1°C for 14 days were presented in Table 6 and Figure 2. It was clear that in fresh products (first day of storage), the addition of wheat or Freek burghul didn't affect the general acceptability of them, whereas the addition of barley burghul had lowering the total score. General acceptability of the cereal fermented dairy products containing Freek; especially when fermented with yogurt culture (S.F.1) treatment; was gained the highest scores in the organoleptic properties followed by their containing of wheat (S.W.2) product. The cereal fermented dairy products containing of Freek were characterized with perfect flavor, body and texture as well as whiteness appearance and colour. On the other hand, the pronounced malt flavor and light brown color or the decline of color to un-natural, were noticed

in cereal fermented dairy products containing of barely; especially at (S.B.2) treatment, these results are in agreement with [20]. During storage at 5 ± 1°C for 14 days, Sensory evaluation was decreased for all products due to the acidity increased in all samples gradually and effect on organoleptic properties and decreased its total score. Vijayalakshmi et al. [36] found that during storage of cereal based low fat fruit yogurt, acidic or malt flavor, firm or ropy body and texture, shrunken or free whey appearance, as well as light brown color were increased in different cereal fermented dairy products at the end of storage.

At the end of storage time, whole wheat burghul treatments had the highest total scores and then, the Freek burghul samples, while the addition of Whole barley burghul had the lowest total score because of high viscosity and unnatural color and still not acceptable at the end of storage. Blanddino et al. [8] mentioned that during cereal fermentations several volatile compounds are formed, which contribute to a complex blend of flavors in the products [37]. Moreover, the presence of aromas representative of Diacetyl acetic acid and butyric acid make fermented cereal-based products more appetizing. Also, Salmeron et al. [38] found that incubation with the probiotic LAB caused a significant change in the aroma profile of the four cereal broths. In barely, considerable amounts of new volatiles were generated after the fermentation. In general, the volatile production depends more on the substrate than on the microorganisms.

Conclusion

From the above results it could be concluded that either Whole

S.W. = Soft kishk like manufactured from fermented Whole Wheat Burghul Skim milk.
S.B. = Soft kishk like manufactured from fermented Whole Barley Burghul Skim milk.
S.F. = Soft kishk like manufactured from fermented Freek Burghul Skim milk.
Starter = 1: (3% Yoghurt starter), 2: (2% Yoghurt starter + 1% Bio-yoghurt starter) and 3: (2% Yoghurt starter + 1% *Lactobacillus plantarum*).

Figure 2: The pictures of soft Kishk like products.

Wheat Burghul, Whole Barley Burghul and Whole Freek Burghul can be used to produce an acceptable product as functional foods suitable for elderly persons or infants weaning foods. This formula has a high nutritional value and fiber content beside the presence probiotic bacteria with a lot of health benefits.

References

1. Metchnikoff E (1907) Essays optimists Paris: The prolongation of life. Optimistic studies. Translated and edited by P Chalmers Mitchell. London Heinemann.

2. Fuller R (1989) Probiotics in man and animals. J Appl Bacterial 66: 365-378.

3. Fuller R (1991) Probiotics in human medicine. Gut 32: 439-442.

4. Jones PJ (2002) Clinical nutrition 7: Functional foods- More than just nutrition. Canadian Med Assoc J (CMAJ) 166: 1555-1563.

5. Charalampopoulos D, Wang R, Pandiella SS, Webba C (2002b) Application of cereals and cereal components in functional foods: a review. Int J food Microbiol 79: 134.

6. Hambraeus L (1992) Nutritional aspects of milk proteins. In Fox PF (Ed) Advanced dairy chemistry 1: proteins. Elsevier Applied Science. London: 457-490.

7. Gomez-Ruiz JA, Ramos, Reciol I (2002) Angiotensin converting enzyme inhibitory peptides in Manchego cheeses manufactured with different starter cultures. Int Dairy J 12: 697-706.

8. Blandino A, AL-Aseeria ME, Pandiella SS, Canterob D, Webba C, et al. (2003) Review cereal-based fermented foods and beverages. Food Res Int 36: 527.

9. Lorri W, Svanberg U (1993) Lactic fermented cereal gruels with improved in vitro protein digestibility. Int J Food Sci Nutr 44: 29.

10. Sanni AI, Ohilude AA, Ibidabpo OT (1999) Biochemical composition of infant weaning food fabricated from fermented blends of cereals and soybean. Food Chem 65: 35.

11. Damiani P, Gobbetti M, Cossignani L, Corsetti A, Simonetti MS, et al. (1996) The sourdough microflora. Characterisation of hetero and homofermentative lactic acid bacteria, yeasts and their interaction on the basis of the volatile compounds produced. LWT Food Sci. and Technol 29: 63.

12. AOAC (2007) Official methods of analysis (18thedn). Association of Official Agricultural Chemists, CH, Washington D.C., USA 34: 72-80.

13. AACC (2003) Approved methods of the American Association of Cereal Chemists (8thedn.). AACC, St Paul, USA.

14. Richardson HG (1986) Standard methods for the examination of dairy products (15thedn.). American Public Health Association Inc, Washington, USA.

15. Clark S, Costello M, Drake M, Body-felt F (2009) The Sensory evaluation of dairy products. Springer.

16. APHA (1992) American Public Health Association standard method for the examination of dairy products (16thedn.). Washington DC, USA.

17. Difco (1984) Manual of dehydrated culture media and reagent for microbiological and clinical laboratories. Detroit, Michigan, USA.

18. Lapierra LP, Undeland P, Cox LJ (1992) Lithium chloride-Sodium propionate agar for the enumeration of Bifidobacteria in fermented dairy products. J Dairy Sci 75: 1192.

19. SAS (2013) Statistical Analysis System User Guide Version 9.3. SAS Institute Inc. Cary, NC, USA.

20. Hussein GAM (2011) Production and properties of some cereal-based functional fermented dairy products. Egyptian J Dairy Sci 39: 89-100.

21. Barrantes E, Tammime AY, Muir DD, Swoed AM (1994) The effect of substitution of fat by micro-particulated whey protein on the quality of set-type natural yoghurt. J Soci Dairy Technol 47: 61.

22. Hesseltine CW (1979) Some important fermented foods of Mid-Asia the Middle East and Africa. J of American Oil Chemists' Soci 56: 367-374.

23. Damir AA, Salama AA, Mohamed MS (1992) Acidity microbial organic and free amino acid development during fermentation of skimmed milk, kishk. Food Chemi. 43: 265-269.

24. Mehanna AS, Hefnawy SA (1990) A Study to follow the chemical changes during processing and storage of zabady. Egypt J Dairy Sci 18: 425-434.

25. Mehanna AS (1991) An attempt to improve some properties of zabadi by applying low temperature long incubation period in the manufacturing process. Egypt Dairy Sci 19: 221-229.

26. Abou-Donia SA, Attia IA, Khattab AA, El-Shenawy Z (1991) Studies on the formation of fermented milk for infantile and geriatric nutrition. Egypt J Dairy Sci 19: 283-299.

27. Kailasapathy K, Rybka S (1997) *Lactobacillus acidophilus* and *Bifidobacterium* spp. their therapeutic potential and survival in yoghurt. Australian J Dairy Tech 52: 28

28. Tamine Ay, Rbinson Rk (1985) Yoghurt Science and technology pergamon press. Publisgar Robert Maxuell MG London.

29. Aklain AS (1996) L (+) D (-) Lactic acid content and aroma profile in bioghurt bifigurt biograde in comparison with yougurt. Egypt J Dairy Sci 24: 227.30.

30. Salama FM (2002) Production of therapeutic and diabetic stirred yoghurt-like fermented milk products. Egypti. J Dairy Sci 30: 177.

31. Medina LM, Jordano R (1994) Survival of constitute microflora in commercially fermented milk containing Bifidobacteria during refrigerated storage. J Food Prot 58: 70-75 Cited from Int Dairy J 7: 349-356.

32. Shah NP, Lankaputhra WEV, Britz ML, kyle WSA (1995) Survival of *lactobacillus* and *Bifidobacterium* bifidum in commercial yoghurt during refrigerated storage. Int. Dairy J 5: 515-521.

33. Krasackoopt W, Bhandari B, Deeth HC (2004) Survival of probiotic encapsulated in chitsason-coated alginate beads in yoghurt from UHT. And conventionally treated milk during storage. LWT Food Sci Technol 39: 177-183.

34. Charalampopoulos D, Pandiella SS, Webba C (2002a) Growth studies of potentially probiotic lactic acid bacteria in cereal based substrates. J Appl Microbiol 92: 851.

35. Vinderola CG, Bailo N, Reinheimer JA (2000) Survival of probiotic microflora in Argentinian yougurts during refrigerated storage. Food Res Int 33: 97.

36. Vijayalakmi R, Nareshkumar C, Dhanalashmi B (2010) Storage studies of cereal based low fat fruit yoghurt. Egypti J Dairy Sci 38: 53-61.

37. Chavan Jk, Kadam SS (1989) Nutritional improvement of cereals by fermentation. Crit Rev Food Sci Nutr 28: 349.

38. Salmeron I, Charalampopoulos D, Pandiella SS (2009) Volatile compounds produced by the probiotic strain Lactobacillus Plantarum NCIMB 8826 in cereal-based substrates. Food chem 117: 265.

16

Effect of Post-harvest Quality Parameters on Ultra-Sonication Treatment of Khoonphal (*Haematocarpus validus*) of Meghalaya, North-East India

Sasikumar R*, Vivek K, Chakaravarthy S and Deka SC

Department of Agri-Business Management and Food Technology, North-Eastern Hill University, Meghalaya, India

Abstract

Freshly harvested khoonphal (*Haematocarpus validus*) were surface cleansed with ultrasonic treatment. The process variables i.e., ultrasonic amplitude, treatment time and temperature selected were optimized using response surface methodology (RSM) by three factor three level Box-Behnken design. Horn type ultra-sonicator with a power density of 460 W/cm^2 with a constant frequency of 30 Hz was used for all the 17 experiments. Optimum independent variables selected by RSM were ultrasonic amplitude (100%), treatment time (5.10 min) and solvent temperature (25°C). The corresponding optimum values for dependent variables obtained were total plate count (2.94 log CFU/cm^2), firmness (66.67 N) and respiration rate (42.32 N). Linear terms for all the dependent variables were found to be significant ($p<0.05$). Similarly, the interaction terms between ultrasonic amplitude and treatment time had showed significant negative effect on total plate count ($p<0.001$) and firmness ($p<0.05$). But significant positive effect was obtained for respiration rate ($p<0.100$). Therefore, from this study it was concluded that the ultra-sonication was found to be an effective technology in reducing surface microbial load. Hence, this may be used for extending the shelf-life and maintaining the quality of freshly harvested khoonphal, while RSM was proven to be an effective technique in controlling and optimizing the factors responsible for ultrasonic treatment.

Keywords: Khoonphal; RSM; Ultrasonication; Total plate count; Firmness; Respiration rate

Introduction

Khoonphal (*Haematocarpus validus*) is one of the rarest rather extinct fruits recoup after a period of 100 years in India. It added a recent disclosure of its habitats in Mawlakhieng village (Meghalaya, India) [1,2]. It is commonly known as Khoonphal/Raktaphal and is a rich source of iron, vitamin A, various alkaloids and functional phytochemicals. This fruit is famous for their traditional medicinal usages for treating jaundice, cancer, hypertension, arthritis, neurological problems, etc. [3-5]. Bioactive compounds present in this fruit neutralize free radical species generated as a part of biochemical reactions in our body system [6]. Accessibility and availability of these bioactive compounds from dietary sources contribute significantly in traditional health system of tribal and rural population of developing world [7].

Haematocarpus validus is an evergreen perennial creeping woody climber capable of growing under extreme conditions, from very dry environments to highly acidic soils. The tiny, odorous, greenish white flowers unisex and dioecious produce fruit. The vines of the tree produce flower in middle of November-January. The fruits mature in April-May and the fruiting season is May to August [8]. This fruit is highly perishable and has short shelf life i.e., 4-5 days at room temperature. It spoils rapidly during harvesting, storage and transportation due to surface bruising, senescence and surface microbial decay. The application of fungicides should be nullified due to the adverse effects on human health and environment [9]. Therefore, there is an urgent need to develop a technology to maintain the quality and freshness of khoonphal for achieving the longer shelf life.

Ultrasound is an effective technology used in many industries including food [10,11]. It is a non-thermal technology which has wide spread applicability in heat-sensitive foods to retain sensory, functional and nutritional characteristics along with enhanced shelf life. Ultrasound is composed of sound waves with a frequency beyond the limit of human hearing. It is considered safe, environmental friendly and nontoxic. Among food industries fruits and vegetable industry has huge scope in using ultrasound to generate contamination-free products. Various authors have showed the effectiveness of ultrasonic cleaners in eliminating the contaminants and microorganisms present on objects, including sludge, mold, bacteria, fungi, worms and agrochemicals [9,12-14]. Other applications of ultrasound include degassing, inactivation of enzymes, crystallization, leaching, extraction, digestion, etc. [15]. Low power (high frequency) ultrasound controls the food properties by monitoring the physicochemical properties and composition during storage and processing, which is also crucial for improving food quality and safety. This technology is relatively simple, cheap and energy saving hence it is applicable for improving the shelf life of post harvested fruits [9,12]. During ultra-sonication treatment various factors viz. frequency, solvent temperature, percentage of amplitude, treatment time, viscosity of the solvent, ultrasonic input and output power, etc. may affect the efficacy of the treatment [12]. Controlling all these variables is difficult therefore; a robust and potent optimization tool is required for determining the effects of both individual operational factors and their interactions [16]. Response surface methodology (RSM) is a simple and widely used technique in food engineering fields for optimizing food manufacturing operations and preservation techniques i.e., fresh cut lettuce, pear, kiwifruit, strawberry [9,12,17-19].

However, there has been no report published on effect and optimization of ultra-sonication on post-harvest khoonphal for maximizing the shelf life. Therefore, our aim of this study was to optimize the ultrasonic treatment and to study the sonication effect

---

*Corresponding author: Sasikumar R, Department of Agri-Business Management and Food Technology, North-Eastern Hill University, Meghalaya, India
E-mail: sashibiofoodster@gmail.com

on total surface plate count, respiration rate, fruit firmness and some selected quality parameters. Independent variables considered for this study includes ultrasonic amplitude, treatment time and temperature.

Materials and Methods

Materials

The freshly ripened fruits (Khoonphal) were harvested and collected from the study area (West Garo Hills, Tura, Meghalaya, North-East India) in the month of May 2016. The good quality fruits were selected for the experimentation i.e., uniform size, absence of defects and visual wounds. Then the fruits were taken to the laboratory within 2 h from the time of harvest. The selected khoonphal had an initial total soluble solid (TSS) of 16 ± 1 °Brix and moisture content (M.C) of $85.00 \pm 1.00\%$ w.b. (wet basis). Khoonphal collected for experiments were shown in the Figure 1.

Optimization of ultrasonic treatment

Probe ultrasonicator (BBI-8535027, Sartorius Labsonic M, Germany) with a constant frequency of 30 kHz having a maximal output power density of 460 W/cm^2 was used for treatment of khoonphal. A 3-mm titanium probe with maximum amplitude of 180 μm was used for experimental purpose. Experimental combination with different independent variables viz. amplitude, time and temperature was set according to design obtained from RSM – Box Benkhan model. The probe was immersed into solvent (distilled water) by 25 mm. While the

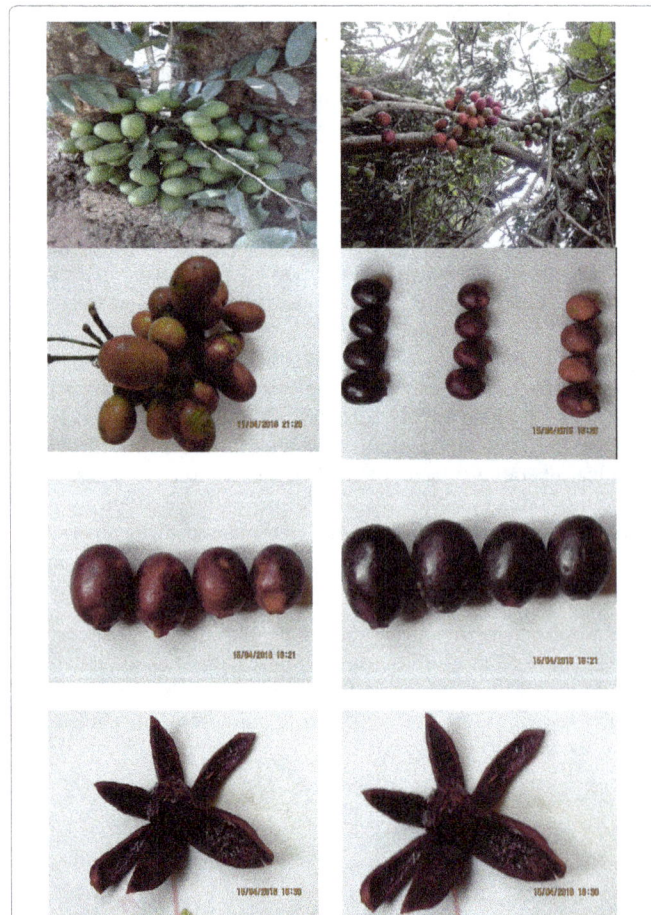

Figure 1: Khoonphal at different stages of growth from West Garo Hills, Meghalaya.

Independent Variables	Level		
	-1	0	1
Ultrasonic amplitude (X1)	60	80	100
Treatment time (X2)	5	10	15
Temperature (X3)	25	37.5	50

Table 1: Independent variables and their level used for central composite design.

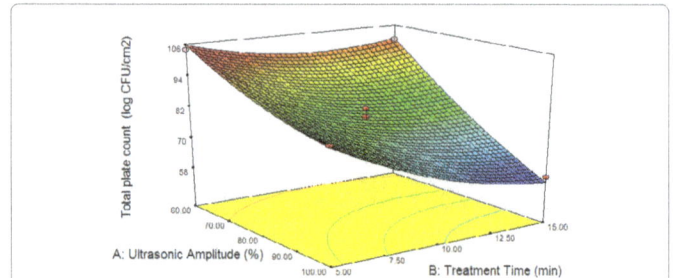

Figure 2: Effect of ultrasonic amplitude and treatment time on total plate count.

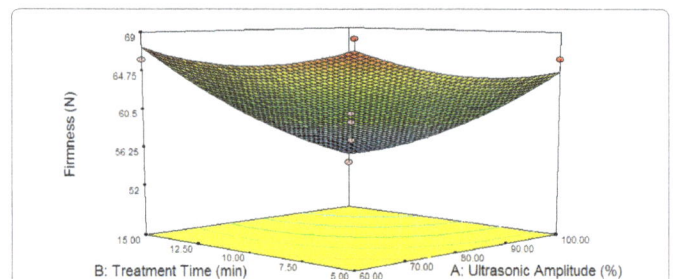

Figure 3: Effect of ultrasonic amplitude and treatment time on firmness.

Cut-off cycle time of 0.5 s was fixed for all the experiment to control the temperature of the solvent. 300 ml of solvent was used to treat 80-90 gram fruits (ratio 1: 3.5). Ultrasonication treatment at maximum amplitude (100%) beyond 15 min would rupture the fruit tissues and severely loosens its firmness. After the treatment fruits were removed immediately and taken for further analyses. Untreated fruit samples are used as a control. All the experiments were conducted twice with two replicates of each treatment per experiment.

Experimental design

Response Surface Methodology (RSM) is considered as an effective optimization tool used to optimize the levels of independent variables. Screening designs were carried out to eliminate the minor variables which are not important during experiment. Therefore, independent variables with major effects on dependent variables (total plate count, respiration rate and firmness) were selected for optimization which includes ultrasonic amplitude, treatment time and temperature. Software design expert (version 7.00, Stat-Ease Inc., Minneapolis, MN) was used to construct model and analyse data. An efficient three-level-three-factor, Box-Behnken design was employed with 17 experimental runs with four replicates at the centre point. The range and centre point values of the independent variables were given in the Table 1. Second order polynomial equation was used to express the dependent variables as a function of independent variables as follows:

$$Y = \beta_0 + \sum_{j=1}^{k} \beta_1 X_j + \sum_{j=1}^{k} \beta_{jj} X_j^2 + \sum \sum_{i<j} \beta_{ij} X_i X_j$$

Where Y is the predicted variable/response, X_i and X_j are the independent variables. While β_0 is the constant coefficient, β_i, β_{ij} and β_{jj} are the regression coefficients for the linear, interaction and quadratic, respectively. The Coefficients obtained were interpreted using the F test. Regression analysis, analysis of variance (ANOVA) was also performed to establish optimum conditions for ultra-sonication treatment for khoonphal. The surface plotting's for the optimized results were shown in Figures 2-4.

Analogy experiment

Optimal conditions (ultrasonic amplitude: 100%, temperature: 25°C and treatment time: 5.10 min) obtained from RSM was used to compare with untreated fruits as a control sample. After ultra-sonication fruits were then immediately analysed for total plate count, firmness, respiration rate and vitamin C. There were three replicates of 80 ± 1 g of fruit each per treatment, and the same experimental combinations was conducted twice. Finally treated fruits were subjected to storage study after vacuum and normal packaging for 15 days (unpublished data).

Total plate count

Total plate count (TPC) was examined according to Pao et al. [20] with minor modification. 100 grams of ultra-sonicated sample was placed into sterilized bags consist of 1 litre of 0.1% (w/v) peptone solution. Then the sterilized sample bags were mixed thoroughly with the help of reciprocal shaker with 100 oscillations/ min at $6 \pm 1°C$ for 3 h. After shaking, the wash solutions obtained were then taken immediately for enumeration of TPC. Appropriate dilutions (1:10) required for sampling (sample plating) were made with 0.1% (w/v)

peptone solution. Each wash solution was then surface plated on plate count agar (PCA) and incubated for 48 h at 35°C. The results were expressed in colony forming unit per square centimeter (CFU/cm^2).

Respiration rate

Respiration rate was measured by sealing 100 ± 1 g fruits into 1 litre plastic container. The container used was fitted with an airtight rubber septum and held at $25 \pm 1°C$ for 1 h. Respiration rate was measured in accordance with Vivek et al. [9] and the experiments were conducted thrice. The ultra-sonicated samples were then taken for measuring respiration rate using gas analyser (Checkmate 2, PBI, dansensor, Ringsted, Denmark). 3 ml head space gas (O_2 and CO_2) in container was taken by the gas analyser for respiration rate analysis. The results were then expressed in mg CO_2 kg^{-1} h^{-1} fresh weight (FW).

Firmness and vitamin C

Firmness of the ultra-sonicated khoonphal was measured accordance with Vivek et al. [9]. Texture analyzer (TA-HD plus, Stable Micro Systems, UK) was used to perform puncture test. The load cell was equipped with a 3 mm diameter stainless steel (SS) probe, at a constant speed of 3 mm s^{-1} to a depth of 0.20 mm. The peak puncture force was treated as firmness of khoonphal and is measured in netwton (N) [21,22]. While the vitamin C content was assayed by titration method (2,6-dichlorophenolindophenol titration method). The results of vitamin C were expressed as mg/100 g of FW [21].

Statistical analysis

All the experimental results obtained were statistically analysed by applying independent sample t-test using SPSS v16 for inspecting the significant differences in the mean values of dependent variables for both the control and ultra-sonicated samples. The mean absolute error (MAE) and root mean square error (RSME) were also calculated to find out the difference between predicted and experimental/observed values for describing the performance of the model. This also shows the deviation of predicted values to the experimental values. The formula used for calculating MAE and RSME were shown in Eqs. (1) and (2).

$$MAE = \frac{1}{N} \sum_{i=1}^{N} |R_{real} - R_p| \tag{1}$$

$$RMSE = \left\{ \frac{\sum_{i=1}^{N} (R_p - R_{real})^2 |}{N} \right\}^{1/2} \tag{2}$$

Where R_p is the predicted value; R_{real} is the experimental/observed value; N is the number of points.

Results and Discussion

Model fitting

Mean values of all the selected dependent/response variables were shown in Table 2. Experimental data was used to obtain all the coefficients of second order polynomial equation for finding the significance of various coefficients of the model. The best fit of the experimental data to the regression model equation was finalised according to coefficients of multiple determinations (R^2), adjusted coefficients of multiple determinations (Adj R^2), mean average error (MAE), root mean square error (RSME), coefficient of variance (CV). Lack of fit for all the dependent variables were found to be insignificant, this indicates the error analysis obtained from RSM among centre points in the experimental combinations was minimum. The linear

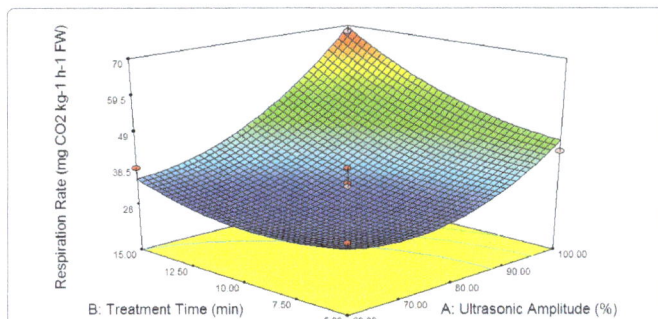

Figure 4: Effect of ultrasonic amplitude and treatment time on respiration rate.

Experimental no	X1	X2	X3	Total plate count (log CFU/cm²)	Respiration Rate (mg CO₂ kg⁻¹ h⁻¹ FW)	Firmness (N)
1	1	0	-1	2.81	58.15	60
2	0	-1	-1	2.95	30.87	65
3	-1	0	-1	2.97	29.08	68
4	0	1	-1	2.85	58.02	61
5	0	1	1	2.88	33.92	59
6	1	0	1	2.77	58.15	54
7	0	0	0	2.90	32.95	60
8	0	0	0	2.91	30.87	61
9	1	-1	0	2.93	43.62	66
10	-1	1	0	3.00	38.77	66
11	-1	0	1	3.00	29.08	64
12	0	-1	1	2.95	38.77	62
13	1	1	0	2.78	67.85	52
14	0	0	0	2.88	32.95	57
15	-1	-1	0	3.02	33.92	69
16	0	0	0	2.92	33.92	63
17	0	0	0	2.89	38.77	59

Table 2: Box behnkan design matrix and response values.

terms of the independent variables for all the dependent variables were found to be significant (p<0.05) which is shown in Table 3. Apart from this few quadratic and interaction terms also showed significant (p<0.05) effect and were shown in Table 3. Similar kind of results were shown by Cao et al. [12] and Vivek et al. [9] for kiwifruit and strawberry fruits, respectively. The number of experimental trails needed to assess and construct the model for multiple variables and their interactions was easily done by using RSM. Finally, the models constructed were statistically measured to describe the deviation in the data.

RSM analysis

Total plate count: The overall model for the total plate count had showed significant difference at p<0.001 with F value 37.38. And the lack of fit for total plate count had showed non-significant difference. The independent variables ultrasonic amplitude and treatment time had showed the negative effect on total plate count. i.e., total plate count decreases with the increase in ultrasonic amplitude and treatment time. Both the independent variables had showed significant difference at p<0.001. The interaction terms between ultrasonic amplitude and treatment time had showed the significant negative effect on total plate count at p<0.001. The other interaction terms between ultrasonic amplitude and temperature had also showed the significant negative effect on total plate count at p<0.100. The quadratic terms for ultrasonic amplitude and treatment time had showed significant positive difference at p<0.100 and p<0.05 respectively. But the quadratic term for temperature had showed the significant negative effect on the total plate count at p<0.100. The coefficient of determination and adjusted coefficient of determination values (R^2=0.98 and Adj R^2=0.95) resulted high for TPC, which indicates the model fits extremely well. The RMSE and MAE values were also calculated (RMSE=1.76 and MAE=1.50), which tells the deviation in the experimental data. The maximum total plate count was observed at 60% ultrasonic amplitude, 5 min and 37.5°C and the minimum total plate count was observed at 100% ultrasonic amplitude, 10 min and 50°C. Similar kind of results were obtained for kiwifruit and strawberry [9,12]. This microbial reduction may be due to cavitation bubbles, localized temperature and pressure occur in solvent during ultrasonication [12].

Constraints Name	Goal	Lower limit	Upper limit	Lower weight	Upper weight	Importance
Total plate count	Minimize	2.77	3.02	1	1	5
Respiration rate	Minimize	52	69	1	1	4
Firmness	Maximize	29.08	67.85	1	1	3

Table 4: Responses and limits of optimizer for optimization using numerical optimization in design expert.

Respiration rate: The overall model for the Respiration rate had showed significant difference at p<0.05 with F value 14.67. And the lack of fit for respiration rate had showed non-significant difference. Independent variables ultrasonic amplitude and treatment time had showed the positive effect on respiration rate. i.e., Respiration rate increases with the increase in ultrasonic amplitude and treatment time. Both the independent variables had showed significant difference at p<0.001. The interaction terms between ultrasonic amplitude and treatment time had showed the significant positive effect on respiration rate at p<0.100. The other interaction terms between treatment time and temperature had also showed the significant negative effect on respiration rate at p<0.05. The quadratic terms for ultrasonic amplitude and treatment time had showed significant positive difference at p<0.050 and p<0.100 respectively. The coefficient of determination and adjusted coefficient of determination values (R^2=0.95 and Adj R^2=0.88) resulted high for respiration rate, which indicates the model fits extremely well. The RMSE and MAE values were also calculated (RMSE=3.50 and MAE=2.20), which tells the deviation in the experimental data. The maximum respiration rate was observed at 100% ultrasonic amplitude, 15 min and 37.5°C and the minimum respiration rate was observed at 60% UA, 10 min and 25°C. Ultrasonication at higher power and longer treatment time for kiwifruits have showed higher respiration rates [9]. This may be due to the due to the rupturing of fruit tissues and cell wall degradation [23,24]. Higher respiration rates indicate a more active metabolism and usually a faster deterioration rate [25]. Heat treated mango fruits showed higher respiration rate [26].

Firmness: The overall model for the firmness had showed significant difference at p<0.05 with F value 6.71. And the lack of fit for firmness had showed non-significant difference. The Independent Variables ultrasonic amplitude and treatment time had showed the negative effect on firmness. i.e., Firmness decreases with increase in ultrasonic amplitude and treatment time. Both the independent variables had showed significant difference at p<0.001. The interaction terms between ultrasonic amplitude and treatment time had showed the significant negative effect on firmness at p<0.05 i.e., combined effect of ultrasonic amplitude and treatment time decreases the fruit firmness. Similar kind of results were reported for strawberry and kiwifruit [9,12]. The other interaction terms showed non-significant difference on fruit firmness. The quadratic terms for ultrasonic amplitude, treatment time and temperature had also showed non-significant difference on fruit firmness. The coefficient of determination and adjusted coefficient of determination values (R^2=0.90 and Adj R^2=0.76) resulted high for fruit firmness, which indicates the model fits well. The RMSE and MAE values were also calculated (RMSE=1.05 and MAE=1.20), which tells the less deviation in the experimental data. The maximum firmness was observed at 60% ultrasonic amplitude, 5 min and 37.5°C and the minimum firmness was observed at 100% ultrasonic amplitude, 15 min and 37.5°C. Various authors have reported that the cell wall degradation and tissue rupturing is mainly due to the disturbances in cell wall constituents i.e., polygalacturonase and pectin methylesterase [9,12,27,28].

Coefficients	Total plate count	Respiration	Firmness
β_0	79.60	33.89	60.00
X_1 (β_1)	-16.00***	12.12***	-4.38***
X_2 (β_2)	-7.87***	6.42***	-3.00***
X_3 (β_3)	0.87	-2.02	-1.88*
X_1X_2 (β_4)	-5.25***	4.85*	-2.75**
X_1X_3 (β_5)	-3.25*	0.00	-0.50
X_2X_3 (β_6)	1.50	-8.00**	0.25
X_1^2 (β_7)	2.70*	7.68**	1.50
X_2^2 (β_8)	4.95**	4.46*	1.75
X_3^2 (β_9)	-3.05*	2.04	0.00
R^2	0.98	0.95	0.90
Adj R^2	0.95	0.88	0.76
Pred R^2	0.88	0.40	0.18
Adeq precision	21.27	13.04	8.51
Std dev	2.93	4.14	2.26
C.V	3.58	10.21	3.67
RMSE	1.76	3.50	1.05
MAE	1.50	2.20	1.20
Lack of fit	N.S	N.S	N.S

* Significant at p<0.1.
** Significant at p<0.05.
*** Significant at p<0.001.

Table 3: Regression coefficient for the responses.

Variables	Optimized conditions
Ultrasonic amplitude (%)	100.00
Treatment time (min)	05.10
Temperature (°C)	25.00
Total plate count (log CFU/cm²)	02.93
Respiration rate (mg CO₂ kg⁻1 h⁻1 FW)	43.16
Firmness (N)	67.06

Table 5: Optimized solution – response optimizer in design expert.

Responses	Ultrasound treated sample (optimized conditions)	Un treated sample	p-value
Total plate count (log CFU/cm²)	2.9433	3.9767	0.00*
Respiration rate (mg CO₂ kg⁻1 h⁻1 FW)	42.3167	33.2733	0.031*
Firmness (N)	66.6667	72.3333	0.003*
Vitamin C (mg/100 g of FW)	70.0633	75.8267	0.416

p-value corresponds to Student's t-test to related samples (paired).
* Significant at p<0.05.

Table 6: Effect of ultrasonic treatment under the optimized conditions on responses.

Optimization of ultrasonic treatment conditions

The desirability function of 0.75 was obtained from numerical optimization using design of expert '7.0'. This approach is considered as an effective technique for the simultaneous determination of optimum settings of input variables that can determine optimum performance levels for one or more responses [29-31]. Importance of '5' was given to total plate count, while importance level of '4' and '3' were given to firmness and respiration rate. Based on the relative contribution to final quality of product importance of '3' was given to all the independent (ultrasound amplitude, treatment time and temperature) variables [9] were shown in Table 4. After giving all the preferences, computer program gives the optimum ultra-sonication conditions for the blood fruit. The optimum values of all the independent variables viz. ultrasound amplitude, treatment time and temperature of the solvent were shown in Table 5 [30,31].

Effect of ultrasonic treatment under optimum conditions on microbial population and quality of blood fruit

The optimum conditions obtained from the RSM [ultrasonic amplitude (%), temperature (25°C) and treatment time (5.10 min)] were considered for determining the microbial population on the surface of fruit. Other quality aspects like firmness and respiration rate for 80 to 100 grams of samples were also evaluated in order to verify and compare the effects of ultra-sonication with the control (Table 6). The mean values of total plate count, firmness, respiration rate and vitamin C for ultra-sonicated samples and control were shown in Table 6. Ultra-sonication significantly (p<0.05) inhibited the microbial growth on the surface by 1 log cycle. It also showed significant (p<0.05) difference for respiration and firmness when compared with the control (Table 6). Insignificant (p>0.05) difference was observed for vitamin C for both the ultra-sonicated and control samples. Similar kind of results was shown for kiwi fruits [9]. The respiration rate for the ultra-sonicated samples were resulted higher (27%) compared to control samples. Total plate count, firmness and vitamin C for the ultra-sonicated samples were decreased (25.9%, 7.8%, 7.6%) compared to control samples. These results confirmed the validity and adequacy of the predicted models. Similar kind of results was shown by Cao et al. [12] where total bacterial count was decreased by 55.2%. Firmness

of ultra-sonicated kiwi fruits were resulted 5.42% less compared to the NaOCl treated samples, while the TSS of the ultrasound treated samples were unaffected [9].

Conclusion

From this study it was concluded that the RSM was proven to be an effective technique in controlling and optimizing the factors responsible for ultrasonic treatment. The application of 100% ultrasonic amplitude for 5.10 min at 25°C was optimum conditions in terms of decreasing the surface microbial load by 1 log cycle and reducing decay by increasing the shelf life of the freshly harvested khoonphal. Ultra-sonication treatment increased the respiration rate by 26.7% (33.27 to 42.32 mg CO₂ kg⁻1 h⁻1 FW), decreases total plate count, firmness and vitamin C by 28% (3.98 to 2.94 log CFU/cm²), 7.8% (72.30 to 66.67 N) and 7.6% (75.83 to 70.06 mg/100 g of FW), respectively compared to control samples.

Acknowledgements

Authors would like to thank Prof. S.C. Deka, Department of Food Engineering and Technology, Tezpur University for his constant encouragement.

References

1. Hanen N, Khaled Z, Sami F, Emna A, Mohamed N (2011) Antioxidant and antimicrobial activities of *Allium roseum* L. "Lazoul," A wild edible endemic species in North Africa. Int J Food Prop 14: 371-380.

2. Odeja O, Obi G, Ogwuche CE, Elemike EE, Oderinlo Y (2015) Phytochemical screening, antioxidant and antimicrobial activities of *Senna occidentalis* L. leaves extract. Clinical Phytosci 1: 1-6.

3. Ahmad N, Zuo Y, Lu X, Anwar F, Hameed S (2016) Characterization of free and conjugated phenolic compounds in fruits of selected wild plants. Food Chem 190: 80-89.

4. Rahim MA, Khatun MJ, Mahfuzur R, Anwar MM, Mirdah MH (2015) Study on the morphology and nutritional status of Roktogota (*Haematocarpus validus*): An important medicinal fruit plant of hilly areas of Bangladesh. Int J Minor Fruits, Med Aromatic Plant 1: 11-19.

5. Singh DR, Singh S, Salim KM, Srivastava RC (2012) Estimation of phytochemicals and antioxidant activity of underutilized fruits of Andaman Islands (India) Int J Food Sci Nutri 63: 446-452.

6. Caprioli G, Iannarelli R, Cianfaglione K, Fiorini D, Giuliani C (2016) Volatile profile, nutritional value and secretory structures of the berry like fruits of *Hypericum androsaemum* L. Food Res Int 79: 1-10.

7. Jin YW, Patricia M, Yasmin BH, Kah FC (2014) Evaluation of antioxidant activities in relation to total phenolics and flavonoids content of selected Malaysian wild edible plants by multivariate analysis. Int J Food Prop 17: 1763-1778.

8. Snigdha R, Uddin MZ, Hassan MA, Rahman MM (2008) Medico botanical report on the Chakma community of Bangladesh. Bangla J Plant Taxonomy 15: 67-72.

9. Vivek K, Subbarao KV, Srivastava B (2016) Optimization of postharvest ultrasonic treatment of kiwifruit using RSM. Ultrasonic Sonochem 32: 328-335.

10. Knorr D, Zenker M, Heinz V, Lee DU (2004) Applications and potential of ultrasonics in food processing. Trend Food Sci Technol 15: 261-266.

11. Vilkhu K, Mawson R, Simons L, Bates D (2008) Applications and opportunities for ultrasound assisted extraction in the food industry-A review. Innovative Food Sci Emerg Technol 9: 161-169.

12. Cao S, Hu Z, Pang B (2010) Optimization of postharvest ultrasonic treatment of strawberry fruit. Postharvest Biol Technol 55: 150-153.

13. Alegria C, Pinheiro J, Gonçalves EM, Fernandes I, Moldão M, et al. (2009) Quality attributes of shredded carrot (*Daucus carota* L. cv. Nantes) as affected by alternative decontamination processes to chlorine. Innovative Food Sci Emer Technol 10: 61-69.

14. Baumann AR, Martin SE, Hao F (2009) Removal of Listeria monocytogenes biofilms from stainless steel by use of ultrasound and ozone. J Food Protection, 72: 1306–1309

15. Jiao Y, Zuo Y (2009) Ultrasonic extraction and HPLC determination of anthraquinones, aloe-emodine, emodine, rheine, chrysophanol and physcione, in roots of Polygoni multiflori. Phytochem Analysis 20: 272-278.

16. Baş D, Boyaci İH (2007) Modeling and optimization I: Usability of response surface methodology. Journal of food engineering, 78(3), 836-845.

17. Ölmez H, Akbas MY (2009) Optimization of ozone treatment of fresh-cut green leaf lettuce. J Food Eng 90: 487-494.

18. Abreu M, Beirao-da-Costa S, Gonçalves EM, Beirão-da-Costa ML, Moldão-Martins M (2003) Use of mild heat pre-treatments for quality retention of fresh-cut 'Rocha' pear. Postharvest Biol Technol 30: 153-160.

19. Beirão-da-Costa S, Steiner A, Correia L, Empis J, Moldão-Martins M (2006) Effects of maturity stage and mild heat treatments on quality of minimally processed kiwifruit. J Food Eng 76: 616-625.

20. Pao S, Brown GE (1998) Reduction of microorganisms on citrus fruit surfaces during packing house processing. J Food Protect 61: 903-906.

21. Pal RS, Kumar VA, Arora S, Sharma AK, Kumar V, et al. (2015) Physicochemical and antioxidant properties of kiwifruit as a function of cultivar and fruit harvested month. Brazilian Archive Biol Technol 58: 262-271.

22. Meng X, Zhang M, Adhikari B (2014) The effects of ultrasound treatment and nano-zinc oxide coating on the physiological activities of fresh-cut kiwifruit. Food Bioprocess Technol 7: 126-132.

23. Gonzalez ME, Barrett DM (2010) Thermal, high pressure, and electric field processing effects on plant cell membrane integrity and relevance to fruit and vegetable quality. J Food Sci 75: 121-130.

24. Pieczywek PM, Kozioł A, Konopacka D, Cybulska J, Zdunek A (2017) Changes in cell wall stiffness and microstructure in ultrasonically treated apple. J Food Eng 197: 1-8.

25. Cantwell MA, Suslow TR (1999) Fresh-cut fruits and vegetables: Aspects of physiology, preparation and handling that affect quality. Annual Workshop Fresh-Cut Products: Maintaining Quality and Safety 5: 1-22.

26. Ketsa S, Chidtragool S, Klein JD, Lurie S (1999) Ethylene synthesis in mango fruit following heat treatment. Postharvest Biol Technol 15: 65-72.

27. Nogata Y, Ohta H, Voragen AGJ (1993) Polygalacturonase in strawberry fruit. Phytochem 34: 617-620.

28. Vicente AR, Saladie M, Rose JK, Labavitch JM (2007) The linkage between cell wall metabolism and fruit softening: looking to the future. J Sci Food Agri 87: 1435-1448.

29. Harrington EC (1965) The desirability function. Industrial Qual Control 21: 494-498.

30. Vivek K, Singh P, Sasikumar R (2016) Optimization of iron rich extruded moringa oleifera snack product for anaemic people using response surface methodology (RSM). J Food Process Technol 7: 639.

31. Ferreira SC, Bruns RE, Ferreira HS, Matos GD, David JM, et al. (2007) Box-Behnken design: An alternative for the optimization of analytical methods. Analytic Chimica Acta 597: 179-186.

17

Optimization of Production of Bread Enriched With Leafy Vegetable Powder

Famuwagun AA[1]*, Taiwo KA[1], Gbadamosi SO[1] and Oyedele DJ[2]

[1]*Department of Food Science and Technology, Obafemi Awolowo University, Ile-Ife, Nigeria*
[2]*Faculty of Agriculture, Department of Soil and Land Management, Obafemi Awolowo University, Ile-Ife, Nigeria*

Abstract

Box-Behnken design was used to study the effect of level of inclusion of dried leafy vegetable powder, mixing time and proofing time, on the weight, volume and specific volume of bread made from composite flour. Data obtained were evaluated using regression analysis. The study revealed all the parameters studied were significant in producing high quality vegetable powder enriched bread. The coefficient determination (R^2) was good for the second-order quadratic model. The study found out those combinations of the parameters; level of vegetable inclusion; 3.65%, proofing time: 90.6 minutes and mixing time: 4.04 minutes were the optimal conditions for the productions of high quality bread enriched with vegetable powder. The study further confirmed through additional analysis that the model is adequate to optimize the process.

Keywords: Vegetable powder; Proofing time; Mixing time; Box-Behnken design

Introduction

Bread is a baked product made from wheat flour. It is one of the most important staple foods in the world and the technology for its production has been in existence for long. There are evidences from food consumption survey in Nigeria of the astronomic rise in the consumption of this baked food [1]. In addition to wheat flour, which is the basic ingredient in bread, yeast, butter, salt, sugar and water are also important ingredients [2]. Other food materials like beverages can also be added based on individual delight. Each of the ingredients has its peculiar purpose in improving the physical characteristics of the final product. Bread, despite lack of some basic nutrients, it is generally accepted and of as such belong to class of food people called 'convenience food' [3].

Leafy vegetables represent inexpensive but high quality nutritional sources, for the poor segment of the population especially where malnutrition is on the increase. They represent a veritable natural pharmacy of minerals, vitamins and phytochemicals [4]. The fibre content of vegetables contributes to the feeling of satisfaction and prevents constipation while the proteins in vegetables are superior to those found in fruits [4]. Fluted pumpkin (*Telfairia occidentalis*) belongs to that class of leafy vegetable that is nutritionally and medicinally beneficial to human health [5]. The bioactive potentials of this vegetable have also been recognized.

The United States Department of Agriculture [6] food pyramids suggest a minimum of three serving of leafy greens per week. Most people fall short of this recommendation. Most health experts agree that the amount of leafy greens that we should consume is likely much higher than just three serving per week. The question that comes to mind is how to meet the recommendations for better health. Improving the nutritional base of a convenience food like bread with highly nutritious vegetable like fluted pumpkin will be a step in the right direction.

Recently, efforts are being made towards improving the quality of wheat bread through wheat flour substitution, probably due to perceived loss in some of nutrient of the wheat flour during milling. Composite flour is a partial replacement of wheat flour with other food materials such as vegetable flour [7]. However, in a bid to enrich the nutrient base of this bread with vegetable powder, some physical characteristics must not be compromised. In preserving the physical characteristics, such as loaf volume, weight and specific volume of bread

enriched with vegetable powder, unit operations such as mixing time, level of vegetable inclusion and proofing time need to be optimized.

Bread, a basic food for human, has no negative health effects when produced in appropriate conditions using appropriate materials [8]. However, improper use of raw materials and non-optimal unit operations such as proofing time, mixing time can affect the physical characteristics of the product and hence, attractions of this product to consumers. Over-mixing or under-mixing of dough during baking process would have negative effect on the quality of the product. Over-proofing or under-proofing also has effect on the actions of the yeast which affect the quality of the final product. Optimization method could be a veritable tool in determining the favourable production conditions for bread enriched with leafy vegetable powder.

The optimization of process parameters can be done by various techniques; one of the effective and commonly used techniques for this purpose is Response Surface Methodology, which is a collection of statistical and mathematical techniques useful for developing, improving and optimizing processes [9]. This technique is a faster and economical method for gathering research results than classic one variable at a time or full factors experimentation [10]. This statistical tool has been used in the optimization of bread by varying the amount of bran, the amount of yeast and the fermentation time on the amount of phytic acid in bread [11]. Mohammed and Sharif [12] optimized composite flour for the production of enhanced storability of leavened flat bread using response surface methodology.

This present study aims at optimizing some critical unit operations in the production of bread enriched with leafy vegetable powder such as mixing and proofing time and the level of inclusion of the vegetable powder on the loaf volume, weight and specific volume of the product. This would ultimately help the industry to gain economic advantage with increased production of high quality vegetable enriched bread.

---

*Corresponding author: Famuwagun AA, Food chemistry and processing Laboratory, Department of Food Science and Technology, Obafemi Awolowo University, Ile-Ife, Nigeria, E-mail: akinsolaalbert@gmail.com

Materials and Methods

Preparation of vegetable powder

The leafy vegetable powder was prepared by modified method of Abraham et al. [13]. Fresh and matured fluted pumpkin (*Telfaria occidentalis*) leaves were obtained from Teaching and Research farms of Obafemi Awolowo University, Ile-Ife, Nigeria. The fresh vegetables were sorted, de-stalked and rinsed with clean water. The cleaned leafy vegetable was sliced to dimensions (0.12 × 0.90 × 0.50 mm) with clean knife. The sliced vegetables were dried using hot air cabinet drier at <60°C for 8 hrs. The dried vegetable leaves were milled to fine powder using Marlex Excella dry mill (Marlex Appliances PVT, Daman).

Preparation of composite flour blends

Wheat flour was blended with 1%, 3% and 5% of dried, milled, fluted pumpkin leaf powder. The composite flour was mixed together to produce 200 g blends of each level of vegetable incorporation. The composite flours were then stored in air-tight container for further use.

Baking of wheat flour bread enriched with vegetable powder

Dough from the composite flour blends were baked using the straight-dough method of Greene and Bovell [14] with some modifications. The formula used in this process is: 200 g of composite flour blends (wheat flour and vegetable powder in different proportions), 6 g yeast, 4.0 g salt, 10 g shortening, 6 g sugar, and 120 ml water. All dry ingredients were weighed and placed in a Kenwood dough mixer (Model A 907 D) set at highest speed and mixed for 50 seconds. Then a suspension of the yeast in water was added. The mixture was further run at high speed for 50 seconds. Water was added to the mixture to make up the required water for the process for the process and the mixture was further mixed for the required length of time. The dough was later kneaded on the kneading table, rounded into balls by hand and placed in lightly greased fermentation bowl and placed in fermentation cabinet (National Company, Lincoln, NE). The dough was then proofed for the required length of time. Baking was done at 250°C for 15 minutes. The baked bread was allowed to cool at room temperature before measurements were taken.

Determination of responses

Volume, weight and specific volume of loaves produced: Loaf volume was measured by small seeds displacement method described by Khalil et al. [15]. Loaf was placed in a container of known volume into which rapeseeds was run until the container was full. The volume of seeds displaced by the loaf was considered as the loaf volumes which were measured in a graduated cylinder. The weight of the loaf was determined using a sensitive weighing balance and the specific volume of the loaf was determined by averaging the loaf volume with loaf weight.

$$\text{Specific volume} \left(cm^3/g\right) = \frac{loaf\ volume}{loaf\ weight}$$

Experimental design: The study employed Box-Behnken designs (BBD) of response surface methodology with three levels. The process was optimized on the basis of three input variables whose interactions were studied as three major responses. These input variables were determined on the basis of preliminary single experimental factor. The levels of various input variables selected are as follows; Inclusion of vegetable powder: 1-5% (w/w), dough mixing time: 3-5 minutes and proofing time; 60-120 minutes. The analyses of all the responses were done in triplicates and the average values reported. Factors and their levels for Box-Behnken design are shown in Table 1. The independent

variables and dependent ones in coded form and the number of experimental runs are presented in Table 2. The selected responses evaluated include; loaf volume, loaf weight and specific volume. Seventeen runs (Table 2) were evaluated to select the best combination of processing conditions that produced the best quality of loaf of bread made from the composite flours.

Statistical analysis of the data: Data were analysed using Response Surface Methodology, Design-expert software version 8.0.3.1 (Stat-Ease Inc., Minneapolis, USA). Data obtained from the experiment were fitted in the second order polynomial model and co-efficient for the regression equations were obtained.

The suitability of the fit was evaluated using analysis of variance (ANOVA). The fitted quadratic response is shown below:

$$Y_i = \beta_0 + \sum_{j=1}^{k} \beta_j X_j \sum_{j=1}^{k} \beta_{jj} X_{jj} + \sum \sum_{i>j=1}^{k} \beta_{ij} X_i X_j + e \ \dots\dots \quad (1)$$

In equation 1 above, Y_i = predicted response, β_0 = a constant, βj = linear coefficient, βjj = squared coefficient, and βij = interaction coefficient, Xi and Xj are the independent variables and e the error, respectively.

Results and Discussion

Model evaluation

The optimization of the loaf volume, loaf weight and specific volume of bread made from composite flours comprising wheat flour

Variables	Code	Range and Level		
		- 1	**0**	**+1**
Level of veg. inclusion (%)	A	1	3	5
Mixing time (min)	B	3	4	5
Proofing time (min)	C	60	90	120

A= Level of veg. powder inclusion (%); B= Mixing time (min); C= Proofing time (min).

Table 1: Experimental ranges and levels of the independent variables for the experimental design.

Runs	Level of veg inclusion (%)	Mixing time (min)	Proofing time (min)	Weight of loaf (g)	Volume of loaf (cm³)	Specific volume (cm³/g)
1	-1	0	0	126.51	400	3.16
2	0	1	1	142.16	415	2.92
3	1	0	1	127.98	380	2.97
4	0	0	0	158.58	635	4.00
5	1	1	0	142.98	470	3.29
6	0	1	-1	140.90	412	2.92
7	1	0	-1	143.09	320	2.24
8	0	0	0	168.80	570	3.39
9	1	0	1	136.90	350	2.56
10	0	1	-1	123.64	430	3.48
11	1	-1	0	143.53	310	2.16
12	1	1	0	140.89	500	3.55
13	0	0	0	165.80	625	3.76
14	1	-1	0	140.30	340	2.42
15	0	-1	1	144.05	420	2.92
16	0	0	0	162.85	620	3.81
17	0	0	0	165.98	618	3.72

Table 2: Box-Behnken experimental design for three factors in coded and uncoded units and the actual responses.

and green leafy vegetable powder was evaluated in this section. The co-efficient of the regression equations for the measured responses, the linear, quadratic and interaction terms of the selected variables were evaluated and values shown in Table 3.

The results of the loaves weight of the bread showed that two linear (A, B), three quadratic (A^2, B^2, C^2) parameters and one interaction (BC) term were significant at $p<0.05$ as shown in Table 3. For the loaf volume, it was shown from the analysis that two linear (A, C) parameters, three quadratic (A^2, B^2, C^2) and one interaction terms were significant at $p<0.05$ and two of the interaction terms were not significant. In the case of the specific volume of the loaves of bread produced, all the three linear (A, B, C) terms and the three quadratic were significant terms. One interaction term (BC) was also significant. Just like the loaf weight and volume, two of the interaction terms were not significant.

Fitting of the quadratic model

The quadrating models fitting are shown in Table 4. From the analysis of variance (ANOVA), it was clear that the model was significant ($p<0.05$) for the predicted loaf volume, loaf weight and the specific volume of the vegetable powder enriched bread. The correlation coefficient (R^2) 0.9239, 0.9602 and 0.9684 for loaf weight, loaf volume,

Parameter	Co-efficient Estimate	Standard Error	95%CI Low	95%CI High	t ratio	Prob > t
Loaf Weight						
Intercept	164.4	2.62	158.2	170.61	62.75	**0.0032**
A	3.19	2.07	-1.71	8.1	1.53	**0.0016**
B	-2.58	2.07	-7.48	2.33	-1.25	**0.025**
C	2.12	2.07	-2.79	7.02	1.02	0.341
AB	-0.31	-7.24	6.63	1	0.042	0.9195
AC	-1.91	-8.85	5.02	1	0.215	0.5348
BC	4.79	2.93	11.72	1	1.635	**0.0468**
A^2	13.28	2.86	-6.52	1.01	4.643	**0.0024**
B^2	9.22	2.86	-2.45	1.01	3.224	**0.0146**
C^2	17.5	2.86	-10.74	1.01	6.119	**0.0005**
Loaf Volume						
Intercept	613.6	15.37	577.25	649.95	39.922	**0.0014**
A	-53.75	12.15	-82.48	-25.02	-4.434	**0.0031**
B	10.38	12.15	-18.36	39.11	0.854	0.4215
C	0.38	12.15	-28.36	39.11	0.031	**0.00097**
AB	0	17.18	-40.64	40.64	0	1
AC	12.5	17.18	-28.14	53.14	0.728	0.4906
BC	-3.25	17.18	-43.89	37.39	-0.189	**0.044**
A^2	-132.67	16.75	-172.28	-93.07	-7.921	**0.0001**
B^2	-75.93	16.75	-115.53	-36.32	4.533	**0.0027**
C^2	-118.42	16.75	-158.03	-78.82	7.07	**0.0002**
Specific Volume						
Intercept	3.74	0.1	3.5	3.98	37.4	**0.0034**
A	-0.45	0.081	-0.64	-0.26	-5.556	**0.0008**
B	0.14	0.081	-0.056	0.33	1.728	**0.0137**
C	-0.054	0.11	-0.24	0.14	-0.491	**0.0025**
AB	0	0.11	-0.27	0.27	0	1
AC	0.13	0.11	-0.14	0.4	1.182	0.3001
BC	-0.13	0.11	-0.41	0.13	-1.182	**0.0258**
A^2	-0.6	0.11	-0.87	-0.34	-5.455	**0.001**
B^2	-0.28	0.11	-0.54	-0.014	-2.545	**0.0415**
C^2	-0.4	0.11	-0.66	-0.14	-3.636	**0.0088**
Values in bold form statistically different at p < 0.05						

Table 3: Regression co-efficient, standard error, t-test confidence level. results of the response surface for loaf weight, loaf volume and specific volume of loaves.

Weight of loaves					
Source	SS	Df	MS	F-value	p-value
Model	2924.01	9	324.89	9.44	**0.0037**
A	81.8	1	81.6	2.37	**0.0016**
B	53.06	1	53.06	1.54	**0.025**
C	35.93	1	35.93	1.04	0.341
AB	0.38	1	0.38	0.011	0.9195
AC	14.67	1	14.67	0.43	0.5348
BC	91.65	1	91.65	2.66	**0.0468**
A^2	743	1	743	21.58	**0.0024**
B^2	357.59	1	357.59	10.39	**0.0146**
C^2	1289.2	1	1289.2	37.45	**0.0005**
Residual	240.99	7	34.43		
Lack of fit	180.9	3	60.3	4.01	0.1064
Experimental Error	60.09	4	15.02		

$R^2 = 92.39\%$; Adjusted $R^2 = 92.60\%$; Predicted $R^2 = 85.58\%$

Volume of loaves					
Source	SS	Df	MS	F-value	p-value
Model	199000	9	22152.51	18.75	**0.0004**
A	23112.5	1	23112.5	19.57	**0.0031**
B	861.13	1	861.13	0.73	0.4215
C	1.13	1	1.13	9.524	**0.00097**
AB	0	1	0	0	1
AC	625	1	625	0.53	0.4906
BC	42.25	1	42.25	0.036	**0.0544**
A^2	74116.44	1	74116.44	62.74	**0.0001**
B^2	24272.02	1	24272.02	20.55	**0.0027**
C^2	59050.44	1	59050.44	49.99	**0.0002**
Residual	8268.95	7	1181.28		
Lack of fit	5719.75	3	1906.58	2.99	0.1587
Experimental Error	2549.2	4	637.3		

$R^2 = 96.02\%$; Adjusted $R^2 = 90.90\%$; Predicted $R^2 = 84.01\%$

Specific Volume of loaves					
Source	SS	Df	MS	F-value	p-value
Model	4.71	9	0.52	10.08	**0.003**
A	1.61	1	1.61	31.02	**0.0008**
B	0.15	1	0.15	2.81	**0.0137**
C	0.023	1	0.023	0.45	**0.0025**
AB	0	1	0	0	1
AC	0.065	1	0.065	1.25	0.3001
BC	0.078	1	0.078	1.51	**0.0258**
A^2	1.54	1	1.54	29.6	**0.001**
B^2	0.32	1	0.32	6.21	**0.0415**
C^2	0.67	1	0.67	12.92	**0.0088**
Residual	0.36	7	0.052		
Lack of fit	0.17	3	0.056	1.14	0.4331
Experimental Error	0.2	4	0.049		

$R^2 = 96.84\%$; Adjusted $R^2 = 93.63\%$; Predicted $R^2 = 91.09\%$
Values in bold form statistically different at p < 0.05

Table 4: Estimated regression model of relationship between responses.

and loaf specific volume respectively were obtained. R-squared value is an indication of the level of responses that can be explained by a particular model. From these results, it could be shown that 92.39%, 96.02% and 96.84% of the responses could be explained by the model. Statistically, the significant level obtained were 0.0037, 0.004 and 0.003 for loaf weight, volume and specific volume respectively. These levels were high and attested to the fitness of the model in evaluating the responses. The results obtained in this study revealed that the model employed is good and could be used for the prediction of the three selected responses (weight, volume and specific volume) from the production of wheat flour bread enriched with leafy vegetable powder.

Using the experimental data in Table 2, second degree polynomial equation model for the loaf weight, loaf volume and specific weight were regressed and the equations are shown below:

$$Weight\ of\ loaf\ = +164.40 + 3.19A - 2.58B + 2.12C - 0.31AB - 1.91AC + 4.79BC - 13.28A^2 - 9.22B^2 - 17.50C^2 \quad (2)$$

$$Loaf\ volume\ = +613.60 - 53.75A + 10.38B + 0.38C + 0.000AB + 12.50AC - 3.25BC - 132.67A2 + 75.93B2 - 118.42C2 \quad (3)$$

$$Specific\ volume\ = +3.74 - 0.45A + 0.14B - 0.054C + 0.000AB + 0.13AC - 0.14BC - 0.60A2 - 0.28B2 + 0.40C2 \quad (4)$$

Response surfaces analysis

The graphical representation of 3 dimensional plots of response surface in Figures 1-3 show the relationships between the dependent and independent variables in the production of bread enriched with green leafy vegetable (fluted pumpkin) powder.

Figures 1(a) - 1(c) are three dimensional plots showing the influence of two of the variables: mixing time and level of vegetable powder inclusion on the weight, volume and the specific volume respectively of the loaves produced while the other variable, proofing time was kept constant. In Figure 1(a), increasing the mixing time of the process from 3-4.5 minutes and the level of vegetable inclusion in the composite flour from 2-4% positively affected the weight of the loaves produced. A decrease in the weight of the loaf was noticed when the mixing time was increased beyond 4.5 minutes and level of vegetable inclusion above 4%. A possible explanation for this trend could be that at higher mixing time, the water absorption capacity of the composite flour protein is damaged due to over-mixing [16]. Also, from the three dimensional plots in Figure 1(a), addition of vegetable contributed immensely to the weight of the loaf probably due to its water absorption capacity.

In Figure 1(b), the effect of the mixing time and level of vegetable inclusion on the volume of the loaf is represented by three dimensional plots. In this Figure, the mixing time was seen contributing more to the volume of the loaf than the level of vegetable powder inclusion. An increase in the volume of the loaf was noticed as the mixing time increased from 3.0-4.8 minutes. The volume decreased as the mixing time approached 5.0 minutes. Mixing is an important stage in bread production. During this process, the activities of the yeast increased, more air is absorbed from the surrounding and hence the volume increased. It is also possible for the yeast to be over worked upon if the mixing time goes beyond the optimal, this might explain the reason for sharp decrease in bread volume as mixing time approached 5 minutes. The contribution of the vegetable powder to the loaf volume was peak at about 3.5%. Addition beyond this level decreased the volume of the loaf. A possible explanation for this might be because of the replacement of the wheat flour with vegetable powder. Composite flour such as the one used in this work is a partial replacement of wheat flour with vegetable powder. One major factor that aids rising of the dough of bread is the gluten which is the protein in wheat flour. High level of vegetable powder inclusion means reduction in the level of wheat gluten and hence the rising ability of dough of bread is affected.

In Figure 1(c), the mixing time and level of vegetable powder inclusion as it affects the specific volume of the loaf is represented. It was evident from Figure 1(c) that increasing the mixing time of the process from 3-5 minutes increased the specific volume of the loaf of bread produced. However, the contribution of level of vegetable powder inclusion to the loaf specific volume of the loaf was peak at about 4.0%.

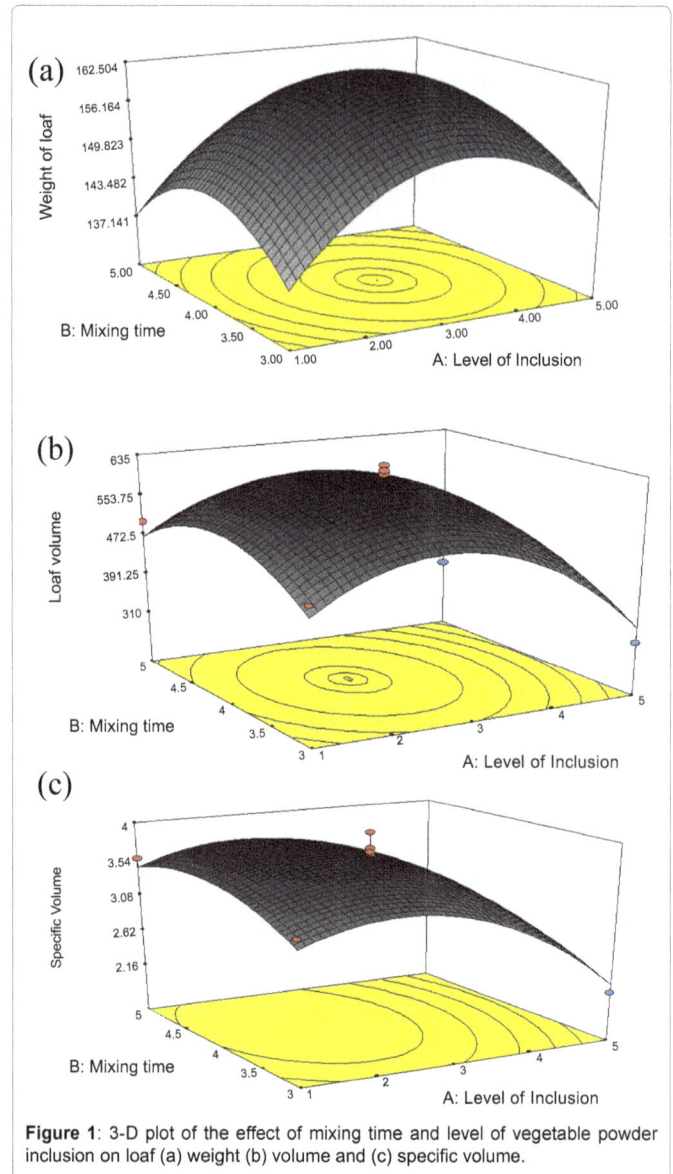

Figure 1: 3-D plot of the effect of mixing time and level of vegetable powder inclusion on loaf (a) weight (b) volume and (c) specific volume.

Figures 2(a) - 2(c) are three dimensional plots showing the loaf weight, volume and specific volume of the loaf of bread produced as functions of two variable parameters; proofing time and level of vegetable powder inclusion while keeping the mixing time of the process constant. In Figure 2(a), the loaf weight as a function of proofing time and level of vegetable inclusion is represented. The results showed increase in the weight of the loaf of bread as the proofing time increased from 60-110 minutes and level of vegetable inclusion increased from 1-4%. Increasing the values of these variables beyond these levels resulted in negative impact on the weight of the loaves of bread. Proofing stage is an important unit operation in bread making. Under humid conditions, the release of CO_2 by the yeast and the trapping of same by gluten take place at this stage. Prolong proofing time might affect the properties of bread as there might be excessive release of the gas and the gluten might be weakened.

Figure 2(b) represents the influence of the proofing time and level of vegetable powder inclusion on the volume of loaves of bread produced. It was evident from Figure 2(a) that increasing the time of proofing 60-105 minutes and the level of vegetable powder inclusion from 1.0-

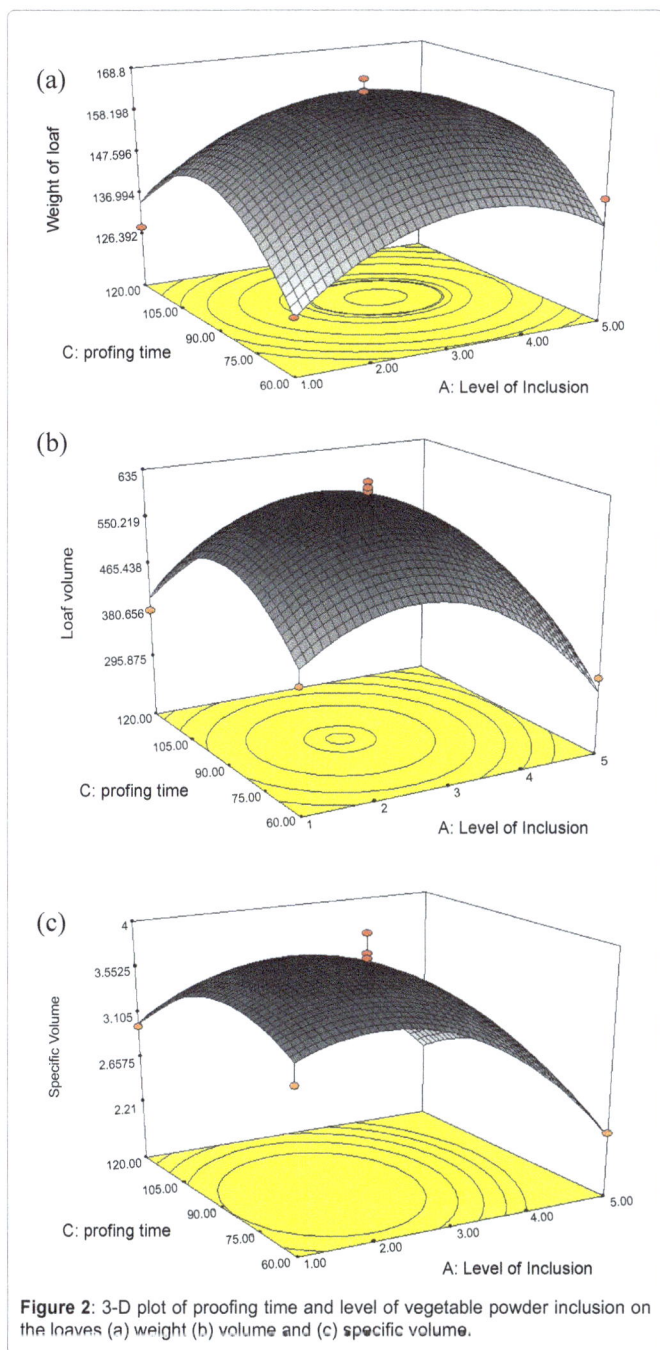

Figure 2: 3-D plot of proofing time and level of vegetable powder inclusion on the loaves (a) weight (b) volume and (c) specific volume.

loaf of bread. The same trend was observed when rice was composited with wheat flour in bread making [17]. Increasing the level of vegetable powder beyond this level would mean further reduction in wheat proteins. Consequently, the protein would be unable to retain gas produced during the fermentation process, resulting in a product that has a low specific volume.

Figures 3(a) - 3(c) represents the influence of proofing time and mixing time on the weight, volume and specific volume of loaf of bread produced. The results showed that proofing time had greater influence on the weight of the loaf than the mixing time as shown in Figure 3(a). The proofing time was peak at 105 minutes, after which there was progressive decrease in the weight of the loaf as the proofing time increased beyond 105 minutes. Mixing time of the process was also peak at 4 minutes. An increase in the mixing time of the process beyond this level decreased the weight of the product.

Figure 3(b) also shows the effect of proofing and mixing time on the loaf volume of the bread. There was noticeable rise in the volume of the loaf of bread when the mixing time was increased from 3 minutes to 4 minutes and the proofing time between 60 minutes to 90 minutes. As the parameters were increased beyond these levels, a decrease in the volume of the loaf was observed. A possible explanation is that over mixing might negatively affect the yeast activity and over-proofing of the dough in the prover which might affect the rising ability of the dough.

The effect of the mixing time and proofing time on the specific volume of the loaf is represented by the three dimensional plot in Figure 3(a). Mixing time of the process was seen having greater effect on the specific volume of the dough. The specific volume of the dough was on the increase as the mixing time increased and peak at about 4.5 minutes. The proofing time was also peak at about 100 minutes. Beyond these levels of proofing and mixing time, the specific volume of the loaf of bread decreased.

Predictive model verification

In verifying the capacity of the model to predict the optimum conditions for the process, maximum desirability was used for the loaf weight, volume and specific volume of the bread enriched with vegetable powder. Optimum level of vegetable powder inclusion, proofing and mixing times were generated by the software and were found to be 3.65%, 4.04 minutes and 90.60 minutes respectively as shown in Table 5. With respect to these optimum conditions, the loaf volume, weight and specific volume produced were 582 cm³, 163.93 g and 3.52 cm³/g respectively. These values were found to be close with the experimental values as shown in Table 5. This shows the reliability of the model in optimizing the process.

Comparison with conventional loaf of bread

Table 6 shows the physical characteristic of the loaf of bread produced without the use of vegetable powder. The loaf weight of the optimized parameters with vegetable powder was more than that obtained in Table 6 without vegetable powder. The values obtained for the volumes were in close range. Comparing Table 5 and Table 6, it is evident that the enrichment of vegetable powder in bread making, in addition to the perceived nutritional benefit, the physical characteristics of the loaf of bread is preserved.

Conclusion

Response Surface Methodology was successfully used to optimize the process condition in the production of wheat flour bread enriched with vegetable powder. The Box-Behnken design of RSM was found to

3.5% increased the loaf volume. Increasing the proofing time and level of inclusion beyond these stages lead to about 16.00% reduction in the volume of the loaf. The result agreed with the observation of Gujral and Rosell [17] on the bread made from flours composed of sorghum and wheat flours, attributing the reason to lower levels of gluten network in the dough and consequently less ability of the dough to rise; due to the weaker cell wall structure.

Specific volume of the loaf of bread as a function of proofing time and level of inclusion of the vegetable powder is shown in Figure 2(c). From the Figure, it could be observed that proofing time influenced the specific volume of the loaf of the bread more than the mixing time. Increase in the proofing time above 105 minutes and level of vegetable inclusion beyond 3.00% lead to a decrease in the specific volume of the

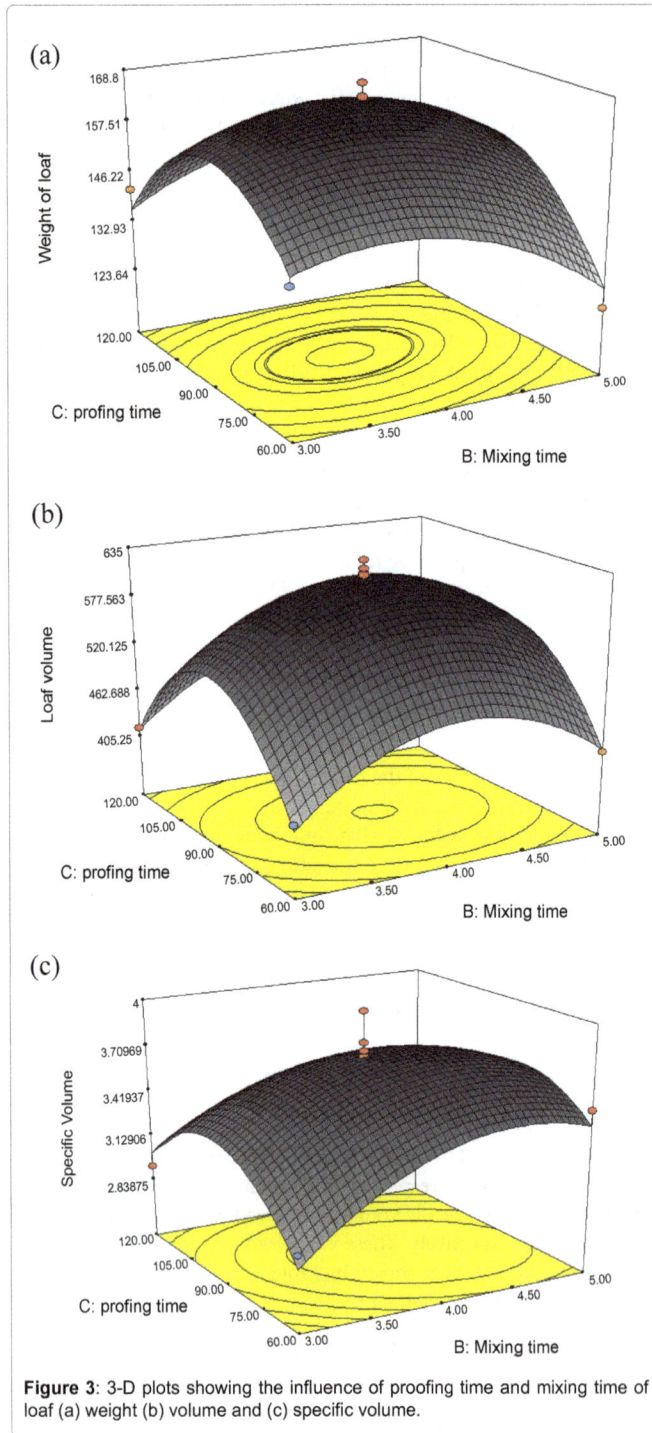

Figure 3: 3-D plots showing the influence of proofing time and mixing time of loaf (a) weight (b) volume and (c) specific volume.

	Level of veg. inclusion (%)	Mixing time (min)	Profing time (min)	Loaf weight (g)	Loaf volume (cm-cube)	Specific volume

Level of veg. inclusion (%)	Mixing time (min)	Profing time (min)	Loaf weight (g)	Loaf volume (cm-cube)	Specific volume
0	3	90	140.75	505	3.58
0	4	90	135.70	585	4.31
0	5	90	138.80	570	4.10

Table 6: Control values of the experiment.

be effective to determine the level at which the vegetable powder will be added to the flour, mixing time and the proofing time of the process that will not negatively affect the physical properties of the product. The optimal conditions of the process parameters can therefore be used in the production of bread enriched with vegetable powder with acceptable physical characteristics.

Acknowledgements

Authors are grateful for funding from IDRC (International Development Research Centre) and the Department of Foreign Affairs, Trade and Development/ Canadian International Food Security Research Fund (DFATD/CIFSRF) through Project 107983 on synergizing indigenous vegetables and fertilizer micro-dosing innovations among West African farmers.

References

1. Anyika JU, Uwaegbute AC (2005) Frequency of consumption and nutrient content of some Snacks eaten by an adolescent secondary and University student in Abia State. Nig J Nutri Sci 26: 10-15.

2. Badifu SO, Chima CE, Ajayi YI, Ogori AF (2005) Influence of Mango mesocarp flour supplement to micronutrient; physical and organoleptic qualities of wheat-based bread. Nig Food J 23: 59-68.

3. Rosales-Juarez M, Gonzalez-Mendoza B, Lopez Guel EC, Lozano-Bautista F, Chanonaperez J, et al. (2008) Changes on dough rheological characteristics and read quality as a result of the addition of germinated and non-germinated soybean flour Food. Bioprocess Technol 1: 152-160.

4. George P (2003) Encyclopedia of foods. Humane Press, Washington 1: 526

5. Ajibade SR, Balogun MO, Afolabi OO, Kupolati MD (2006) Sex differences in biochemical contents of Telfairia occidentalis Hook F. J Food Agri Environ 4: 155-156.

6. USDA, USDHS (2010) Dietary Guidelines for Americans (7thedn.). United States Department of Agriculture and United States Department of Health and Human Services. Washington DC, USA.

7. Shittu T, Sanni LO, Raji AO (2007) Bread from composite cassava-wheat flour: Effect of baking time and temperature on some physical properties of bread loaf. Food Res Int 40: 280-290.

8. Tekindal MA, Bayrak H, Ozkaya BGY (2012) Box- Behnken experimental design in factorial experiments: the importance of bread for nutrition and health. Turk J Field Crop 17: 115-123.

9. Carley MK, Kamneva YN, Reminga J (2004) Response surface methodology. CASOS Technical report, CMU- ISRI.

10. Li F, Yang L, Zhao T, Zou J, Zou Y, et al. (2012) Optimization of enzymatic pre-treatment for n-hexane ex- traction of oil from Silybummarianum seeds using response surface methodology. Food Bio-prod Process 90: 87-94.

11. Zubair F, Salim R, Muhammad A (2013) Application of response surface methodology to optimize composite flour for the production and enhanced storability of leavened flat bread (Naan). J Food Process Preserv 40: 32-36.

12. Mohammed MI, Sharif N (2011) Mineral composition of some leafy vegetables consumed in Kano Nigeria. Niger J Basic Appl Sci 19: 208-211.

13. Abraham IS, Joseph OA, Dick IG (2012) Effect of Moringa oleifera leaf powder supplementation on some quality characteristics of wheat bread. Food Nutri Sci 4: 270-275.

14. Greene JL, Bovell-Benjamin AC (2004) Macroscopic and sensory evaluation of bread supplemented with sweet potato flour. J Food Sci 69: 167-173.

15. Khalil AH, Mansour EH, Dawood FM (2000) Influence of malt on rheological and baking properties of wheat-cassava composite flours. Lebens Wissen Technol 33: 159-164.

	Optimum conditions	Modified conditions
Mixing time (min)	4.04	4
Vegetable powder inclusion (%)	3.65	3
Proofing time (min)	90.6	90
Weight(g)	163.94 ± 1.50	164.4
Volume (cm³)	582.17 ± 1.16	613.6
Specific volume (cm³)	3.52 ± 0.10	3.74

Table 5: Optimum conditions, modified conditions and the experimental value of response for the process.

18

Structure and Function of Storage Pit, *Polota*, for Long-Term Storage of Sorghum-A Case Study of Storage Pit in Dirashe Special Worenda, Ethiopias

Yui Sunano*

School of Agricultural Sciences, Graduate School of Bio-agricultural Sciences, Nagoya University, Japan

Abstract

"Storage pits" with bag-like, tubular or flask-like shapes had been utilized for storing grains and nuts all around the world until several centuries BC. Storage pits were, however, mostly replaced with aboveground storehouse once the cultivation of rice, wheat and barley, which were not suitable for underground storage, became widespread. Nevertheless, such storage pits are now still being used locally in some rural villages in Ethiopia and Sudan. Storage pits can prevent losses due to weather, mice, sparrows, fire, water, and theft. However, storage pits are highly humid inside, which leads fungal and bacterial proliferation. Stored grains tend to severely deteriorate within several months, which is often before the next harvest season comes. However, the local people in the Dirashe area in Southern Ethiopia state that underground storehouses called *polota* with a flask-like shape can store sorghum for a maximum of 20 years. This study investigates the location, structure and storage function of *polota* in order to understand the reason why *polota* is capable of long-term storage while such storage pits highly humid inside. First, soil samples were collected from the locations where *polota* were constructed. The x-ray fluorescence analysis was conducted on these samples to analyze their chemical compositions. Then, the rates of iron (g)/aluminum (g), aluminum (g)/titanium (g), silicon (mol)/aluminum (g) were calculated. The result indicates that *polota* were built in the areas where basalt layers were chemically weathered. Based on the actual measurement of *polota*, all the *polota* are shaped like a flask, about 1.5 m in diameter and about 2 m in depth. The chemically weathered basalt makes it easier to work on while maintaining its dense composition, thus it can be easily formed into a flask-like shape. Also, the airtight characteristic can maintain the temperature and humidity inside stable. The measurement of hygro-thermal properties performed inside *polota* in which sorghum was stored shows the stable temperature at 31 Celsius and relative humidity at 92%. A low concentration of oxygen (O_2), 2.7%, and high concentration of carbon dioxide (CO_2), 1,60,000 ppm, were also measured inside *polota*. While *polota* is as high in humidity as other storage pits, thus not suitable for storage, it now revealed that a low O_2 concentration prevents propagation of noxious insects and a high CO_2 concentration induces a state of quiescence to the stored grains, inhibiting deterioration and enabling long-term storage.

Keywords: Underground storage; Local storage; Longer preservation; Sorghum; Africa; Self sufficient

Introduction

Until several centuries BC, various kinds of underground storage pits with bag-like, tubular and flask-like shapes had been used to store grains such as millet (*Setariaitalica*) and Japanese millet (*Echinochloa esculenta*), nuts such as beech (Fagaceae) and horse chestnut (*Aesculus turbinata*) and other plants throughout the world [1-4]. As Shazali et al. [5] states, "Underground grain storage in pits is practised in hot dry climates and may have developed independently in many similar places around the world". Remains of storage pits were discovered at the sites from the Iron Age in UK [6] and in Hungary [7] as well. However, storage pits can rarely be discovered at the post Iron Ages sites. As the wheat and barley cultivation became widely introduced, storage pits were replaced by aboveground storage methods as these grains cannot be preserved in humid storage pits. In China, grains were stored in storage pits in the same manner. Such books as "Nousho (Book on Agriculture)," "Noseizensho (Complete Book on Agricultural Policy)," and "Keijohen (Collection of Articles on Miscellaneous Affairs)" describe in detail how storage pits became less and less widespread. According to these records, the usage of storage pits began in the early Neolithic Age, accelerated by the development of agriculture and promotion by the government, and continued to store millet until the Sui and Tang Periods [4]. Yet, after the Song Period, the cultivation of rice in the Yangtze River areas and introduction of raised-floor storehouse across China caused storage pits to pass out of existence.

While storage pits can prevent stored grains from being damaged due to wind and rain, sparrows, mice, fire and water, and from being stolen [4], they tend to be highly humid inside [3], which degraded stored grains. In "Nousho (Book on Agriculture)," the Northern part of China with solid ground soil and cold climate was stated to be suitable for constructing storage pits [4]. Also, the traces of backing the wall of

***Corresponding author:** Yui Sunano, School of Agricultural Sciences, Graduate School of Bio-agricultural Sciences, Nagoya University, Furocho, Chikusa-ku, Nagoya, Aichi 464-8601, Japan, E-mail: yui.sunano@gmail.com

storage pits with basket were discovered at the sites in United Kingdom [6] and traces of fumigation at the sites in Cyprus and Hungary [7,8], both of which were for the purpose of controlling humidity inside. On the other hand, storage pits are still being used to store sorghum in such African countries as Ethiopia and Sudan [9-14]. However, such storage pits tend to be very high in humidity inside because, when the grains make contact with the moist inner wall, it often leads entry of moisture into the space among grains, resulting in increasing the moisture content of grains and relative humidity inside [15].

Furthermore, respiration by insect pests, mainly Sitophilus, Sitotroga and Tribolium species, micro-fungi, other micro-organisms and the living grain itself produces high temperature and water activity [16-19]. Fungi and bacteria grown in storage are known as the major cause for degradation of grains, pulses, and oilseeds [17,20] and Aspergillus and Penicillium are particularly well known. Such an environment in storage pits described thus far causes grains to be molded and deteriorated [21,22]. In the Hararghe region in Eastern Ethiopia, storages pits are also being used. Still, storage loss is too large to sufficiently store grains for a year until the next harvest season. It is estimated that 13% of grains weight are lost in 7-8 months storage period with its maximum weight loss at 24% in the Hararghe region. Similarly, during the same storage period, seed germination declined from 83-27% on average [15]. Grain deterioration of sorghum in storage pits in Alemaya in the state of Hararghe is at 2-25% when the pits are completely filled and 7-35% when half-filled [11-13]. The similar pattern can be observed in Sudan as well [5]. Yet, such weight losses in underground storehouses can be efficiently reduced by de-humidifying, as in experiments using a system of polythene sacks [23] or polyethylene lining [5]. On the other hand, since the 16th century, the Dirashe people living in Southern Ethiopia have managed to store sorghum for the extended period of several years in the storage pits called *polota*. *Polota* is a flask-like shaped underground storehouse with the depth of about 2 m and width of 1.5 m, and has a capacity of storing 2 tons of sorghum. A male interviewee in the Dirashe area stated, "When I witnessed the *polota*, which had been left for 20 years, opened several years ago, the stored sorghum inside was not damaged at all." *Polota* can maintain the quality of sorghum for up to 20 years, which successfully prepared people for a year of crop failure. Storehouses like *polota* that can store sorghum without causing much damage for several years are quite rare if seen globally, even including aboveground storehouses.

This study focuses on the materials, structure, and function of *polota*. The purpose of this study is, then, to identify the major causes for *polota*'s characteristic of long-term storage whereas it is categorized into a storage pit known for a large storage loss.

Research Methods

Participatory observations and interviews on the field were performed for a total of 13 months (January to March 2009, June to August 2009, January to February 2011, December 2011 to January, 2012, January to February 2013). Amharic language, the official language of Ethiopia, was mainly used, supplemented with the language used in the Dirashe area. Most people except for the elderly women and small children were educated in Amharic language, thus fluent, but the Dirashe people talk to each other in Dirashe language. Due to the limited linguistic capability of the author, a male Dirasha assistant was hired to translate the conversation in Dirasha language that the author was unable to comprehend. The following analyses were carried out in order to understand the quality, structure and function of *polota*.

The method of data collection will be described later, and methods of analyses, the main data for this study, are stated here.

Firstly, *onga* in 'A' village, 'O' village and 'At' village, *sogita* above *onga* in 'A' village, gravel stones of basalt in 'A' village were collected and brought back to Japan. The samples were outsourced for grain refining, then for x-ray fluorescence analysis in order to analyze their chemical compositions. Then, the rates of iron (g)/ aluminum (g) and of aluminum (g)/ titanium (g) were calculated to identify their base materials. The quantity of silicon in these samples was converted into molar mass (mol) so that the rate of silicon (mol)/aluminum (g) was calculated to analyze the level of weathering. It should be also noted that onga in 'A' village and 'O' village were collected from two different locations.

Secondly, the assistant measured the size of *polota* from within: 4 in 'A' village, 6 in 'O' village and 8 in 'W' village. Then, the illustration of *polota* was produced based on the actual measurement of *polota* located in 'A' village.

Thirdly, the temperature and humidity inside *polota* in 'A' village, which was filled with sorghum up to two thirds, were measured for two days for the purpose of investigating its storage function. In comparison, the same process was carried out in a house with thatched roof and mud wall in 'A' village. Furthermore, an atmospheric gas was collected into a gas barrier vacuum-sealed bag through a silicon tube, which was inserted through the slightly opened lid into the same *polota*. As a comparison data, an atmospheric gas was collected from an empty *polota*. After returning to Japan, oxygen concentration (%) and carbon dioxide concentration (ppm) of the collected atmospheric gas were quantitated.

Overview of Research Sites

Gidole, the administrative center of the Dirashe special woreda, is located 550 km southeast of the capital, Addis Ababa, and 50 km south of Arbaminch, the center of the neighboring Gomo Gofa Zone (Figure 1). It is geographically located on Segen valley plateau at about 1,000 m above sea level connected to the lakeshore of Lake Chamo, and on the slope and plateau of Mount Gadolla with its elevation of 2.561 m (Figure 1). The statistical data by the Dirashe special worenda shows that Gidole has an area of 1,500 square kilometers with a population of approximately 130,000 of which mostly consisted of the ethnically Dirashe people. They are mainly dependent on agriculture for their living.

In consequence of the rule by Ethiopian Empire from the late 19th to the late 20th Century, the Dirashe people differentiate the plateau near the mountain top, which is above about 1,800 m elevation, and the slope and valley plain area, which is about 1,000-1,800 m elevation, calling the former dega and the latter kola respectively. Agricultural fields are mostly located in kola area. Sorghum and *Zea mays* are cultivated in untilled kola area. Kola has a dry climate with its highest average temperature at 31°C and its lowest average temperature at 17°C. In kola area, there are two planting seasons in a year. The precipitation rates differ significantly in those two seasons: The rainfall in the first planting season called kasyana (February to July) is 500 mm on average, while it is 230 mm on average in the second planting season called hagayte (August to November) with large year-to-year differences (Figure 2). Therefore, sorghum, which is draught-resistant, has been cultivated as a staple crop in the region. In the Dirashe area, the yield amount of sorghum per each unit area is the highest in Southern Ethiopia, and its cultivation dates back to over 400 years ago. In the villages along main

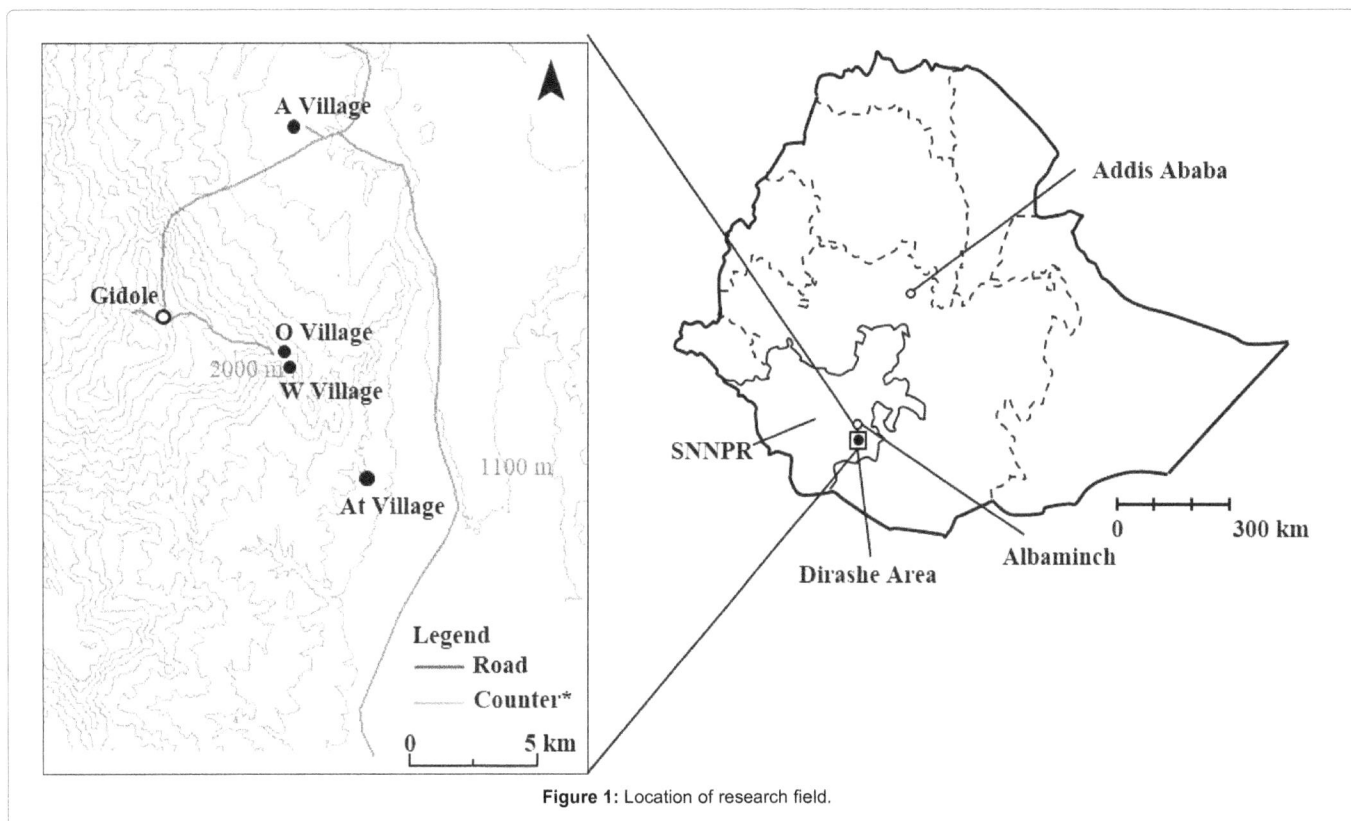

Figure 1: Location of research field.

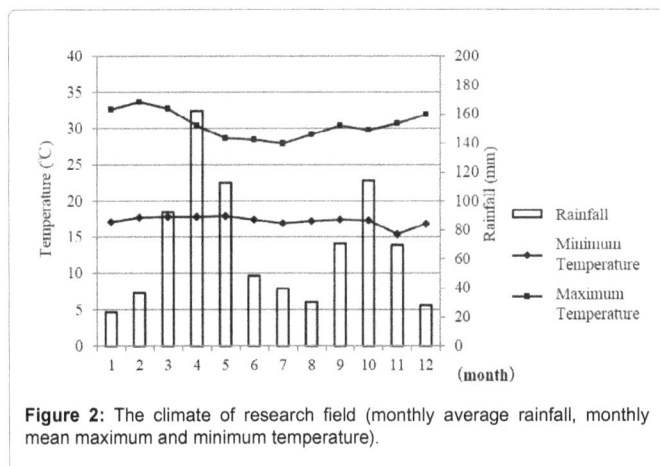

Figure 2: The climate of research field (monthly average rainfall, monthly mean maximum and minimum temperature).

roads, due to its recent integration into a market-oriented economy, people cultivate *Zea mays* for cash during kasyana season. However, due to its low precipitation rate, sorghum is cultivated in hagayte season. The Dirashe people have stored the harvested sorghum in the underground storehouse, *polota*, and used them accordingly. *Polota* is only constructed in valley plain and kola areas where their elevations are below 1,800 m. Therefore, the research activities were conducted only in *kola* area. Two kinds of villages with different characteristics were selected from kola area: 1) the traditional villages built in the 16th Century ('O' village, 'W' village and 'At' village), and 2) the newly developed village along main roads in the late 19th Century due to the government's rural development plan ('A' village). 'O' village and 'W' village are located approximately 5 km away from Gildole and on the steep slope of Mount Gadolla at about 1,800 m above sea level. These

villages are far from roads and no transportation means such as bus and taxi are available. Thus, owing to a lack of communication with urban areas, integration into a market-orient economy progresses only slowly. The local people continue their traditional lifestyle based on cultivation of sorghum while they also cultivate *Zea mays* for cash and food.

'A' village is at about 1,200 m above sea level and located along main roads that connect Gidole to the neighboring city of Arbaminch. The village is thus mostly integrated into a market-oriented economy and people are more likely to use cash compared to 'O' village and 'W' village. Therefore, the residents of 'A' village cultivate *Zea mays* with a large yield and high market demand during kasyana season for cash conversion and self-consumption. They store *Zea mays* with insecticide in the aboveground storehouse called gotera, and sell them for cash when needed. They also cultivate sorghum during hagayte season and store them in *polota* for self-consumption.

Results and Discussion

Location of *Polota*

Visual aspects of location of *Polota*: When constructing *polota*, a certain requirement has to be fulfilled at a construction site. An aquiclude or what the local people call onga must be situated near ground level. An onga layer is hardly water-permeable and exists in every village located in kola area, but the color, texture and hardness of onga are not uniform. Onga from 'A' village at about 1,200 m above sea level is of reddish brown color and smooth texture, and can hardly be crushed even when hit against a rock. On the other hand, onga from 'O' village and 'W' village at about 1,800 m above sea level is of white color and has a polished surface. When scratching its surface with fingernails, it can be slightly scraped off. Onga from 'At' village with its elevation of 1,200 m resembles a marble with various colors of crème, yellow, black, green and white, and has an uneven surface. While it was not damaged at all when hit with a rock, it broke into pieces when pierced strongly with a sharp spear.

The permeable layer developed above onga layer also has various soil characteristics with village-to-village differences. It is of yellow colored hard soil in 'A' village, but of white color and consisted of clay and sand in 'O' village, 'W' village and 'At' village. The former is called sogita while the latter is called pusuka. A pusuka layer can be easily dug with a shovel, but a sogita layer can be crushed only when pierced with a digging bar. Neither of them is suitable for building *polota* as they cannot sustain a flask-like complex shape.

Identification of base material and degree of weathering of Onga layer: Onga in 'A' village, 'O' village and 'At' village and *sogita* above *onga* in 'A' village, and basalt gravel in 'A' village were collected and taken to Japan, then x-ray fluorescence analysis was performed on these samples in order to identify their chemical compositions. Then, the rates of iron (g)/ aluminum (g) and of aluminum (g)/ titanium (g) were calculated based on the results of analyses. All the samples of onga and sogita had less than 10 g% of the rate of iron/ aluminum and above 0.6 g% of aluminum/ titanium, resulting in identifying their base material as basalt. The result of basalt gravel collected in 'A' village also fit between these ranges (Figure 3). The quantity of silicon in these samples was converted into molar mass (mol), and accordingly, the rate of silicon (mol)/ aluminum (g) was calculated. The average rates of silicon/ aluminum were 2.1 mol% for onga from 'A' village, 4.0 mol% for onga from 'O' village, 4.5 mol% for onga from 'At' village and 2.1 mol% for sogita from 'A' village. All the results obtained were lower than that of basalt gravel, which was 5.8 mol%. The results indicate that the base material of these samples, basalt, was weathered. The above stated results of iron/aluminum rates to identify a base material and of silicon/aluminum rates to identify a degree of weathering are jointly shown in Figure 4. The closer the iron/aluminum rates in vertical axis are to each other, the more similar the qualities of base material are among the same basalt category. The lower the rate of silicon/ aluminum in horizontal axis is, the more the weathering is progressed. Although all the samples are identified as basalt-derived, the characteristics of base material and degree of weathering are dissimilar, thus resulting in the different colors, texture and hardness.

A number of onion-shaped weathered rocks can be seen around onga in 'O' village (① in Figure 4). Based on the surrounding environment and ratio of iron, aluminum and titanium, onga in this village is basalt whose iron composition was eliminated due to a redution by weathering. Holistically judging from the white-colored appearance and smooth texture, onga can be identified as kaolinite $(Al_2(Si_2O_5)(OH)_4)$. Kaolinite is derived from decomposition of plagio

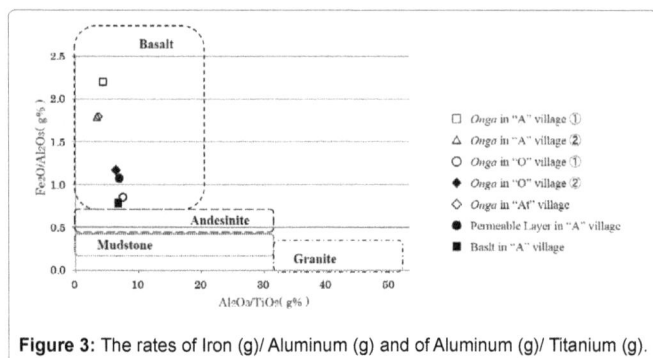

Figure 3: The rates of Iron (g)/ Aluminum (g) and of Aluminum (g)/ Titanium (g).

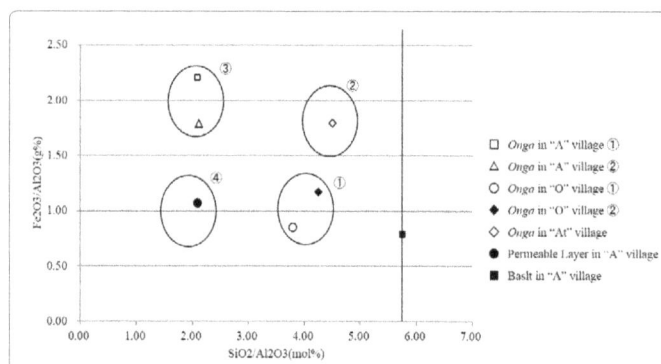

Figure 4: The rates of Iron (g)/ Aluminum (g) and of silicon (mol)/ aluminum (g).

clase in basalt because of weathering. Such rocks and stones generated due to weathering are composed of fine particles and hardly water permeable, and therefore suitable for building *polota*.

The analysis of onga from 'At' village (② in Figure 4) indicates that, among its chemical composition, Ca is as high as 46.7%. In the Great Rift Valley, there are volcanic rocks called carbonatite in places, of which more than 50% of its major constituent is carbonite mineral, and the area near Lake Chamo is also dotted with carbonatite [24]. The rocks found in the vicinity, Ca content and visual aspect of onga indicate that a part of Ca in carbonatite was leached due to the weathering and reducing action, which consequently generated onga. As onga in 'A' village (③ in Figure 4) is of reddish brown color, it is presumably high in iron. Such a reddish brown soil is generated as a result of leaching of Bases and silica, chemical elements that are easily to migrate, and concentration of iron and aluminum that do not usually migrate, and often seen in Africa [25]. It can be also presumed that onga in 'A' village was generated by the similar process.

On the other hand, sogita from 'A' village (③ in Figure 4) is also derived from the same sort of basalt as onga from 'O' village based on the ratio of iron, aluminum and titanium. However, it is highly water-permeable, which means that the original dense structure of basalt has been dissolved. This sample was collected from an area where *polota* were heavily concentrated in 'A' village. The area is low-lying depression into which rainfall is flowed. Presumably, the base material of basalt was physically weathered, refined into fine particles, transformed into silt, channeled into the low-lying depression due to rainfall from the peripheral areas, accumulated on the top of onga, resulting in consisting the current soil layers.

As stated above, onga is either basalt or what the base material with the similar composition to basalt was weathered. It is also assumed that the process and degree of weathering bring about the differences in appearance, texture and hardness. Yet, all the onga layers share a common characteristic of being hardly water-permeable. Weathering can be categorized into "physical weathering" and "chemical weathering". Physical weathering causes grain refining of objects, thus transforms its physical features, whereas chemical weathering refers to a transformation in chemical composition by leaching or weathering and reducing action. By chemical weathering, the elemental composition of object is altered, but the volume remains the same. As onga is water-impervious in the same manner as rock, it is assumed that onga is chemically weathered, but not physically. Therefore, though the composition of onga remains dense, it is possible to shape it with a sharp-edged bar. On the other hand, since the soil layer above onga is what basalt is transformed due to physical weathering, it is water-permeable. *Polota* is staunch in structure, and therefore once it is built, it can be passed down for generations.

Structure of *Polota*

A cross-section diagram of *polota* and names of each section in the Dirashe language are shown in Figure 5. The diagram is drawn based on the actual measurement of *polota* by the research assistant, which was used by a family in 'A' village. *Polota* is of a rounded flask-like shape with a circular-cylinder-shaped section on the top of it as seen in the diagram. The top of the circular cylinder section is led to ground surface and the flask-like shaped section is dug underground. The circular cylinder section is called wake or waketa (hereinafter, wake). The flask-shaped section is called makara or matameta (hereinafter, makara), and it can be further divided into smaller sections with different names as follows. The mouth section of makara is named matabu, which leads to kabada, a neck-shaped section below. The main flask-like section with its enlarging diameter as it goes deeper is called talashade. People place sorghum in the flask-shaped makara and close matabu with a disk type stone as a lid. Then, the lid will be tightly sealed with clay and wake will be entirely covered with mud. As a result, the sealed *polota* cannot be found from the ground. In *polota* shown in Figure 5, the diameter of *wake* is 80 cm and the depth, 50 cm. The depth of makara is 200 cm with matabu, 40 cm in diameter. The diameter of matabu remains consistent and it goes deeper to form kabate. Talashade has a maximum diameter of 160 cm at the depth of 180 cm from matabu. This *polota*'s volumetric capacity is approximately 2,200 liters and it can store as much sorghum as 1,700 kg. As *polota* has a small entry channel despite of an extended storage space below, rainfall permeated from ground surface can rarely get inside. Yet, the structure allows *polota* to keep a high volumetric capacity.

Furthermore, it should be noted that the flask-like structure of *polota* also plays a significant role in providing a high level of safety when in use. Two men in a pair are required for the procedure to take out sorghum from *polota*: one stands at the edge of matabu, the other goes inside makara. A man who goes inside puts sorghum in a bamboo basket called kunna of 40 cm in diameter and passes it to the other outside. They will repeat this process a number of times. After taking out the required amount of sorghum, the man stationed outside will bring up the other inside by pulling up his hands. Then, *polota* will be once again sealed with a stone lid, mud and clay. If the diameter drastically extends as going downwards, the weight of the man standing at the edge of matabu can possibly cause the entrance of *polota* to collapse when carrying out the aforementioned procedure.

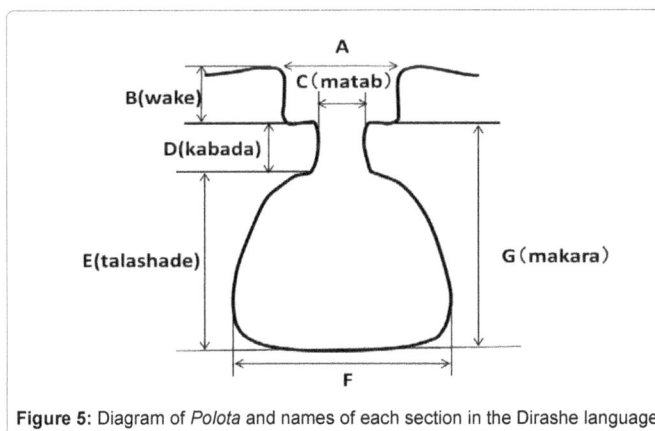

Figure 5: Diagram of *Polota* and names of each section in the Dirashe language.

The flask-like shape with a neck part like matabu can disperse load from the above, increasing safety when in use. The author and research assistant measured the size of *polota* from inside: 4 in 'A' village, 6 in 'O' village and 8 in 'W' village (Table 1). The result is that, as for makara, the average depth is about 190 cm, the average diameter of entrance part is about 45 cm and the maximum diameter is about 170 cm. The *polota* that the author investigated in 'A' village basically had the average size. However, focusing on wake, the diameter is about the same in all *polota*, which is approximately 90 cm, but the depth varies significantly from 30 cm to 100 cm. This variation is attributed to the thickness of water-permeable layers where *polota* are constructed in each location as stated below.

A construction plot for *polota* is required to have a water-permeable soil layer of tens of centimeters on the surface as well as a thick onga layer. Firstly, the water-permeable layer is dug down in a circular cylinder shape with its diameter at 90 cm in order to form wake. When reaching to the hardly water-permeable onga layer, the diameter is reduced to about a half. After digging farther to a certain extent, the flask-like makara will be formed, gradually extending its diameter, thus forming a shallow curve. That is to say, the thickness of water-permeable layer in each location causes differences in the depth of wake of 18 *polota*. Makara is, on contrary, situated as deep as 160 cm to 190 cm and in the hardly water-permeable layer of *onga*. Some construction plots have the water-permeable layer below onga layer, when dug farther down, but makara is only constructed within onga layer. The flask-like space to store grains can only be built in onga layer. However, *polota* is never built in such a location where onga layer is surfaced on the ground. The water-permeable layer above onga layer is soft soil with lots of voids, so it can be easily and evenly backfilled even after digging over. Therefore, this layer plays such roles as reducing a temperature change in *polota*, lessening an impact of human or car passing over, preventing rainfall from flowing into *polota* in a rainy season and hiding the location of *polota* from the sight of thieves.

Function of *Polota*

Temperature and humidity in *Polota*: One of the notable characteristics of *polota* is its high airtightness as it is only built in the hardly water-permeable onga layer of dense structure, and as the result, a specific environment can be created when sorghum is stored inside. Temperature and humidity of a *polota*, which was filled with sorghum up to two thirds, in 'A' village were measured for two days. By slightly opening a lid, a thermo-hygrometer was inserted and placed inside *polota*, and then it was completely sealed again. For comparison, a thermo-hygrometer was also placed in a house with a thatched roof and mud wall for two days. The results are shown in Figures 6 and 7. The temperature and humidity remained the same in *polota*, 31°C and 92% respectively. On contrary, the temperature in the house varied from 20°C to 30°C and the relative humidity varied from 45% to 80%, fluctuating from the daytime to the night (Figures 6 and 7). It can be stated that *polota* can achieve an excellent level of airtightness since both the temperature and humidity inside were kept stable. Insect development is accelerated with temperature up to 35°C, but thereafter, insects become sluggish or immobile as the temperature increases and they will perish at 42°C and above. However, fungi and bacteria can continue to proliferate until the temperature hits about 80°C. Therefore, this biological activity needs to be regulated in order to successfully preserve the stored materials [26]. Judging from the results collected, the temperature and humidity inside *polota* provide the environment suited for proliferation of insects, bacteria and fungi.

Part	Minimum (cm)	Average (mm)	Maximum (mm)
A	70	95	115
B	40	60	90
C	40	45	55
D	30	45	60
E	160	190	230
F	140	170	220
G	190	210	260

Table1: Minimum, average and maximum for each part size of *Polota*.

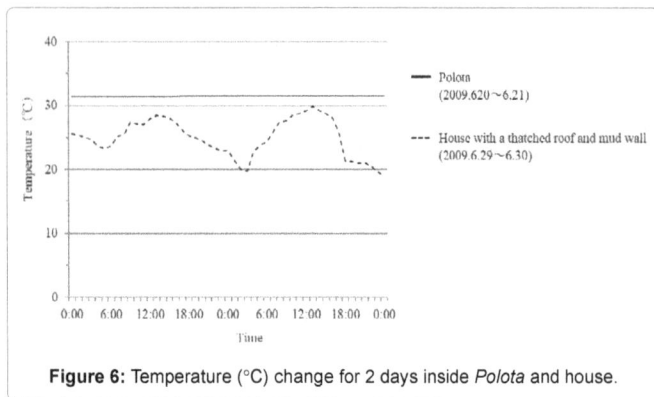

Figure 6: Temperature (°C) change for 2 days inside *Polota* and house.

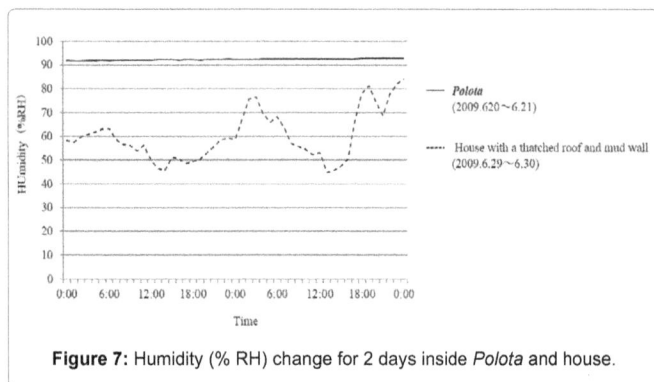

Figure 7: Humidity (% RH) change for 2 days inside *Polota* and house.

Oxygen concentration in *Polota*

Atmospheric gases were collected from a *polota* filled with sorghum up to two thirds and from an empty *polota*, and their oxygen concentrations were measured (Table 2). The oxygen concentration of the *polota* with sorghum inside was 2.7% while that of the empty *polota* was 17%. Both of the results are below the oxygen concentration of standard atmosphere at 21%, and that of *polota* with sorghum is drastically lower. Human suffer from oxygen deficiency when the oxygen concentration falls under 18%, and they will pass away within 6 to 8 minutes at 8% and instantly at 6% and below [27]. The oxygen concentration inside *polota* is normally maintained at the low level, and storage of sorghum causes it to be lower, creating the environment with the extremely low oxygen concentration. The Dirasha are well aware of the dangerous condition of newly opened *polota*. They leave the lid of *polota* open for two hours up to two days for ventilation before going inside *polota* to take out the stored materials. In dry season, many *polota* are left open, but nobody ever tries to go inside to steal the stored materials even when no one is around. In some rare cases, people go inside *polota* before the ventilation inside is completed and pass out owing to oxygen deficiency. When the author visited 'O'

village in February 2011, a father and son passed out in *polota* owing to the deficiency of oxygen when they were taking out the stored grains. The atmospheric condition lacking in oxygen inside *polota*, which can impact human activity, is highly hazardous to insects as well. While it seems that *polota* provide a preferred environment for proliferation of insects due to high levels of temperature and humidity, the low oxygen concentration prevents growth of insects and therefore prevents damage of sorghum. Furthermore, the low oxygen concentration inside *polota* prevents aerobic bacteria from growing inside as well.

Carbon dioxide concentration in *Polota*

In the same manner as the measurement of oxygen concentration stated above, atmospheric gases were collected from the same *polota* filled with sorghum up to two thirds and from the same empty *polota* in order to quantitate the carbon dioxide concentration (ppm) (Table 2). The carbon dioxide concentration of the *polota* with sorghum was 160,000 ppm and that of the empty *polota* was 40,020 ppm. Normally, the atmospheric gas has carbon dioxide concentration of 380 ppm [28], which is exponentially lower than these results. How was such a space as this created? While the empty *polota* had a relatively high level of carbon dioxide concentration at 40,020 ppm, which was higher than the standard rate at 380 ppm, it was not as high as that of *polota* filled with sorghum up to two thirds. *Polota* is air tight as previously stated. Yet, in rainy season, a small amount of rainfall can get inside through gaps around lid or the water-permeable layer above *onga* layer. The temperature inside *polota* is maintained at 31°C, which is suitable for germination. A small fraction of seeds can germinate with scarcely available oxygen and water. When opening *polota*, the upper portion of stored sorghum is covered with a styrene-foam-like seedcake called putcot and kama. At the same time, a portion of stored sorghum at the bottom is mixed with mud and degenerated into a seedcake called mutaitade. It is considered that anaerobic bacteria called spergillus and penicillium living in *polota* attach to the surface of germinated seeds and multiply proliferously [29,30], and transform the part of sorghum into seedcake-like condition [12,16,31]. Seeds consume oxygen and emit a large amount of carbon dioxide in the process of germination. This is possibly causing and maintaining the environment inside *polota* to be low in oxygen and high in carbon dioxide [32-36].

As for some of grains and pulse, protein in seeds and carbon dioxide are physically fused in a high carbon dioxide environment, resulting in preventing deterioration in quality [37]. Prevention of quality deterioration specifically means reducing seed respiration, preventing B1 contents in seeds from degenerating, maintaining a peroxidase activity, controlling formation of free fatty acid caused by oxidation of lipid component and generation of carbonyl compound [37,38]. Such a method of storing grains as this by physically fusing protein in seeds and carbon dioxide in the artificially created environment with a high level of carbon dioxide is typically called "carbon dioxide hibernation storage method." In Japan, rice stored by this method is named "Touminmai (hibernated rice)." Rice is placed in a gas-barrier film bag whose carbon dioxide concentration is artificially increased by injecting carbon dioxide in order to hibernate rice for long-term storage [39-44]. To put it briefly, the quality deterioration of sorghum

	Oxygen Concentration (%)	Carbon Dioxide Concentration (ppm)
Polota filled with Sorghum up to Two	17.0	40020.0
Empty *Polota*	2.7	160000.0

Table 2: Oxygen Concentration (%) and carbon dioxide concentration (ppm) inside *Polota* filled with Sorghum up to two thirds and from an empty *Polota*.

stored in *polota* is also most likely to be controlled by the physical fusion of protein in seeds and carbon dioxide in the environment of the high carbon dioxide level.

Conclusion

Most of farmers in developing nations including Africa utilize local storage methods without air conditioning. These local storage methods are conventional and usually built with woods, plant fibers and argilliferous soil. The storage method can be categorized into three systems; 1) open storage system, 2) semi-open storage system and 3) sealed storage system. If harvested crops are still moist, open storage system should be suitable. The framework of the open storage system is made with woods and the roof with straws. Spikes and rachis are stratified on the wooden foundation for drying. Alternatively, crops are hung from the roof, and in some cases, above fire. This drying process can control fungal and bacterial growth, but will be more likely to risk crops damaged by mice and birds. Semi-open system is particularly prevalent in semi-arid regions. The wooden framework is covered with matted walls made out of either twigs or straws. Whereas it is rather resistant to rain and wind, insects can easily proliferate due to its lower level of ventilation. The third category of sealed storage system is built either by daubing the walls of semi-open storage system with argillaceous soil or by piling stone materials or baked clay materials. This type of storage is most widely used around the world. Storage pits are also under this category.

However, storage losses by mice, birds and insects in these conventional storage facilities are significantly large, and among those, insect damage is especially severe. In Africa, sorghum is widely cultivated as a main crop due to its high plasticity to the environment. At the same time, however, the problem is that storage loss of sorghum is quite large. For example, about 15% weight losses and 38.7% grain damage were reported on sorghum that was stored in traditional storage facilities in Ethiopia. And, also in Sudan, storage losses of 2.5-7.6% was recorded; whereas weight losses ranged from 6.1-14.3% were reported on sorghum stored for 4 months in traditional granaries in Kenya.

When stored, grains are internally deteriorated by the increment of seed respiration, and also externally damaged by insects, fungi, rodents and birds. Internal deterioration means that the quality and nutrition quantity decrease because seed respiration breaks down starchy components in seeds into oxygen, carbon dioxide, water and heat. External deterioration refers to grain weight loss by vermin or putrefaction due to fungal and bacteria proliferation. In Africa, grain damages caused by maize weevil (*Sitophilus zea mays*), grain moth (*Sitotroga cerealella*) and rust-red flour beetle (*Tribolium castaneum*) are particularly severe and sorghum is particularly prone to the damages by these insects. Additionally, sorghum grains generate carbon dioxide due to grain heating when placed in a closed space and it increases temperature and humidity inside. This consequently promotes growth of fungi and bacteria. The proliferation of fungi and bacteria inhibits germinating capacity of grains or causes discoloration of germs. Furthermore, it also reduces nutrition in grains, changes quality, causes grains to have musty or bad smell or oxidative rancidity, generates toxic substances, transforms grains into a seedcake-like form and reduces quantity. In order to prevent the proliferation of fungi and bacteria, it is a prerequisite that a storehouse is well ventilated and maintains the temperature and humidity inside at the low levels. However, the more gaps a storehouse structure has, the more the nutritional decomposition progresses as wind and rain cause grains to repeat swelling and drying processes. Besides, rodents and birds can easily come inside and damage the stored grains. Therefore, it is necessary to manage hygiene, temperature, humidity and air circulation in a storehouse, and exterminate vermin in order that the aforementioned multiple factors do not affect the stored grains. Nevertheless, it is still difficult for the local storage systems without air-conditioning facilities in developing nations to preserve grains for more than one year.

While *polota* in Dirashe special worenda, in which the temperature is maintained at 31°C and the relative humidity at 92%, can most likely induce both internal and external deteriorations, *polota* can store sorghum for about 20 years. The high storage performance of *polota* is attributed to the dense structure of layer in which *polota* is constructed since it can maintain the low concentration of oxygen and high concentration of carbon dioxide. The *polota* with sorghum inside has only 2.7% of oxygen concentration, and not only human, but also insects are unable to survive or grow. As a consequence, insects will not damage stored sorghum. In addition, the low level of carbon dioxide inside *polota* prevents aerobic bacteria from proliferating. The *polota* with sorghum inside also maintains the high level of carbon dioxide concentration at 160,000 ppm. If sorghum is placed in such a space of high carbon dioxide concentration as *polota*, protein in grains and carbon dioxide in air are physically fused, and it will prevent grain deterioration by reducing seed respiration, preventing degradation of B1 contents in seeds, sustaining a peroxidase activity, preventing oxidation of lipid compound to form free fatty acid, and preventing generation of carbonyl compound. This storage technique generally called "carbon dioxide hibernation storage method" is applied not only to the storage of grains in the Dirashe area, but also to the storage of rice in Japan.

The *polota*'s characteristics of low oxygen and high carbon dioxide concentration are realized due to the highly dense structure of onga layer in which *polota* is constructed. Onga is a hardly water-permeable layer generated as a consequence of chemical weathering of basalt. Owing to its dense structure, it is possible to retain the condition of low concentration of oxygen and high concentration of carbon dioxide generated by seed respiration of sorghum. This condition vests *polota* a long-term storage performance capability that other sort of storage pits lacks. In most of rural areas in developing nations, the local residents do not have many means of obtaining cash other than agricultural activity. As they do not have cash to spare, they have difficulty in purchasing food at the time of lean crop, and therefore prone to food insecurity. In the semi-arid Dirashe region, *polota* is indispensable as the people can ensure the reliable amount of food. Even today, *polota* is still preferred over pesticide and aboveground storage due to its long-term storage capacity without causing much loss in both quality and quantity compared to other kinds of storage pits, which has become a thing of the past in many other parts of the world.

In most of the rural areas in developing nations, people live a self-sufficient lifestyle, and particularly in Africa, it is the role of small-scale farmers to produce most of food. In such areas, people are reliant on the effective storage method for their staple food, which enables to minimize loss in quality and quantity, until the next good harvest, possibly more than a year, in the case of bad harvest.

Acknowledgements

Firstly, I would like to convey my gratitude to the village residents in Dirashe special woreda who warmly accepted me as one of them and took the most integral part in helping my research. I also would like to thank Prof. Juichi Itani and Prof. Gen Yamakoshi, Prof. Shigeru Araki and Prof. Masayoshi Shigeta at Graduate School of Asia and African Area Studies, Kyoto University (ASAFAS) for their advice and guidance which led me through this study. This study was supported by Sasakawa Scientific Research Grant of the Japan Science Society (2009) and JSPS KAKENHI Grant Number 15K16188 and 25300012.

References

1. Driver HE, Massy WC (1958) Comparative Studies of North American Indians. American Anthropologist 60: 1204-1206.

2. Testart A, Richard G, Forbis, Hayden B, Ingold T, et al. (1982) The significance of food storage among hunter-gatherers: residence patterns, population densities, and social inequalities. Current Anthropology 23: 523-537.

3. Sakaguchi T (2003) Uncompleted Archaeological Library. The Study of Storage Pits in the Jomon Period of Japan, Umpromotion-214, Tokyo, Japan.

4. Bray F (2007) Needham's Science and Civilisation in China: Agriculture in China. Translated by Hisao Furukawa, Kyoto University, Kyoto.

5. Shazali MH (1992) Matmura (underground pit) storage of sorghum in the Sudan. Bull Grain Technol 30: 207-212.

6. Reynolds PJ (1974) Experimental Iron Age Storage Pits: An Interim Report. Proceedings of the Prehistoric Society 40: 118-131.

7. Bersu G (1940) Excavations at Little Woodbury, Wiltshire. Proceedings of the Prehistoric Society 6: 30-111.

8. Abdalla AT, Ali KH, Stigter CJ, Adam IA, Mohamed HA, et al. (2002) Impact of Soil Types on Sorghum Grain Stored in Underground Pits in Central Sudan. J Agric Eng Res 11: 219-229.

9. Abdalla AT, Ali KH, Stigter CJ, Bakhiet NI, Gough MC, et al. (2002) Traditional Underground Grain Storage in Clay Soils in Sudan Improved by Recent Innovations. Tropicultura 20: 170-175.

10. Blum A, Bekele A (2002) Storing Grains as a Survival Strategy of Small Farmers in Ethiopia. J Int Agr Ext Ed 9: 77-83.

11. Gilman GA (1968) Storage Problems in Ethiopia with Special Reference to Deterioration by Fungi. Publ Trop Prod Inst 48: 1-50.

12. Niles EV (1976) The Mycoflora of Sorghum Stored in Underground Pits in Ethiopia. Trop Sci 18: 115-124.

13. Lynch B, Reichel M, Salomon Berhane, Kuhne W (1986) Examination and Improvement of Underground Grain Storage Pits in Alamaya Wereda of Hararghe Region, Ethiopia. Alamaya: Alamaya University of Agriculture.

14. Lemessa F (2008) Under and above ground storage loss of sorghum grain in eastern hararge, Ethiopia. Agricultural Mechanization in Asia, Africa, and Latin America 39: 49-52.

15. Dejene M (2004) Grain Storage Methods and Their Effects on Sorghum Grain Quality in Hararghe, Ethiopia. Swedish University of Agricultural Sciences.

16. Neergaard P (1979) Seed Pathology, Vol. I and II. Revised Edition. The MacMillan Press, London.

17. Bothast RJ (1978) Fungal deterioration and related phenomena in cereals, legumes and oilseeds. In: Hultin HO, Milner M (eds.) Post-harvest biology and biotechnology. Food and Nutrition Press, Westport, Connecticut, USA, pp: 210-243.

18. Meronuck RA (1987) The significance of fungi in cereal grains. Plant Disease 71: 287-291.

19. Copeland LO, McDonald MB (1995) Principles of seed science and technology. Chapman & Hall, New York.

20. Janicki LJ, Jr Green VE (1976) Rice losses during harvest, drying and storage. II Riso 25: 333-338.

21. Christensen CM, Kaufmann HH (1974) Mycroflora. In: Christensen CM (ed.) Storage of Cereal Grains and Their Products. American Association of Cereal Chemists, Minneapolis, MN, pp:158-192.

22. Shashidhar RB, Ramakrishna Y, Bhat RV (1992) Moulds and mycotoxins in sorghum stored in traditional containers in India. J Stored Prod Res 28: 257-260.

23. Boxall RA (1974) Underground storage of grain in Harar Province, Ethiopia. Trop Stored Prod Info 28: 39-48.

24. George RM, Rogers NW (2002) Plume Dynamics Beneath the African Plate Inferred from the Geochemistry of the Tertiary Basalts of Southern Ethiopia. Contributions to Mineralogy and Petrology 144: 286-304.

25. Araki S (1996) Tsuchi to Miomborin-Bemba no Yakihatanoukou to Sono Henbou (Soil and Miombo Woodland-Burn Agriculture and its Change in Bemba). In: Tanaka, Jiro, Mokoto Kakeya, Mitsuo Ichikawa, Itaru Ohta (eds) Ecological Anthropology in African Societies 2. Academia, pp: 306-338.

26. Smith CV, Gough MC (1990) Meteorology and grain storage. Technical note 101, WMO, Geneva

27. Ministry of Health, Labour and Welfare (2008) Sansoketsuboushou tou no Roudousaigai Hasseijoukyou no Bunseki nitsuite (Analysis on Occupational Accidents Occurrence including Oxygen Deficiency). Circular Notice from Safety and Health Department, Labour Standards Bureau, No. 0701001.

28. WMO (2014) WMO Greenhous Gas Bulletin. No. 10.

29. López LC, Christensen CM (1963) Factors influencing invasion of sorghum seed by storage fungi. Plant Dis Rep 47: 597-601.

30. López LC, Christensen CM (1967) Effect of moisture content and temperature on invasion of stored corn by Aspergillus flavus. Phytopathology 57: 588-590.

31. Brown RL, Cleveland TE, Payne GA, Woloshuk CP, Campbell KW, et al. (1995) Determination of resistance to aflatoxin production in maize kernels and detection of fungal colonization using an Aspergillus flavus transformat expressing Escherichia coli ß-glucuronidase. Phytopathology 85: 983-989.

32. Bekele AJ, Ofori DO, Hassanali A (1997) Evaluation of Ocimum kenyense London, (Ayobangira) as Source of Repellents, Toxicants and Protectants in Storage against Three Major Stored Product Insect Pests. J Appl Entomology 121: 169-173.

33. Eticha F, Tadesse A (1999) Insect pests of farm-stored sorghum in the Bako area. Pest Manage J Ethiopia 3: 53-60.

34. Fields RW, King TH (1962) Influence of storage fungi on deterioration of stored pea seed. Phytopathology 52: 336-339.

35. Gwinner J, Harnisch R, Mück O (1996) Manual of the Prevention of Post Harvest Grain Losses. GTZ-Postharvest Project.

36. Hill DS (1990) Pests of Stored Products and Their Control. Belhaven Press, London.

37. Mitsuda H, Kawai F, Kuga M, Yamamoto A (1972) Carbon Dioxide Gas Adsorption by the Cereal Grains and its Application to Packaging (Part 1). J Japanese Soc Food Nutri 25: 627-631.

38. Mitsuda H, Kawai F, Yamamoto A, Kimura Y (1971) Changes of Qualities in Rice during Storage: Studies on Under-water Storage of Cereals (Part 4). J Japanese Soc Food Nutri 24: 216-226.

39. Abdalla AT, Stigter CJ, Mohamed HA, Mohammed AE, Gough MC, et al. (2001) Effects of wall linings on moisture ingress into traditional grain storage pits. Int J Biometeorol 45: 75-80.

40. Nyambo B T (1993) Post Harvest Maize and Sorghum Grain Losses in Traditional and Improved Stores in South Nyanza district, Kenya. Int J Pest Manag 39: 181-187.

41. Poswal MT, Akpa AD (1991) Current trends in the use of traditional and organic methods for the control of crop pests and diseases in Nigeria. Trop Pest Manage 37: 329-333.

42. Qasem SA, Christensen CM (1958) Influence of moisture content, temperature, and time on the deterioration of stored corn by fungi. Phytopathology 48: 544-549.

43. Seifelnasr YE (1992) Stored grain insects found in sorghum stored in the central production belt of Sudan and losses caused. Trop Sci 32: 223-230.

44. Sinha KK, Sinha AK (1992) Impact of stored grain pests on seed deterioration and aflatoxin contamination in maize. J Stored Prod Res 28: 211-219.

19

Physicochemical Characterization of Gum of Some Guar (*Cyamposis tetragonoloba L. Taup*) Lines

Mawada E Yousif[1]*, Babiker E Mohamed[1] and Elkhedir AE[2]

[1]*Department of Food Science and Technology, Faculty of Agriculture, University of Khartoum, Sudan*
[2]*Food Industry Department, Industrial Research and Consultancy Center (IRCC), Khartoum, Sudan*

Abstract

The objective of the present investigation was to determine the physical and chemical properties of five guar (*Cyamposis tetragonoloba L. Taup*) lines, planted in the experimental from of the Faculty of Agriculture, University of Khartoum. The results obtained from all lines studied, GM2, GM6, GM8, GM9 and GM34 were compared those from a known control cultivar, L53. Physical characterization of the guar gum extracted included the determination of pH, relative viscosity, refractive index, solubility and optical density, while the chemical studies involved the determination of the proximate chemical composition of the tested Guar lines namely: moisture, ash, oil, fiber, protein and carbohydrate. The physical properties showed significant differences ($P \leq 0.05$) in optical density, solubility and pH among lines and between lines and control, while no significant differences ($P \leq 0.05$) were observed in refractive index and viscosity. The thousand-seed weight ranged from 30.21 g-30.75 g, pH 7.16 to 7.40%, relative viscosity 0.1000 to 0.197 cps, refractive index 1.34% to 1.35%, solubility 76.67% to 89.83% and optical density 0.035 to 0.047. The results showed significant differences ($P \leq 0.05$) among lines and between lines and control in all physical parameters studied. The moisture content ranged from 8.37% to 8.80%, ash content 3.33% to 4.96%, fat content 1.70% to 2.47%, fiber content 10.53% to 11.83%, protein 25.80% to 30.52%, carbohydrate 43.80% to 48.77%. The results showed significant differences ($P \leq 0.05$) among lines and between lines and control in the various levels of guar lines chemical component. All five lines are characterized by possessing physical and chemical properties related to the control sample and to those from previous findings lines GM2 and GM6 proved to possess the best physical and chemical properties a among others and those reported from previous works.

Keywords: *Cyamposis tetragonoloba L. Taup*; Guar gum; Physicochemical properties

Introduction

Guar gum is one of the outstanding representatives of new generation of plant gums. It's source is an annual pod-bearing, drought resistant plant, called Guar, or cluster bean (*Cyamopsiste tragonolobuosr L. Taup*), belonging to the family Leguminosae, Genus Cyamposis and Species *Cyamposiste tragonoloba*. It has been grown for several thousand years in India and Pakistan as a vegetable, and a forage crop [1]. In the Sudan, guar plant Was known as a wild plant in the Red Sea mountains and Arashekol mountains of White Nile state. The advent of guar gum production in Sudan will pressurize gum Arabic to better commercial production and quality in order to compete in the national and international markets. The gum is contained within a portion of the seed known as the endosperm which is 35% to 42% of seed weight. The endosperm is ground into powder by the usual mechanical processing technique that does not produce completely pure endosperm. Therefore, the gum is not perfectly pure but contains small amounts of hull and germ, this contamination lower the gum quality, but does not harm its suitability as food additives [2]. Guar gum is a white to yellowish white powder and is nearly odorless. Guar gum is a natural high molecular weight hydro colloidal polysaccharide composed of galactan and mannan units combined through glycosidic linkages, which may be described chemically as galactomannan [3]. Guar gum is an economical thickener and stabilizer. The special properties of guar gum known in India make it most suitable for various industrial applications. Chemically, guar gum is a polysaccharide composed of the sugars galactose and mannose. The backbone is a linear chain of β 1, 4-linked mannose residues to which galactose residues are 1, 6-linked at every second mannose, forming short side-branches. Guar Gum is known as one of the best thickening additives, emulsifying additives and stabilizing additives [4], it has a polymeric structure, containing several hydroxyl groups. The various derivatives or industrial grades of guar gum are manufactured by reaction of these hydroxyl groups with chemicals [5]. There are more than 300 industrial applications of guar gum it is mainly used as natural thickener, emulsifier, stabilizer, bonding agent, hydrocolloid, gelling agent, soil stabilizer, natural fiber, flocculants and fracturing agent. Guar gum is soluble in cold water but insoluble in most organic solvents and has strong hydrogen bonding properties [1]. It has excellent thickening, emulsion, stabilizing and film forming properties. It is compatible with a variety of inorganic and organic substances including certain dyes and various constituents of food [6]. The seeds of guar are split and the endosperm and germ can be separated from the endosperm by sieving. Through heating, grinding and polishing process the husk is separated from the endosperm halves and the refined guar gum split are obtained. The refined guar splits are then treated and converted into powder [1]. The objective of this study was to extend the knowledge concerning the physical and chemical properties of guar gum. The attainment of this objective required the determination of the physical and chemical properties of five guar lines (GM2, GM6, GM8, GM9 and GM34) as compared to a known control sample (L53).

*Corresponding author: Mawada E Yousif, Department of Food Science and Technology, Faculty of Agriculture, University of Khartoum, Sudan
E-mail: mawadelfatih15@gmail.com

Materials and Methods

Materials

Guar samples: Seeds of five guar lines (GM2, GM6, GM8, GM9 and GM34) together with a known control (L53) sample were obtained from the Department of Agronomy, Faculty of Agriculture, University of Khartoum. These lines were planted in the experimental farm of the faculty of Agriculture university of Khartoum (Shambat area). After harvesting, guar seeds were sieved to remove broken seeds, soil particles and foreign material.

Methods

Preparation of the samples: Guar seeds were ground to fine particle size, using milling machine, sieved by 0.4 mm mesh sieve, stored in polyethylene bags, and kept in a refrigerator at 4°C for further analysis.

Guar gum extraction: Gum extraction was carried out according to AOAC [7]. Guar seeds were soaked in distilled water for about 10 hours. The seeds were swollen and the outer layer (Hull) was removed easily. It was observed that twenty-four hours are required for the seed coat to be removed from the seed. The hull was opened into two separate parts the medium layer (endosperm) and the inner portion (germ). Then these extracts were put in an oven and dried at a temperature of 100°C for 20 minutes. After extraction, the endosperm was ground to fine particle size using milling machine sieved by 0.4 mm mesh sieve and stored in polyethylene bags.

Determination of physical properties: The determination of the physical properties namely pH, relative viscosity, refractive index, and optical density were done according to AOAC [7], while solubility was determined accord to the procedure of Osman [8].

Proximate chemical composition: The proximate chemical composition from the guar gums were carried out according to AOAC [7]. Carbohydrates were determined by difference.

Statistical analysis: Data were subjected to Duran's Multiple Range Test to evaluate the statistical significance using Analysis of Variance (ANOVA), and the significance was established at P ≤ 0.05.

Results and Discussion

General properties of guar seeds

Thousand kernel weight (gram): The 1000 kernel weight (gram) of the five samples under study as well as of the control sample was determined. The obtained results are presented in Table 1. The values of thousand kernel weight of the five guar lines ranged from 30.21 g to 30.75 g, while that of the control sample (L53) was 30.61g, lying within same range. The values obtained from all samples, including the

Guar Lines	1000 kernel weight (g)
GM2	30.25^e (± 0.31)
GM8	30.51^c (± 0.29)
GM34	30.75^a (± 0.27)
GM9	30.37^d (± 0.30)
GM6	30.21^f (± 0.34)
L53 (control)	30.61^b (± 0.23)
L.S.D	0.09
*Means not sharing the same letters in the same Column are significantly different (P ≤ 0.05). *Each value in the table is mean of three replicates ±S.D.	

Table 1: Thousand kernel weight (gram) of guar lines as compared to the control sample.

control, were significantly different (p ≤ 0.05). The values of the 1000 kernel weight of the five guar lines, under study, and the control sample are in agreement with the lower limit of the range from 30.75 to 31.57 g reported by Eldirany [8], while they were lower than the range 35.6 g- 35.7 g reported by Sabah Elkhier [9].

Physical properties of guar gum: The physical properties of the guar samples investigated were viscosity, pH, refractive index, optical density and solubility. The results obtained are presented in Table 2.

pH: The results showed that the pH-values of the solution prepared from five lines investigated ranged from 7.16 to 7.40. The highest value (7.40) was obtained from line GM9, while the lowest value (7.16) was from line GM2. There were no significant differences (P ≤ 0.05) in pH-value observed among the tested lines except in line GM2 which was significantly different from all other lines. The pH-value of the control sample was significantly different (P ≤ 0.05) from all samples. Results obtained in this study are higher than the values of 4.07 to 5.99 reported by Sabah Elkhier [9], and lower than the range from 7.5 to 10.5 and 8.00 to 9.00 reported by Loggale [10], and Eldow [11]. The variation of pH values might be attributed to genetical variation.

Relative viscosity: The results showed that the relative viscosity of the five lines investigated ranged from 0.100 cps-0.197 cps. All samples showed the same relative viscosity (0.197 cps) with the exception of line GM9 which gave 0.1000 cps. The relative viscosity of the standard commercial cultivar was 0.197cps, a value which was similar to values obtained from four of the lines studied. Relative viscosity obtained from all samples was not significantly different (P ≤ 0.05). The relative viscosity values obtained in this study were lower than the value 1.30 cps reported by Loggale [10]. The variation in relative viscosity might be attributed to genetical variations.

Refractive index: A narrow range from 1.34 to 1.35 refractive index was obtained for the five lines under study. The control sample gave a refractive index of 1.34. GM2 and GM8 were not significantly different (P ≤ 0.05) in their refractive index. GM34, GM9 and GM6 were also not significantly different, however these were significantly different (P ≤ 0.05) from the two lines, GM2 and GM8. The results obtained are close to the same value reported by both Eldirany [8] and Sabah Elkhier [9] which was 1.3237.

Solubility: The results showed that solubility of the five lines investigated ranged from 76.67 to 89.83%. The highest value (89.83) was obtained from line GM34, while the lowest value (76.67) was found in line GM2. There was no significant difference (P ≤ 0.05) in solubility among the tested samples except that from line GM2 which is significantly different (P ≤ 0.05) from all other tested lines. The solubility of the standard commercial line was 86.51; a value which is not significantly different (P ≤ 0.05) from the values obtained from the five lines investigated. The results of this study showed higher values than those ranging from 70.53 79.1% reported by Eldirany [8]. Earlier finding showed solubility in guar gum to range from 75.23 to 85.55 and 72.67 to 81.88 as reported by loggale [10] and Eldow [11] respectively. The small variation in solubility between the extracted guar gum and the values reported earlier might be attributed to genetical factors.

Optical density: The results showed that optical density of the five lines investigated ranged from 0.035 to 0.047. The highest value (0.047) was found in line GM2, while the lowest value (0.035) was obtained from line GM34. There were no significant difference (P ≤ 0.05) in optical density observed among all tested samples. The optical density of the standard commercial cultivar was 0.037, a value which is lower than that obtained from the five lines, under study with no significant

Lines	pH	Relative Viscosity (cps)	Refractive index	Solubility (%)	Optical density
GM2	7.16ª (± 0.07)	0.197ª (± 0.04)	1.35ª (± 0.30)	76.67ª (± 0.21)	0.047ª (± 0.20)
GM8	7.24ªᵇ (± 0.05)	0.197ª (± 0.06)	1.35ª (± 0.16)	80.00ᵈ (± 0.23)	0.042ᶜ (± 0.34)
GM34	7.25ᵇᶜ (± 0.04)	0.197ª (± 0.07)	1.34ᵇ (± 0.07)	89.83ª (± 0.09)	0.035ᶠ (± 0.31)
GM9	7.40ᵇᶜ (± 0.07)	0.1000ª (± 0.03)	1.34ᵇ (± 0.20)	83.38ᶜ (± 0.15)	0.040ᵈ (± 0.24)
GM6	7.21ᵇᶜ (± 0.10)	0.197ª (± 0.02)	1.34ᵇ (± 0.13)	79.97ª (± 0.14)	0.045ᵇ (± 0.21)
L53(**Control**)	7.32ᶜ (± 0.05)	0.197ª (± 0.03)	1.34ᵇ (± 0.14)	86.51ᵇ (± 0.12)	0.037ᵉ (± 0.31)
L.S.D	0.1196	0.1196	0.01	0.473	0.0014

*Mean not sharing the same letter in the same Colum is significantly different (P ≤ 0.05)
*Each value in the Table is a mean of three replicates ± S.D
* Viscosity = 0.1 % solution

Table 2: Physical characteristics of guar gum as lines as compared to the control.

Genotype	Moisture %	Ash %	Oil %	Fiber %	Protein %	Carbohydrate %
GM2	8.80ª (± 0.07)	3.33ᵉ (± 0.04)	1.93ᵇᶜ (± 0.30)	10.67ᶜ (± 0.21)	30.52ª (± 0.20)	44.74ᵈ (± 0.41)
GM8	8.63ᵇ (± 0.05)	3.72ᵇᶜ (± 0.06)	2.30ªᵇ (± 0.16)	11.37ᵇ (± 0.23)	25.80ᵉ (± 0.34)	4818ᶜ (± 0.36)
GM34	8.53ᶜ (± 0.04)	4.96ª (± 0.07)	2.27ªᵇ (± 0.07)	11.83ª (± 0.09)	28.60ᵇ (± 0.31)	43.80ᵉ (± 0.42)
GM9	8.50ᶜ (± 0.07)	3.43ᵈ (± 0.03)	2.47ª (± 0.20)	11.33ᵇ (± 0.15)	25.97ᵈ (± 0.24)	48.31ᶜ (± 0.35)
GM6	8.37ᵈ (± 0.10)	3.80ᵇ (± 0.02)	1.70ᶜ (± 0.13)	10.53ᵈ (± 0.14)	26.83ᶜ (± 0.21)	48.77ᵇ (± 0.51)
L53 (Control)	8.23ᵉ (± 0.05)	3.63ᶜ (± 0.03)	2.30ªᵇ (± 0.14)	10.73ᶜ (± 0.12)	25.30ᶠ (± 0.31)	49.80ª (± 0.43)
L.S.D	0.0636	0.0935	0.3723	0.1344	0.1261	2.22814

*Means not sharing the same letter in the same Colum are significantly different (P ≤ 0.05).
*Each value in the table is a mean of three replicates ± S.D
*Carbohydrate by difference

Table 3: Chemical composition of guar seed from different lines as compared to control sample.

differences (P ≤ 0.05) between all tested samples and the control. The results obtained in this study lie within the range from 0.020 to 0.095, reported by Sabah Elkhier [9], however they are comparable to the values ranging from 0.031 to 0.044 reported by Eldirany [8]. The small variations in optical density of extracted guar gum might be attributed to genetical factors.

Chemical composition of guar gums: The chemical components determined were moisture, protein, oil, ash, crude fiber and carbohydrate. The values obtained for these components in each of the five lines studied and the control sample are presented in Table 3.

Moisture content (%): Table 2 shows that moisture content of all lines investigated ranged from 8.37 to 8.80%. The highest value of moisture content (8.80%) was found in line GM2, while the lowest value (8.37%) was found in line GM6. There was no significant difference (P ≤ 0.05) in moisture content among the tested lines, however lines GM34 and GM9 did not show significant difference (P ≤ 0.05) in their moisture content. The moisture content of the standard commercial cultivar was 8.23%, a value lower than those obtained for each of the five lines and was significantly different (P ≤ 0.05) from all lines under study. The values of moisture obtained in this study are higher than the range of 7.10% to 8.19% reported by Sabah Elkhier [9] in guar gum and the values from 5.5% to 5.9% lower than values obtained by Elsiddig and Khalid [12].

Ash content: Table 3 shows that the ash content of the five lines ranged from 3.33% to 4.96%, with line GM34 having the highest value and line GM2 the lowest value. Significant differences (P ≤ 0.05) in ash content were observed among the tested lines. The ash content of the control was 3.63%, which was not significantly different from all lines. Lower values ranging from 0.5% to 1% were reported by Elsiddig and Khalid [12]. Values within the range obtained in this study were found by Eldow [11] who reported ash content ranging from 3.25% to 3.75% in guar seed. Higher values of ash were found by Elsiddig and Khalid [12] who reported a range from 5% to 6.5%. Higher ash content ranging from 5% to 5.54% were also reported by Eldirany [8]. The variation in the

ash content may be due to genetic factors and environmental factors under which plant was grown.

Oil content: Oil content ranging from 1.70% to 2.47%, were obtained from the five lines. The highest value 2.47% was from line GM9 while the lowest value 1.70 was from line GM6 (Table 3). The difference in oil content was significant (P ≤ 0.05), among the lines, however lines GM8 and GM34 did not show significant difference (P ≤ 0.05) between them. The oil content of the control was 2.30% a value relatively like that obtained from the five lines under study. GM8 and GM34 were not significantly different (P ≤ 0.05), however a significant difference (p ≤ 0.05) was found between GM2 and the other five lines. The results of oil content obtained are comparable to the values of 1.47 to 2.2% reported by Elsiddig and Khalid [12]. The obtained values were lower than the values found by Saba Elkhier [9] who reported ranges of 3.04% to 3.27% and 0.87% to 5% respectively. The variation in the oil content may be affected by genetic factors and environmental condition.

Crude fiber content: The results (Table 3) showed that crude fiber content of all lines investigated ranged from 10.53% to 11.83%. The highest value of fiber content (11.83%) was found in line GM34, while the lowest value (10.53%) was found in line GM6. Significant differences (P ≤ 0.05) in crude fiber content were observed among the tested lines, however lines GM8 and GM9 showed no significant difference (P ≤ 0.05) in their crude fiber content. On the other hand the control sample crude fiber was significantly different (P ≤ 0.05) from all tested samples with the exception of line GM2. Earlier findings showed crude fiber content in guar seeds to range from 12% to 13.8%, 9.03% to 10.1%, 8.48% to 9.37%, and 7.78% to 9.56% [8,9,11] respectively. The variation in the crude fiber content among genotypes might be attributed to genetic variation.

Protein content: The results showed that protein content of the five lines ranged from 25.80% to 30.52%. The highest value of protein (30.52%) was found in line GM2, while the lowest value (25.80%) was in line GM8. Significant differences (p ≤ 0.05) were observed among all lines and between lines and control while gave the lowest protein content. The results obtained in this study are lower than value of protein ranging from 37.6% to 42.80% reported by Eldirany [8] and higher than the values ranging from 16.6% to 20.5% reported by Elsiddig and Khalid [12]. Sabah Elkier [9] found crude protein of guar seed in the range from 25.3% to 26.62%. Values ranging from 28.17% to 29.62% from guar seed were reported by Eldaw [11]. The variation in protein content obtained from the five lines and those reported from each line wish could be attributed to genetical variation.

Carbohydrate content: The carbohydrate content of the lines investigated ranged from 43.80% to 48.77%. The highest value of carbohydrate content (48.77%) was found in line GM6, while the lowest value (43.80%) was found in line GM34 (Table 3). significant differences (P ≤ 0.05) carbohydrate content were observed among the tested lines. However, lines GM8 and GM9 did not show significant difference (P ≤ 0.05) in their carbohydrate content. The carbohydrate content of the control was 49.80%, a value which was higher than the tested lines carbohydrate content. The results obtained in this study concerning carbohydrate content were higher than the values ranging from 30.25% to 38.57% reported by Eldirany [8] in guar gum and is lower than the values ranging from 58.5% to 60.7% reported by Elsiddig and Khalid [12]. Eldaw [11] reported carbohydrate content in guar seed to range from 44.8% to 47.1%. The variation in carbohydrate (as glactomanan) content among the different genotypes might be also attributed to genetical variation [13-16].

Conclusion

In conclusion, all five lines investigated are characterized by possessing physical and chemical properties related to those of the control sample and those found in earlier reports. Line GM2 and GM6 showed the best physical and chemical properties related to those of the control sample and those reported earlier from previous findings.

References

1. Sharma BR, Cechani V, Dhuldoya NC, Merchant UC (2007) Guar Gum. J Science Tech Entrepreneur, Lucid Colloids Limited, Jodhpur, Rajasthan, India.

2. Heye E, Whistler RL (1984) Chemical composition and properties of guar polysaccharides. J A soc 70: 249-252.

3. Abuelgasim EH (1985) Guar variety trial. Annual Report, El Obeid Research Station.

4. Taha MB (1993) Effects of intra row spacing and seed/hole on rain grown guar. 1992-93 Annual Report, Gedarif research station, ARC, Sudan.

5. Baker CW, Whistler RL (1975) Distribution of D-galactosyl group in guaran and locust bean gum carbohydrate. Res 45: 237-243.

6. Fox JE (1992) Seed Gums: Thickening and gelling agents for food. Glasgow, Scotland.

7. AOAC (2001) Association of official analytical chemists: Official methods of analysis. (17th edn), Assoc of Analytical Chemists, Washington, DC, USA.

8. Eldirany AA (2009) A study of physicochemical and functional properties of four new genotypes of guar (*Cymposis Tetragonoloba*). Faculty of Agriculture, University of Kingdom Sudan.

9. Sabah Elkhier MK (1999) Improvement of yield and quality of guar (*Cymposis Tetragnoloba*). Faculty of Agriculture, University of Kingdom Sudan.

10. Loggale LB (2001) Response of guar to plant spacing and number of plants/ Hole. 2000-01 Annual report, Food Legume Program, Kenana Research Station Abu Naama, ARC, Sudan.

11. Eldaw GE (1998) A study of guar seed and guar gum properties (*Cymposis Tetragnoloba* L) Faculty of Agriculture, University of Kingdom Sudan.

12. El-Siddig AE, Khalid AI (1999) The effect of bradyrhizobium inoculation on yield and seed quality of guar (*Cyamposis Teteragonoloba* L.). J Food chem 19: 8-19.

13. Osman AK (2001) Guar within row spacing trial. 2000-01 Annual report, Elobeid research station, ARC, Sudan.

14. Bureng PL (1996) Internal report. Food research centre, Shambat, Khartoum, Sudan.

15. Dickinson E (2003) Hydrocolloids at interface and influence on the properties of dispersed system. J Food Hydrocolloids 17: 25-39.

16. Glicksman M (1969) Gum technology in the food industry. Academic press, New York.

20

Product Development on Goulash Sided by Parsley-Potatoes and Almond-Broccoli

Mohammad H Rahman*, Nasim Marzban, Theresa Schmidl, Saiful Hasan and Caroline Nandwa

Department of Organic Agricultural Sciences, Kassel University, Hesse, Germany

Abstract

The team consisting of Mohammad Habibur Rahman, Nasim Marzban, Saiful Hasan, Caroline Nandwa, and Theresa Schmidl developed a traditional German Goulash half pork and half beef sided by parsley-potatoes and almond-broccoli. In the first stage of the project, product ideas were identified and product concept was developed. In three laboratory sessions, a prototype for the Goulash was created focusing on the needs of Geman seniors, >80 years, showing an early onset of dementia. Based on this prototype, the report was written considering important aspects to launch a product in a catering service in future: Product characteristics, marketing design specifications, product design specifications, quality and regulatory aspects. For this module, the project ended in the phase of the prototype.

The aim to provide a lunch for German seniors could be fulfilled as a successful prototype was developed. It is expected to be accepted by the German seniors as it reflects German traditions and seniors' childhood memories.

Keywords: Goulash; Product development; German seniors; Innovation; Healthy and nutritious

Introduction

In this paper, a traditional German Goulash–half pork and half beef–sided by parsley potatoes and almond-broccoli is being developed in teamwork by Mohammad Habibur Rahman, Nasim Marzban, Saiful Hasan, Caroline Nandwa and Theresa Schmidl. The product's marketing name is: "Goulash sided by parsley-potatoes and almond-broccoli". The aim of the product is to fulfil the needs and demands of German seniors, >80 yrs., showing an early onset of dementia.

Product Concept

In the following chapter, the product concept to the dish "Goulash sided by parsley-potatoes and almond-broccoli" is presented to the reader. The concept deals with the ingredients, the processing and serving recommendations, the nutritional aspects, the sensory and physical attributes, as well as the target group and the packaging, delivery and marketing.

Ingredients, processing and serving recommendation

The aim of group E is to produce a tasty, healthy and fresh goulash sided by parsley-potatoes and almond-broccoli. Therefore, the following ingredients are needed: Lean pork and beef, onions, tomato paste, native sunflower seed oil, and spices (red pepper spice, chili, garlic powder, pepper and salt), potatoes, parsley, broccoli and almonds (Appendix 1).

For the preparation, the group needs to cut the meat into small pieces cutting off all possible fat and filaments. As well, the onions and the parsley need to be chopped. The lentils, set aside for swelling on the previous day, are heated and mashed. The potatoes are peeled and boiled in a microwave-stable Tupperware. For the Goulash: The meat is roasted in native sunflower seed oil using a pressure cooker. As soon as the meat is well done, the onions are roasted as well. Then the meat is put back into the cooker and the tomato paste as well as some water is added. The pressure cooker is now used for approx. 20-25 min. to reach the ideal tenderness of the meat before spicing the Goulash with the above-mentioned spices. For the side-dishes: The cooked potatoes are cut into adequate-sized squares and mixed with chopped parsley.

The cleaned and cooked broccoli is refined using thin slices of roasted almonds (according to an old family recipe).

The food is served in a metal plate with three sections–a bigger one for the Goulash and two smaller ones for the side-dishes. This allows the seniors to choose how to eat the Goulash, it keeps the heat easily and it is fast and hygienic to clean.

The big advantage of Goulash is that the consistence of the meat sauce is not tremendously altered when the dish is sitting aside waiting until lunch time. The potatoes and the broccoli can be served with adequate properties as well. The dish should be served warm to hot. It can be reheated in a microwave or oven, if the senior feels the heat not to be sufficient. The Goulash can be stored in the fridge for one to two days from the delivery day on.

Nutritional aspects

The dish will contain between 400 kcal-600 kcal which is appropriate for a senior's main meal. It will also provide the seniors with many added nutritional values. As meat is one of the main ingredients, it provides a high content of protein with the full count of amino acids. Generally red meat is an important source of minerals, including iron, phosphorus, zinc, copper and vitamins such as vitamin A and B. Due to the high content of saturated fat and cholesterol in red meat, less fatty cuts are preferred for the dish from a nutritional point of view. Pork is a good source of thiamin converting carbohydrate into energy in the body. It is also a good source of zinc. Potatoes are a good source of energy, vitamin B_6, vitamin C, and potassium. Almonds' calcium content is higher than the one of any other nut and they are

*Corresponding author: Mohammad H Rahman, Department of Organic Agricultural Sciences, Kassel University, Hesse, Germany
E-mail: tuki.st07@gmail.com

an excellent source of iron, riboflavin, and vitamin E. Adding herbs such as parsley to the food is a good replacement for fat and salt and it provides a healthier diet. Broccoli is high in vitamin A (beta-carotene) and C. Additionally; it has anti-cancer properties due to various phytochemicals. Also onions contain phytochemicals with antioxidant capacity. It is recommended to serve a fruit yoghurt or quark dessert to the Goulash to provide the important calcium, as well as vitamin A, B, and D. It will also increase the number of the bacterial beneficial flora in the intestine. As the sense of taste declines when aging, spices will improve the appetite, though not adding much nutritional value [1].

Sensory and physical attributes

Malnutrition in geriatrics is mostly attributed to the increased need of energy and the inadequate food intake, and often due to loss of appetite and alteration of taste and smell perceptions. Medical conditions, prescribed drugs, nutritional status, cognitive status, mood and feelings are also few factors that will affect the sensory and physical attachment to food intake [2]. Although these chemo-sensory deficits are generally not reversible, sensory and physical interventions including intensification of taste and odor can compensate for perceptual losses [3].

Amplification of flavor and taste can improve food palatability and acceptance, increase salivary flow and immunity, and reduce oral complaints in both sick and healthy elderly [3]. The flavors and spices must however be mildly use so that they do not irritate the mouth or stomach [4].

In the case of Goulash stew, the use of spices (red pepper, chili), herbs (garlic, parsley) and salt) will be used to enhance the olfactory as well as the gustatory perceptions. The wide array of the dish and the preparation method (boiling, pressure cooking and stewing), will also provide the right food texture suitable for geriatrics [4].

In order to enhance the physical attributes, plate size and food in the plate arrangement, color of the dish (potatoes being whitish, the goulash brown and the broccoli green) and use of almonds to garnish (including the nutritional benefits it supplies), is aimed to stimulate the appetite and willingness to take in the meal [3,4].

Target group

The target group is German seniors, >80 yrs., showing an early onset of dementia. Most of these seniors are distressed from muscle function, anemia, oral health, reduced cognitive function, higher hospital and readmission rate, mortality or dementia. Often nutritional aspects such as malnutrition, BMI, dieting intake and dieting habits are responsible for biological and physiological changes [5].

The 80+ seniors in Germany need to take in carbohydrate (k/cal), protein (g/day), dietary fiber, calcium, folate, iron, minerals (K, Na), vitamin- A, E, C, D and B_{12} [6]. Omega-3 fatty acids are also very essential for seniors showing an early onset dementia. Research shows that cold water fishes like mackerel, salmon, tuna etc. reduce the risk of increasing dementia/Alzheimer by over 50 percent [7].

Goulash can provide the seniors with all the essential energy compactness and macro-/micro-nutrients required at their life phase. As seniors often show dehydration, the juicy and slightly spicy Goulash can activate the water consumption. A limited sensory perception decreases the appetite of seniors. By creating a color-balanced dish the desire to eat should be animated. Often seniors only eat small portions for each meal, so the Goulash dish is kept in adequate dimensions. Lastly, seniors are mostly fighting problems with chewing and swallowing. To decrease this problem, the meat, potato and broccoli pieces are kept in an easy-to-eat size. Though, it is important to not only serve mashes to the seniors. The pieces should be in a size which allows eating without prior cutting, as seniors often lose the capability to manage the hand/arm-coordination [8].

Other than that, the dish is kept lightly spicy by adding herbs and spices to enhance the flavor perception. Salt is only lightly used as it supports dehydration. As the product is intended for German seniors (80+ yrs., early onset of dementia) and is thought to provide them with a feeling of their early childhood as Goulash is a common German dish, especially in the Eastern parts of Germany and the former German regions of Poland and Czech Republic.

Delivery, packaging and marketing

"Food packaging can retard product deterioration, retain the beneficial effects of processing, extend shelf-life, and maintain or increase the quality and safety of food [9]. The golupack consists of a metal bowl keeping the heat, a three-section metal plate and a hard plastic cover.

All parts are made out of hygienic and easily cleanable materials. The golupack keeps the Goulash dish hot for more than 90 minutes. It is a truly eco-friendly solution to hot food transport and storage. The golupack is suitable for the microwave so if the seniors want to reheat their food, they can do so later. The delivery system includes a thermal food carrier to store the golupacks during the delivery to the seniors' homes keeping the dish hot (Figure 1).

As the package is the face of a product it is very important for a new product to be advertised by unique and innovative packaging concept. This helps to attract consumers and increases the sales [10]. The golupack is designed to increase the product performance and to give good product image. Nutritional value, ingredient declaration, net weight, manufacturer information, brand identification, and pricing are given on the product's declaration. Additionally serving instructions may impact the consumer attraction and satisfaction.

The marketing of the Goulash is in a line with all the other dishes of the caterer "Gerichtebringer". The description of the Goulash in the menu is as the following: "Homemade traditional Goulash (half/half) sided by parsley-potatoes and almond-broccoli, heartwarming dish for cold winter days".

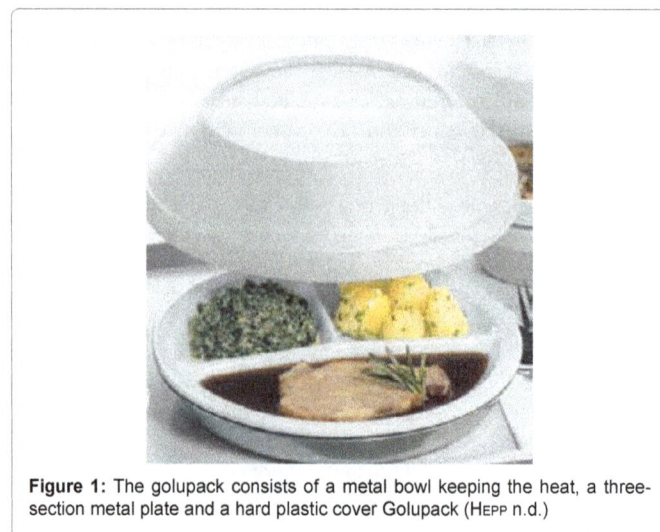

Figure 1: The golupack consists of a metal bowl keeping the heat, a three-section metal plate and a hard plastic cover Golupack (Hepp n.d.)

Product Characteristics

Sensory characteristics

When looking at the sensory characteristics provided by the Goulash, two different aspects shall be analysed: Taste and appearance [11]. The taste of the dish is described as following. The dish tastes as a traditional Goulash, but is only little salted in order to meet the nutritional requirements of seniors [12]. The onion and red pepper provide sweetness to the Goulash. Bitter and sour tastes cannot be found in the dish [11].

The parsley provides an herbal freshness as well as certain spiciness to the product underlining the potatoes in taste and color. Lastly, the chili and spicy paprika powder enhance the subtle spiciness of the dish. This will enhance the flavor perception (Trigeminus) of the seniors and by that, influences the food intake. Here it is important to not over-spice the dish not to irritate the taste buds and the food acceptance of the seniors [11].

For the appearance, next to the reddish-brown color of the Goulash and the yellowish-white potatoes, green broccoli was chosen. This should provide a feeling of freshness and remembering the earthiness of the vegetable. Additionally, the application of parsley and almonds further enhances the color experience when serving the dish [11,13].

Nutritional characteristics

The nutritional characteristics are based on the ingredients found in Appendix 1. It is important to notice that the list of ingredients includes the dessert recommendation given in the product concept. The strawberries, the sugar and the low-fat yoghurt are blended together to a small dessert providing additional protein to the dish (Appendix 1).

For the calculation of the nutritional values with PRODI, a meal calculation program, the setting 'Mittagsverpflegung stat. Senioreneinrichtung (>65 Jahre PAL 1.2)' was chosen. This is a calculation suggestion for public senior institutions serving lunch to seniors older than 65 years with a low mobility. According to PRODI, the amount of fat in the dish is 26 percent, the carbohydrates are 35 percent and the protein is 35 percent. The amount of dietary fibers is 4.3% (Appendix 2).

When looking in Appendix 3 and 4, it can be found that the kilocalories are nearly ideal for a seniors' lunch. The protein content of the dish is relatively high (186%), while the fat content (90%) and the carbohydrates content (70%) are lower than the recommended serving. The dish is high in fibers (120%). The micro nutrients are very well provided by the Goulash. The vitamins B_1, B_2, B_6, C, D and E are as well given in a satisfactory amount as vitamin B_{12} which is important for seniors. Calcium is provided to 66% – including the dessert. Folic acids, important for senior nutrition, are sufficiently provided (114%).

Physical characteristics

Several physical characteristics of the dish can be identified. The texture of the Goulash is soft and juicy, as the vegetables and meats are pressure-cooked forming a soft consistency. The red pepper and the onions do not only provide nutritional value and certain sweetness to the product but also form a juicy basis for the Goulash. The broccoli is cooked to a point where a distinct crispness is sensed in the eating experience. The potatoes offer a creamy component to the dish, which makes it more pleasant to consume.

In general, the texture of a dish can influence its taste, aromas and the mouth-feel. Other components such as almonds provide a crunchy, nutty, roasted taste to the dish enhancing the appetite of the elderly [11].

Chemical characteristics

The chemical characteristics of the Goulash are changed during the warm-storage between the delivery and the consumption of the meal. Here, the structure of the molecules in the food is easily changed. Vitamin losses can be a result. The main chemical effects on the Goulash could be caused by non-enzymatic browning or enzymatic browning, for example affecting the broccoli. Also the physical characteristics, such as the texture are affected by the change in the chemical characteristics [14].

Microbiological characteristics

The Goulash could potentially be affected by several microbes. Mycotoxines might be found in the Goulash. Especially the almonds processed in the dish could be infected with aflatoxin. According to the food law, only certain amounts of aflatoxins are allowed in products for human consumption. Almonds should only be purchased in small amounts allowing quick turn-over and they should be stored in a dry and cool place [15].

Clostridium perfringens, a anaerobic bacterium, could affect consumers when deficient hygiene was practiced during the dish's preparation phase. Especially meals containing meat which were kept warm are affected. This risk can be minimized by storing the food at more than 70°C. If this is not possible, the food should be stored in an environment colder than 5°C and it should be reheated just before the consumption [15].

Salmonella could affect the Goulash through cross-contamination in the processing area or by keeping food warm at a temperature between 20°C and 45°C. In order to reduce the risk of salmonellosis, the food products should at least be cooked at 80°C core temperature for ten minutes. Additionally, all highly perishable ingredients need to be stored in a refrigerator. Cutlery and crockery should be cleaned properly after processing meat products and other risky ingredients such as eggs [15].

Staphylococcus aureus is a bacterium which could also affect the Goulash. Through deficient personal hygiene or cross-contamination in the preparation phase as well as incorrect storage of ingredients, an outbreak of *staphylococcus aureus* could be fostered. The bacterium can be killed by heat application, though its toxins are highly heat-resistant. Cooking processes of 90 minutes at 100°C are needed to destroy the toxins [15].

Processing

The processing of the 'Goulash sided by parsley-potatoes and almond-broccoli' can be seen as three parallel preparation parts which stem from an old family recipe.

Firstly the parsley-potatoes are prepared by rinsing the parsley to get rid of the soil attached to the leafy greens. After cutting off the stems of the parsley, the leaves are finely chopped. The potatoes are washed, peeled and cut into small squares of approximately two times two centimeters. The potatoes are cooked in preheated water for approx. 20 minutes depending on the potato variety. When they are tender, the cooked potatoes are given into a bowl. They are sprinkled with small butter flakes. A little salt and the chopped parsley are added. All the ingredients are mixed and filled into a small section of the serving plate.

For preparing the almond-broccoli, the almonds are roasted until they have a fine brown tan. The broccoli is cut into small flowers in a size adequate for eating without prior cutting. The broccoli is washed carefully and the flowers are steamed in a sieve over a pot with boiling water. This requires approx. nine minutes. After the steaming, the broccoli is mixed together with the rest of the butter, a little salt and the roasted almonds. This part of the dish is also filled into a small section of the serving plate.

The third preparation part is the Goulash. Here, the meat is washed, dried and fat is cut off. Then, the meat pieces are cut into convenient to eat two times two centimeters pieces. The onions are cleaned and chopped. The red pepper is washed and sliced into thin (0.5 cm) and short (3 cm) cuts. The garlic is cleaned and finely chopped. Extra native oil is added into a pre-heated pressure cooker first browning the onions, followed by the meat which is seared. When the meat is well done, garlic, red pepper and water are added for the cooking process. In the pressure cooker, the dish cooks on medium heat with a medium pressure for 24 minutes. After this cooking process, tomato mark and spices (pepper [2 g], salt [0.5 g], paprika powder sweet [3 g], paprika powder spicy [1 g] and chili powder [2.5 g]) are added to the mixture. After letting the Goulash sit for two minutes, it can be served in the largest section of the plate (According to the family recipe).

Packaging

The Goulash is served on a three-section plate which is closed with a lit on top. This serving crockery is placed inside heater boxes which can restore the product's heat until being consumed. On top of the plates lit, the nutrition facts and product specifications can be presented to the customer.

Storage

The Goulash can be stored in the heated box until it is consumed. If it is not eaten on the day of the production and delivery, the Goulash should be stored in the fridge for maximum one day. It can be reheated on low heat in any oven or microwave. The dish can be frozen as well. In this case, the microwave's defrosting component should be used to get a wishful state of the dish. Plate and lit of the crockery are both microwavable [16].

Marketing Design Specifications

The following attributes should be promoted. Promotion is one of the marketing's four Ps (Price, product, promotion and place). According to Perreault and Carthy [17], promotion is sharing information between seller and buyer. Its main objective is to tell the consumer about product, price and place. According to Kotler and Armstrong, promotional activities have five tools: Advertising, personal selling, sales promotion, public relations and direct marketing [18].

Advertising is the paid presentation and promotion of goods, ideas or services by an identified sponsor. The following media is attempted to influence the Goulash's consumers: Television, newspapers, magazines and radio. The personal selling is an oral communication with the consumer with the purpose of selling products and building relationships. The sales promotion is providing short-term incentives to consumers and the distribution channels encourage the purchase or sale of a product [18].

The public relations are communicating with the target audience directly or indirectly through the media aiming to create and maintain a positive image and create a strong relationship with the audience. Examples are press releases, newsletters and public appearances. Direct marketing communicates with targeted consumers directly to get immediate responses for a product by using e-mail, telephone and other tools [18].

For the previously selected target customer group (German seniors, >80 yrs., showing an early onset of dementia) advertisement, personal selling and direct marketing could be possible options for the Goulash's marketing and promotion.

According to Kotler and Armstrong [18] the "Distribution channel is a set of interdependent organization involved in the process of making a product or service available of use or consumption by the consumer and service sector". The distribution method depends on the individual product. The Goulash's distribution is complex as the target consumers are German seniors. This is why a very crucial distribution channel is needed which supports the customer's demand and the product flow. According to Shepherd [19] there are some possible distribution channels:

a. Direct to consumers (Suitable option for covering small areas)

b. Retailers (If interested to sell the product)

c. Supermarkets (If product acceptable and sufficient quantities can be delivered)

d. Wholesalers (Suitable for larger processors and large consumer demand)

e. Institutions and the catering trade

The direct distribution to the target group is here the favored alternative for the Goulash dish [19].

Product Design Specifications

Product design specifications: The product design and process includes integration, creativity, systematic planning and monitoring. The product design is one of the most important parts of the product development work. As the project progresses, the product is more clearly Defined and the study of the variables of the process becomes more important allowing achieving the optimal product [20].

Product formulation

The raw materials for the Goulash sided by parsley-potatoes and almond-broccoli can be found in Appendix 1.

The utensils applied during the processing are listed in Table 1. The process flow chart can be found in Appendix 5.

Marketing package design

Golupack is served in a three sectional metal bowl with a plastic cover. The logo of the catering service "Gerichtebringer" is broadly designed on the packaging to coin the caterer with high quality, fresh, healthy and good tasting food. The description of the Goulash in the menu is as the following: "Homemade traditional Goulash (half/half) sided by parsley-potatoes and almond-broccoli, Heartwarming dish for cold winter days". The following information will be printed with attractive coloring on the package's wrapping paper as declaration for the consumers (Table 2).

Regulatory Aspects

To successfully introduce a new food product or ingredient, it is important to consider the regulatory environment and aspects applicable to its markets, especially at the early stage of the product development [21]. The General European Union Food Law Regulation (EC) 178/2002 aims at ensuring high levels of protection of human life

Utensils	
➤	Pressure cooker
➤	Pots
➤	Kitchen towels
➤	Jars or containers
➤	Stove
➤	Spoons and knives
➤	Cutting board
➤	Steam cooker
Source: Own design	

Table 1: Processing utensils.

Product sheet	
Article description	Goulash sided by parsley-potatoes and almond-broccoli
Manufacturer information	Gerichtebringer GmbH, Examplestraße 1, 12345 Example
Pricing	?
Net weight	400 g for the main meal + 50 g for the recommended dessert
Nutritional value	Caloric value 408.26 kcal Protein 35.43% Carbohydrates 34.25% Sugar 0% Fat 26.04% Fiber 4.28%
Raw material requirements	All ingredients must be impeccable with a high quality and freshness
Ingredients declaration	Lean pork, beef, onions, tomato paste, native sunflower seed oil, red pepper spice, chilli, garlic powder, pepper, salt, potatoes, parsley, broccoli, almonds
Instructions for reheating	Oven: 180°C for 5 minutes; Microwave: high for 1.5 minutes
Recycling information	

Table 2: Wrapping paper as declaration for the consumers.

and health, both at national and EU level, and established the rights of consumers to safe food and accurate and honest information (EU Food Law). That is, "food must not be injurious to health and must be fit for consumption" (EU REGULATION 178/2002). In the EU, a set of regulations, directives, scientific opinions and guidance documents have been established which assist producers and manufactures during the development of products [21]. This regulatory framework covers a wide array of issues. However, for the purposes of this paper, the Goulash dish, the applicable regulatory frameworks to it shall be considered.

Food hygiene

Food hygiene results from the implementation of prerequisite requirements and procedures by food businesses based on the HACCP principles [22]. In line with this, the implementation of certain provisions of Regulation (EC) 852/2004 on the hygiene of food stuffs, some prerequisite requirements to be observed include: The safe handling of food, food waste handling, sanitation procedures and personal hygiene.

Labelling

The EU Regulation 1169/2011, on the provision of food information to consumers, (Also Directive 2000/13/EC on labelling, presentation and advertising of foodstuffs and Directive 90/496/EEC on nutritional

labelling for foodstuffs), states that consumers must not be misled by inadequate labelling on weight, constituents and additives, shelf life, nutritive value and the production method of products. Food labelling is thus also seen as a food safety measure, since information on allergens, the proper handling of the dish (Refrigeration, shelf life, cooking) and any food intolerance is provided (EU Food Law). In Germany, in line with this Regulation, the "Lebensmittel-und Futtermittelgesetzbuch" and the "Lebensmittelkennzeichnungs-Verordnung" also provide information on how food must be labelled. For the Goulash dish, the following labelling information shall be included: The name of the food, the list of ingredients, allergens (In this case almonds), quantity of certain ingredients, net quantity of the dish, the date of minimum durability, storage conditions (Refrigeration), and the instructions of use.

Nutritional and health claims

According to EU Regulation (EC) 1924/2006 foreseeing implementing measures to ensure that claims made on food labels are clear and based on evidence. Low fat, not more than 3 g fat per 100 g solids, and low salt, 0.12 g per 100 g, as stated in the regulation were considered in labelling the Goulash "low in fat and salt". For this case, low fat ingredients (Lean beef, lean pork and a small quantity of sunflower seed oil), and small amount of iodized Dead Sea salt was used. No health claim is intended to be used on this product.

Allergens

The EU Regulation 1169/2011 lists nuts and almonds as allergen-commodity examples listed in the Regulation Annex II. The Goulash dish is garnished with roasted almonds. Therefore it must be clearly labelled for individuals who cannot tolerate them or who may have allergic reactions.

Food for particular nutritional uses

The Goulash dish is ideal for seniors over the age 65 (geriatric nutrition). Therefore, all the nutritional and sensory aspects have been put in consideration in its preparation.

Genetically modified ingredients

Based on the Regulation (EC) 1829/2003 on GMO`s, and in order to surpass this risk, all the ingredients used are natural ingredients, grown/produced organically and ecologically.

Food additives/flavorings/enzymes

As stated in the annex list of food additives, flavorings and enzymes in the Regulation (EC) 1333/2008; no synthetic flavorings, additives and enzymes were used in the preparation of the Goulash. All-natural flavors from garlic and red bell pepper were used to enhance the flavors.

Food packaging

The EU Directive 2000/13 on "presentation of food stuffs" aims to also provide information to consumers, through labelling, in a better legible text including minimum size of text. The material of the packaging was considered to prevent cross-contamination of substances from the packaging into the food.

Food waste

The EU states its goal to reduce food waste by at least 30 percent by 2025 (EU Food Law). This targets sectors including manufacturing, retail, food service/hospitality and households. One of the ways in achieving this is avoiding food surpluses. The preparation and the serving of the Goulash dish, aims to attain this through using the exact amount of ingredients and correct portioning of the meal. This attempts to ensure that almost all the dish is consumed by an individual with minimal waste.

Quality aspects

Quality is the combined knowledge of how to use good hygiene practices (General and specific hygiene requirements) and hazard analysis as well as critical control points (HACCP) principles [23]. To get and sustain quality standards for Goulash, the focus lies on all production activities and responsibilities such as purchasing (Fresh lean meat, organic ingredients etc.), personal hygiene, food preparation, food processing, sensory tests and packaging which need to be kept up to a desired level. With the concern on quality aspects of the Goulash, basic hygiene practices were followed in all stages of the processing. This allows serving the dish with little food hazards, less product failure and high product nutrient value [24-27]. There are no chemical preservatives added during the preparation of the Goulash. The time-temperature combination during stewing and cooking was focused upon.

Packaging also a very important part of the food quality aspects. It can play a vital role in the food processing industries or food catering services. It makes food more suitable, convenient and gives the food safety assurance from microorganisms and biological, chemical and physical changes. Sometimes also taste depends on packaging [9]. The packaging keeps the Goulash dish hot for more than 90 minutes and extends the shelf life. The storage of the cooked Goulash after its packaging is an important critical control point (CCP). All microorganisms need to be killed in this stage to extend the product's shelf life.

As the product is intended for German seniors (>80 yrs., early onset of dementia) thought to provide them with a feeling of their early childhood, the researchers tried to maintain all the quality procedures during the Goulash preparation. The end product had the benefit that it did not contain artificial additives, chemicals or genetically modified organisms (GMO).

Conclusion

When reflecting on the project process, it was found that the teamwork, especially in the laboratory was very good as many preparation steps needed to be undertaken. Also the idea and the recipe of the Goulash were quickly found and the product concept was developed within the group. In general, when problems or difficulties occurred during the processing stage, immediate actions were taken to correct or stabilize individual processes.

On the other hand, several setbacks were experienced as the initially planned use of the microwave for steaming the potatoes did not work out. Also one of the pressure cookers was not fully functioning (Not allowing closing lit properly). These limitations were due to technical problems in the kitchen which could not be solved. Solved limitations were the steaming of the broccoli which was first too long altering its color and the buttering of the potatoes to make them lustrous as well as to make the parsley adhere to the surface of the potatoes.

Other improvements such as time and temperature, which could be carried out in the future are the following. The beef could be firstly roasted in order to get a softer consistence compared to the pork. Additionally, simmering instead of boiling the Goulash for a relatively long time under low heat might optimize the taste and texture of the dish. Lastly, the Goulash could be thickened with cornstarch or flour if a short processing time does not allow the dish to boil down.

Concluding, the project's aim to provide a lunch for German seniors, >80 years, showing an early onset of dementia, could be fulfilled as a promising prototype was developed (Appendix 6-8). It is expected to be accepted by the German seniors as it reflects the German traditions and with it the seniors' childhood memories.

References:

1. Friedman MI, Tordof MG, Kare MR (1991) Chemical senses. Appet Nutri 4: 341.

2. Savina C, Donini LM, Anzivino R, De Felice MR, Canella C, et al. (2003) Administering the "AHSP questionnaire" (Appetite, hunger, sensory perception) in geriatric rehabilitation care. J Nutrition Health Aging 7: 385-389.

3. Rolls BJ (1998) Do chemosensory changes influence food intake in the elderly? Elsevier J physiol Behav 66: 193-197.

4. Schiffman SS (2000) Intensification of sensory properties of foods for the elderly. J Nutrition 130: 927-930.

5. Stratmann M (2013) Nutrition of the elderly: Fresenius conference shines a light on products requirements for the "silver generation". Die Akademie Fresenius. Dortmund/Mainz, Germany.

6. Volkert D, Kreuel K, Heseker H, Stehle P (2004) Energy and nutrient intake of young-old, old-old and very old elderly in Germany. European J Clinic Nutri 58: 1190-2000.

7. Food for the brain (2010) Championing optimum nutrition for the mind.

8. Engel M (2004) Too little? Too much? The wrong? Dossier zur Seniorn- ernährung in Deutschland. Verbraucherzentrale Bundesverband.

9. Ansari IA, Dattaan AK (2003) Overview of sterilization methods for packaging materials used in aseptic packaging systems. Food Bioprod Process 81: 57-65.

10. Marsh K, Bugusu B (2007) Food packaging-roles, materials and environmental issues. J Food Sci 72: 39-55.

11. Vierich TA, Vilgis TA (2013) Aroma: The art of seasoning (2ndedn). Stiftung Warentest, Berlin.

12. Bremen G (2001) Diet and age: Senior nutrition in focus.

13. Deutsche Gesellschaft V (2014) Healthy eating, better living.

14. Rubin D (2011) Vitamins-the greatest errors. Deutsche Gesellschaft für Ernährung MDR–Hauptsache Gesund.

15. Schlieper CA (2010) Basic issues of food. Verlag Dr. Felix Büchner - Handwerk und Technik, Hamburg.

16. Hepp (2014) Hospitalia.

17. Perreault DW, McCarthy JE (2000) Essential of marketing. A global-managerial approach (8thedn). McGraw-Hill Higher Education.

18. Kotler P, Armstrong G (2000) Marketing an Introduction (5thedn.). Library of Congress. Cataloging-in-Publication Data, New Jersey.

19. Shepherd WA (2003) Market research for agroprocessors. Food and Agriculture Organization of the United Nations (FAO), Rome.

20. Earle M, Earle R, Anderson A (2001) Food product development. CRC Press. New York, Washington DC.

21. TNO Innovations for life (2014) Food safety and innovation: Safety and regulatory strategy.

22. Codex Alimentarius (1997) Food Hygiene: Basic texts. Agriculture and Consumer Protection.

23. Verma E (2014) Quality management. Simplilearn.com. The turning point. How to improve quality management consistently.

24. European Commission (2014) Food pages. Food safety: overview.

25. European Commission (2004) Health & Consumer Protection Directorate-General: Guidance Document-Implementation of certain provisions of Regulation (EC) No 852/2004 on the hygiene of food stuffs.

26. European Commission (2006) Health & Consumer Protection Directorate-General: Guidance Document-Implementation of procedures based on the HACCP principles and facilitation of the implementation of the HACCP principles in certain food businesses.

27. University of California Los Angeles and Dole Food Company (2002) Encyclopedia of foods A guide to healthy nutrition. Prepared by medical and nutrition experts from Mayo Clinic, University of California Los Angeles and Dole Food Company Inc, Academic press, California.

21

Squash from Tamarind Pulp by Blending with Mango Pulp

Kiranmai E*, Uma Maheswari K and Vimala B

Department of Food Processing Engineering, Sam Higginbottom Institute of Agriculture, Technology and Sciences, Allahabad, UP, India

Abstract

A study was conducted on development of squash with tamarind by blending with mango pulp at different levels (10%, 20% and 30%) and different sugar concentrates. all the treatments were kept for three months' storage period to evaluate their storage stability. During the storage period, all the treatments were evaluated for the physico-chemical, microbial and sensory quality. The results revealed that among all the treatments highest acceptability observed in squash prepared with 80% tamarind pulp and 20% mango pulp (T6) during the storage period. No microbial growth was observed in all the treatments. The products were stored without any deterioration in physico-chemical, sensory quality and microbial count up to 3 months of storage period.

Keywords: Tamarind; Mango; Squash; Overall acceptability; Storage

Introduction

Tamarind is native fruit of Africa. It belongs to Leguminosae family with botanical name *Tamarindus indica*. L. The tamarind is prized for its shade and shelter [1]. It is one of the important tropical fruit tree and is widely grows in India. There are only a few varieties of tamarind grown in India, some are sweet and some are sour. Fruit is the most important part of the tree and it is the most acidic of all fruits and contains an uncommon plant acid i.e., tartaric acid 8% to 18% [2]. India is the chief producer and consumer of tamarind in the world. It is estimated that India produces 3,00,000 MT of fruits and export tamarind products worth about Rs. 50.0 crores per annum. Tamarind pulp is the chief agent for souring food products like sauces, chutneys, sambar, rasam and beverages. The fruit pulp is the important raw material for the manufacture of tamarind pulp concentrate and soft drinks. The pulp of fruit is used extensively in the local confectionary industry in several developed countries [3]. Due to high acidity in the tamarind fruit, the utilization of these fruits for preparation of various processed products is limited. Tamarind also has hypoglycemic and hypocholestrolemic effect and it helps in reducing obesity. Blending of fruits like mango will be helpful to enhance the sensory quality characteristics such as color, flavor, taste and overall acceptability of the prepared products. Keeping the above facts in view, tamarind squash could be prepared by blending with mango pulp for better utilization of tamarind.

Materials and Methods

Tamarind was procured from local market and seeds were removed and cleaned properly. Then the tamarind was soaked in water in 1:1.5 ratios, heated up to 100°C, then cooled and crushed. After crushing it was passed through a siever to obtain pulp. The pulp so obtained was used for the preparation of squash. Simultaneously mangoes were procured and cleaned. Tamarind squash prepared by blending with mango pulp (10%, 20% and 30%) and different sugar concentrates (45°B, 46°B and 47°B) was used in different treatments. Sugar syrup was prepared; juice was added to the cooled syrup and mixed thoroughly. Potassium Meta bisulphate was added as a preservative. Filled in sterilized bottles and capped. Squash was diluted (juice 1: water 4) before serving. The flow diagram depicting preparation of squash was given in Figure 1.

The products so prepared were evaluated for physico-chemical parameters such as total soluble solids (TSS) [4], Acidity (%), Reducing sugars (%), Total sugars (%) [5]. Sensory evaluation was done by the sensory scoring by a panel of 10 members in the laboratory of PGRC, using a score card developed for the purpose. Descriptive terms were given to various quality attributes like appearance, color, flavor,

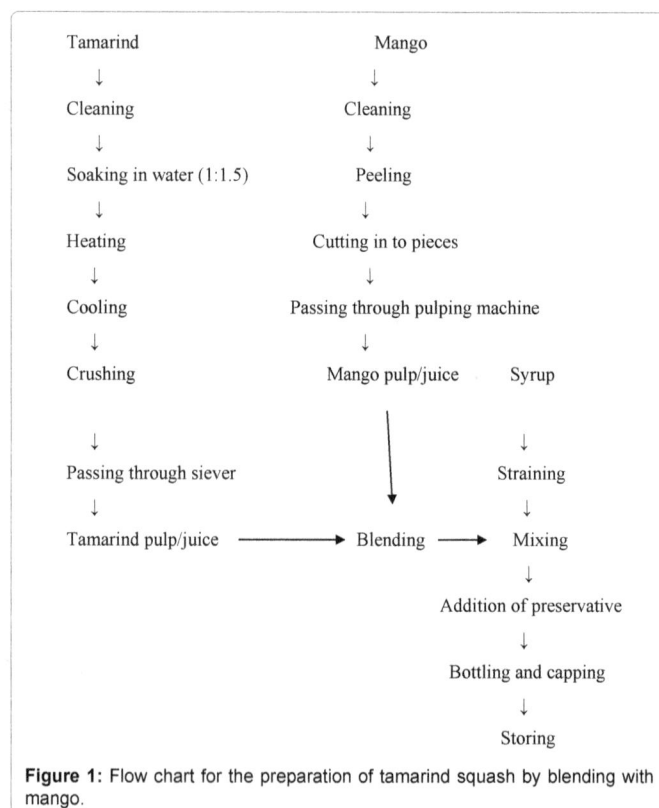

Tamarind	Mango
↓	↓
Cleaning	Cleaning
↓	↓
Soaking in water (1:1.5)	Peeling
↓	↓
Heating	Cutting in to pieces
↓	↓
Cooling	Passing through pulping machine
↓	↓
Crushing	Mango pulp/juice Syrup

Figure 1: Flow chart for the preparation of tamarind squash by blending with mango.

consistency, taste and overall acceptability (Figure 2). Numerical scores were assigned to each attribute. A five-point scale was adopted to score each of the attributes, while scoring, highest score (5) was assigned to

*Corresponding author: Kiranmai E, Department of Food Processing Engineering, Sam Higginbottom Institute of Agriculture, Technology and Sciences, Allahabad, UP, India, E-mail: kiranfoodtech@gmail.com

most preferred characteristic and least score (1) to the least designed characteristics. For estimating microbial count (bacteria, Yeast and moulds) population in different samples, dilution plate method was followed [6]. The data was subjected to statistical analysis as per the procedure described by Panse and Sukhatme [7]. The experimental design was complete randomized design with factorial concept.

Results and Discussion

Total soluble solids (TSS) recorded in different treatments and days of storage were given in Table 1. No significant change in total soluble solids during the storage period was observed. Treatments recorded significant differences, where as interactions were found non-significant. Among the treatments employed for preparation of tamarind squash initially T4, T5, T6 and T7 47°B recorded highest TSS values in comparison with T1, T2 and T3. During storage, there was

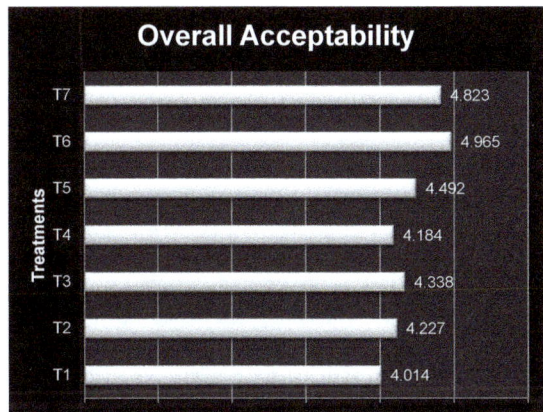

T₁: tamarind pulp 100%+45°Brix; T₂: tamarind pulp 100%+46°Brix; T₃: tamarind pulp 100%+46°Brix;

T₄: tamarind pulp 100%+47°Brix; T₅: tamarind pulp 90%+ mango pulp 10%; T₆: tamarind pulp 80%+mango pulp 20%; T₇: tamarind pulp 70%+mango pulp 30%.

Figure 2: Mean values of overall acceptability of sorghum squash at room temperature during storage period.

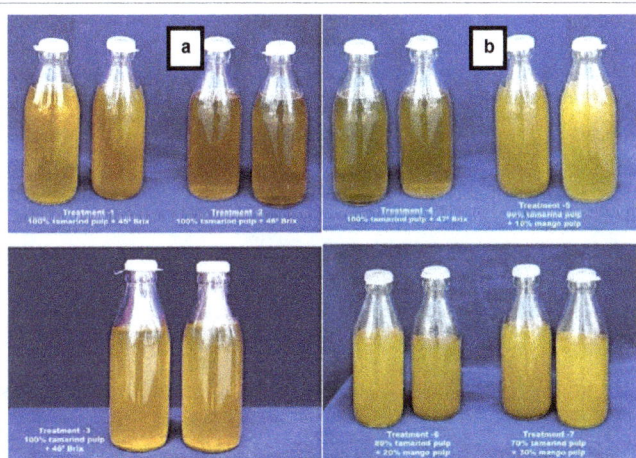

a) T₁: tamarind pulp 100%+45°Brix; T₂: tamarind pulp 100%+46°Brix; T₃: tamarind pulp 100%+46⁰Brix

b) T₄: tamarind pulp 100%+47°Brix; T₅: tamarind pulp 90%+ mango pulp 10%; T₆: tamarind pulp 80%+mango pulp 20%; T₇: tamarind pulp 70%+mango pulp 30%.

Figure 3: Tamarind squash by blending with mango pulp using different treatments.

Treatments	Storage Period	TSS (°B)	Acidity (%)	Reducing Sugars (%)	Total Sugars (%)
T1	0 day	45	0.407	8.44	16.816
	90 days	45.03	0.404	15.68	16.516
T2	0 day	46	0.423	9.56	17.24
	90 days	46.03	0.427	16.35	17.023
T3	0 day	46	0.446	11.36	20.474
	90 days	46.06	0.451	16.98	20.133
T4	0 day	47	0.475	12.95	22.22
	90 days	47.06	0.481	17.35	22.056
T5	0 day	47	0.495	14.65	26.656
	90 days	47.09	0.497	17.35	26.333
T6	0 day	47	0.517	15.56	27.97
	90 days	47.09	0.521	19.77	27.65
T7	0 day	47	0.508	15.15	27.853
	90 days	47.09	0.489	19.55	27.533

T₁: tamarind pulp 100%+45°Brix; T₂: tamarind pulp 100%+46°Brix; T₃: tamarind pulp 100%+46°Brix; T₄: tamarind pulp 100%+47°Brix; T₅: tamarind pulp 90%+ mango pulp 10%; T₆: tamarind pulp 80%+mango pulp 20%; T₇: tamarind pulp 70%+mango pulp 30%.

Table 1: Effect of storage period on physico-chemical parameters in tamarind squash at room temperature.

no significant increase in mean TSS content of the tamarind squash from 0 day (46.42°B) to 90 days (46.48°B) of storage (Figure 3). The interaction effects between days of storage and treatments were also not significant. However, a slight increase in TSS was observed among all treatments during the storage period. This may be due to conversion of polysaccharides in to sugars. Similar observations were reported by Saikia et al. [8] in ou-tenga fruit squash. Acidity values recorded in different treatments and days of storage are given in Table 1. No significant change in acidity was observed during the storage period. Treatments recorded significant differences, where as interactions were found to be non-significant. Among the different treatments, initially T6 (0.517%) recorded significantly higher acidity value and least recorded in T1 (0.407%). During storage, there was no significant change in acidity from 0 day (0.467%) to 90 days (0.467%) of storage. T6 recorded Maximum acidity value (0.521%), and least acidity value was recorded in T1 (0.404%) at 90 day of storage. Similar findings were reported in guava and papaya RTS beverage [9] and in blends of mango nectar [10]. Among treatments, significant changes found in acidity might be due to initial differences maintained during processing in acidity. Reducing sugars of tamarind squash recorded in different treatments and days of storage is given in Table 1. There was significant change in reducing sugars during the storage period, among the different treatments and interactions. All treatments differed significantly from one another. Among the different treatments employed for tamarind squash, initially T6 recorded significantly highest reducing sugar content (15.56%) and least was recorded in T1 (8.44%). During storage, there was a significant increase in mean reducing sugar content of the samples from 0 days (12.52%) to 90 days (17.57%) of storage period. The interaction effects of treatments and days of storage were also found to be significant. T6 recorded the maximum reducing sugar content (19.77%) at 90 days of storage. Increase in reducing sugar content may be due to hydrolysis of total sugars by acid present in fruit, which might have resulted in degradation of disaccharides to monosaccharides [11]. Similar observations were made by Farheen [12] in guava-grape and guava-pineapple nectar blends and in watermelon nectar prepared from different blends of watermelon with other fruits.

Treatments (F1)	Overall acceptability			
	0	45	90	Mean
T_1	4.02	4.02	4.003	4.014
T_2	4.33	4.236	4.116	4.227
T_3	4.6	4.34	4.216	4.338
T_4	4.323	4.22	4.01	4.184
T_5	4.72	4.42	4.336	4.492
T_6	4.966	4.966	4.963	4.965
T_7	4.91	4.826	4.733	4.823
Mean	4.532	4.432	4.34	--
	F value	Sed ±	CD at 5%	--
Treatments (F_1)	**	0.005	0.001	--
Periods (F_2)	NS	0.007	NS	--
F_1* F_2 interaction	NS	0.013	NS	--

T_1: tamarind pulp 100%+45°Brix; T_2: tamarind pulp 100%+46°Brix; T_3: tamarind pulp 100%+46°Brix; T_4: tamarind pulp 100%+47°Brix; T_5: tamarind pulp 90%+ mango pulp 10%; T_6: tamarind pulp 80%+mango pulp 20%; T_7: tamarind pulp 70%+mango pulp 30%.

Table 2: Effect of storage period on overall acceptability in tamarind squash at room temperature.

Treat-ments	Microbial load (Colony forming units/gm)							
	0 Days		30 Days		60 Days		90 Days	
	Bacteria	Y & M	Bacteria	Y & M	Bacteria	Y & M	Bacteria	Y & M
T_1	-	-	-	-	-	-	3×10^1	7×10^1
T_2	-	-	-	-	-	-	3×10^1	7×10^1
T_3	-	-	-	-	-	-	2×10^1	5×10^1
T_4	-	-	-	-	-	-	2×10^1	5×10^1
T_5	-	-	-	-	-	-	1×10^1	3×10^1
T_6	-	-	-	-	-	-	1×10^1	3×10^1
T_7							1×10^1	3×10^1

T_1: tamarind pulp 100%+45°Brix; T_2: tamarind pulp 100%+46°Brix; T_3: tamarind pulp 100%+46°Brix; T_4: tamarind pulp 100%+47°Brix; T_5: tamarind pulp 90%+ mango pulp 10%; T_6: tamarind pulp 80%+mango pulp 20%; T_7: tamarind pulp 70%+mango pulp 30%.

Table 3: Effect of storage period on microbial load (colony farming units/gm) of tamarind squash at room temperature.

Total sugars recorded in different treatments and days of storage are given in Table 1. No significant change in total sugar content was observed during the storage period. Treatments recorded significant differences, where as interactions were found non-significant. Among the treatments, initially T6 recorded highest (27.970%) total sugar content and least was in T1 (16.81%). During the storage, there was no significant decrease in the mean content of total sugars in squash from 0 day (22.74%) to 90 days (22.46%) of storage periods. The interaction effects of treatments and days of storage were also found to be non-significant during different storage period. Decrease in total sugars may be attributed to the increase in the bacterial count, which might have utilized for their survival. These findings were in conformity with the results reported by Sheeja and Prema [13] in papya squash, Chahal et al. [14] in watermelon juices and Krishnaveni et al. [15] in jack fruit RTS beverage. Of all the treatments, the overall acceptability score (Tables 2 and 3) was significantly highest for T6 (4.97) followed by T2 (4.33), T3 (4.6), T4 (4.32), T5 (4.72), T7 (4.91) and least overall acceptability score

was observed in T1 (4.02). There was decrease in all sensory scores for the products during storage. Decrease in colour of the products may be due to browning of the products. Similar findings were reported by Ranganna [4] in phalsa and litchi squashes. Decrease in flavor and taste upon storage may be due to the loss of volatile aromatic substances responsible for flavor. Temperature also plays an important role on the biochemical changes in the products, which leads to the formation of flavor and discoloration, masking the original flavor of the products with the storage period [16]. Similar findings were reported by Kaur et al. [10] in plum nectars, Sogi et al. [17] in kinnow squash.

The microbial examination showed (Table 3) that no yeast and mold count was observed till 60 days of storage. T1, T2 recorded higher load (7×10^{-1}) followed by T3, T4 (5×10^{-1}), and T5, T6, T7 (3×10^{-1}) at the end of 90 days storage period. The bacterial growth was observed at 90 days only. T1 and T2 recorded higher bacterial count (3×10^{-1}), followed by T3 (2×10^{-1}), T4 (2×10^{-1}) and the least were observed for T5, T6, and T7 (1×10^{-1}) at the end of 90 days' storage. However, the increase in microbial growth was negligible and within the permissible limits of squash. Application of heat during processing reduced microbial load [18]. This has been reported in watermelon nectar and in mixed fruit RTS beverage by Bidyut et al. [19].

Conclusion

The overall acceptability was highest in squash prepared with 80% tamarind pulp and 20% mango pulp (T6). Negligible growth of microbes was observed in all the treatments. The products stored without any deterioration in physico-chemical, sensory quality and microbial count and are consumer acceptable up to 3 months of storage as per the study. Profit estimated for 1 litre of tamarind squash Rs. 25.00 when compared with locally available products. Hence it can be concluded that blending with mango pulp can bring value addition to tamarind and increase in appearance and taste.

References

1. Chaturvedi MD (1956) The tamarind is prized for its shade and shelter. Indian farming.

2. Duke JA (1981) Handbook of legumes of world economic important. Plenum press, New York.

3. Lewis YS, Neelakantan S (1964). The chemistry, biochemistry, and technology of tamarind. J Sci Industry Res: 23: 204-206.

4. Ranganna S (1986) Handbook of analysis and quality control for fruits and vegetable products. Tata MC Graw Hill, New Delhi.

5. AOAC (1975) Association of official analytical chemists. Official methods of analysis, Geneva, Switzerland.

6. Cruikshank R, Durgid JP, Masmion BP, Sirion RHA (1975) Hedonic scale method of Measuring Food Preferences. Food Technol 11: 9-14.

7. Panse VG, Sukhatme PV (1985) Statastical methods for agricultural workers. ICAR, New Delhi.

8. Saikia L, Saikia J (2002) Processing of ou-tenga (Dillenia indica) fruit for preparation of squash and its quality changes during storage. J Food Sci Technology 39(2): 149-151.

9. Tiwari RB (2000) Studies on blending on guava and papaya pulp for RTS beverage. Indian Food Pack 54: 68-72.

10. Kaur C, Khurdiya DS (1993) Improvement in the quality of fruit nectar. Beverage Food World 23: 15-18.

11. Aruna K, Vimala V, Giridhar N, Rao DG (1997) Studies on preparation and storage of nector prepared from papaya (Carica papaya L.). Beverage Food World 28: 29-30.

12. Farheen F (2004) Studies on the preparation of squash, nectar, RTS and juice blends from watermelon fruits. Acharya NG Ranga Agricultural University, Hyderabad.

13. Sheeja N, Prema L (1995) Impact of pre-treatments on the shelf life quality of papaya squash. South India Horticultur 43: 49-51.

14. Chahal GS, Sain SPS (1999) Storability of juice from new hybrid watermelon variety. Indian. Food Pack 53: 12-17.

15. Krishnaveni A, Manimegalai G, Saravana KR (2001) Storage stability of jack fruit (Artocarpus Heterophyllus) RTS beverage. J Food Sci Technol 38: 601-602.

16. Doodnath L, Badriel N (2000) Processing quality evaluation of ready-to-serve watermelon nectars. J Food Sci Technol 38: 495-498.

17. Sogi DS, Singh S (2001) Studies on bitterness development in Kinnow juice, RTS beverage, squash, jam and candy. J Food Sci Technol 38: 433-438.

18. Srivastava P, Sanjeev K (2002) Fruits and vegetable preservation and principles and practices. International Book Distribution Company, Lucknow.

19. Bidyut CD, Sethi V (2001) Preparation of mixed fruit juices spiced RTS beverage Indian. Food Pack 55: 58-61.

22

Optimization of Operational Parameters in a Cyclone Type Pneumatic Rice Polisher

Samadder M*, Someswararao CH and Das SK

Department of Agriculture and Food Engineering, Indian Institute of Technology, Kharagpur, West Bengal, India

Abstract

A pneumatic rice polishing system has been developed to minimize broken content (B_p) and maximize the degree of polishing (D_p). This polisher consists of a metallic cyclone coated with hard abrasive material, a blower, a collection system. Brown rice is feed vertically at the rate of 1.5 kg/min that flows towards the abrasive cyclone through horizontal air flow at 72.2 m/s. Experiments were carried out at different moisture content (M) and grit size of abrasive material (E) at fixed number of passes. Degree of polishing and broken content varied with both M and E. System showed D_p from $4.224 \pm 0.02\%$ (13% M and 60 E) to $13.250 \pm 0.56\%$ (12% M and 36 E) whereas B_p from $2.146 \pm 0.14\%$ (10% M and 60 E) to $49.717 \pm 2.64\%$ (13% M and 36 E). Effect of M and E and their square were found significant ($p < 0.05$ and $p < 0.01$) on D_p and B_p. Optimum values of D_p and B_p were achieved at 100 E and 11.70% M (wet basis) respectively with a value of 10.359% and 0.476%.

Keywords: Pneumatic polishing system; Degree of polishing; Broken content; Rice; Moisture content

Abbreviations: Cm: Centimeter; dia: Diameter; M: Meter; min: Minute; mm: Millimeter; S: Second; wb: Wet Basis

Introduction

Rice (*Oryza sativa*) is the staple food of almost half of the world population. Asia accounts for 92% of world's rice production. India is the second largest rice growing country in the world [1]. Rice is the main source of nutrition for majority (65%) of people in India. Thus, rice milling becomes the largest agro-based industry in India. Polishing of brown rice is necessary to improve cooking quality, digestibility, and extension of storage period. Ideally, polishing should be within 5% to 6% of the weight of brown rice. However, commercially excessive polishing (10% to 12%) with removal of deeper starchy endosperm layer is a common practice because of consumer preference with white shiny grains. Most of the nutrients such as dietary fibers, essential amino acids, minerals, proteins and vitamins (B & E) present in the outer layers of the kernel and the germ are removed during polishing, which considerably reduces the nutritional value of rice. This result in loss of beneficial dietary fiber, large yields of broken (20% to 50%) and less oil content in the resultant bran mixed with starchy endosperm fraction [2]. Polishing is the most energy intensive operation among all other operations in rice milling process [3]. The major part of the supplied energy in polishing gets converted into heat energy; rises grain temperature significantly that leads to induce fissure development and ultimately cracking of the grain. Generally, two types of rice polishers are used–friction type and abrasion type. Friction type machines have metallic polishing rollers, which remove bran by inter-grain friction under relatively high pressure while the abrasion type polishers accomplish it by cutting of surface layers of grain by sharp edges of abrasive particles, such as carborundum, at relatively low pressure. Both operations involve high shearing action on the rice kernels under high pressure (5000-10000 Pa), involving inter-granular attrition and high rate friction with abrasive surface [4]. Large amount of the input energy to polishers is utilized for its moving parts to overcome friction and inertial forces, resulting high specific energy consumption. The moving parts of these heavy rollers also suffer excessive wear and tear that leads to high maintenance cost.

Prakash et al. [5] developed a horizontal abrasive pipe pneumatic rice polisher without any moving part and with negligible thermal and compressive stresses, was found successful in carrying out control level of polishing of rice grains with yield of broken as low as 8.52% compared to 20% in laboratory abrasive polisher. Restriction of flow with the formation of dunes; large pressure drop with excessive friction in the pipe are some of the drawbacks of this system. To overcome this, a new technique of cyclone type pneumatic rice polishing system has been developed. In this new technique of rice polishing, the modification of cyclone separator has been tried. The inner surface of the cyclone separator is lined with abrasive surface (Caborundum grit size 100, 60 and 36, Concord, India). Mixture of air and rice particles enters tangentially into the system near the top of the cylindrical section. The grains follow swirling path inside the polisher and come in contact with the hard-abrasive emery surface and get polished. The mixture of bran and milled rice was collected at the bottom of the polisher. The inlet air velocity and feed rate were maintained at 72.2 m/s and 1.5 kg/min respectively. Abrasion under moderate pressure results in effective polishing of grains under low mechanical and thermal stress. Absence of moving part might be beneficial also to reduce breakage percentage of kernels. The present study is aimed to optimize the moisture content of brown rice and type of abrasive surface for maximizing the degree of polishing with minimum broken content.

Materials and Methods

Pneumatic rice polishing system

The improved pneumatic rice polishing system consisted of no moving part. Basically, it comprised of two parts - an upper cylindrical part and a conical bottom part; similar to that of a cyclone separator as shown in Figure 1. In the present study, a system having diameter of

---

*Corresponding author: Samadder M, Department of Agriculture and Food Engineering, Indian Institute of Technology, Kharagpur-721302, West Bengal, India, E-mail: samadderagril@gmail.com

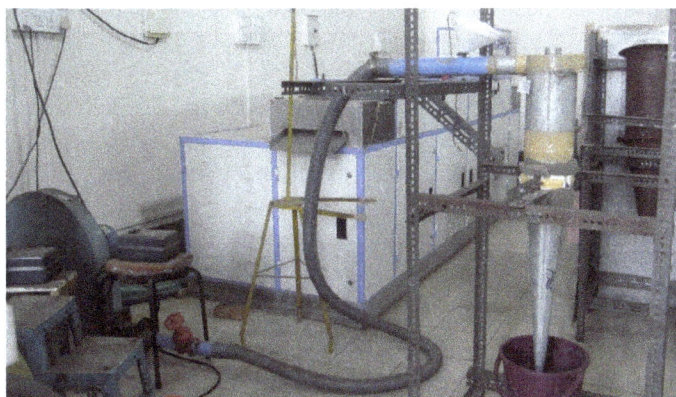

Figure 1: Laboratory pneumatic rice polishing systems.

Figure 2: Picture showing (a) Paddy (b) brown rice (c) polished rice.

the cylinder and height of the polisher as 0.2 m and 1.06 m respectively was selected. The cylindrical part consisted of an inlet port (feed inlet) provided with 180° scroll. This facilitated two-phase tangential flow of air and brown rice at the entry. The inside surface of the polisher was layered with three grades of hard abrasive particles (Concord, India). These were categorized as fine (FN), medium (MD) and coarse (CR) depending upon their grit sizes as commonly expressed in industry; like 100, 60 and 36, respectively. Corresponding average particle dimension were 122, 254 and 483 m. Air velocity of 30 m s^{-1} was used. This flow rate was adequate for generating high centrifugal force on the rice grains (terminal velocity of rice is around 6.0 ms^{-1}) for its movement along the abrasive surface. The final stream of bran and polished rice mixture at the outlet passes through the bran separator where bran particles escape through the perforations and deposited in the outer chamber of the cylinder. Partially polished rice is collected at the bottom of the system. Due to less residence time of rice inside the polisher, required degree of polishing cannot be achieved in a single pass. The collected rice was recycled for 60 times and the broken content and degree of polishing was measured after every 10 passes. The air escaped at the top and carried very fine bran particles with it. Experiments were conducted with brown rice of 10, 11, 12, and 13% (wb) moisture content.

Raw materials

To ascertain uniformity in polishing of grains, visual examination was the first choice. For this study, "Annapurna" a pigmented variety of paddy, was procured from Hatigeria village (Kharagpur, West Bengal). These variety exhibit a whitish endosperm covered with 10.5% purple color bran (Figure 2 shows the picture of paddy, kernel, and polished kernel). So, the changes of color (purple to white) due to removal of bran can be easily ascertained qualitatively by visual inspection. The paddy was procured with initial moisture content around 11.62% (wb) and dried with conventional sun drying method up to moisture content of 10.35% (wb) before de-husking it (Figure 1).

For de-husking the paddy, a laboratory rubber roll sheller (Satake

Corporation, Model THU35A, Japan) was used. For using the whole kernels for polishing with the developed polisher, broken kernels were separated from brown rice by using laboratory grader (Burrows equipment Company, Illinois, USA) (Figure 2).

Preparation of pre-conditioned rice (Hydration or dehydration)

The polishing operations were designed to perform at different wet basis moisture levels (10, 11, 12 and 13). The initial moisture content of the sample was determined by using air oven method [2]. About 5 g brown rice were taken and kept in a hot air oven maintained at 105 ± 10°C for 24 h, the moisture content, 11.82 ± 0.09% (wb) was determined gravimetrically by mean of four replications. Brown rice with desired moisture levels was obtained by the following procedure.

Decreasing moisture content (Dehydration): A lump of silica gel was placed inside a hot air oven maintained at 90°C up to an extent until the color become dark blue. The completely moisture free silica gel was filled up to the screen glass desiccator. A known quantity of brown rice was then kept inside the desiccator. The sample was then weighed periodically until the silica gel absorbed the excess moisture.

Increasing of moisture content (Hydration): A bottom of the desiccator was filled with water up to 2-3 cm below the screen. A known quantity of brown rice was then spread uniformly over the screen fitted at the middle of the desiccator. Weight of the grain was then taken periodically until the grain absorbed the desired moisture.

Polishing of rice with the pneumatic polishing system

Each experiment was carried out with 180 g brown rice at 10%, 11%, 12% and 13% moisture content. Only whole kernels were taken for polishing as stated above. The grains were fed into the system from the hopper. The grains, collected carefully were recycled 60 times. After 60 recycling of grains (pass), the degree of polishing was evaluated gravimetrically. To study the effect of abrasive material of abrasive cyclone on degree of polishing and yield of broken, experiment was

M, % wb	Coarse Emery	Medium Emery	Fine Emery
10 ± 0.15	D_p = 9.848, Bp = 26.816	D_p = 5.2511, Bp = 2.146	D_p = 7.3695, Bp = 2.403
11 ± 0.15	D_p = 11.440, Bp = 9.722	D_p = 7.774, Bp = 4.576	D_p = 10.438, Bp = 5.299
12 ± 0.15	D_p = 13.250, Bp = 12.169	D_p = 6.753, Bp = 4.441	D_p = 9.805, Bp = 4.511
13 ± 0.15	D_p = 12.675, Bp = 49.717	D_p = 4.224, Bp = 15.789	D_p = 9.103, Bp = 14.341

Figure 3: Surface view of rice after polishing (60 passes) at different moisture content using different emery grit size.

carried out with 3 different abrasive material (36E (Coarse Emery), 60E (Medium Emery) and 100E (Fine Emery), degree of polishing was determined for samples after 60 passes as stated above. After 60 passes, broken rice was separated from the sample using laboratory grader, and total breakage of kernels was estimated gravimetrically.

Estimation of degree of polishing

Gravimetric method Juliano and Bechtel [6] was adopted for measuring the degree of polishing. An accurately weighed amount of the sample was subjected to polishing operation. After each experiment the milled rice samples were aspirated by using a laboratory aspirator (Bates Aspirator, USA) out to remove any adhering bran sticking to sample surface followed by weighting it using electronic balance (Sartorius, Model BT 323 S, accuracy 0.001 g) and change of weight was recorded to calculate the degree of polishing, D_p using Eq. (1).

$$D.O.M.(\%) = \left(\frac{w}{W_i}\right) \times 100 \qquad (1)$$

Where, w and W_i are weight of removed bran after 60 passes and initial weight of brown rice respectively.

Estimation of broken content (%)

In each set of experiment with varying moisture content and grit size of abrasive material, broken content was determined after 60 passes. After 60 passes, the polished rice put on the laboratory grader to separate broken kernels. Both these streams were collected, weighted and B_p was estimated from Eq. (2).

$$\text{Broken Content: } B_p = \left(\frac{w_{bk}}{W_i}\right) \times 100 \qquad (2)$$

A designated program (Design Expert 7.0) was used to obtain regression equation correlating independent and dependent parameters. The significant terms in the model were determined by analysis of variance (ANOVA) for each of the response parameters. The goodness of fit was evaluated with several statistical parameters like probability of failure (P), adjusted R^2, predicted R^2, predicted error sum of squares (PRESS) and adequate precision. A good model should have large predicted R^2 and a low PRESS.

Optimization of the process parameters

For optimization of process parameters, experiments were conducted using three levels of emery paper, four level of moisture content at fixed number of passes (60). The general factorial option for experiment design was applied for optimizing the process in Design Expert 7 software (Design expert version 7.0.0, Stat {Ease INC., 2009, USA). The number of distinct combinations of the experiments was worked out to be 12 (with replication it was 36). Twelve combinations came for each level of emery paper combination with moisture content. To find out the effect of independent variable (X) on dependent variable (response, Y), the following quadratic regression equation (model) was fitted.

$$Y = b_0 + b_1X_1 + b_2X_2 + b_{12}X_1X_2 + b_{11}X_1^2 + b_{22}X_2^2 \qquad (3)$$

Optimization of independent parameters, moisture level (M), and emery paper (coarse, medium, and fine) was carried out for maximization of degree of polishing and minimizing of percentage of broken. Optimum conditions were finally selected considering the feasibility of moisture content and grade of emery paper following the desirability function approach which is one of the most widely used

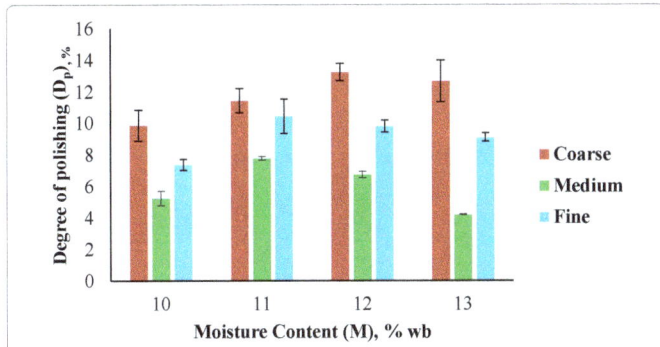

Figure 4: Degree of polishing at different moisture content using different emery grit size.

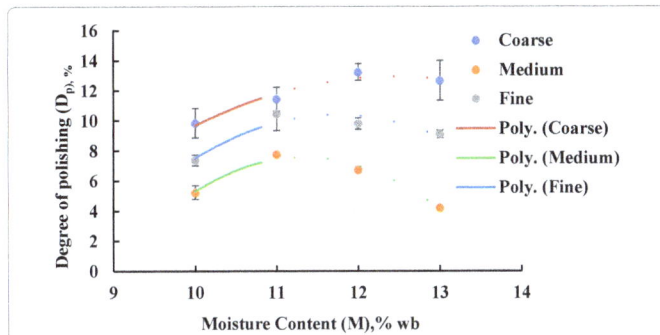

Figure 5: Effect of moisture content and emery grit size on degree of polishing.

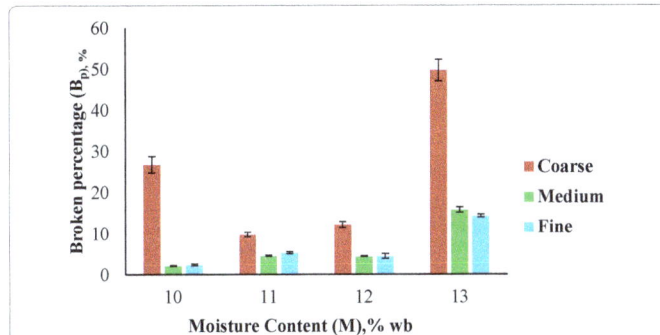

Figure 6: Broken percentage at different moisture content using different emery grit size.

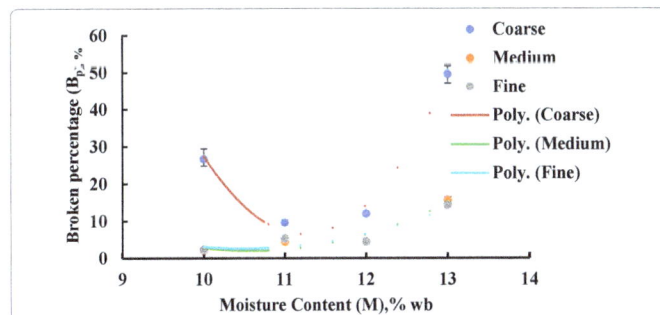

Figure 7: Effect of moisture content and emery grit size on broken content of brown rice.

industry methods for the optimization of multiple response processes. The method finds operating conditions (x) that provide the "most desirable" response

For each response, $Y_i(x)$ a desirability function $d_i (Y_i)$ assigns numbers between 0 (completely undesirable) and 1 (completely desirable or ideal response value). Based on desirability function of each of the parameters involved [$d_i (Y_i)$], the overall desirability (D) was obtained from geometric mean of the individual desirability as:

$$D = [d_1(Y_1) \times d_2(Y_2) .... d_k(Y_k)]1/k \qquad (4)$$

Where, k denotes the number of responses. The designated software (Design Expert 7) was employed for optimization of parameters.

Results and Discussion

Figure 3 shows the image of polished rice at different moisture content after 60 passes in a pneumatic polisher with different abrasive surfaces. Visual assessment with all these samples revels that the rice kernels were polished uniformly with all the moisture content and emery papers. Figures 4 and 5 shows the variation of D_p after 60 passes for different moisture content and grit size of emery. Changes of D_p were found to vary with moisture content and grit size of emery paper. D_p was found to vary from 4.224 ± 0.02% (13% M and Medium emery) to 13.250 ± 0.56% (12% M and Coarse emery). Figures 6 and 7 show the variation of B_p after 60 passes for different moisture content and grit size of emery. Changes of B_p were found to vary with moisture content and grit size of emery paper. D_p was found to vary from 2.146 ± 0.14% (10% M and Medium emery) to 49.717 ± 2.64% (13% M and Coarse emery).

Analysis of quality parameters of polished rice

From the foregoing discussion, it is apparent that both degree of polishing and broken percentage are affected by the combined effects of moisture content brown rice and grit size of abrasive materials. This will lead to obtaining an optimum condition for maximizing the effects. The correlation between moisture content (M) and polishing characteristics are shown in Table 1.

Degree of polishing: Analysis of variance for this regression equation showed close agreement between adjusted R^2 and predicted R^2 (0.8628 and 0.8124) with the estimated values of coefficient of R^2 and the coefficient of variance (C.V.) are 0.8794 and 11.81% respectively. This indicates that the regression equation fits best with the experimental results. Further all the linear terms, M and E significant at p < 0.05 and M^2 and E^2 significant at p < 0.01 on degree of polishing. A generalized regression equation for D_p (Eq. (5)) has been obtained using specific software (Design Expert 7.0) (Figures 6 and 7).

$$D_p = -92.73317 -0.72355 \times E + 21.41956 \times M + 5.01878 \times 10^{-3}E^2 - 0.91569 \times M^2 \qquad (5)$$

Figure 8 shows the response surface plot for degree of polishing as a function of moisture content and emery grit size. Degree of polishing changed with both moisture content and emery grit size where it increased from 10 to 12% moisture content and then decreased at 13% moisture content (Table 2).

Emery	Degree of Polishing (D_p)	Broken Content (Bp)
Coarse	$D_p = 0.541M^2 + 13.489M - 71.001$ $R^2 = 0.9502$	$B_p = 8.0025M^2 - 183.73M + 1063.5$ $R^2 = 0.9904$
Medium	$D_p = 1.2629M^2 + 28.637M - 154.73$ $R^2 = 0.972$	$B_p = 2.2294M^2 - 47.198M + 251.88$ $R^2 = 0.9126$
Fine	$D_p = 0.9425M^2 + 22.134M - 119.54$ $R^2 = 0.8745$	$B_p = 1.7335M^2 - 36.368M + 193.45$ $R^2 = 0.8776$
R^2 = Coefficient of determination		

Table 1: Relationship between moisture content (M) and polishing characteristics.

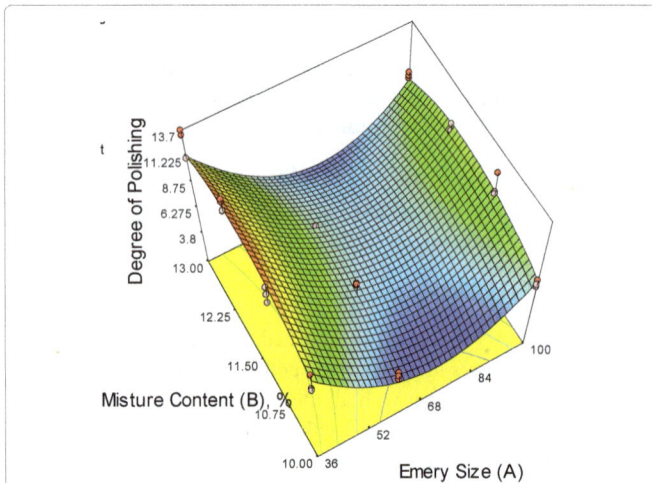

Figure 8: Response surface plot showing the effect of moisture content and emery grit size on degree of polishing.

Source	Sum of Squares	D_f	Mean Square	F-value	p-value
Model	238.7	5	59.64	52.87	0.0001
E*	41.31	1	41.31	36.62	0.0001
M**	5.76	1	5.79	5.13	0.0312
E2*	181.92	1	181.92	161.25	0.0001
M2*	30.19	1	30.19	26.76	0.0001

$R^2 = 0.8794$, R^2(adjusted) = 0.8628 and R^2 (predicted) = 0.8124 *Significant at p<0.01%, **significant at p<0.05

Table 2: Analysis of variance (ANOVA) for degree of polishing.

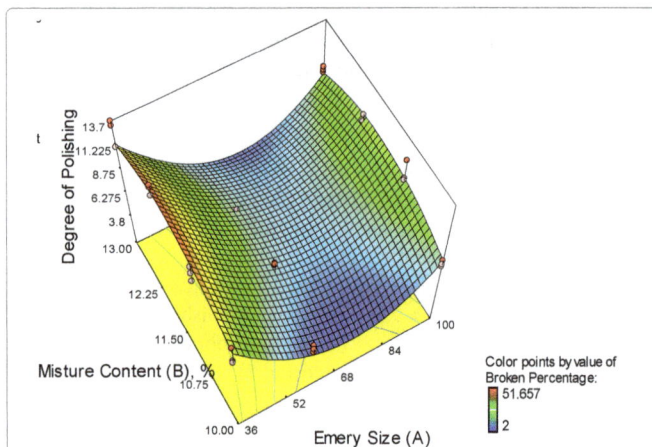

Figure 9: Response surface plot showing the effect of moisture content and emery grit size on broken content.

Source	Sum of Squares	d_f	Mean Square	F Value	p-value
Model	4891	4	1222.75	26.75	0.0001
E*	1937.74	1	1937.74	42.38	0.0001
M*	1097.77	1	1079.77	23.62	0.0001
E2*	971.13	1	971.13	21.24	0.0001
M2*	1241.86	1	1241.86	27.16	0.0001

$R^2 = 0.7867$, R^2 (adjusted) = 0.7573 and R^2 (predicted) = 0.6701 *Significant at p<0.01%,

Table 3: Analysis of variance (ANOVA) for broken content.

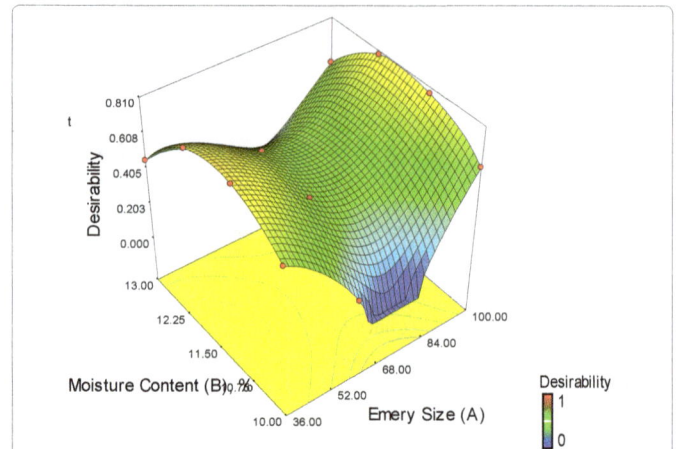

Figure 10: Response surface plot of desirability as a function of moisture content and emery grit size.

Number	E, grit	M, % wb	$B_{P,\%}$	$D_{P,\%}$	Desirability
1	100	11.70	0.476	10.359	0.806
2	100	11.68	0.343	10.359	0.806
3	36	11.32	16.578	12.854	0.803
4	36	11.39	16.785	12.895	0.803

Table 4: Optimization value of moisture content and emery grit size.

Broken content: Analysis of variance for this regression equation showed close agreement between adjusted R^2 and predicted R^2 (0.7573 and 0.6701) with the estimated values of coefficient of R^2 and the coefficient of variance (C.V.) are 0.7876 and 53.40% respectively. This indicates that the regression equation fits best with the experimental results. Further all the linear terms, M and E and M^2 and E^2 significant at p < 0.01 on broken content. A generalized regression equation for B_p (Eq. (6)) has been obtained using specific software (Design Expert 7.0)

$$B_p = 789.53708 - 1.85782 \times E - 130.18820 \times M + 0.011596 \times E + 5.87333 \times M^2 \tag{6}$$

Figure 9 shows the response surface plot for broken content as a function of moisture content and emery grit size. Broken content changed with the increase of moisture content and decrease of emery grit size. Brocen content changed rapiedly from 12% to 13% moisture content (Table 3).

Optimization of pneumatic polishing of rice

Neumarical optimization of the above Eqs. (3) and (4) were carried out using a designated software programe (Design Expert 7) as stated in the previus section. The response citeria for optimization were chosen for maximization of degree of polishing and minimization broken content. On the basis of highest desirability value of 0.806, the optimized values of independent parameters A and B were 100 and 11.70. Which gave broken content 0.4755% while degree of polishing 10.3588% (Figure 10) (Table 4).

Conclusion

The developed pneumatic rice polishing system was found to give uniform polishing and yielded less breakage compared to abrasive polishers ($D_p \approx 10\%$ causes around 20% B_p; Model TM 05, Japan) currently used. Linear increasing trend of degree of polish with moisture content reveals possibility of this system with higher grit size. However, energy consumption needs to be evaluated with further study

References

1. FAO (2014) Food and Agriculture Organization: Statistical Data base. Rome.

2. AOAC (2005) Official methods of analysis. (18th edn), Association of Official Analytical Chemists Inc, Washington DC.

3. Mohapatra D, Bal S (2010) Optimization of polishing conditions for long grain basmati rice in a laboratory abrasive mill. Food Bioprocess Technol 3: 466-472.

4. Satake RS (1994) New methods and equipment for processing rice. Rice Science and Technology, New York, USA.

5. Prakas KS, Someswarrarao CH, Das SK (2014) Pneumatic polishing of rice in a horizontal abrasive pipe: A new approach in rice polishing. Innovative Food Sci Emerg Technol 22: 175-179.

6. Juliano BO, Bechtel DB (1985) The rice grain and its gross composition. Rice: Chemistry and technology, American Association of Cereal Chemists, St. Paul Minnesota, USA pp: 17-50.

23

Studies on the Proximate, Anti-Nutritional and Antioxidant Properties of Fermented and Unfermented *Kariya* (*Hildergardia barterii*) Seed Protein Isolates

Gbadamosi SO and Famuwagun AA*

Department of Food Science and Technology, Obafemi Awolowo University, Ile-Ife, Nigeria

Abstract

The study prepared protein isolates from fermented *Kariya* seeds. Nutritional, anti-nutritional and antioxidant properties of the fermented (FKI) and unfermented (UKI) isolates were evaluated. Results showed that fermentation increased the protein content of the isolates were between 90.71% to 93.91%. The processing treatments was found to reduce the levels of some anti-nutrients in the protein isolates from 3.29 mg, 1.26 mg and 0.05 mg/100 g in unfermented isolate to 1.32 mg, 0.55 mg and 0.02 mg/100 g in fermented isolate for oxalate, tannin and saponin respectively. The result of antioxidant properties revealed that FKI had better antioxidant properties than UKI and the anti-oxidative properties of the samples increased with increasing sample concentration. The study concluded that fermented *Kariya* seeds protein isolates could find applications as potential food ingredient.

Keywords: Anti-nutrients; Fermentation; Isolates; Antioxidants

Introduction

A wide range of oil-bearing seeds exist in the forest of many African countries which are underutilized. Some of these oil seeds have been shown to be functional foods. Functional foods are important ingredients of a balanced human diet in many parts of the world due to their high protein content [1]. Plant proteins play significant roles in human nutrition, particularly in developing countries where average protein intake is less than the required [2]. It is worthy of note that plant protein products are gaining increasing interest as ingredients in food systems in many parts of the world and the final success of utilizing plant proteins as food additives depends greatly upon the functional characteristics that they impart to foods [3]. Since oil-seeds are valuable sources of proteins, many studies on protein functionality of major and minor oilseeds such as soybean [4] peanut, winged bean and ground nut have been reported [5]. Many of the vegetable proteins require processing techniques to provide food material with acceptable functional properties, such as emulsification, fat and water absorption, texture modifications, colour control and whipping properties, which are attributed primarily to the protein characteristics.

Kariya seeds (*Hildergardia barterii*) are consumed mostly in West African countries as raw or roasted nuts having a flavour like that of peanuts and it is grown for the ornamental nature. According to Hildergaia [6], the flowers, which are usually borne on leafless branches, mature into one-seeded pods, each about 50 mm in length, having a peanut–like seed in a nutshell. When the pods are completely matured and dry, they drop from the tree and are disposed as refuse in many parts of the world where they are found. Only in few West African countries are the kernels used in preparing traditional foods as condiments, eaten raw or roasted like peanut. Studies by Ogunsina et al. [7] showed that *Hildergardia barterii* kernel contains 17.5, 37.5, 2.8 and 6.5% of crude protein, crude fat, ash, and crude fibre respectively. In view of the high level of crude protein (17.5%) in *Kariya* seed, processing the whole flour to protein rich products such as protein isolate could enhance its utilization as a food ingredient.

However, consumption of most of the oil seeds found in the world is limited because in their raw state, they contain high levels of anti-nutrients which are potentially toxic [8]. *Kariya* seed is not an exception, the concentration of these anti-nutrients in plant protein sources vary with the species of plant, cultivar and post-harvest treatments (processing methods) [9]. Khare [10] revealed that

processing treatments such as soaking, cooking and fermentation are capable of reducing the anti-nutrient in legumes and oilseeds. The nutritional values of many plant foods can be enhanced through fermentation as it improves the nutritional properties of plant foods, prolongs shelf life and increases the protein content and carbohydrate accessibility and reduce anti-nutrients of plant foods [8]. Adebayo et al. [11] worked on the physicochemical and functional properties of *Kariya* flours. Previous studies have also suggested that fermentation improved the properties of oil seeds [12]. There is no reported work in the literature on the protein isolates of fermented *Kariya* seeds. This work therefore, aimed at fermenting *Kariya* seed; isolate its proteins and then evaluating the effect of fermentation on the physic-chemical, functional anti-nutritional and anti-oxidant characteristics of the protein isolate with a view to increasing its utilization as food ingredients.

Materials and Methods

Collection and preparation of plant materials

Dried *Kariya* pods were gathered from various ornamental *Kariya* trees in Obafemi Awolowo University, Ile-Ife, Nigeria. The nuts extracted from the pods were sorted to remove extraneous materials such as stones and leaves. The kernels were obtained by shelling the nuts manually which were cleaned to remove chaffs and immature kernels.

Fermentation and preparation of samples

Kariya kernels were rinsed with tap water and drained. The samples were divided into two portions; a portion was soaked for 24 h with warm water at 50°C and the water was changed every 6 h interval. The

---

***Corresponding author:** Famuwagun AA, Department of Food Science and Technology, Obafemi Awolowo University, Ile-Ife, Nigeria
E-mail: akinsolaalbert@gmail.com

soaked seeds were then transferred into different calabash pots, lined uniformly with banana leaves (up to 5 layers) and allowed to ferment inside the incubator (30°C). The fermented seeds were taken out after 96 h and oven dried at 60°C to terminate the fermentation process. The second portion was neither soaked nor fermented. The fermented and the unfermented samples were milled separately using Kenwood' grinder (PM-Y44B2, England) and sieved through 200 µm sieve. The resulting flours of the two samples were subsequently defatted using n-hexane in a sohxlet extraction apparatus. The defatted flours were desolventized by drying in a fume hood and the dried flours finely ground to obtain homogenous defatted flours. The flour samples were packaged in an air-tight polythene bags for further processing.

Preparation of protein isolates

Kariya protein isolate was prepared by the method described by Gbadamosi [13]. A known weight (100 g) of the defatted flours (fermented and non-fermented) was dispersed in 1000 ml of distilled water to give a final flour to liquid ratio of 1:10 in separate containers. The suspension was gently stirred on a magnetic stirrer for 10 min. The pH of the resultant slurry was adjusted to the point at which the protein was most soluble (pH 10.0) and the extraction was allowed to proceed with gentle stirring for 4 h keeping the pH constant. Non-solubilized materials were removed by centrifugation at 3500 × g for 10 min. The proteins in the extracts were then precipitated by drop wise addition of 0.1 N HCl with constant stirring until the pH was adjusted to the point at which the protein was least soluble (pH 4.0). The mixture was centrifuged (Harrier 15/80 MSE) at 3500 × g for 10 min in order to recover the protein. After separation of proteins by centrifugation, the precipitate was washed twice with distilled water. The precipitated protein was re-suspended in distilled water and the pH was adjusted to 7.0 with 0.1 M NaOH prior to freeze-drying. The freeze-dried protein was later stored in air-tight plastic container at room temperature for further use.

Proximate composition of fermented and unfermented *Kariya* isolates

Moisture content determination: Moisture content was determined by the standard [14] official method by weighing 1 g (W_1) of the samples in moisture cans and drying in a hot air-oven (Uniscope, SM9053, England) at 105 ± 1°C until to constant weight (W_2) was obtained. The samples were removed from the oven, cooled in a desiccator and weighed. The results were expressed as percentage of dry matter as shown in the equation below:

$$Moisture\ content\ (\%) = \frac{W_1 - W_2}{W_1} \times 100$$

Where,

W_1 = Weight of flour before drying,

W_2 = Weight of flour after drying,

Ash content determination: Ash content was determined by the official [14] method using muffle furnace (Carbolite AAF1100, UK). Two grams (W_3) of the sample were weighed into already weighed (W_2) ashing crucible and placed in the muffle furnace chambers at 700°C until the samples turned into ashes within 3 h. The crucibles were removed, cooled in a desiccator and weighed (W_1). Ash content was expressed as the percentage of the weight of the original sample.

$$Ash\ content\ (\%) = \left(\frac{W_1 - W_2}{W_3}\right) \times 100$$

Where,

W_1 = Weight of crucible + ash

W_2 = Weight of empty crucible

W_3 = Weight of sample

Protein content determination: The total protein content was determined using the Kjeldahl method [14]. The protein isolates (0.20 g) was weighed into a Kjeldahl flask. Ten milliliter of concentrated sulphuric acid was added followed by one Kjeltec tablet (Kjeltec-Auto 1030 Analyzer, USA). The mixture was digested on heating racket to obtain a clear solution. The digestate was cooled, and made up to 75 ml with distilled water and transferred onto kjeldahl distillation set up followed by 50 ml of 40% sodium hydroxide solution, the ammonia formed in the mixture was subsequently distilled into 25 ml, 2% boric acid solution containing 0.5 ml of the mixture of 100 ml of bromocresol green solution (prepared by dissolving 100 mg of bromocresol green in 100 ml of methanol) and 70 ml of methyl red solution (prepared by dissolving 100 mg of methyl red in 100 ml methanol) indicators. The distillate collected was then titrated with 0.05M HCl. Blank determination was carried out by excluding the sample from the above procedure

$$(\%)\ protein = \frac{1.401 \times M \times F\ (ml\ titrant - ml\ blank)}{sample\ weight}$$

Where,

M = Molarity of acid used = 0.05

F = Kjeldahl factor = 6.25

Carbohydrate content: Carbohydrate was expressed as a percentage of the difference between the addition of other proximate chemical components and 100% as shown in equation below;

Carbohydrates = 100 - (protein crude fat + ash + fibre + moisture)

Anti-nutritional properties of *Kariya* protein isolates

Determination of tannins: The concentration of tannin in the kariya protein isolates was determined using the modified vanillin–hydrochloric acid (MV – HCl) method of Price [15] was used.

Various concentrations (0.0, 0.1, 0.2, 0.4, 0.6, 0.8 and 1.0 mg/ml) of the catechin standard solution was pipetted into clean dried test tubes in duplicate. To one set was added 5.0 ml of freshly prepared vanillin – HCl reagent prepared by mixing equal volume of 4% (w/v) vanillin/MeOH and 16% (v/v) HCl/ MeOH and to the second set was added 5.0 ml of 4% (v/v) HCl/methanol to serve as blank. The solutions were left for 20 min before the absorbance was taken at 500 nm. The absorbance of the blank was subtracted from that of the standards. The difference was used to plot a standard graph of absorbance against concentration.

Kariya protein isolate was extracted separately with 10 ml of 1.0% (v/v) HCl–MeOH. The extraction time was 1 hour with continuous shaking. The mixture was filtered and made up to 10 ml mark with extracting solvent. Filtrate (1.0 ml) was reacted with 5.0 ml vanillin–HCl reagent and another with 5.0 ml of 4% (v/v) HCl–MeOH solution to serve as blank. The mixture was left to stand for 20 min before the absorbance was taken at 500 nm.

$$Tannin\ (mg/g) = \frac{x\ (mg/ml) \times 10 ml}{0.2\ (g)} = 50x\ (mg/g)$$

Where,

x = value obtained from standard catechin graph

Determination of oxalate: Oxalate was determined using titrimetric method by Falade [16]. Two grammes of the sample was weighed in triplicate into conical flasks and extracted with a 190 ml distilled water and 10 ml 6M HCl. The suspension was placed in boiling water for 2 h and filtered and made up to 250 ml with water in a volumetric flask. To 50 ml aliquot was added 10 ml of 6M HCl and filtered and the precipitate washed with hot water. The filtrate and the wash water combined and titrated against conc. NH_4OH until the salmon pink colour of the methyl red indicator changed to faint yellow. The solution was heated to 90°C and 10 ml 5% (w/v) $CaCl_2$ solution was added to precipitate the oxalate overnight. The precipitate was washed free of calcium with distilled water and then washed into 100 ml conical flask with 10 ml hot 25% (v/v) H_2SO_4 and then with 15 ml distilled water. The final solution was heated to 90°C and titrated against a standard 0.05M $KMnO_4$ until a faint purple solution persisted for 30 s. The oxalate was calculated as the sodium oxalate equivalent.

1 ml of 0.05M KMnO₄ =2 mg sodium oxalate equivalent/g of sample

Determination of saponin: The spectrophotometric method of Brunner [17] was used for saponin analysis. 1 g of finely ground sample was weighed into 250 ml beaker and 100 ml of isobutyl alcohol was added. The mixture was shaken for 2 h to ensure uniform mixing. Thereafter the mixture was filtered through a Whatman No. 1 filter paper into a 100 ml beaker and 20 ml of 40% saturated solution of magnesium carbonate was added and the mixture made up to 250 ml. The mixture obtained with saturated $MgCO_3$ was again filtered through a whatman No. 1 filter paper to obtain a clear colourless solution. 1 ml of the colourless solution was pipette into a 50 ml volumetric flask and 2 ml of 5% $FeCl_3$ solution was added and made up to mark with distilled water. It was allowed to stand for 30 min for blood red colour to develop. Saponin stock solution was prepared and 1-10 ppm standard saponin solutions were prepared from saponin stock solution. The standard solution was treated similarly with 5% of $FeCl_3$ solution as done for 1 ml of sample above. A dilution of 1 to 10 was made from the prepared solution. The absorbances of the samples as well as that of the standard solution were read after colour development in a 752S Spectrum lab UV, VIS Spectrophotometer at a wavelength of 380 nm.

$$Saponin = \frac{absorbance\ of\ sample \times dil.\ factor \times gradient\ of\ standard\ graph}{sample\ weight \times 10,000.} (mg/g)$$

Antioxidant properties of protein isolates

DPPH radical scavenging activity assay: The free radical scavenging ability of the extract was determined using the stable radical DPPH (2, 2-diphenyl-2-picrylhydrazyl hydrate) as described by Pownall [18]. To 1 ml of different concentrations (0.5, 1.0, 1.5, 2.0 and 2.5 mg/ml) of the extract or standard (vitamin C) in a test tube was added 1 ml of 0.3 mM DPPH in methanol. The mixture was mixed and incubated in the dark for 30 min after which the absorbance was read at 517 nm against a DPPH control containing only 1 ml methanol in place of the extract. The percent of inhibition was calculated from the following equation:

$$Inhibition(\%) = \frac{(Acontrol - Asample)}{Acontrol} \times 100$$

Where $A_{control}$ is the absorbance of the control reaction (containing all reagents except the test compound) and A_{sample} is the absorbance of the test compound. Inhibition concentration leading to 50% inhibition (IC_{50}) was calculated from the graph plotting inhibition percentage against extract concentrations.

Metal chelating ability assay: The metal-chelating activity of the isolates was carried out according to the method described by Singh [19]. Solutions of 2 mM $FeCl_2 \cdot 4H_2O$ and 5 mM ferrozine was diluted 20 times (1 ml of each of the solutions made up to 20 ml with distilled water separately). An aliquot (1 ml) of different concentrations (6.25, 12.5, 25.0, 50.0 and 100.0 mg/ml) of sample extract was mixed with 1 ml $FeCl_2 \cdot 4H_2O$. After 5 min incubation, the reaction was initiated by the addition of ferrozine (1 ml). The mixture was shaken vigorously and after a further 10 min incubation period the absorbance of the solution was measured spectrophotometrically at 562 nm. The percentage inhibition of ferrozine–Fe^{+2} complex formations was calculated using the formula:

$$Chelating\ effect = \frac{Acontrol - Asample}{Acontrol} \times 100$$

Where,

$A_{control}$ = absorbance of control sample (the control contains 1 ml each of $FeCl_2$ and ferrozine, complex formation molecules) and

A_{sample} = absorbance of a tested samples.

Determination of ferric reducing antioxidant power (FRAP): The FRAP assay uses antioxidants as reductants in a redox-linked colorimetric method with absorbance measured with a spectrophotometer The principle of this method is based on the reduction of a colourless ferric-tripyridyltriazine complex to its blue ferrous coloured form owing to the action of electron donating in the presence of antioxidants [20]. A 300 mmol/L acetate buffer of pH 3.6, 10 mmol/L 2,4,6-tri-(2-pyridyl)-1,3,5-triazine and 20 mmol/L $FeCl_3 \cdot 6H_2O$ was mixed together in the ratio of 10:1:1 respectively, to give the working FRAP reagent. A 50 µl aliquot of the extract at concentration (0.0, 0.2, 0.4, 0.0.6, 0.8 and 1 mg/ml) and 50 µl of standard solutions of ascorbic acid (20, 40, 60, 80, 100 µg/ml) was added to 1 ml of FRAP reagent. Absorbance measurement was taken at 593 nm exactly 10 minutes after mixing against reagent blank containing 50 µl of distilled water and 1 ml of FRAP reagent.

The reducing power was expressed as equivalent concentration (EC) which is defined as the concentration of antioxidant that gave a ferric reducing ability equivalent to that of the ascorbic acid standard.

Statistical analysis: All the analyses were conducted in triplicate and subjected to statistical analysis using analysis of variance (ANOVA). Means were separated using Duncan's multiple range test.

Results and Discussion

Proximate composition of fermented and unfermented *Kariya* protein isolates

The results of the effect of fermentation on the proximate composition of Fermented kariya isolates (FKI) and unfermented kariya protein isolates (UKI) presented in Table 1. The results showed that UKI had higher moisture content of (3.39 ± 0.09) than FKI (2.35 ± 0.11). The protein content of FKI was (93.91 ± 1.93) was higher than the value obtained for UKI (90.71 ± 1.61) and this values was significantly different ($p > 0.05$) from each other. Ash content of 0.49 ± 0.03 was recorded for FKI and the value was lower than the value (0.67 ± 0.06) obtained for UKI. Crude fibre was not detected in the samples and the value obtained for carbohydrate content in sample UKI was higher than the value obtained for FKI. The values obtained for UKI and FKI in this work were higher than the value reported for conophor nut isolates (80.00%) by Gbadamosi [13] and compared favourably with the values reported by Samruan et al. [21] for sunflower protein isolates (90.1%).

Proximate composition	UKI	FKI
Moisture content	3.39 ± 0.09^a	2.35 ± 0.1^b
Crude fibre	ND	ND
Ash	0.67 ± 0.06^a	0.49 ± 0.03^b
Crude fat	0.98 ± 0.02^b	1.09 ± 0.21^a
Protein	90.71 ± 1.61^b	93.91 ± 1.93^a
Carbohydrates	4.43 ± 0.98^a	1.98 ± 0.51^b

Values reported are means ± standard deviation of triplicate determinations. Mean values with different superscript within the same row are significantly ($P<0.05$) different.

Table 1: Proximate compositions of fermented and unfermented *Kariya* seed protein isolates.

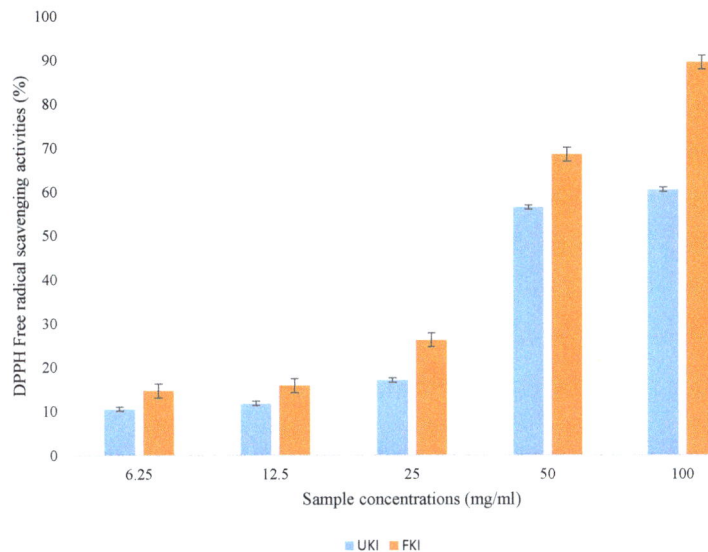

Figure 1: DPPH free radical scavenging abilities of fermented and unfermented *Kariya* protein Isolates. Error bars showing the standard deviation of triplicate determinations.

The results reveal that *Kariya* seed protein isolates has the potential to satisfy the protein needs of the ever increasing population. Ash content is an indication of the mineral contents of the samples. The ash content of the samples indicated the usefulness of this under-utilized seed to satisfy the macro and micro elements need of the consuming populace in the developing world.

Anti-nutritional properties of fermented and unfermented *Kariya* seed protein isolates

The effect of fermentation on some anti-nutritional properties (oxalate, tannin and saponin) of fermented *Kariya* isolates (FKI) and unfermented *Kariya* isolates (UKI). The study showed that UKI contained 3.29 mg, 1.26 mg and 0.05 mg/100 g for oxalate, tannin and saponin respectively while FKI contained 1.32 mg, 0.55 mg and 0.02 mg/100 g for oxalate, tannin and saponin respectively and these values were significantly different ($p<0.05$). The values represented about 61.70%, 56.35% and 60% reduction in the levels of the oxalate, tannin and saponin, respectively in unfermented *Kariya* protein isolates. Defatting was employed as processing technique on UKI while soaking, fermentation and defatting were employed as processing techniques on FKI. The significant reduction in the levels of the anti-nutrient could be attributed to soaking and fermentation processes carried out during the processing of FKI. Factors such as soaking, defatting and fermentation applied during sample preparation could be responsible for degrading the anti-nutrients in these samples. Oxalates bind minerals like calcium and magnesium and interfere with their metabolism, which leads to muscular weakness and paralysis [22]. Tannins have been reported

to affect nutritive value of food products by chelating metals such as iron and zinc and reduce the absorption of these nutrients and also forming complex with protein thereby inhibiting their digestion and absorption [22]. Saponins have been found to cause haemolytic activity by reacting with the sterols of erythrocyte membrane The levels of these tested anti-nutrients in UKI and FKI were low and were within the tolerable (safe) levels for man (12.0, 1.5 and 100 mg /100 g, for oxalate, tannin and saponin respectively) [23]. This study however revealed that soaking, fermentation and defatting could be employed separately or in combination in the processing of *Kariya* seeds to significantly reduce the levels of anti-nutrients isolates in *Kariya* protein isolates.

Antioxidant properties of fermented and unfermented *Kariya* protein isolates

The effect of fermentation on the 1,1-diphenyl-2 picrylhydrazyl (DPPH) free radical scavenging abilities of FKI and UKI is shown in Figure 1. The results showed that the free radical scavenging capacities of the samples as measured by DPPH assay increased as the concentration of the sample extract increased from 0.5-2.5 mg/ml. The increase was significant when the FKI was compared with UKI at each of the concentration considered. At the highest concentration of the sample extracts (2.5 mg/ml), the inhibition percentage of the sample extracts for FKI was 69.44% and this value was higher than 61.35% obtained for UKI at the same concentration (2.5 mg/ml). The result clearly showed that fermentation enhanced the free radical scavenging capability of the isolates by about 13.00% than the unfermented isolates. Table 2 shows the potency of the samples in terms of IC_{50} value. The results

SAMPLE	DPPH IC$_{50}$ (mg/ml)	MC IC$_{50}$ (µg/ml)	FRAP (AAEµg/g)
FKI	1.73 ± 0.2[b]	1.85 ± 0.17[b]	0.93 ± 0.3[b]
UKI	2.21 ± 0.77[a]	1.93 ± 0.04[a]	1.50 ± 0.12[a]
Ascorbic acid	0.67 ± 0.03[c]	-	-
EDTA	-	0.08 ± 0.06[c]	-

Values reported are means ± standard deviation of triplicate determinations. Mean values with different superscript within same column are significantly (P<0.05) different. FKI: Fermented *Kariya* Isolate, UKI: Unfermented *Kariya* isolate.

Table 2: IC$_{50}$ values fermented and unfermented *Kariya* seed protein isolates.

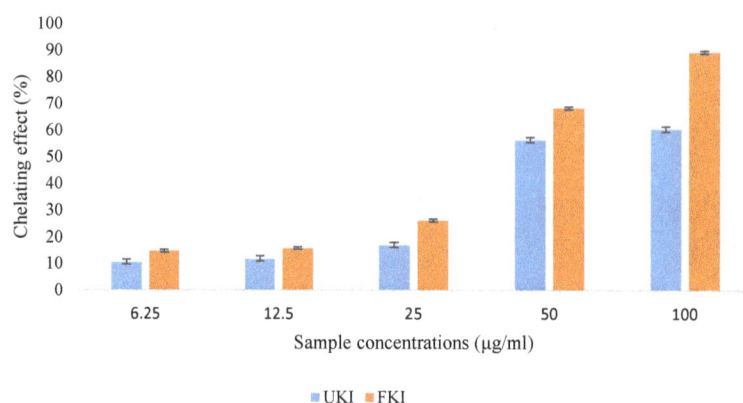

Figure 2: Metal chelating effect of fermented and unfermented *Kariya* protein Isolates. Error bars showing the standard deviation of triplicate determinations.

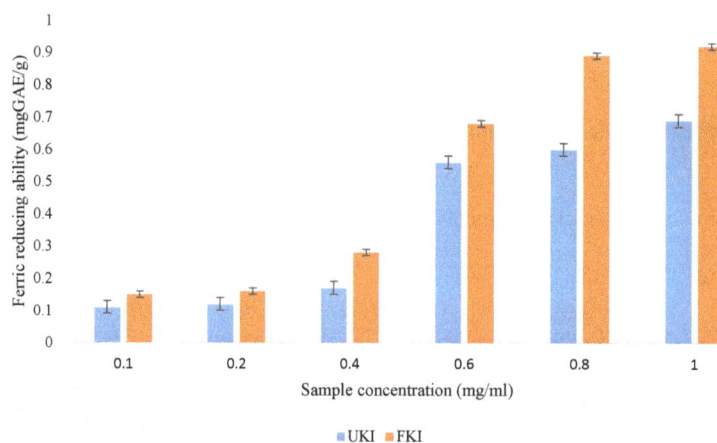

Figure 3: Ferric reducing power of fermented and unfermented *Kariya* protein isolate. Error bars showing the standard deviation of triplicate determinations.

revealed lower value of FKI when compared with UKI. The lower value of FKI indicates better radical scavenging than UKI. Similar results were reported by Je [23] on the fermentation of soybeans proteins.

The chelating effects of the isolates as influenced by fermentation are shown in Figure 2. The results showed that sample FKI had better chelating effect than sample UKI at each of the concentration considered. Just like the DPPH, the chelating effect of the samples increased with an increase in the sample concentration. At the highest concentration of 100 µg/g, chelating percentage for sample FKI was 78.69. The value was higher than 70.9% obtained for UKI. Considering the IC$_{50}$ values of the samples, it was observed that 1.93 was obtained for FKI. The value was higher than 1.85 recorded for sample unfermented kariya isolate 1.95. The trend observed in this work was in agreement with the observation of Je [24] on the fermentation of soy proteins.

Ferric reducing effect of the samples as a function of fermentation is presented in Figure 3. The results revealed the positive influence of fermentation on the ferric reducing ability of the samples. Results showed progressive increase in the ferric reducing abilities as sample concentration increased. Isolates produced from fermented kariya seed (FKI) was found to have higher chelating effect than the unfermented sample (UKI). The results also revealed the potential of FKI as having better ferric reducing than UKI. The result was in line with the observation of Samruan [21] on the ferric reducing abilities of rapeseed proteins.

The results of the anti-oxidant properties of the fermented isolate clearly showed the beneficial effects of fermentation in positively influencing the free radical scavenging abilities, chelating metals and in reducing ferric ions of *Kariya* protein isolates.

Conclusion

The study investigated the effect of fermentation on physicochemical, functional, anti-nutritional properties of *Kariya* protein isolates. Fermentation increased emulsifying properties, water and oil absorption capabilities, *in-vitro* protein digestibility of *Kariya* seed isolates. On the other hand, fermentation decreased bulk density and foaming properties. The processing methods employed (fermentation) significantly reduced the levels of tested anti-nutrients (tannin, saponin and oxalate) below the tolerance levels. Fermentation was also observed to increase the levels of some antioxidant properties isolates produced from fermented *Kariya* seed. The study revealed that fermented *Kariya* isolates could find application as functional ingredient in food systems.

References

1. Ezeagwu IE, Gopalakrishna AG, Khatoon S, Gowda LR (2004) Physicochemical characterization of seed oil and nutrient assessment of *Adenantherapavonina*. Ecology. Food Nutri 43: 295-305.

2. Yoshie-Stark Y, Wada Y, Wäsche A (2000) Chemical composition, functional properties, and bio-activities of rapeseed protein isolates. Food Chem 107: 32-39.

3. Yu J, Ahmedna M, Goktepe I (2008) Peanut protein concentrates: Production and functional properties as affected by processing. Food Chem 103: 121-129.

4. Molina Ortiz SE, Puppo MC, Wagner JR (2004) Relationship between structural changes and functional properties of soy protein isolates. Food Hydrocol 18: 1045-1053.

5. Lawal O, Adebowale K, Adebowale Y (2007) Functional properties of native and chemically modified protein concentrates from bambarra groundnut. Food Res Int 40: 1003-1011.

6. Hildergaia (2012) The Hildegardia.

7. Ogunsina BS, Olaoye IO, Adegbenjo AO, Babawale BD (2011) Nutritional and physical properties of *Kariya* seeds (*Hidergardia bateri*). Int Agrophys 25: 97-100.

8. Fasoyiro SB, Ajibade, SR, Omole AJ, Adeniyan ON, Farinde EO, et al. (2000) Proximate, minerals and anti-nutritional factors of some underutilized grain legumes in the South West Nigeria. J Nutri Food Sci 36: 18-23.

9. Akande KE, Doma UD, Agu HO, Adamu HM (2009) Animal protein. Niger Pak J Nutri 8: 827-832.

10. Khare SK (2000) Application of immobilized enzymes. The Third International Soybean Processing and Utilization (ISPCRC III): Innovation era for mycotoxins. Farm animal metabolism and Soybeans, Tsukuba, Ibaraka, Japan. 1: 381-382

11. Adebayo WA, Ogunsina BS, Gbadamosi SO (2013) Some physicochemical and functional properties of kariya (*Hildedardia barterii*) kernel flours. Ife J Sci 15: 3.

12. Igbabul BD, Bello FA, Ani EC (2014) Effect of fermentation on the proximate composition and functional properties of defatted coconut (*Cocosnucifera L.*) flour. J Food Sci 3: 34-40.

13. Gbadamosi SO, Abiose SH, Aluko RE (2012) Amino acid profile, protein digestibility thermal and functional properties of Conophor nut (*Tetracarpidium conophorum*) defatted flour, protein concentrate and isolates. Int J Food Sci Technol 47: 731-739.

14. Association of Analytical Chemists (2000) Official methods of analysis of AOAC, (17thedn). AOAC International, Gaithersburg, MD, USA.

15. Price ML, Van Scoyoc S, Butler LG (1974) A critical evaluation of vanillin reaction as an assay for tannin in sorghum grain. J Agri Food Chem 26: 1214-1218.

16. Falade OS, Dare AF, Bello MO, Osuntogun BO, Adewusi SRA, et al. (2004) Varietal changes in proximate composition and the effect of processing on the ascorbic acid content of some Nigerian vegetable. J Food Technol 2: 103-108.

17. Brunner JH (1984) Direct spectrophotometric determination of saponin. J Analytic Chem 34: 1314-1326.

18. Pownall TL, Udenigwe CC, Aluko RE (2010) Amino acid composition and antioxidant properties of pea seed (*Pisum sativum L.*) enzymatic protein hydrolysate fractions. J Agri Food Chem 58: 4712-4718.

19. Singh N, Rajini PS (2004) Free radical scavenging activity of an aqueous extract of potato peel. Food Chem 85: 6-11.

20. Benzie IFF, Strain JJ (1999) Ferric reducing ability of plasma (FRAP) as a measure of antioxidant power: The FRAP assay. Analytic Biochem 239: 70-76.

21. Samruan W, Oosivilai A, Oosivilai R (2012) Soybean and fermented Soybean Extract Antioxidant Activities. World Acad Sci Eng Technol 72: 2-6.

22. Soetan KO, Oyewole OE (2009) The need for adequate processing to reduce the Anti-nutritional factors in plants used as human foods and animal feeds. Afri J Food Sci 3: 223-232.

23. Health and Safety Publications (2011) Permissible levels of anti-nutrients in: Series on the safety of novel foods and feeds and Environment. Health and Safety Publications.

24. Je JY, Park PJ, Kim SK (2005) Antioxidant activity of a peptide isolated from Alaska Pollack (*Theragra chalcogramma*) frame protein hydrolysate. Food Res Int 38: 45-50.

24

Quality and Antioxidant Properties of Apricot Fruits at Ready-to-Eat: Influence of the Weather Conditions under Mediterranean Coastal Area

Bartolini Susanna[1]*, Leccese Annamaria[1] and Viti Raffaella[2]

[1]Institute of Life Science, Scuola Superiore Sant'Anna, Piazza Martiri della Libertà 33, Pisa, Italy

[2]Department of Agriculture, Food and Environment-Interdepartmental Research Center, Nutrafood 'Nutraceuticals and Food for Health', University of Pisa, Via del Borghetto, Pisa, Italy

Abstract

The effect of different weather conditions on fruit quality of 'Pisana' apricot cultivar (*Prunus armeniaca* L.) was evaluated over seven consecutive harvesting seasons in central Italy. The main physical-chemical traits, total antioxidant capacity and total phenols of fresh apricots at ready-to-eat were studied. The fruit quality showed a high variability in relation to the climatic conditions, particularly due to the summer rainfall. The most influenced quality parameters were TSS, TA and antioxidant levels: under wet seasons a huge reduction was observed, while strong drought conditions increased these chemical compounds. To improve fruit quality, 'Pisana' cultivar benefits of environmental conditions typical of temperate and semi-temperate regions, where water is usually limited.

Keywords: *Prunus armeniaca* L.; Climatic conditions; Fruits; Physical-chemical traits; Total antioxidants; Total phenols

Introduction

Apricot (*Prunus armeniaca* L.) is a fruit species with a nutritional value of interest having a good source of fiber, minerals (especially potassium but also calcium, iron, magnesium, zinc, phosphorus and selenium) and vitamins such as vitamin A, vitamin C, thiamin, riboflavin, niacin and pantothenic acid [1]. Moreover, apricots contain a number of main secondary metabolites such as polyphenols, carotenoids, fatty acids, volatiles and polysaccharides whose biological activities are considered useful for exerting various biological activities desirable for human health [2]. In particular, phenolic compounds are one of the main sources of antioxidant activity which are able to prevent oxidative stress scavenging free radicals and nitrogen species [3]. Antioxidant properties and quality traits of fruits are strictly related to genetics (species and cultivars) whose are influenced by geographic area, environment and cultivation techniques [4-6]. Climate has an important role on quality, affecting the nutritional value of vegetables and fruits [7]. In particular, light intensity and temperature together with water availability are related to the antioxidant activity in different fruit species [8]. Furthermore the deficit irrigation, as well, influence the phenol content in fruit as reported by several authors [9-11]. In apricot, recent studies showed that the bioactive compounds, such as the antioxidant content, are mainly related to the genotype and pedo-climatic conditions [12]. A screening among several apricot cultivars from international and Italian germplasm revealed some interesting varieties showing a high antioxidant capacity of fruits [13]. In particular, 'Pisana' cultivar stood out for its excellent pomological and antioxidant fruit properties, so as it is also appreciated in non-EU countries such as Latin America [14]. 'Pisana' was patented in Italy by the University of Pisa's breeding programmes and it is characterized by late blooming and ripening time, and strong fruit attractiveness for the fresh market [15,16]. In a two-year experimental trial, it has been showed that some fruit quality traits of 'Pisana', mainly total antioxidant capacity and total phenols, can be influenced by climatic conditions during the fruit growth and ripening [17]. Consequently, more studies about the effect of annual climatic variability on apricot quality are needed. In particular, researches combining environmental conditions, pomological traits and nutraceutical properties of apricot fruits over many years are rare. The aim of this research was to assess the influence of temperature and rainfall on fruit quality of 'Pisana' cultivar, over seven consecutive harvesting seasons. In particular, the effect of these climatic factors on the antioxidant potential of apricot was determined.

Materials and Methods

Plant material

The research was conducted over several harvesting seasons (2005-2012) on full bearing apricot trees of cultivar 'Pisana'. This cultivar i s characterized by late blooming and ripening time, +8 days from 'San Castrese' (last decade of June), reference cultivar for the Italian apricot ripening calendar. Trees, grafted onto Myrabolan 29C rootstock, were grown at the experimental Station of the Department of Agriculture, Food and Environment of Pisa University located in a coastal area of Tuscany (Italy, altitude 6 m a.s.l., lat. 43.02 N, long. 10.36 E). The site is characterized by mild-winters and annual average rainfall is about 600 mm; the soil in orchard is loam, moderately deep, medium texture, slightly alkaline, non-calcareous. Trees, trained to a free palmette system (4 m × 4.5 m) with rows facing east-west, were not irrigated and routine conventional horticultural management (pruning, thinning, fertilization, pest and disease protection) was performed. The experimental design was established in a completely randomized design. The main climatic parameters were acquired. Hourly temperatures were registered by an automatic data-loggers (Tynitag Plus', West Sussex , UK, 2003) and rainfall data were provided by the Regional Agro-meteorological Service of Florence (ARSIA,

*Corresponding author: Bartolini Susanna, Institute of Life Science, Scuola Superiore Sant'Anna, Piazza Martiri della Libertà 33, Pisa, Italy
E-mail: susanna.bartolini@sssup.it

'Agenzia Regionale per lo Sviluppo e l'Innovazione nel settore Agricolo Forestale' Tuscany, Italy).

Crop entity and physical-chemical fruit parameters

Crop entity and apricot fruit quality were assessed at physiological maturity (ready-to-eat stage). The crop entity average was evaluated on ten trees (kg/tree) and expressed as crop index (CI) related to 5 yield classes: < 1 kg (class 1); 1-5 kg (class 2); 5.1-10 kg (class 3); 10.1-20 kg (class 4); > 20 kg (class 5). Samples of 30 fruits were randomly collected to determine the main physical-chemical parameters, total antioxidant capacity, and total phenol content. From each fruit, measurements of fresh weight, peel and flesh color, pulp firmness, total soluble solids (TSS), and titratable acidity (TA) were determined. The skin color of the un-blushed side was evaluated using a color chart for apricot fruit according to Lichou et al. [1] by 10 shades of growing intensity from 1 (green) to 10 (red-orange) through different categories (1-4: yellow-green; 5-8: yellow-orange; 9-10 red-orange). The skin color of the blushed side ranges from pink to red. The area of the blushes was evaluated visually by classifying the red area according to the following classes: SC-b: < 15% (class 1); 15.1-25% (class2); 25.1-35 (class 3); > 35% (class 4). Firmness (kg 0.5 cm^2) was evaluated with a manual penetrometer (Model 53200SP TR, TR-Turoni & C. Inc Forlì, Italy) on two opposite sides at the equatorial region of the apricot, using an 8-mm-wide plunger. TSS was measured using a refractometer (Model 53015C TR, TR-Turoni & C. Inc Forlì, Italy) and expressed in °Brix at room temperature. TA was determined in fruit juice by titrating known volume of juice with 0.1 N sodium hydroxide (NaOH) to an end point of neutral pH (8.1). TA was expressed as milliequivalents per 100 grams of fresh weight (meq 100 g^{-1} FW).

Antioxidant proprieties

The Total Antioxidant Capacity (TAC) and Total Phenols (TP) analyses were carried out on the same fruits that had been previously subjected to the physical and chemical determinations. Samples of fresh material (3 g, in triplicate), homogenized using an ultra-Turrax T25 (Ika, Staufen, Germany) at 4°C to avoid oxidation, were performed in 80% ethanol for 1 h in a shaker in the dark and subsequently centrifuged at 2600g for 10 min at 2-4°C.

TAC assay

Total antioxidant capacity was evaluated using the improved Trolox Equivalent Antioxidant Capacity (TEAC) method [18]. The TEAC value was calculated in relation to the reactivity of Trolox, a water-soluble vitamin E analogue, which was used as an antioxidant standard. In the assay, 40 µl of the diluted samples, controls, or blanks were added to 1960 µl ABTS$^{•+}$ solution, which resulted in a 20-80% inhibit ion of the absorbance. The decrease in absorbance at 734 nm was recorded at 6 min after an initial mixing, and plotted against a dose-response curve calculated for Trolox (0-30 µM). Antioxidant activity was expressed as micromoles of Trolox equivalents per gram of fresh fruit weight (µmolTE g^{-1} FW). Trolox was purchased from Sigma Chemical Co. (St. Louis, MO, USA).

TP assay

Total phenolic content was determined according to the improved Folin- Ciocalteu (F-C) method [19]. The assay provides a rapid and useful indication of the antioxidant status of the studied material and has been widely applied to different food samples. Gallic acid (GA; Sigma Chemical Co., St. Louis, MO, USA) was used as a standard compound for the calibration curve. Total phenol content was

calculated as milligrams of GA equivalent (GAE) per gram of fresh fruit weight (mgGAE g^{-1} FW). The absorbance of the blue colored solutions was read at 765 nm after incubation for 2 h at room temperature.

Statistical Analysis

Data, reported as means ± standard errors (SEM), were analyzed by one-way analysis of variance (ANOVA), and differences were considered statistically significant at $p \leq 0.05$ according to Tukey test. Pearson's correlation and regression analysis were performed in order to determine relationships between pomological and antioxidant properties, cumulative rainfall and TAC-TP levels, respectively.

Results

Climatic conditions

Average monthly maximum and minimum temperatures and the amount of rainfall from March to June, over a 7-year period (2005-2012), are shown in Figures 1 and 2. During the final stages of fruit growth, the average (AVG) of minimum and maximum temperatures in the last years have been 13.5-24.2°C (May) and 16.2-26°C (June), respectively. In particular, 2007 and 2009 showed the highest maximum temperatures, 3-4°C more than to the AVG of the considered climatic area; as a consequence, the highest max-minimum temperature fluctuations occurred. These two years were also characterized by rainy spring-early summer seasons and the cumulative precipitations were unusually high (380 mm in 2007 and 420 mm in 2009), against the average value of the last ten year period (155 mm); both years recorded rainfalls over May and June. The driest year was 2006, when the cumulative rainfall was only 65 mm, mainly occurred in March; this year was also characterized by minimum temperatures higher than the AVG.

Physical-chemical fruit parameters

The main pomological traits of fruits are reported in Table 1. Among the tested years, the average of fruit size by weight was about 72 g, differing from 65.2 g (2006) to 82.9 g (2009), independently from the crop index which was similar in these years (class 2). Considering the physical traits of fruit, the skin color of the un-blushed side was yellow-orange (7-8 intensity) and the area of the blushes was between class 2 and 3, denoting a moderate cover color, always below 35% over the years. Fruit flesh firmness, TSS and TA showed significant variations among years. Flesh firmness varied from 1.7 kg 0.5 cm^{-2} (2011) to 2.9 kg 0.5 cm^{-2} (2009); soluble sugars (TSS) ranged from 11.9 °Brix in 2009 to 16.3 °Brix in 2006, the wettest and driest years, respectively. TA values changed between 9.0 meq 100 g^{-1} FW in 2011 and 17.3 meq 100 g^{-1} FW in the wettest year 2009. As a consequence, the TSS/TA ratio changed and it was particularly low (0.8 and 0.7) in the wettest years 2007 and 2009.

Antioxidant proprieties

A significant years's effect was observed on the on the antioxidant levels. The total antioxidant capacity (TAC) ranged from 2.69 to 9.01 µmolTE g^{-1} FW (Figure 3A). In the wettest years (2007 and 2009) the lowest values were recorded, while in the years 2006, 2011 and 2012, values were 3-fold higher than 2007 and 2009. Total phenol content (TP), similar to TAC, showed a significant variability related to the years, and values ranged from 0.55 to 1.53 mgGAEg^{-1} FW (Figure 3B); the lowest and the highest values were recorded in the wettest and driest years, respectively. A high correlation was found between antioxidant proprieties (TAC and TP) and rainfall occurred during the spring-early

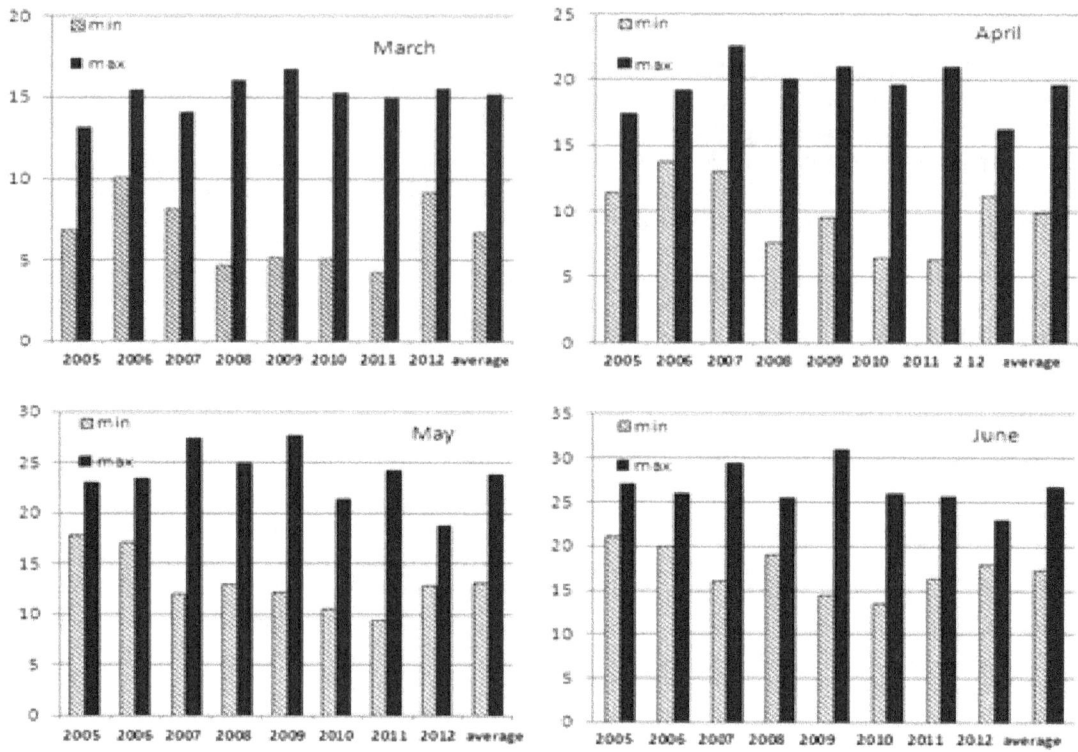

Figure 1: Average monthly minimum and maximum temperatures (°C) from March to June (2005-2012). The relative average of years is also showed

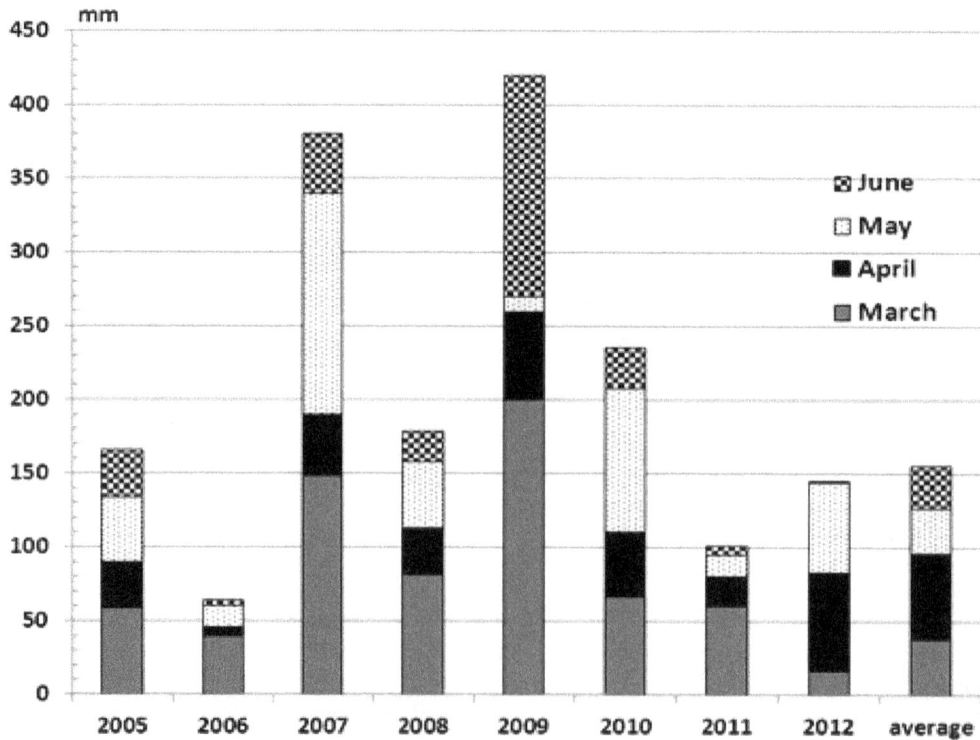

Figure 2: Cumulative monthly rainfall (mm) from March to June (2005-2012).

Year	C.I	F.W.	SC-b	F.F	TSS	TA	TSS/TA
2005	3	71.4 ± 3.6ab	2	2.5 ± 0.4ab	16.3 ± 0.6c	10.5 ± 0.2a	1.5
2006	2	65.2 ± 4.0a	3	2.7 ± 0.3ab	16.1 ± 0.4c	14.4 ± 0.6b	1.1
2007	4	69.0 ± 1.7ab	2	1.6 ± 0.1a	13.1 ± 0.5ab	15.6 ± 0.3b	0.8
2009	2	82.9 ± 3.8b	2	2.9 ± 2.0b	11.9 ± 0.4a	17.3 ± 0.6b	0.7
2010	3	73.8 ± 3.6ab	2	2.8 ± 0.4b	12.2 ± 0.6a	11.0 ± 0.4a	1.1
2011	4	72.8 ± 3.4ab	2	1.7 ± 0.1a	14.9 ± 0.3bc	9.0 ± 0.2a	1.6
2012	2	69 ± 1.9ab	2	2.4 ± 0.2a	15.9 ± 0.6bc	16.4 ± 0.3b	1
Avg	2.9	72.1 ± 1 0.2	3.2	2.4 ± 0.1	14.4 ± 0.6	13.1	1.1

Table 1: Main physical-chemical traits from apricots, cv 'Pisana', recorded over a 7-year period: C.I. (crop index, kg/tree), F.W. (fruit weight, g), SC-b (skin colour of the blushed side), F.F. (flesh firmness, kg 0.5 cm^{-2}), TSS (total soluble sugars, °Brix), T.A. (titratable acidity, meq 100 g^{-1} FW), TSS/TA (sugars/acids ratio). Mean ± standard error. Means within the same column followed by the same letter, do not differ significantly according to Tukey test at $p \le 0.05$.

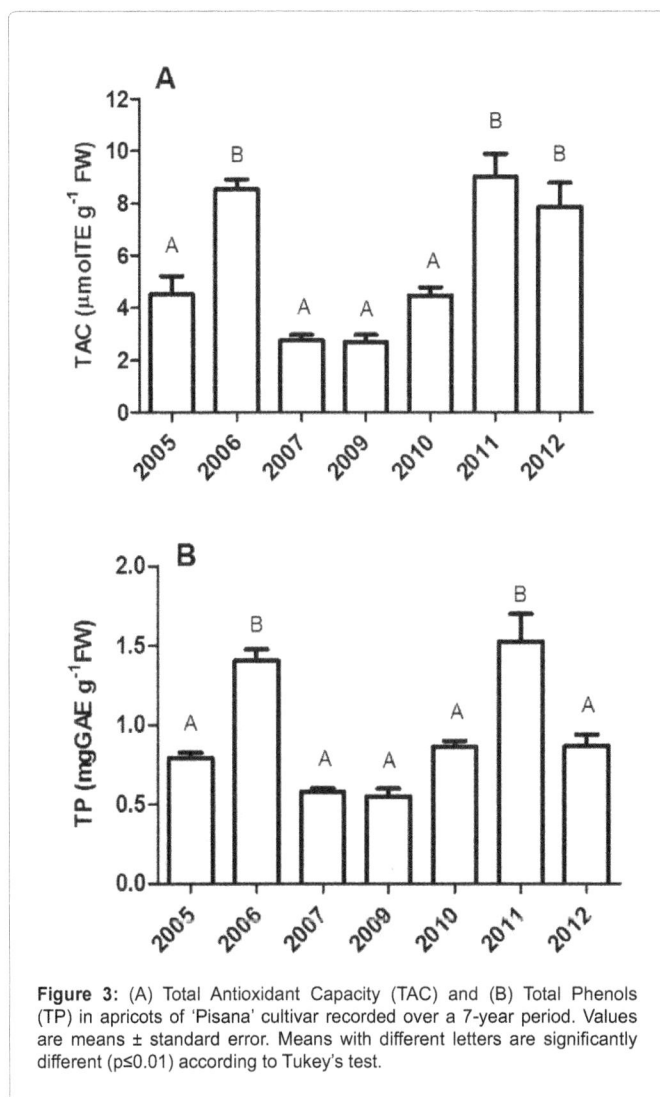

Figure 3: (A) Total Antioxidant Capacity (TAC) and (B) Total Phenols (TP) in apricots of 'Pisana' cultivar recorded over a 7-year period. Values are means ± standard error. Means with different letters are significantly different ($p \le 0.01$) according to Tukey's test.

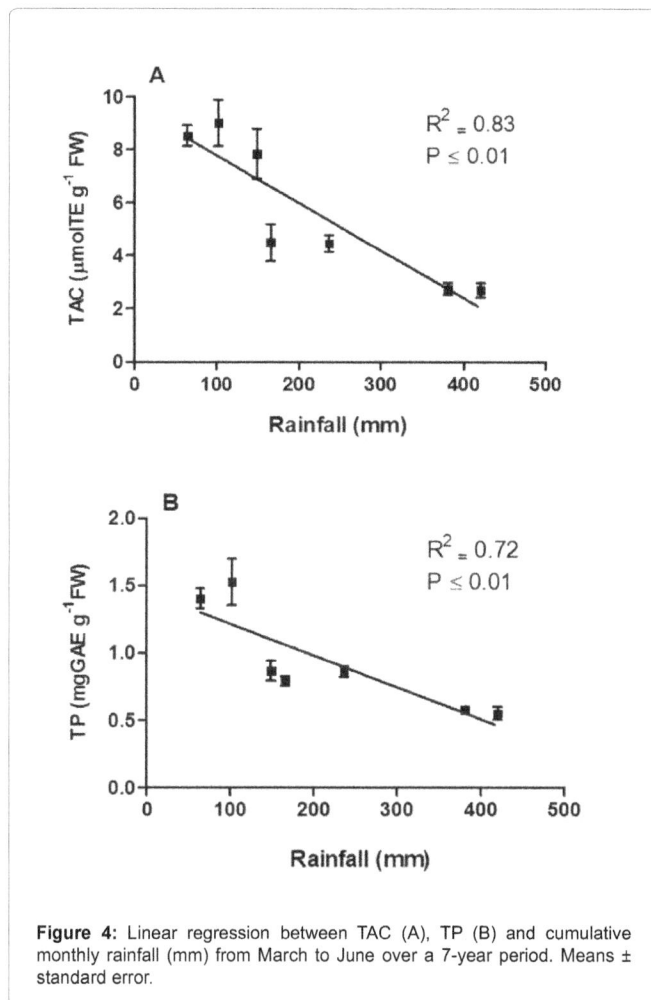

Figure 4: Linear regression between TAC (A), TP (B) and cumulative monthly rainfall (mm) from March to June over a 7-year period. Means ± standard error.

summer seasons (Figures 4A,4B). The correlation coefficient was 0.83 for TAC and 0.72 for TP.

Discussion

The main quality traits of 'Pisana' apricots at ready-to-eat were characterized by fluctuations in relation to the studied years. However, the average physical-chemical data were similar to those recorded in previous researches; TAC and TP levels were inside the interval of 'Pisana' which define this cultivar with a good antioxidant power, when compared to a wide number of commercial apricot genotypes [20]. In Table 2 are reported the Pearson's coefficients among different fruit pomological and chemical properties to find out possible relations between pomological traits and antioxidants. The results over a 7-year period showed a general negative correlation between fruit weight and firmness with chemical parameters, such as sugar content, total antioxidant capacity and total phenols. A weak relation between TSS and TA was observed, not fully in agreement with other researches where these parameters were well linked, denoting them

	FW (g)	Firmness (kg0.5cm²)	TSS (°Brix)	TA (meq100g⁻¹ FW)	TSS/TA	TAC (μmolTEg⁻¹ FW)	TP (mgGAEg⁻¹ FW)
FW	1						
Firmness	0.121	1					
TSS	-0.621	-0.459	1				
TA	0.165	0.622	0.092	1			
TSS/TA	-0.314	**-0.805**	0.173	**-0.903**	1		
TAC	-0.51	-0.751	**0.873**	-0.341	0.576	1	
TP	-0.393	-0.7	0.656	-0.555	0.649	**0.877**	1

Table 2: Pearson's coefficients among fruit weight (FW), flesh firmness, total soluble sugar (TSS), tritatable acidity (TA), sugar/acid ratio (TSS/TA), total antioxidant capacity (TAC) and total phenols (TP) for 'Pisana' cultivar over a 7-year period. Bold coefficients are significant at p ≤ 0.05.

as determinant to define the fruit gustative quality [21]. Positive correlation coefficient was found between TAC and TP ($r = 0.877$), which confirms a significant contribution of polyphenols to the total antioxidant capacity as reported for several fruit species and apricot too [4,12,22,23]. TAC was also significantly correlated with TSS ($r = 0.873$).

During the considered 7-year period, the final stage of fruit growth in 2006, 2007 and 2009 was characterized by climatic conditions which differed from the average of the last years. In May and June of 2007 and 2009 years, high temperatures associated to unusual and heavy rainfall events occurred; on the other hand, the year 2006 was particularly warm and dry. These climatic disorders determined an anomalous microclimate which also differed from the seasonal averages of the past 20 years for the same cultivation area [24]. During the growth-ripening period, the different weather conditions strongly influenced the quality parameters of fruits. In the warm-wet 2007 and 2009 years, 'Pisana' apricots showed the lowest TSS , highest TA and a consequent very low TSS/TA ratio, key parameters related to the eating quality for consumer preference [25,26]. These traits were markedly modified in the dry year (2006) which led to a smaller fruit size characterized by a TSS increase, and a TA decrease with a more balanced TSS/TA ratio. Analogous results were also obtained by several authors, confirming the positive influence of dry conditions on apricot quality traits during the intensive ripening period [27,28]. In particular, the accumulation of sugars in fruit, by conversion of starch, can be enhanced by water stress as a result of reduced irrigation [29-32]. In spite of this, the different climatic conditions had little effect on other physical quality attributes, such as the cover color, confirming this trait as genetic imprint of a genotype [33]. On the other hand, climatic conditions strongly influenced the antioxidant proprieties expressed by TAC and TP values, which markedly changed among the analyzed years. A linear significant inverse relationship between cumulative monthly rainfall and TAC-TP levels was found (Figure 4): in the years characterized by high rainfalls and concomitant warm temperatures, fruits had the lowest antioxidant levels, while in the driest year they reached the highest TAC and TP values. These results are in accordance with recent works showing the key role of water availability on fruit quality traits. These investigations, involving pomological properties, phenolic composition and volatile compounds of different species, have found an inverse relationship between water status and antioxidant content [10,11]. In apricot, a relationship between antioxidant levels and drought conditions was found comparing the autochthonous cultivar 'San Castrese' under different growing sites [34]. Moreover, variability between harvest seasons and antioxidant values was recently found in several apricot genotypes [20]. A number of experimental researches have been addressed on the application of regulated deficit irrigation as a strategy to be applied in areas where water resources are limited. It has been found that deficit irrigation during the fruit growth period might have a positive effect on fruit quality by improving taste, associated with an increase in soluble solids content [27,35,36]. For certain fruit

species, such as pomegranate, this agronomical practice is considered as determinant to enhance fruit composition and postharvest performance [11]. Alternatively, excessive watering may have adverse effects on fruit quality, since it increases tree vegetative growth, which promotes a nutritional imbalance and decreases fruit dry mass [37].

Conclusions

From the results presented in this study, 'Pisana' cultivar confirmed to have an excellent fruit qualitative profile which combines good source of antioxidant compounds and pomological traits. These appeals could drive this cultivar as possible genetic source for breeding programs addressed to produce new genotypes which associate the best agronomic performance to fruit quality traits. The analysis carried out over seven consecutive harvesting seasons allowed establishing that fruit quality showed a high variability in relation to the climatic conditions, particularly due to summer rainfall. The most influenced quality parameters were TSS, TA and antioxidant levels as a physiological response to abiotic stresses; under wet seasons a huge reduction was observed, while strong drought conditions increased these chemical compounds. From the agronomical point of view, to improve fruit quality, 'Pisana' cultivar benefits of environmental conditions typical of temperate and semi-temperate regions where water is usually limited, which seems to be a condition able to enhance the antioxidant level whose importance is strictly related to human health.

References

1. Lichou J, Jay M, Vaysse P, Lespinasse N (2003) Recognizing variety apricot. Paris, France.

2. Erdogan-Orhan I, Kartal M (2011) Insights into research on phytochemistry and biological activities of *Prunus armeniaca* L. (apricot). Food Res Intern 44: 1238-1243.

3. Singleton VL (1981) Naturally occurring food toxicants: phenolic substances of plant origin common in foods. Adv Food Res 27: 149-242.

4. Scalzo J, Politi A, Pellegrini N, Mezzetti B, Battino Mm, et al. (2005) Plant genotype affects total antioxidant capacity and phenolic contents in fruit. Nutrition 21: 207-213.

5. Dragovic-Uzelac V, Levaj B, Mrkic V, Bursac D, Boras M, et al. (2007) The content polyphenols and carotenoids in three apricot cultivars depending on the stage of maturity and geographical origin. Food Chem 102: 966-975.

6. Roussos PA, Gasparatos D (2009) Apple tree growth and overall fruit quality under organic and conventional orchard management. Sci Hort 123: 247-252.

7. Weston LA, Barth MM (1997) Preharvest factors affecting postharvest quality of vegetables. Hortsci 32: 812-816.

8. Lee SK, Kader AA (2000) Preharvest and postharvest factors influencing vitamin C content of horticultural crops. Postharv Biol Technol 20: 207-220.

9. Terry LA, Chope GA, Bordonaba JG (2007) Effect of water deficit irrigation and inoculation with Botrytis cinerea on strawberry (*Fragaria x ananassa*) fruit quality. J Agric Food Chem 55: 10812-10819.

10. Navarro JM, Pérez-Pérez JG, Romero P, Botia P (2010) Analysis of the

Quality and Antioxidant Properties of Apricot Fruits at Ready-to-Eat: Influence of the Weather Conditions under...

139

changes in quality in mandarin fruit, produced by deficit irrigation treatments. Food Chem 119: 1591-1596.

11. Laribi AI, Palou L, Intrigliolo DS, Nortes PA, Rojas-Argudo C, et al. (2013) Effect of sustained and regulated deficit irrigation on fruit quality of pomegranate cv. 'Mollar de Elche' at harvest and during cold storage. Agric Water Manag 125: 61- 70.

12. Leccese A, Bartolini S, Viti R (2008) Total antioxidant capacity and phenolics content in fresh apricots. Acta Alim 37: 65-76.

13. Leccese A, Bartolini S, Viti R (2012) From genotype to apricot fruit quality: the antioxidant properties contribution. Plant Foods Hum Nutr 67: 317-325.

14. Seibert E, Rubio P, Infante R, Nilo R, Orellana A, et al. (2010) Intermittent warming heat shock on 'Pisana' apricot during postharvest: sensorial q uality and proteomic approach. Acta Hort 862: 599-604.

15. Guerriero R, Monteleone P (1992) 'Pisana'. Fruttico ltura 6: 8-11.

16. Guerriero R, Massai R, Canterella F, Remorini D (2006) Agronomic behaviour of 'Pisana' cultivar on several rootstocks in dry, sandy hills. Acta Hort 717: 163-167.

17. Bartolini S, Leccese A, Iacona C, Andreini L, Viti R, et al. (2014) Influence of rootstock on fruit entity, quality and antioxidant properties of fresh apricots (cv. 'Pisana'). New Zeal J of Crop and Hort Sci 42: 265-274.

18. Arts MJTJ, Dallinga JS, Voss HP, Haenen GRMM, Bast A, et al. (2004) A new approach to assess the total antioxidant capacity using the TEAC assay. Food Chem 88: 567-570.

19. Waterhouse AL (2001) Determination of total phenolics: Current Protocols in Food Analytical Chemistry. John Wiley & Sons, New York.

20. Leccese A, Bartolini S, Viti R (2012b) Genotype, harvest season and cold storage influence on fruit quality and antioxidant proprieties of apricot. Intern J Food Prop 15: 864-879.

21. Bassi D, Selli R (1990) Evaluation of fruit quality in peach and apricot. Adv Hort Sci 2: 107-111.

22. Kalt W, Forney CF, Martin A, Prior RL (1999) Antioxidant capacity, vitamin C, phenolics, and anthocyanins after fresh storage of small fruits. J Agric Food Chem 47: 4638-4644.

23. Kim D, Jeong SW, Lee CY (2003) Antioxidant capacity of phenolic phytochemicals from various cultivars of plums. Food Chem 81: 321-326.

24. Guerriero R, Viti R, Iacona C, Bartolini S (2010) Is apricot germplasm capable of withstanding warmer winters? This is what we learned from last winter. Acta Hort 862: 265-272.

25. Shaw DV (1990) Genotypic variation and genotypic correlations for sugars and organic acids in strawberries. J Am Soc Hort Sci 115: 839-843.

26. Biondi G, Pratella GC, Bassi R (1991) Maturity indexes as a function of quality in apricot harvesting. Acta Hort 293: 667-671.

27. Perez-Pastor A, Ruiz-Sanchez MC, Martınez JA, Nortes PA, Artes F, et al. (2007) Effect of deficit irrigation on apricot fruit quality at harvest and during storage. J Sci Food and Agric 87: 2409-2415.

28. Milinovic B, Jelacic T, Halapijakazija D, Cicek D, Vujevic P, et al. (2012) The effect of weather conditions on fruit skin colour development and pomological characteristics of four apricot cultivars planted in Donja Zelina. Agric Conspectus Scientificus 77: 191-197.

29. Kramer PJ (1983) Water Relations of Plants. Academic Press, San Francisco.

30. Ebel RC, Proebsting EL, Patterson ME (1993) Regulated deficit irrigation may alter apple maturity, quality, and storage life. HortSci 28: 141-143.

31. Irving DE, Drost JH (1987) Effects of water deficit on vegetative growth, fruit growth and fruit quality in Cox's Orange Pippin apple. J Hort Sci 62: 427-432.

32. Marsal J, Lopez G, Mata M, Girona J (2012) Postharvest deficit irrigation in 'Conference' pear: effects on subsequent yield and fruit quality. Agric Wat Manag 103: 1-7.

33. Ruiz D, Egea J (2008) Phenotypic diversity and relationships of fruit quality traits in apricot (Prunus armeniaca L.) germplasm. Euphytica 163: 143-158.

34. Leccese A, Bartolini S, Viti R, Pirazzini, P (2010) Fruit quality performance of organic apricots at harvest and after storage from different environmental conditions. Acta Hort 873: 165-172.

35. Crisosto CH, Johnson RS, Luza JG, Crisosto GM (1994) Irrigation regimes affect fruit soluble solids concentration and rate of water loss of 'O'Henry' peaches. HortSci 29: 1169- 1171.

36. Torrecillas A, Domingo R, Galego R, Ruiz-Sánchez MC(2000) Apricot tree response to withholding irrigation at different phenological periods. Scientia Hort 85: 201-215.

37. Herrero A, Guardia J (1992) Conservación de Frutos, Manual Técnico. Madrid, Spain.

25

Study on the Effect of Thermal Processing on Ready-To-Eat Poultry Egg Keema

Ashraf G*, Sonkar C, Masih D and Shams R

Department of Food Process Engineering, Sam Higginbottom University of Agriculture, Technology and Sciences, Allahabad, UP, India

Abstract

Study was conducted to analyze the characteristics of thermally processed ready-to-eat poultry egg keema. Egg keema was prepared as per standardized procedure i.e., utilizing gravy from vegetables i.e., tomato, onion and green peas along with spices and boiled eggs cut into cubes. The product made was filled in pre-sterilized tin cans, sealed hermetically and then thermally processed at different time and temperature combinations viz., 110°C,115°C and 120°C for 10 minutes, 15 minutes and 20 minutes respectively in order to interpret the effects of thermal processing. Samples were evaluated initially and after that at regular intervals of 20 days for physico-chemical, microbiological, sensory and statistical analysis during the entire storage period of 60 days. It was found that the ready-to-eat poultry egg keema samples processed at 120°C for 10 minutes had adequate fat and ash content while as samples processed at 110°C for 10 minutes had adequate protein content. Poultry egg keema samples processed at 120°C for 20 minutes had adequate and standard amount of moisture content, desired pH, reduced microbial loads and maximum consumer acceptability in all respects.

Keywords: Thermal processing; Ready-to-eat foods; Shelf-life; Canning; Hermetic seal

Introduction

Poultry eggs play an important role in the human diet and nutrition as an affordable nutrient-rich food commodity that contains highly digestible proteins, lipids, minerals, and vitamins [1]. Over the past 35 years, global egg production has grown to 203.2%, due to rapid increasing demand for proteins in the developing world [2]. Today, eggs remain a staple food within human diet consumed by people throughout the world. They are recognized by consumers as a versatile and wholesome as they have natural balance of essential nutrients. The proximate composition of egg as given by Watkins [3] Whole egg, egg white, egg yolk contains solid content of 24.5%, 12.1%, 51.8% respectively; protein content of 12%, 10.2%, 16.1% respectively; carbohydrate content of 1%, 1%, 1% respectively; ash content of 1%, 0.68%, 1.69% respectively and lipid content of whole egg and egg yolk is 10.9% and 34.1% respectively. Eggs contain approximately 75% water, 12% each lipids and proteins, ~1% carbohydrates and minerals Burley and Vadehra [4] and Li-Chan et al. [5]. Avian eggs are an excellent source of nutrients, particularly high-quality proteins, lipids, minerals, and vitamins [6-8]. There is a growing consumer interest in related eating patterns such as the avoidance or reduced consumption of red meat. An estimated 7 million people currently either avoid red meat or are vegetarians, compared to 2 million in 1984 [9] and the current state of art to produce thermally processed ready-to-eat egg products has increased the commercial value and offered a level of quality, safety and convenience to consumers. Egg products will continue to be an important part of our daily diet as the primary animal protein in many parts of the world, thus new technologies and methods of egg processing will see applications to improve nutrition, safety, shelf-life and taste of egg products, or to create new egg products as functional foods, nutraceuticals and other non-food uses.

Over the centuries world has witnessed change in the pattern of consumption, from raw to cooked to ready-to-eat foods; emergence of it in the global food industry and now to the Indian markets. Ready-to-eat foods are foods that are offered for sale without additional cooking or preparation, packaged on the premises where they are being sold and are ready for consumption. Canned foods, convenience foods, fast foods, frozen foods, instant products, dried foods, preserved foods, etc. all come under the category of ready-to-eat foods [10]. The increasing demand by consumer for high quality convenient ready-to-eat foods has led to an increase in the commercial production of these products [11] for example Indian curries, desserts, frozen heat and eat products like chicken and mutton curry, canned beef meat etc. which are generally retorted and are shelf stable [12]. Unlike pasteurized products where the survival of heat resistant microorganisms is accepted, the main aim of sterilization of egg products is the destruction of all microorganisms including their spores. Thermal treatment of ready-to-eat egg products must be intense enough to inactivate/kill the most heat resistant bacterial spores of *Bacillus* and *Clostridium*. Temperatures above 100°C, usually ranging from 110°C to 121°C are applied to achieve the goal [13].

Materials and Methods

Preparation of ready-to-eat poultry egg keema

Poultry eggs properly washed were boiled for about 20 minutes, cooled, shelled and cut into cubes keeping the yolk intact. Gravy was made as per the standardized procedure by frying vegetables in oil along with the ground dried spices until golden brown. Then water, mint, coriander leaves and green peas were added and cooked until all water was evaporated and thick curry left behind. Finally chopped eggs were added to the curry.

Thermal processing of ready-to-eat poultry egg keema

The poultry egg keema was filled into pre-sterilized cans maintaining a proper head space (1.10 cm). After filling, exhausting was done in hot water bath at 82°C for 10 minutes. Cans were then sealed with double seaming machine after which they were thermally processing in an autoclave at different time and temperature combinations. After

*Corresponding author: Ashraf G, Department of Food Process Engineering, Sam Higginbottom University of Agriculture, Technology and Sciences, Allahabad, UP, India, E-mail: gousiamir123@gmail.com

thermal processing cans were cooled in a vessel containing tap water to reduce the temperature, dried and finally stored in cool and dry place at ambient temperatures (25°C to 30°C).

Proximate and chemical analysis

Moisture, protein, fat, ash, pH of the thermally processed samples was determined initially and at the regular intervals of 20 days for 60 days storage period as per AOAC.

Microbiological analysis

Thermally processed ready-to-eat poultry egg keema was analyzed for commercial sterility at the regular intervals of 20 days for 60 days of entire storage period. Standard plate count was determined using nutrient agar for incubation at 37°C for 24 hours while as yeast and mold count was determined using Potato Dextrose Agar (PDA) for incubation at 37°C for 48 hours to 72 hours.

Sensory analysis

Sensory evaluation of thermally processed ready-to-eat poultry egg keema was done using 9-point Hedonic scale which is the basis to differentiate between pleasurable and un-pleasurable experiences to determine the effect of different thermal treatments on the organoleptic properties like color, taste, flavor, aroma and overall acceptability.

Statistical analysis

Statistical analysis was conducted as per the data obtained from three levels of temperatures for three levels of time periods i.e., from 9 treatments and 3 replications during trial and was analyzed statistically by Analysis of Variance technique, 3-way classification considering the effect of time, temperature and days on the product. The significant effect of treatments i.e., time, temperature and days was judged with the help of 'F' (variance ratio). Calculated values of F were compared with the table value of F at 5% level of significance (0.05) (Table 1).

Results and Discussion

Effect of thermal processing on ready-to-eat poultry egg keema

The product standardization of poultry egg keema was done in cans filled to constant weight i.e., 500 g maintaining 1.10 cm headspace and then adequacy of thermal processing was maintained at different time and temperature combinations i.e., 110°C, 115°C and 120°C for 10 minutes, 15 minutes and 20 minutes. The cans were stored at ambient temperature i.e., 25°C to 30°C. The sensory evaluation of the ready-to-eat poultry egg keema was done on the basis of organoleptic characteristics to indicate its definite maturation and improvement during storage and it was observed that the samples thermally processed

at 120°C for 20 minutes were acceptable in all respects. Similar results were interpreted during the chemical, microbial and sensory analysis of canned meat based mutton and beef curry Madhwaraj [14], Rajkumar et al. [15] evaluated the appearance, color, flavor, juiciness, texture and overall acceptability of Chettinad goat meat curry showing scores of 8.0 to 8.4 on a 9-point hedonic scale after thermal treatment.

Composition of ready-to-eat poultry egg keema in cans

Poultry eggs: 65% moisture, 12% to 14% protein, 10% to 12% fat, 1% ash

Boiled egg cubes considered for each can: 250 g

Gravy made with vegetables and spices for each can: 250 g

Weight of empty can: 100 g

Weight of each packed can: 600 g (i.e., 500 g egg keema and 100 g weight of can)

Chemical Characteristics of Ready-To-Eat Poultry Egg Keema

Effect of thermal processing on moisture percentage of ready-to-eat poultry egg keema

Mean value of moisture percentage at 0 days of storage at 120°C i.e., 69.50 was significantly superior to the mean value at 115°C i.e., 69.49 which again was found significantly superior to 110°C i.e., 69.45. For time treatments, it was observed that the mean value at 20 minutes i.e., 69.34 was significantly superior to the mean value at time 15 minutes i.e., 69.54 which again was significantly superior to 10 minutes i.e., 69.57. Also during the interaction analysis (both time and temperature) mean moisture percentage of samples processed at 120°C for 20 minutes i.e., 69.48 was significantly superior to the interaction at 115°C for 20 minutes i.e., 69.31 which was again superior to 110°C for 20 minutes i.e., 69.24. Similar results were depicted after the analysis of 20 days, 40 days and 60 days of storage. It was observed from the findings that the moisture percentage of poultry egg keema slightly decreased during canning, cooking as the temperature increased. High moisture content results in microbial growth during canning of meat by using different gravy's. Moisture content decreases as the temperature increases [16]. Statistical analysis showed that the calculated value of F due to treatments (time, temperature and days) is greater than the tabulated value at 5% probability level concluding the significant effect of treatments on moisture content of samples (Figure 1).

Effect of thermal processing on protein percentage of ready-to-eat poultry egg keema

Mean value of protein percentage at 0 days storage at 110°C i.e.,

Ingredients	Weight
Poultry Eggs (Nos.)	100
Onion (kg)	2.5
Tomato (kg)	1.5
Green peas (g)	500
Soya-bean oil (ml)	300
Ginger garlic paste (g)	120
Chilli powder (g)	70
Turmeric powder (g)	10
Salt (g)	180
Water (L)	2.5

Table 1: The ingredients used in the preparation of poultry egg keema.

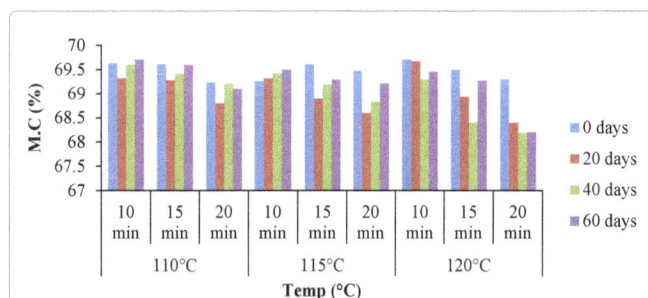

Figure 1: Effect of thermal processing on moisture percentage of ready-to-eat poultry egg keema.

15.35 was significantly superior to the mean value at 115°C i.e., 15.29 which again was significantly superior to 120°C i.e., 15.17. For time treatments, it was observed that the mean value at time 10 minutes i.e., 15.38 was significantly superior to the mean value at 15 minutes i.e., 15.28 which again was significantly superior to 20 minutes i.e., 15.14. Also during the interaction analysis (both time and temperature) mean protein percentage of samples processed at 110°C for 10 minutes i.e., 15.43 was significantly superior to the interaction at 115°C for 10 minutes i.e., 15.42 which was again superior to 120°C for 10 minutes i.e., 15.31. Similar results were depicted after the analysis of 20 days, 40 days and 60 days of storage. It was observed from the findings that proteins present in poultry egg keema denatured at higher temperatures thus affecting their physical properties and decreasing the biological value. Morgan and Kern [17] found that during the canning of meat, the biological activity of the proteins decreased as the severity of heat treatment increased. Processes like canning and roasting may adversely affect the meat protein properties by bringing changes in their linkages and making them unsusceptible to enzymatic digestion [18]. Statistical analysis showed that the calculated value of F due to treatments (time, temperature and days) is greater than the tabulated value at 5% probability level concluding the significant effect of treatments on protein content of samples (Figure 2).

Effect of thermal processing on fat percentage of ready-to-eat poultry egg keema

Mean value of the fat percentage at 0 days storage at temperature 120°C i.e., 18.66 was significantly superior to the mean value at temperature 110°C i.e., 18.36 which again was found significantly superior to 115°C i.e., 18.24. For time treatments, it was observed that the mean value at time 20 minutes i.e., 18.50 was significantly superior to the mean value at time 15 minutes i.e., 18.43 which again was significantly superior to 10 min i.e., 15.33. Also during the interaction analysis (both time and temperature) mean fat percentage of samples processed at 120°C for 10 minutes i.e., 18.82 was significantly superior to the interaction at 110°C for 10 minutes i.e., 18.12 which was again superior to 115°C for 10 minutes i.e., 18.07. Similar results were depicted after the analysis of 20 days, 40 days and 60 days of storage.

The higher temperature and time combinations decreased the fat percentage of the product during storage due to the possibility of breakage of long chain fatty acid chains into individual fatty acid moiety. Increased temperatures influence the physical properties of meat fat [16]. Similar results were obtained during the determination of fat in canned meats [14]. Statistical analysis showed that the calculated value of F due to treatments time and temperature is less than the tabulated value at 5% probability level but reverse is observed for the

effect of days, concluding the significant effect of days and insignificant of time and temperature on fat content of samples (Figure 3).

Effect of thermal processing on PH of ready-to-eat poultry egg keema

Mean value of the pH at 0 days storage at temperature 120°C i.e., 5.61 and 115°C i.e., 5.60 was found significantly superior to 110°C i.e., 5.57. For the time treatments, it was observed that mean value at time 20 minutes i.e., 5.96 was significantly superior to the mean value at time 15 minutes i.e., 5.59 which again was superior to the mean value at 10 minutes i.e., 5.57. Also during the interaction analysis (both time and temperature) mean pH of samples processed at 120°C for 20 minutes i.e., 5.63 was significantly superior to the interaction at 115°C for 20 minutes i.e., 5.63 which was again superior to 110°C for 20 minutes i.e., 5.59. Similar results were depicted after the analysis of 20 days, 40 days and 60 days of storage.

The pH of egg never reaches such high acid values but the higher temperatures of about 110°C to 115°C favors the reaction. The results are in accordance with Morgan and Kern [17] who stated that during meat roasting and canning, if the pH of the product lies between 5.4-5.8, it will be edible. The results are in conformation with the study on the effect of thermal processing on shelf stable canned salted beef with tomato gravy [19]. Statistical analysis showed that the calculated value of F due to treatments (time, temperature and days) is greater than the tabulated value at 5% probability level, concluding the significant effect of treatments on pH of samples (Figure 4).

Effect of thermal processing on ash percentage of ready-to-eat poultry egg keema

Mean value of the ash percentage at 0 days storage at temperature

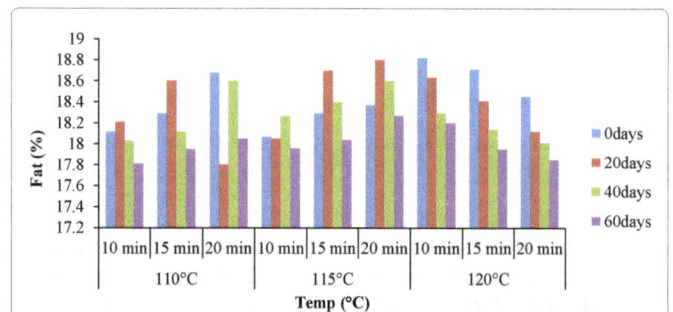

Figure 3: Effect of thermal processing on fat percentage of ready-to-eat poultry egg keema.

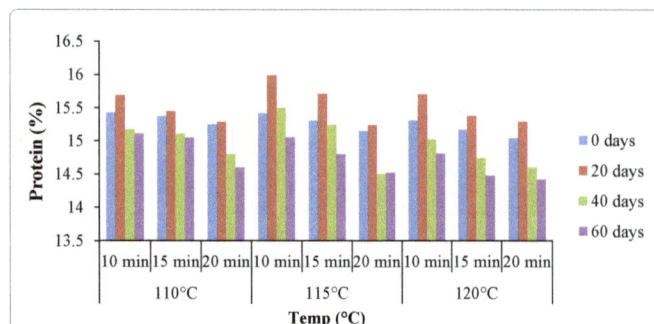

Figure 2: Effect of thermal processing on protein percentage of ready-to-eat poultry egg keema.

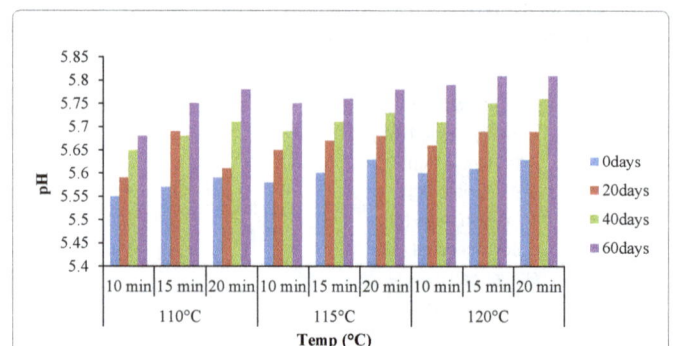

Figure 4: Effect of thermal processing on pH of ready-to-eat poultry egg keema.

120°C, 115°C and 110°C were comparable i.e., 1.83. For the time treatments, it was observed that mean value at time 10 minutes i.e., 2 was significantly superior to the mean value at time 15 minutes and 20 minutes i.e., 1.83. Also during the interaction analysis (both time and temperature) mean ash percentage of samples processed at 120°C for 10 minutes, 115°C for 10 minutes and 110°C for 10 minutes is 2. Similar results were depicted after the analysis of 20 days, 40 days and 60 days of storage.

The processing and cooking methods had little or no effect on the mineral elements of food products as suggested by Ackurt [20]; Gall et al. [21]; Steiner-Asiedu et al. [22]. Cooking is responsible for mineral losses due to their sensitivity to heat, oxygen, pH of solvent or combination of these as described by Harris [23]. Potassium is probably the most sensitive mineral lost during cooking [24]. Statistical analysis showed that the calculated value of F due to treatments (time and days) is greater than the tabulated value at 5% probability level but the reverse is observed due to effect of temperature concluding the significant effect of time and days and insignificant effect of temperature on ash content of samples (Figure 5).

Effect of thermal processing on microbiological characteristics of ready-to-eat poultry egg keema

Mean value of standard plate count (SPC) at temperature 120°C i.e., 30.33 was significantly superior to the mean value at temperature 115°C i.e., 44.66 which was again significantly superior to 110°C i.e., 54. For time treatments, it was observed that the mean value at 20 minutes i.e., 26.66 was significantly superior to 115 minutes i.e., 43 which was again significantly superior to 10 minutes i.e., 59.33. Also during the interaction analysis (both time and temperature) mean SPC of samples processed at 120°C for 20 minutes i.e., 16 was significantly superior to interaction at 115°C for 20 minutes i.e., 30 which again was significantly superior to 110°C for 20 minutes i.e., 34. Similar results were depicted after the analysis of 20 days, 40 days and 60 days of storage.

The results are in accordance with Kumar et al. [25], Agathian et al. [26] who studied retort processed ready-to-eat foods and found commercial sterility after retort processing and the entire period of the storage under different temperature. Rajkumar et al. [15] determined total viable, anaerobic, coliform, *staphylococcal, streptococcal, clostridial* and yeast and mold counts of Chettinad goat meat curry retorted to an F₀ value of 12.1 minutes and showed that the product was commercially sterile. Statistical analysis showed that the calculated value of F due to treatments (time, temperature and days) is greater than the tabulated value at 5% probability level, concluding the significant effect of treatments on the microbial count of samples (Figure 6).

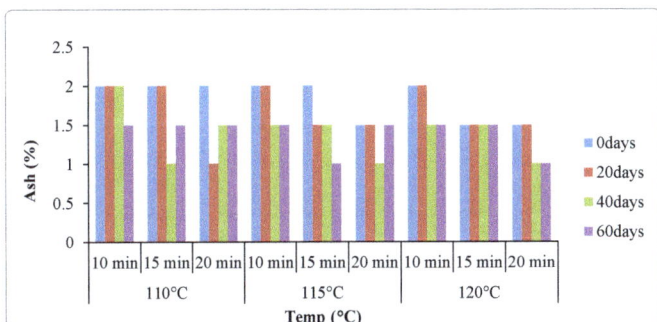

Figure 5: Effect of thermal processing on ash percentage of ready-to-eat poultry egg keema.

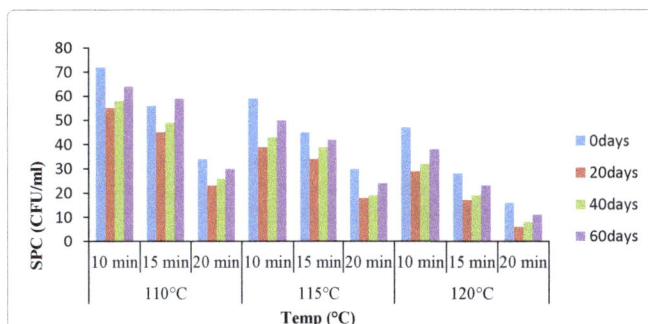

Figure 6: Effect of thermal processing on microbiological characteristics of ready-to-eat poultry egg keema.

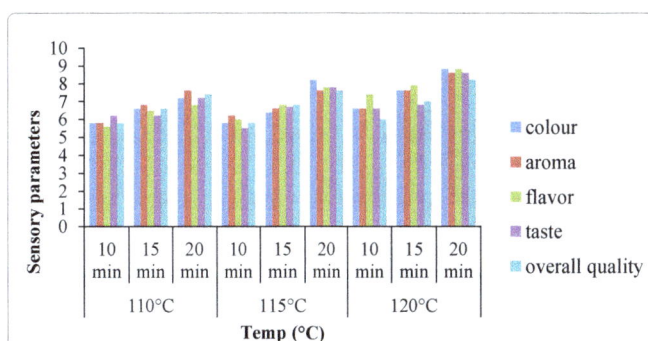

Figure 7: Effect of thermal processing on overall acceptability of ready-to-eat poultry egg keema.

Effect of thermal processing on sensory characteristics of ready-to-eat poultry egg keema

Mean value of sensory scores showed that the ready-to-eat poultry egg keema thermally processed at 120°C for 20 minutes had significant mean sensory score as compared to that of control which clearly depicts its shelf stability and maximum acceptability [27-29].

Effect of thermal processing on overall acceptability of ready-to-eat poultry egg keema

Mean value of sensory scores for overall acceptability at temperature 120°C i.e., 7.06 was significantly superior to mean value at temperature 115°C i.e., 6.73 which was again significantly superior to 110°C i.e., 6.60. For time treatments, it was observed that the mean value at 20 minutes i.e., 7.73 was significantly superior to 15 minutes i.e., 6.80 which was again significantly superior to 10 minutes i.e., 5.86. Also during the interaction analysis (both time and temperature) of mean sensory scores of overall acceptability of samples processed at 120°C for 20 minutes i.e., 8.2 was significantly superior to the interaction at 115°C for 20 min i.e., 7.6 which again was significantly superior to 110°C for 20 minutes i.e., 7.4. Similar results were depicted after the analysis of 20 days, 40 days and 60 days of storage and also during analysis of color, taste, aroma and flavor. The results are in accordance with Gopal et al. [28] who evaluated Kerala style fish curry and showed an overall acceptance of 8.0 on a 9-point scale rating after heat treatment, which then decreased to 7.5 after 12 months of storage (Figure 7).

Conclusion

The present study revealed that due to application of thermal processing at different time and temperature combinations, the microbial stability, sensory attributes as well as the nutritive characteristics of the ready-to-eat poultry egg keema were retained. Poultry egg keema samples thermally processed at 120°C for 20 minutes had significantly superior acceptability, adequate moisture content and pH. Samples processed at 110°C for 10 minutes had adequate protein content as compared to samples processed at other temperature and time combinations. Samples processed at 120°C for 10 minutes had adequate fat and ash content as compared to samples processed at other temperature and time combinations. Similarly, the maximum decline in the microbial load was depicted after the poultry egg keema was thermally processed at 120°C for 20 minutes. Results from the temperature and time measurements along with microbiological tests showed that the product was commercially sterile and acceptable throughout the storage period of 60 days.

References

1. Fisinin VI, Papazyan TT, Surai PF (2008) Producing specialist poultry products to meet human nutrition requirements: Selenium enriched eggs. World's Poultry Sci J 64: 85-98.

2. Windhorst HW (2007) Changes in the structure of global egg production. World's Poultry Sci J 23: 24-25.

3. Watkins BA (1995) The nutritive value of the egg. The Haworth Press Inc, New York, USA.

4. Burley RW, Vadehra DV (1989) The avian egg: Chemistry and biology. Wiley publications, New York, USA.

5. Li-Chan E, Powrie WD, Nakai S (1995) The chemistry of eggs and egg products. Food Products Press, New York, USA.

6. Herron KL, Fernandez ML (2004) Are the current dietary guidelines regarding egg consumption appropriate? J Nutri 134: 187-190.

7. Kovacs-Nolan J, Phillips M, Mine Y (2005) Advances in the value of eggs and egg components for human health. J Agri Food Chem 53: 8421-8431.

8. Li-Chan E, Kim HO (2008) Structure and chemical composition of eggs. Egg Biosci Biotechnol pp: 1-65.

9. Gregory S (2001) Changing attitudes to meat consumption. Real eat survey.

10. Selvarajn PRM (2012) Consumer attitudes towards ready-to-eat packed food items. The Seventh International Research Conference on Management and Finance pp: 322-332.

11. Kumar R, Johnsy G, Rajamanickam R, Lakshmana JH, Kathiravan T (2013) Effect of gamma irradiation and retort processing on microbial, chemical and sensory quality of ready-to-eat chicken pulav. Int Food Res J 20: 1579-1584.

12. Mann JE, Brashears MM (2004) Contribution of humidity to the lethality of surface-attached heat-resistant salmonella during the thermal processing of cooked ready-to-eat roast beef. J Food Protect 70: 762-765.

13. Han JM, Ledward DA (2004) High pressure/thermal treatment effects on the beef muscle. Meat Science 68: 347-355.

14. Madhwaraj MS, Kadkol SB, Nair PR, Dhanraj S, Govindarajan VS (1979) Effect of thermal processing on shelf stable canned salted beef with tomato gravy. Central Food Technological Research Institute, Mysore, India.

15. Rajkumar D, Dushyanthan K, Das AK (2010) Retort pouch processing of Chettinad style goat meat curry a heritage meat product. J Food Sci Technol 47: 372-379.

16. Reiser R, Shorland FB (1990) Meat fats and fatty acids. J Food Sci Technol 5: 21-62.

17. Morgan AF, Andg EK (1934) The effect of heat upon the biological value of meat protein. J Nutrition 7: 367.

18. Howker JJ, Shults GW, Wierbicki E (1976) Effect of combined irradiation and thermal processing on canned beef. Army Natick Research and Development Command Mass Food Engineering Lab. Agri Res Rev 30: 44-48.

19. Singh A, Genitha TR, Singh R, Shakya BR (2012) Effect of thermal processing on shelf stable canned salted beef with tomato gravy. IOSR J Environmental Sci Toxicol Food Technol (IOSR-JESTFT) 1: 11-18.

20. Ackurt F (1991) Nutrient retention during preparation and cooking of meat and fish by traditional methods. Gida Sanayii 20: 58-66.

21. Gall KL, Otwell WS, Koburger JA, Appledorf H (1983) Effects of four cooking methods on proximate, mineral and fatty acid composition of fish fillets. J Food Sci 48:1068-1074.

22. Steiner AM, Julshamn K, Lie Ø (1991) Effect of local processing methods (cooking, frying and smoking) on three fish species from Ghana: Part I, Proximate composition, fatty acids, minerals, trace elements and vitamins. Food Chem 40: 309-321.

23. Harris RS (1988) General discussion on stability of nutrients. In: nutritional evaluation of food processing. Van Nostrand Reinstein Co, New York, USA.

24. Adams CE, Erdman JW (1988) Effects of home food preparation practices on nutritional content of foods. Van Nostrand Reinstein Co, New York, USA.

25. Kumar R, Nataraju S, Jayaprahash C, Sabhapathy SN, Bawa AS (2007) Development and evaluation of retort pouch processed ready-to-eat coconut kheer. India Coconut J 37: 2-6.

26. Agathian G, Nataraj S, Sashikanth S, Sabapathy SN, Bawa AS (2009) Development of shelf stable retort pouch processed ready-to-eat dal makhni. Indian Food Packer 63: 55-62.

27. AOAC (2000) Official methods of analysis. Association of Official Analytical Chemists International. Maryland, USA.

28. Gopal TKS, Vijayan PK, Balachandran KK, Madhavan P, Iyer TSG (2001) Traditional Kerala style fish curry in indigenous retort pouch. Food Control 12: 523-527.

29. Singh A, Genitha TR, Singh R, Shakya BR (2012) Effect of thermal processing on shelf stable canned salted beef with tomato gravy. IOSR J Environmental Sci Toxicol Food Technol (IOSR-JESTFT) 1: 11-18.

26

Physico-Chemical Properties Chemical Composition and Acceptability of Instant 'Ogi' from Blends of Fermented Maize, Conophor Nut and Melon Seeds

Oluwabukola Ojo D* and Ndigwe Enujiugha V

Federal University of Technology, Akure, School of Agriculture and Agricultural Technology, Department of Food Science and Technology, Akure, Ondo State, Nigeria

Abstract

'Ogi' is a popularly known fermented starch of staple cereala such as maize, sorghum and millet. Instant'ogi' production ariises from the need to meet the demand for 'ogi' among the urban populace in Nigeria and other developing countries. This study was carried out to evaluate the proximate, functional, pasting, mineral, anti - nutrient content and consumer acceptability of 'ogi' from blends of fermented maize, conophor nut and melon seed flours (90:5:5, 80:10:10, 70:15:15, 100:0:0). Pasting properties were determined by use of the rapid visco analyser. The mineral elements Ca, Mg, Fe, Zn, Cu, were determined by atomic absorption spectrophotometry while Na and K values were determined by flame photometry. Consumer acceptability of the instant 'ogi' was rated best at 5% supplementation level with conophor and melon seed flours (90:5:5) when compared with the control (100% fermented maize).

Keywords: Fermented maize; Instant ogi; Supplementation; Acceptability; Nutritional value

Introduction

'Ogi', a fermented maize gruel, is a popular breakfast cereal and infant weaning food in West Africa. Maize (*Zea mays*) also referred to as corn, is the most important cereal in the world after wheat and rice with regard to cultivation areas and total production Osagie and Eka [1]. Apart from being consumed by humans, it is also used to prepare animal feeds, and useful in the chemical industry. Maize can be cooked, roasted, fried, ground, pounded or crushed Abdulraharan, and Kolawole [2].

The conophor nut plant (*Tetracarpidum conophorum*) commonly called the African Walnut, is a perennial climbing shrub found in the moist forest zones of Sub-Saharan Africa. It is cultivated principally for the nuts, which are cooked and consumed as snacks, along with boiled corn [3]. Conophor nut commonly called 'Ukpa', 'asala', 'awusa' in some parts of southern Nigeria is one of the several high nutrient dense food with the presence of protein, fibre, carbohydrate and vitamins [4]. Conophor nut is a rich source of minerals such as calcium, magnesium, sodium, potassium, and phosphorus [5] A bitter after taste is usually observed upon drinking water immediately after eating conophor nut and this could be attributed to the presence of alkaloids and other anti-nutritional and toxic factors. Ripe conophor nuts are mostly consumed in the fresh or toasted form or used in cakes, desserts and confectionaries.

Melons are food crops with several varieties which serve as a major food source. Melon seeds are generally rich in oil and are a good source of protein. The seed contains about 44% oil and 32% protein [6]. It has both nutritional and cosmetic importance and is rich in vitamin C, riboflavin and carbohydrates. Melon seed is a good source of amino-acids such as isoleucine and leucine [7]. It also contains palmitic, stearic, linoleic and oleic acids important in protecting the heart. It can serve as an important supplementary baby food, helping to prevent malnutrition. The objective of our study was to determine the effect of supplementing fermented maize flour at different levels with conophor nut and melon seed flours in the production of instant 'ogi'.

Materials and Methods

Raw material source and collection

White maize (*Zea mays*), melon seed (*Citrullus lanatus*) and conophor nut (*Tetracarpidum conophorum)* were obtained from the local 'Oba' market in Akure, Ondo State, Nigeria.

Processing of fermented maize flour

The maize grains were cleaned and sorted by removing the pest-infested grains and discoloured ones. It was then steeped for 72 h at room temperature and the steep water was decanted while the fermented grain was washed with portable water and wet-milled. It was then wet-sieved and the slurry was allowed to ferment for 24 h. It was afterwards decanted, dried at 70°C for 4 h and milled using hammer mill. The fermented maize flour was then sieved to obtain a finer particle (630 μm mesh size) and packaged in air-tight containers prior to analysis. The production chart is presented in Figure 1.

Processing of melon seed flour

The melon seeds were sorted to remove the discoloured ones and then dried at 65°C for 6 h, milled with hammer mill and was defatted using n-hexane as the solvent for 6 h. The defatted melon was air dried and milled using hammer mill. The melon flour was then sieved to obtain a finer particle and packaged in air-tight containers prior to analysis. The production chart is shown in Figure 2.

Processing of conophor nut flour

The conophor nuts were cleaned to remove debris and dirt and cooked at 100°C for 1 h. It was then shelled to obtain the kernels. The kernels were dried at 50°C for 8 h, milled with hammer mill and defatted using n-hexane as solvent for 6 h. The defatted conophor nut

*Corresponding author: Oluwabukola Ojo D, Federal University of Technology, Akure, School of Agriculture and Agricultural Technology, Department of Food Science and Technology, PMB 704, Akure, Ondo State, Nigeria E-mail: doojo@futa.edu.ng

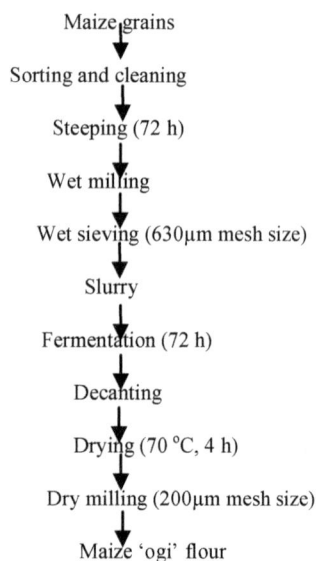

Figure 1: Flow chart for the production of fermented maize 'ogi flour.

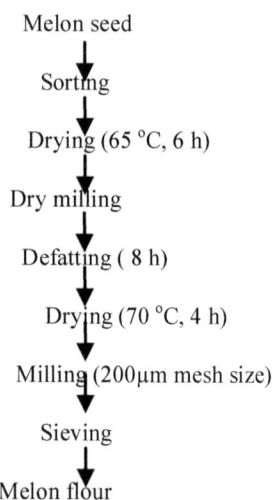

Figure 2: Flow chart for the production of melon seed flour.

cake was afterwards air dried at 70°C for 4 h and milled using hammer mill. The conophor nut flour was then sieved to obtain a finer particle and packaged in air- tight containers prior to analysis. The production chart is shown in Figure 3.

Blend formulation

Four formulations were made in the following proportions (maize: melon seeds: conophor nut); 90:5:5, 80:10:10, 70:15:15, 100:0:0. The sample consisting of 100% 'ogi' flour was used as the control. A hand mixer (model: Kenwood, UK.) was used for mixing samples for 5 min to achieve uniform blending.

Analysis

Proximate chemical composition analysis

Proximate chemical composition of the samples were determined using the methods of AOAC [8]. Carbohydrate content was determined by substracting the sum of the percentage weight of crude protein, crude fibre, ash, fat from 100 percent.

Functional properties analysis

For the determination of functional properties, the method of Onwuka [9] was used for the determination of Water/Oil absorption capacity. Bulk density and Swelling index were determined by the method described by Ukpabi and Ndimele. The rotating spindle method described in the Encyclopedia of Industrial Chemical Analysis E.I.C.A [10] was employed in viscosity determination.

pH was determined using a pH meter (Model: S90526, Fischer Scientific Ltd., Singapore). The method of Jitngarmkusol et al. [11] with some slight modifications was used for the determination of the foaming capacity and stability of the instant 'ogi' flour blends. Emulsion capacity was determined by the method of Yasumatsu et al. [12] Least gelation concentration (LGC) of the flour blends was determined using the modified method of Coffman and Garcia [13] Solubility index were determined as described by Takashi and Sieb [14] using a SPECTRA UK (Merlin 503) centrifuge. Reconstitution index were also determined as described by Banigo and Akpapunam [15].

Pasting properties analysis

The pasting properties of the samples were determined using a rapid visco–analyser (Model: NEWPORT SCIENTIFIC, NSW, Australia) as described by Adeyemi et al. [16] The peak, viscousity, trough, breakdown, final viscosity, set back, peak time and pasting temperature were read with the aid of Thermocline for Windows Software connected to a computer.

Mineral elements analysis

The mineral elements Ca, Mg, Fe, Zn, Cu, were determined using atomic absorption spectrophotometer (Model: PYE UNICAM SP9, Cambridge, UK). Flame photometer (JENWAY PFP7, Bibby Scientific Ltd, Staffordshire, U. K.) was also used to measure the values of Na and

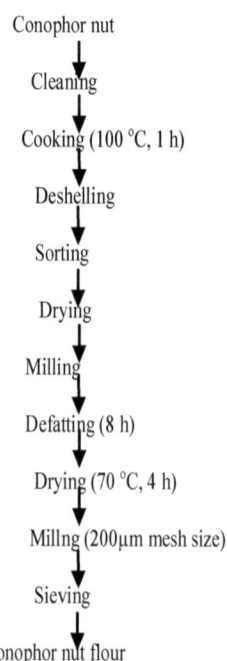

Figure 3: Flow chart for the production of conophor nut flour.

K in all the samples and Phosphorus (P) was determined as described by AOAC [8].

Anti- nutrients content analysis

For anti- nutrients content analysis, tannin content was determined by the method of Makkar and Goodchild [17] Oxalate content was determined by the method of Nwika et al. [18] and phytate content was determined by the method of Latta and Eskin [19].

Sensory evaluation

The instant 'ogi' was made into slurry by adding water till it formed a paste and boiled water was added to it and stirred continuously till it became viscous and formed a gruel. The products were evaluated for taste, appearance, aroma, and overall acceptability by a panel of ten members using a 9-point Hedonic scale. The rating of the samples ranged from 1 (dislike extremely) to 9 (Like extremely).

Statistical analysis

The data obtained were analyzed using a one-way Analysis of Variance and the means separated by Duncan New Multiple Range Tests (DMNRT) at 5% significance level (SPSS version 19 computer software) [20].

Results and Discussion

Proximate chemical composition of the instant 'ogi' flour blends

The proximate composition of instant 'ogi' from blends of fermented maize, conophor nut and melon seed flours is presented in Table 1. The increase in the protein value of the flour was due to the supplementation of the maize flour with melon seeds and conophor nuts. Low fat content of the flour coupled with the low moisture content of the flour blends is an indication that the samples will be stable during storage. According to Adeyeye and Adejuyo [21], the low moisture content of the samples would hinder the growth of micro-organism and increase the shelf life of the samples. Sample SPB had the highest crude fibre content. According to Norman and Joseph [22], fibre has an important role in providing roughage or bulk that aids in digestion, softens stool and lowers plasma cholesterol level in the body. Increased melon/conophor nut flour substitution gave progressively higher protein, crude fibre and ash contents of the samples while fat and carbohydrate contents were reduced. Crude protein, ash and crude fibre were significantly different in the four samples; however, there were no significant differences in the fat content of samples BPO and SPB.

Functional properties of the instant 'ogi' flour blends

According to Oyerekua and Adeyeye [22] high water absorption capacity (WAC) is desirable for the improvement of mouth feel and viscosity reduction in food products. According to Afoadek and Sefa-Dedeh [23], WAC and OAC in the blended flour might be due to the thickness of interfacial bi-layer model of protein to protein interaction. Sample BPC had the lowest oil absorption capacity. The reduced value of OAC in Sample BPC might be due to collapse of the flour blend proteins thereby increasing the contact between protein molecules leading to coalescence and thus reduce stability of the samples. Bulk density is an important factor in food products handling, packaging, storage, processing and distribution. It is particularly useful in the specification of products derived from size reduction or drying processes.

Bulk densities of the samples were similar to that reported by Adeyemi and Becky [24]. The bulk densities ranged from 0.66-0.90 g/ml with sample POS having the highest value which indicate that its packaging would be economical. Plaami [25] reported that higher bulk density is desirable, since it helps to reduce the paste thickness which is an important factor in convalescent and child feeding. Viscousity ranged from 0.61-0.70 dPa with samples SPB and BPC having the highest value. pH is important in determining the acid factor which is an indicator of the rate of conversion of starch to dextrin. The pH value ranged from 5.70-6.60. The foaming capacity ranged from 1.38%-10.00% with BPO having the highest value. The increase in foaming capacity with melon and conophor nut supplementation might be due to soluble proteins and higher emulsion capacity; this might make it a better flavor retainer and enhance mouth feel [26]. It has also been reported that foam capacity is related to the rate of decrease of the surface tension of the air/water interface caused by absorption of protein molecules [27]. The foaming stability of the flour increased with increment in the supplementation level of the flour though sample SPB had the highest value. Sample BPC had the highest value for least gelation capacity. The emulsion capacity ranged from 50.20%-78.15%, with sample BPO having the highest value. High level of least gelation capacity means less thickening capacity of food; the contents ranged from 6.0%-18.0%. Reconstitution index ranged from 3.61-5.05 ml/g with Sample POS (control) having the highest value. The functional properties of the instant 'ogi' flour blends are shown in Table 2.

Pasting characteristics of the instant 'ogi' flour blends

Table 3 shows the results of the pasting characteristics of the instant 'ogi' flour blends. The pasting properties of the samples BPO, SPB, BPC and control (POS) were significantly different (p < 0.05). Peak viscosity of the instant 'ogi' samples ranged from 133.51-213.83 RVU, the values were observed to reduce with increase in supplementation levels. Final viscousity is a measure of stability of the cooked sample [28]. The final viscousity ranged from 145.67-243.59 RVU with BPO having the highest value; this implies that highly viscous paste can be formed during cooking. The setback value is a measure of retrogadation the cooked sample and it ranged from 60.17-108.58 RVU with BPO having the highest value. Pasting temperature is also a measure of the temperature at which flour viscousity begins to rise during cooking, it provides information on the cost of energy required to cook the instant 'ogi' [24]. The pasting temperature of the instant 'ogi' ranged from 83.65°C to 94.75°C with BPC having the highest value. The pasting time ranged from 5.36-5.85 sec, with POS having the highest value.

Mineral content of the instant 'ogi' flour blends

Table 4 shows the mineral contents of the instant 'ogi' flour

Samples	BPO	SPB	BPC	POS
Moisture	4.73 ± 1.09[b]	4.73 ± 1.09[b]	4.73 ± 1.09[b]	4.73 ± 1.09[b]
Ash	9.01 ± 0.72[a]	1.80 ± 0.00[b]	2.44 ± 0.00[a]	2.98 ± 0.00[c]
Crude fibre	5.94 ± 0.87[b]	1.14 ± 0.12[c]	3.96 ± 0.01[c]	3.96 ± 0.01[b]
Fat	5.55 ± 1.11[b]	0.44 ± 0.12[d]	5.20 ± 0.29[c]	5.20 ± 0.29[a]
Crude fibre	9.01 ± 0.72[a]	1.80 ± 0.00[b]	2.44 ± 0.00[a]	2.98 ± 0.00[c]
Carbohydrate	5.94 ± 0.87[b]	1.14 ± 0.12[c]	3.96 ± 0.01[c]	3.96 ± 0.01[b]

Values with different superscript in a row are significantly different (p < 0.05). Values are means ± standard deviation of triplicate determinations.
Key: BPO = 70% fermented maize flour, 15% melon flour, 15% Conophor nut flour; SPB = 80% fermented maize flour, 10% melon flour, 10% conophor nut flour; BPC = 90% fermented maize flour, 5% Melon flour, 5% Conophor nut flour; POS = 100% fermented maize flour.

Table 1: Percentage proximate composition of instant 'ogi' from blends of fermented maize, conophor nut and melon seed flours.

Samples	BPO	SPB	BPC	POS
Ph	6.60 ± 0.00[a]	6.20 ± 0.00[b]	5.70 ± 0.00[c]	6.20 ± 0.00[b]
WAC (%)	660.00 ± 0.70[a]	680.00 ± 0.28[a]	660.00 ± 0.56[a]	665.00 ± 0.63[a]
OAC (%)	830.00 ± 0.70[a]	870.00 ± 0.21[a]	800.00 ± 0.28[a]	820.00 ± 0.00[a]
EC (%)	78.15 ± 0.21[a]	50.20 ± 0.28[b]	75.25 ± 0.35[a]	53.25 ± 0.47[b]
Viscosity (dPa)	0.61 ± 0.01[b]	0.70 ± 0.01[a]	0.70 ± 0.01[a]	0.61 ± 0.01[b]
RI (ml/g)	3.61 ± 0.02[c]	3.62 ± 0.03[c]	4.35 ± 0.00[b]	5.05 ± 0.07[a]
BD (g/ml)	0.66 ± 0.00[d]	0.71 ± 0.00[c]	0.76 ± 0.00[b]	0.90 ± 0.00[a]
FC (%)	10.00 ± 0.01[a]	9.85 ± 0.03[b]	4.28 ± 0.02[c]	1.38 ± 0.01[d]
FS (%)	4.28 ± 0.03[a]	5.63 ± 0.02[a]	1.42 ± 0.01[c]	1.38 ± 0.01[d]
SI (v/v)	3.61 ± 0.02[c]	3.62 ± 0.03[c]	4.35 ± 0.07[b]	5.05 ± 0.07[a]
LGC (%)	18.00 ± 0.00[b]	14.00 ± 0.00[c]	20.00 ± 0.00[a]	6.00 ± 0.00[d]

Seed flours
Values with different superscript in a row are significantly different (p < 0.05).
Values are means ± standard deviation of triplicate determinations.
Abbrevations: Water absorption capacity: WAC; Oil absorption capacity: OAC; Emulsion capacity: EC; **Reconstitution index:** RI; Bulk density: BD; Foaming capacity: FC; Foaming stability: FS; Swelling index: SI; Least gelation capacity: LGC
Key: BPO = 70% fermented maize flour, 15% Melon flour, 15% conophor nut flour; SPB = 80% fermented maize flour, 10% melon flour, 10% conophor nut flour; BPC = 90% fermented maize flour, 5% melon flour, 5% conophor nut Flour; POS = 100% fermented maize flour

Table 2: Functional properties of instant 'ogi' from blends of fermented maize, conophor nut and melon.

Samples	BPO	SPB	BPC	POS
Peak viscosity (RVU)	163.17 ± 0.07[b]	161.17 ± 0.17[c]	133.51 ± 0.14[d]	213.83 ± 0.21[a]
Trough (RVU)	135.00 ± 0.14[a]	123.25 ± 0.01[b]	85.08 ± 0.07[d]	107.08 ± 0.14[c]
Breakdown (RVU)	28.17 ± 0.07[d]	37.93 ± 0.14[c]	48.41 ± 0.07[b]	106.76 ± 0.14[a]
Final viscosity (RVU)	243.59 ± 0.14[b]	247.34 ± 0.07[a]	145.25 ± 0.07[d]	196.67 ± 0.10[c]
Setback (RVU)	108.58 ± 0.14[b]	104.09 ± 0.10[a]	60.17 ± 0.10[d]	89.57 ± 0.07[c]
Pasting time (sec)	5.36 ± 0.10[d]	5.68 ± 0.07[b]	5.58 ± 0.07[c]	5.85 ± 0.07[a]
Pasting temperature (°C)	86.05 ± 0.14[b]	85.95 ± 0.28[c]	94.75 ± 0.07[a]	83.65 ± 0.10[d]

Values with different superscript in a row are significantly different (p < 0.05).
Values are means ± standard deviation of triplicate determinations.
Key: BPO = 70% fermented maize flour, 15% melon flour, 15% conophor nut flour; SPB = 80% fermented maize flour, 10% melon flour, 10% conophor nut flour; BPC = 90% fermented maize flour, 5% melon flour, 5% conophor nut flour; POS = 100% fermented maize flour

Table 3: Pasting characteristics of instant 'ogi' from blends of fermented maize, conophor nut and melon seed flours.

blends. Calcium value decreased with increase in supplementation level but magnesium content increased. Magnesium is well known to be important in cellular energy production and enzyme activity; its value ranged from 106.06-126.03 mg/100 g. Iron (Fe) ranged from 9.87-11.70 mg/100 g with the sample BPC having the lowest value. The instant 'ogi' flour blends provides a good amount of iron that is needed in the production of haemoglobin which carries oxygen in the blood. Zn ranged between 2.25 mg/100 g and 2.91 mg/100 g. The potassium content ranged from 195.68-198.37 mg/100 g and there were no significant differences between samples BPO, BPC and POS (p < 0.05). A major function of potassium is to maintain the excitability of nerve and muscle tissue.

Anti-nutrient content of the instant 'ogi' flour blends

Table 5 shows the level of anti–nutrients in the instant 'ogi' flour blends. The phytate content of the ogi flour blends ranged from 5.25-5.96 mg/g. Phytate are known to form complexes with iron, zinc, calcium

and magnesium making them less available and thus inadequate in food samples especially for children however, the phytate content of the ogi flour blends are far lower than the minimum amounts of phytic acid reported by Siddhuraju and Becker [29] to hinder the absorption of iron and zinc. Oxalates are also known to make complexes with calcium to form an insoluble calcium–oxalate salt. Siddhuraju and Becker [29] reported a safe normal range of 4-9 mg/ 100 g for oxalates. The oxalate content of the samples which range from 2.48 - 2.67 mg/100 g is quite lower than the reported value. Tannin content range from 4.65–5.85 mg/100 g. Tannins have been implicated in the interference of iron absorption; it usually forms insoluble complexes with proteins, thereby interfering with their bioavailability [30-32].

Sensory evaluation of gruel of instant 'ogi' from blends of fermented maize, conophor nut and melon seed flours

Table 6 shows the sensory evaluation results of the instant 'ogi' flour blends. Sensory evaluation was carried out by ten (10) untrained panelists and the parameters evaluated were taste, flavor, appearance and overall acceptability. Consumer evaluation of taste showed that

Samples	BPO	SPB	BPC	POS
Iron	11.70 ± 0.63[a]	11.53 ± 0.58[a]	9.87 ± 0.17[b]	10.87 ± 0.40[ab]
Zinc	2.91 ± 0.10[a]	2.64 ± 0.21[b]	2.25 ± 0.14[ab]	2.30 ± 0.35[ab]
Calcium	140.68 ± 0.45[d]	144.05 ± 0.63[c]	145.77 ± 0.26[b]	150.46 ± 0.37[a]
Magnesium	126.03 ± 0.09[a]	123.66 ± 0.43[b]	120.06 ± 0.12[c]	106.06 ± 0.38[d]
Potassium	196.59 ± 0.50[b]	198.37 ± 0.53[a]	195.83 ± 0.25[b]	195.68 ± 0.37[b]
Sodium	111.88 ± 0.24[d]	115.90 ± 0.14[c]	122.65 ± 0.35[b]	145.84 ± 0.44[a]
Copper	1.87 ± 0.11[a]	2.27 ± 0.35[a]	2.07 ± 0.10[a]	0.97 ± 0.35[b]
Phosphorus	64.97 ± 0.25[a]	63.67± 0.60[b]	58.89 ± 0.21[c]	56.57 ± 0.81[d]

Values with different superscript in a row are significantly different (p < 0.05).
Values are means ± standard deviation of triplicate determinations.
Key: BPO = 70% fermented maize flour, 15% melon flour, 15% conophor nut flour; SPB = 80% fermented maize flour, 10% melon flour, 10% conophor nut flour; BPC = 90% fermented maize flour, 5% melon flour, 5% conophor nut flour; POS = 100% fermented maize flour

Table 4: Mineral composition of instant 'ogi' from blends of fermented maize, conophor nut and melon seed flours (mg/100 g).

Samples	Phytate (mg/100 g)	Oxalates (mg/100 g)	Tannin (mg/100 g)
BPO	5.64 ± 0.01[c]	2.66 ± 0.01[a]	4.66 ± 0.02[d]
SPB	5.76 ± 0.01[b]	2.59 ± 0.02[b]	5.37 ± 0.03[a]
BPC	5.96 ± 0.03[a]	2.48 ± 0.01[c]	4.86 ± 0.01[b]
POS	5.26 ± 0.02[b]	2.67 ± 0.02[a]	4.78 ± 0.01[c]

Values with different superscript in a row are significantly different (p < 0.05).
Values are means ± standard deviation from triplicate determinations.
Key: BPO = 70% fermented maize flour, 15% Melon flour, 15% conophor nut flour; SPB = 80% fermented maize flour, 10% melon flour, conophor nut flour; BPC = 90% fermented maize flour, 5% melon flour, 5% conophor nut flour; POS = 100% fermented maize flour

Table 5: Anti – nutritional properties of instant 'ogi' from blends of fermented maize, conophor nut and melon seed flours (mg/100 g).

Samples	Taste	Appearance	Aroma	Overall acceptability
BPO	4.50 ± 1.26[c]	6.10 ± 0.99[b]	6.30 ± 0.94[b]	5.30 ± 0.67[c]
SPB	6.50 ± 0.70[b]	6.10 ± 1.19[b]	7.30 ± 0.94[b]	6.30 ± 0.82[b]
BPC	5.80 ± 0.42[b]	7.00 ± 0.94[a]	7.10 ± 0.31[a]	6.80 ± 0.42[b]
POS	7.50 ± 0.52[a]	7.50 ± 1.26[a]	7.70 ± 0.94[a]	7.70 ± 1.15[a]

Values are means ± standard deviation from triplicate determinations.
Values with different superscript in a row are significantly different (p < 0.05).
Key: BPO = 70% fermented maize flour, 15% melon flour, 15% conophor nut flour; SPB = 80% fermented maize flour, 10% melon flour, 10% conophor nut flour; BPC = 90% fermented maize flour, 5% Melon flour, 5% conophor nut flour; POS = 100% fermented maize flour

Table 6: Sensory evaluation of gruel of instant 'ogi' from blends of fermented maize, conophor nut and melon seed flours.

there were no significant differences between sample BPC and SPB. Sample BPC were also rated higher than SPB in terms of appearance and aroma. In terms of overall acceptability, Samples BPC and SPB compared favourably with the control (POS) and there were no significant differences between them.

Conclusion

The study has shown that fermented maize- conophor nut- melonogi flour with improved nutrient s, pasting properties and sensory quality compared to the traditional fermented maize 'ogi' flour can be obtained up to 80:10:10 ratio. Conophor and melon seeds which are under-utilized are suitable for use in instant 'ogi' flour production. The production of instant 'ogi' from fermented maize- conophor nut- melon flour makes the the local food product 'ogi' readily available for consumption, increases its variety, hence, consumer choice of 'ogi'.

References

1. Osagie AU, Eka OU, Igodan VO (1998) Nutritional quality of plant foods. Post harvest research unit University of Benin Benin L 34-41.

2. Abdulraharan AA, Kolawole OM (2006) Traditional preparations and uses of maize in Nigeria. Ethno botanical leaflets 10: 219-227.

3. Enujiugha VN (2008) Tetracarpidium conophorum: Conophor nut In the encyclopedia of fruit and nuts. CABI Publishing, Oxfordshine, UK: 378-379.

4. Savage GP, Mc Niel, DL Dutta PC (2001) Some nutritional advantage of walnuts. J Acta Horticulture 544: 557-563.

5. James NR (2009) Volatile components of green walnuts husks. J Agri Food Chem 48 : 2858-2861.

6. Enujiugha VN, Ayodele-Oni O (2003) Evaluation of nutrients and some anti-nutrients in lesser known, underutilized oil seeds. Int J Food Sci Technol 38: 525-528.

7. Olaofe O, Adeyemi FO, Adediran GO (1994) Amino acid, Mineral composition, functional properties of some oil seeds. J Agri Food Chem 42: 878-884.

8. AOAC (2012) Association of official analytical chemists. Official Methods of Analysis.

9. Onwuka GI (2005) Food analysis and instrumentation theory and practice. Naphtali prints Lagos Nigeria.

10. EICA (1971) Encyclopedia of industrial chemical analysis. Interscience publishers, New York, London, Sydney, Toronto 12: 2.

11. Jitngarmkusol S, Hongsuwankul J,Tananuwong K (2008) Chemical compositions, functional properties and microstructure of defatted macadamia flours. J Food Chemistry 110 : 23-30.

12. Yasumatsu K, Sawada K, Maritaka S, Mikasi M, Toda J, et al. (1972) Whipping and emulsifying properties of soybean products. J Agricultural and Biological Chemistry. 36: 719-727.

13. Coffman CW, Garcia VV (1977) Functional properties of flours prepared from Chinese indigenous legume seed. J Food Chem 61: 429-433.

14. Takashi S, Sieb PA (1988) Paste and gel properties of prime corn and wheat starches with and without native lipids. J Cereal Chemistry 65: 474-480.

15. Akpapunam B (1987) Physico-chemical and nutritional evaluation of protein-enriched fermented maize flour. Niger Food J 5: 30-36.

16. Adeyemi IA, Adabiri BO, Afolabi OA, Oke OL (1992) Evaluation of some quality characteristics and baking potentials of Amaranth flour. Niger Food J 10: 8-15.

17. Makkar AOS, Goodchild S (1996) Quantification of tannins. A laboratory manual Internal centre for Agricultural Research in Dry Areas (ICARDA) Aleppo 25.

18. Nwika N, Ibe G, Ekeke G (2005) Proximate composition and level of toxicants in four commonly consumed spices. J Applied Science and Environmental Management 9: 150-155.

19. Latta, M, Eskin M (1980) A simple and rapid colorimetric method for phytate determination. J Agricultural and Food Chemistry 28: 1313-1315.

20. Steel R, Torrie J, Dickey D (1997) Principles and procedures of statistics: A biometrical approach.3rd edn, Mc Graw Hill Book Co New York USA.

21. Oyerekua MA, Adeyeye EI (2009) Comparative evaluation of the nutritional quality, functional properties and amino acid profile of co-fermented maize/cowpea and sorghum/cowpea ogi as infant complementary food. Asian J clinical nutrition 1 : 31-39.

22. Norman NP, Joseph HH (1995) Food Science (5thedn.) Chapman and Hall Publishers, New York (USA) 55.

23. Afoakwa EO, Sefa -Dedeh S (2002) Viscoelastic properties and changes in pasting characteristics of trifoliate yam (Dioscorea dumentorum pax) starch after harvest. J Food Chem 77: 85-91.

24. Adeyemi IA, Beckley O (1986) Effect of period of maize fermentation and souring on chemical properties and amylograph viscousity of Ogi. J Cereal Sci 4: 353-360.

25. Plaami SP (1997) Content of dietary fibre in foods and its physiological effects. Food Rev Int 13 : 27-76.

26. Adeyeye EI, Ayejuyo OO (1994) Chemical composition of cola accuminata and grarcina kola seed grown in nigeria. Int J Food Sci Nutri 45: 223-230.

27. Kiin Kabari OB, Eke Ejiofor J, Giami SY (2015) Wheat/Plantain flour enriched with bambara groundnut protein concentrate. Int J Food Sci Nutri Eng 5: 75-81.

28. Chinma CE, Adewuyi AO, Abu JO (2009) Effect of germination on the chemical, functional; and pasting properties of flour from brown and yellow varieties of tigernut. Food Res Int 42 : 1004-1009.

29. Siddhuraju P, Becker K (2001) Effect of various domestic processing methods on anti-nutrients and in-vitro protein and starch digestibility of two indigenous varieties of Indian tribal pulse (Mucuna pruriens var. utilis). J Agri Food Chem 49: 3058-3067.

30. Enujiugha, Agbede (2000) Nutritional and anti-nutritional characteristics of African oil bean (Pentaclethra macrophylla) seeds. Appl Tropic Agri 5: 11-14.

31. Enujiugha VN (2003) Chemical and functional characteristics of conophor nut. Pak J Nutri 2: 335-338.

32. Ukpabi UJ, Ndimele C (1990) Evaluation of the quality of garri produced in Imo state. Niger Food J 8: 105-108.

27

Optimization of Baking Temperature, Time and Thickness for Production of Gluten Free Biscuits from Keyetena Teff (*Eragrostis tef*) Variety

Teshome E, Tola YB* and Mohammed A

Department of Post-Harvest Management, Jimma University College of Agriculture and Veterinary Medicine, Jimma, Ethiopia

Abstract

The demand for gluten-free foods is certainly increasing. Teff was becoming popular crop noticeably due to its very attractive nutritional profile and gluten-free nature. This study aims to develop Teff biscuits as an alternative food source for gluten intolerant people. Optimization of three independent variables, baking temperature (174,180 and 186°C), baking time (4, 8 and 12 min), and biscuit thickness (4.5, 5.5 and 6.5 ml) were taken as important factor to determine physical and nutritional quality of biscuits. There were twenty combinations created by Central Composite Design with the aid of Design-Expert software to get best quality Teff product. Moisture, protein, ether extract, fiber, ash, carbohydrate, and gross energy in biscuit samples were found in the range of 4.20-6.98%, 14.59-18.14%, 13.63-14.80%, 3.93-4.08%, 3.38 to 3.65%, 52.4-60.25%, and 414.22-422.15 Kcal/100 g respectively. Results showed that baking temperature and time were the most important factors that significantly affected (p <0.01) rehydration ratio, biscuit hardness, water activity, bulk density, protein, moisture content, carbohydrate and gross energy. Biscuit diameter, thickness, ash, fat, fiber, and mineral were not significantly affected by interaction of baking conditions. Based on all parameters with exception of sensory analysis the best treatment combination of temperature, time, and biscuit thickness were; 174°C, 9 min, and 4.5 mm. Biscuits sensory score shows variation in terms of colour and crispness. Biscuits baked at low temperature slighter thickness got better overall acceptability. Generally, result of this study confirmed that the possibility of production of gluten free Teff based biscuit as an alternative food source for gluten intolerant or tolerant consumers.

Keywords: Biscuit; Celiac disease; Gluten free diet; Teff; Sensory evaluation

Introduction

Teff (*Eragrostis tef* (Zucc.) Trotter) is a tropical cereal used throughout the world as grain for human consumption and as forage for livestock [1]. Teff grain flour is widely used in Ethiopia for making injera (staple spongy like bread made from fermented dough of Teff flour), sweet unleavened bread, local spirit, porridges, and soups [2]. Hopman et al. [3] and Dekking et al. [4] investigated the presence or absence of gluten in pepsin and trypsin digests of 14 Teff varieties. The digests were analyzed for the presence of T-cell-stimulatory epitopes. In contrast to known gluten containing cereals; no T-cell stimulatory epitopes were detected in the protein digests of all the Teff varieties assayed, thus confirming the absence of gluten in Teff. Because of this and nutritional merits, there is a growing interest on Teff grain utilizations like Quinoa. Due to this Teff has an increasingly important grain for individuals who suffer from gluten intolerance [5], and hence it is recommended as functional food for celiac patients.

Celiac Disease (CD) is a chronic entheropathy produced by gluten intolerance, more precisely to certain proteins called prolamines, which causes atrophy of intestinal villi, malabsorption and clinical symptoms that can appear in both childhood and adulthood [6]. Prolamins of wheat, barley, and rye are characterized by high proline content [7]. These proteins are the main constituents of gluten; contain toxic that can trigger celiac disease. Consequently, inadequate intakes of essential nutrients such as folate and vitamin B_{12} [8] calcium, iron, and fiber have been observed in those with CD [9] due to fear to eat products of those cereal crops. The recommended treatment for those with CD available to date is to follow a strict gluten-free diet [10].

Recent data indicated that the average worldwide prevalence of celiac patient is estimated as high as 1:266 (one out of 266 individuals). In Ethiopia, Celiac disease kills many children each year, mainly because it usually goes undiagnosed and untreated [11]. To alleviate the problem associated with gluten, value added gluten free product diversifications at national and international level are the only way to mitigate the problem. Biscuit produced from Teff could be used by celiac disease patients and those people who need to eat gluten free foods. In addition, there have been numerous studies on the effects of process conditions such as baking temperature, types of oven used and baking time to the final product qualities of biscuit made from wheat flour. However, so far there is no study conducted to investigate effect of baking conditions on physical and nutritional qualities of Teff biscuits. The study aimed to investigate and optimize certain baking processing conditions to produce better quality of Teff based biscuit.

Materials and Methods

Description of the study area

The experiment was conducted at Jimma University College of Agriculture and Veterinary Medicine (JUCAVM), Post-Harvest Management, Animal nutrition laboratory, and Ethiopian Public Health Institution (EPHI) laboratories for different parameters.

Experimental material collection and preparation

Teff variety used for preparation of biscuits was selected based upon market demand and colour. DZ-01-1681(Keyetena variety) was selected among red Teff varieties. Teff grain was brought from Debrezeit Agricultural Research Center (DZARC). Grain was manually cleaned, milled, and sieved using 0.5 mm sieve and stored in polyethylene bag till further use. Other ingredients like sugar, milk powder, mayonnaise,

*Corresponding author: Tola YB, Department of Post-Harvest Management, Jimma University College of agriculture and veterinary medicine, Jimma, Ethiopia
E-mail: yetenayet@gmail.com

salt and baking powder was purchased from Jimma local market. Egg was obtained from JUCAVM Animal Science Department. Biscuits were prepared according to method (Method No.10.52) described in American Association of Cereal Chemists [12] with minor modification as indicated in Table 1.

Experimental design

The study was conducted in two separate phases. In first phase, optimization study was conducted to determine better baking temperature, time, and thickness of Teff biscuit. To accomplish this, treatment combinations were designed using statistical software package (Design-Expert*, version 6.02, Stat-Ease (Minneapolis USA). Response Surface Methodology Specifically Central Composie Design (CCD) was used to conduct experiment. Sensory analysis was conducted in subsequent phase.

Data collected

Determination of physical and functional properties: Bulk density was determined according to a method indicated in Adeleke and

Odedeji [13]. Diameter and thickness were measured by two ways with digital Venire caliper (Fowler, US) and calibrated ruler as described in Ayo et al. [14] and AACC [1]. Rehydration ratio (RR) according to Yu et al. [15], Hardness of biscuit was determined with method of Nath and Chattopadhyaya [16], with stable micro system Texture Analyzer (TA-XT plus, 2012, UK). Water activity (aw) was measured with method of Hematian Sourki et al. [17] using Lab Master aw (Novasina AG, CH-8853 Lachen Sprint Switzerland) water activity meter.

Proximate composition analysis: Proximate composition for flours and biscuits were done in duplicate and average value was taken. Moisture content of the samples were determined by air oven (Model: Leicester, LE67 5FT, England) method AOAC (Method 925.10) [18];, Ash (Method 923.03) [18], Crude protein content was determined by Kjeldahl method AOAC (Method 988.05) [18], Ether extract with Soxhlet method (Model: SZC-C fat determinate, China) (Method 45.06) [18], Crude fiber (AOAC method 962.09) [19], Total percentage carbohydrate was determined by the difference as reported in Ponka et al. [20] and gross energy according to Osborne and Voogt [21]. Mineral analyses were accomplished using Atomic Absorption Spectrophotometer (AAS) (AA 6800, Japan) method as per the AOAC method, 985.35 [19].

Sensory evaluation: Sensory analysis was conducted for biscuits baked at optimized baking conditions (baking temperature, time and biscuit thickness). Sensory attributes measured were colour, taste, aroma, crispness, and overall acceptability using a five-point Hedonic scale consisting of 1 (extremely dislike), 2 (dislike moderately), 3 (neither like nor dislike), 4 (like moderately) and 5 (extremely like) [22].

Data analysis

A statistical software package (Design-Expert*, version 6.02,

Ingredients	Amount	Percentages (%)
Teff flour (g)	250	52.7
Mayonnaise (g)	45	9.5
Sugar (g)	60	12.7
Salt (g)	2.5	0.56
Baking powder (g)	5	1.08
Milk powder (g)	30	6.38
Whole egg (g)	50	10.6
Deionized Water (ml)	30	6.38

Table 1: The proportion of various ingredients for preparation of gluten free biscuit [56].

Run	A	B	C	D (mm)	DT (mm)	BD (g/ml)	Hardness (N)	aw	RR (%)	M.C (%)	Ash (%)	CP (%)	EE (%)	CF (%)	CHO (%)	Energy kcal/100g	Fe mg/100 g	Ca mg/100 g	Zn mg/100 g
1	186	6	6.5	53.4	0.3	0.7	85.4	0.6	12.9	6.4	3.6	17.2	14.3	4.0	54.5	415.7	46.6	229.9	6.2
2	186	12	6.5	52.7	0.0	0.7	95.3	0.4	15.4	5.0	3.5	15.6	13.9	4.0	58.2	420.2	45.9	229.1	5.6
3	180	9	3.8	53.0	0.2	0.7	93.1	0.5	15.8	5.3	3.4	15.8	13.8	4.0	57.7	418.4	45.8	229.4	5.9
4	180	9	5.5	53.6	0.4	0.7	92.7	0.5	15.1	5.7	3.5	16.8	14.2	4.1	55.7	418.0	46.2	229.3	5.8
5	190	9	5.5	52.6	-0.1	0.8	96.0	0.4	15.1	4.8	3.5	15.3	13.9	4.0	58.6	420.6	46.0	229.1	5.6
6	174	6	6.5	54.0	0.6	0.7	80.8	0.6	12.1	6.9	3.7	18.1	14.8	4.1	52.4	415.4	47.0	230.7	6.7
7	180	9	5.5	53.5	0.3	0.7	92.6	0.5	15.4	5.8	3.5	16.9	14.3	4.0	55.4	417.9	46.3	229.3	6.0
8	180	9	5.5	53.2	0.3	0.7	92.6	0.5	15.1	5.8	3.6	17.0	14.3	4.0	55.3	418.1	46.3	229.4	5.9
9	180	9	5.5	53.4	0.2	0.7	92.6	0.5	15.2	5.9	3.5	17.0	14.3	4.1	55.3	417.6	46.2	229.3	5.8
10	174	6	4.5	53.8	0.3	0.7	83.8	0.6	14.0	6.0	3.6	17.5	14.4	4.1	54.4	417.5	46.7	230.0	6.5
11	180	9	7.2	53.6	0.5	0.7	86.8	0.6	13.2	6.7	3.5	17.5	14.2	4.0	54.1	414.2	46.3	229.8	6.3
12	180	9	5.5	53.5	0.3	0.7	92.6	0.5	15.1	5.8	3.6	17.1	14.3	4.0	55.2	418.1	46.3	229.6	6.1
13	174	12	6.5	53.4	0.3	0.7	83.8	0.6	13.5	6.2	3.5	17.2	14.2	4.0	55.0	416.2	46.2	229.5	6.0
14	186	12	4.5	52.5	-0.1	0.8	100.2	0.4	16.2	4.2	3.4	14.6	13.6	3.9	60.3	422.2	45.7	228.9	5.5
15	180	14	5.5	52.8	0.0	0.8	92.7	0.4	16.0	4.7	3.4	15.2	13.8	4.0	58.9	420.8	45.8	229.0	5.5
16	174	12	4.5	53.2	0.2	0.7	88.9	0.5	15.6	5.1	3.5	15.9	14.0	4.0	57.6	419.7	46.2	229.4	5.9
17	186	6	4.5	53.4	0.3	0.8	87.9	0.6	13.6	6.0	3.5	16.7	14.2	4.0	55.5	417.0	46.3	229.5	6.0
18	180	9	5.5	53.5	0.2	0.7	92.7	0.5	15.2	5.7	3.6	16.9	14.3	4.0	55.4	418.2	46.3	229.4	5.9
19	180	4	5.5	53.8	0.5	0.7	79.9	0.6	12.3	7.0	3.6	18.0	14.6	4.1	52.7	414.3	46.8	230.0	6.6
20	170	9	5.5	53.8	0.3	0.7	82.8	0.5	13.8	6.0	3.6	17.2	14.4	4.1	54.7	417.6	46.5	229.9	6.4
Pure Teff flour										10.2	2.7	9.5	2.9	4.1	70.6	346.6	44.4	111.0	4.0
Mayonnaise (Herman Soybean oil based)										0	0	1	71.9	0	3.6	653.6	0.5	16.5	0.0
Milk powder (NIDO brand)												24.0	28.2	0	37.4	499.4	460.0	10	4.5

Where: A: Baking temperature; B: Baking time; C: Biscuit thickness before baking; D: Diameter; DT: Difference in thickness; BD: Bulk density; aw : Water activity; RR: Rehydration Ratio; MC: Moisture content; CP: Crude protein; EE: Ether extract; CF: Crude fiber; CHO: Carbohydrate

Table 2: Baking parameters and response variables for physical; functional and nutritional properties of Teff based biscuit.

(Minneapolis USA) was used to generate test of factor combination for better quality biscuit. Response surface methodology which involves design of experiments, fitting mathematical models and finally selecting levels of variables by optimizing the response [23] was employed in the study. The combinations were obtained based on a CCD. The statistical significance of terms in the regression equations were examined by analysis of variance (ANOVA) for each response and the significance test level was set at 5%. Data obtained from the sensory evaluation of biscuits were analyzed using CRD with three replicates by Minitab version 16.0 statistical software.

Results and Discussion

Effects of baking conditions on physical and functional properties

Diameter difference: Before baking biscuits had equal diameter but after baking their diameter varied from 52.5-54.0 mm with no significance difference. However, when average values are compared, bigger diameter (54 mm) was measured for biscuit baked at 174°C, 6 min, 6.5 mm thick and smaller one (52.5 mm) at 186°C, 12 min, 4.5 mm thick (Table 2). This implies that relatively higher temperature, longer baking time and slightly thick dough contributed less for expansion of the dough during baking. The bigger diameter from commercial point of view is the better quality as indicated Labuschagne et al. [24].

Thickness biscuit difference: Difference in thickness implies the difference in thickness of biscuit after and before baking. Since Teff

flour is free from gluten, almost no significant variation was observed on thickness. The value ranged from-0.1 to 0.6 mm as shown on Table 2. The bigger (0.6 mm) difference obtained from biscuit baked at 174°C, 6 min, 6.5 mm thickness and the smaller in reduction (-0.1 mm) from biscuit at 190°C, 9 min, 5.5 mm and 186°C, 12 min, 4.5 mm thickness (Table 2). An increase in thickness of biscuits after baking is one of important features required by commercial producers for an increase in volume of production. However, like diameter of Teff based biscuits, a small increase in thickness of biscuits was observed at relatively lower baking temperature and time. In similar work, Biniyam [25] reported that, the average cookie thickness was not significantly reduced but a decrease in thickness of the cookies was observed as the baking temperature and time were increased. This might be associated with significant decrease in moisture content of biscuits which results in for shrinkage of the thickness.

Bulk density of biscuit: Bulk density of Keyetena Teff flour was 0.62 g/ml and that of biscuits baked at different conditions showed significant (P<0.01) difference. The results ranged from 0.66 g/ml (186°C, 12 min, 4.5 mm) to 0.77 g/ml (174°C, 6 min, 6.5 mm) (Table 2). Bulk density shows significant (P<0.05) difference as temperature and time interacts as well as temperature and thickness interacts, but time and thickness showed non-significant effect. As indicated in Figure 1A as baking time and temperature increases there is an increment of bulk density on biscuit. Similarly, as indicated in Figure 1B with an increase in baking temperature and decrease in biscuit thickness an increment in bulk density is observed. This implies that a denser packaging material may be required for this type product [26] (Figure 1).

Hardness: Hardness showed highly significant difference in main, quadratic and interaction of model terms at (p<0.05) (Table 3). Result ranged between 80 N (180°C, 4 min, 5.5 mm) to 100 N (186°C, 12 min and 4.5 mm (Table 2). As expected, an increase in hardness with higher temperature, longer bakes time and thinner biscuit. The increase in baking time and higher temperature has shown a decrease in moisture loss of breads [27] and this might be true also for biscuits to have a harder texture due to higher temperature and longer baking time. Biscuits with relatively higher moisture content generally had a softer texture and therefore, there seems to be a good relation between the hardness of the biscuits and their moisture contents. Lower hardness in low temperature and shorter baking time indicates less brittle (but crunchy) biscuits with greater internal cohesiveness and springiness.

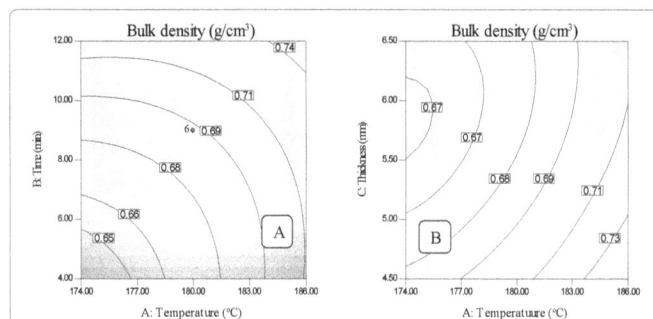

Figure 1: Effects of baking temperature and time (A) baking temperature and thickness (B) on bulk density of biscuits.

Source	D (mm)	D (mm)	Hardness (N)	aw	BD (g/cm³)	RR (%)	MC (%)	Ash (%)	CP (%)	EE (%)	CF (%)	CHO (%)	Energy K.cal /100g	Fe mg /100g	Ca mg /100g	Zn mg /100g
Model	***	***	***	***	***	***	***	***	***	***	***	***	***	***	***	***
Int	1.55	0.33	91.12	0.54	0.68	14.78	6	3.56	17.21	14.37	4.05	54.8	417.39	46.35	229.5	6.02
A	-0.31*	-0.103*	3.9***	-0.03**	0.03**	0.28**	-0.28**	-0.04**	-0.54*	-0.16*	-0.02**	1.03**	0.54*	-0.18*	-0.27**	-0.21**
B	-0.4**	-0.18**	3.79**	-0.09**	0.029**	1.74**	-0.87**	-0.07**	-0.99*	-0.3**	-0.04**	2.29**	2.38**	-0.45*	-0.524*	-0.45*
C	0.11*	0.078*	-1.9*	0.037*	-0.013*	-0.70*	0.376*	0.026*	0.40**	0.12**	0.017*	-0.94*	-1.08*	0.143*	0.179*	0.105*
A2	-0.08*	-0.063*	-1.2*	-0.012*	0.010*	-0.30*	-0.139*	-0.008*	-0.2**	-0.03ns	-0.01*	0.41**	0.47*	0.014ns	0.064ns	0.027ns
B2	-0.08ns	-0.02ns	-2.3**	0.006*	0.012*	-0.7*	0.020ns	-0.006ns	-0.17*	-0.03ns	-0.02**	0.20*	-0.14ns	0.071ns	0.105ns	0.082ns
C2	-0.04ns	0.025ns	-1.0**	0.011*	0.010*	-0.3**	0.064*	-0.02**	-0.10*	-0.08*	-0.015*	0.15*	-0.53*	-0.04ns	0.101*	0.066*
A*B	-0.07ns	-0.05ns	1.78**	-0.03**	-0.012*	0.36**	-0.262*	0.002ns	-0.19*	0.01ns	-0.02ns	0.44**	1.10**	0.002ns	0.087ns	0.008ns
A*C	-0.03ns	-0.04ns	0.10**	-0.013*	-0.006*	0.32**	-0.10**	-0.004ns	-0.06ns	-0.02ns	0.001ns	0.19*	0.31*	0.039ns	-0.047ns	-0.014ns
B*C	0.03ns	-0.02ns	-0.6**	0.003ns	0.002ns	-0.06ns	0.078*	0.005ns	0.19*	0.003ns	0.005ns	-0.28*	-0.34*	-0.06ns	-0.110ns	-0.018ns
LoF	0.6468	0.764	0.1816	0.222	0.1152	0.1614	0.2752	0.093	0.1995	0.0699	0.332	0.3406	0.0602	0.1176	0.3178	0.9424
Adj R2	0.9098	0.879	0.9999	0.9877	0.971	0.9865	0.9922	0.9663	0.9857	0.9491	0.9534	0.9917	0.9756	0.9398	0.9045	0.9289

Where: A: baking temperature; B: baking time; C: biscuit thickness before baking; ns: non-significant difference; DT: Difference in thickness; BD: Bulk density; aw : water activity; RR: Rehydration ratio; MC: moisture content; CP: crude protein; EE: Ether Extract; CF: Crude fiber; CHO: Carbohydrate; P>0.05; *P<0.05; **P<0.01 and *** P< 0.001

Table 3: Estimated regression coefficients; degree of significance and lack of fit of parameters for physical; functional and nutritional parameters of biscuit model equations.

When the effects of temperature and time were compared, baking time was more effective for increment of biscuit hardness (Table 2).

But when compared to commercially available wheat based biscuits, Teff based biscuit was a bit harder to break. This might be due to lack of gluten in Teff flour to give the desired textural property related to hardness. In addition to absence of gluten, hardness level might be influenced by high iron and crude fiber contents of Keyetena Teff Umeta and Parker [28] as compared to commercially available wheat flour based biscuits. Similar increase in hardness of cookies was also reported in Singh et al. [29] on incorporation of sweet potato flour in wheat and concluded that an increase in fiber content in the formulation was the reason for variation (Figure 2) (Table 3).

Water activity: Correlating water activity with shelf-life is of critical importance in work with biocontrol formulations [30]. Typical of biscuits is not only low in moisture content but also has low value of aw. Foods with aw <0.60 are considered as microbiologically stable [31]. Water activity of biscuits in this study as indicated on Table 2 ranged from 0.39 to 0.64 with significant effects (P<0.05) in main, quadratic, and interaction of model terms (Table 3). From 20 treatment combinations, 3 of them exhibited a water activity above 0.6 (Table 2) which might not recommended for extended storage life of biscuit. Higher aw was obtained at 174°C for 6 min and 6.5 and lower at 186°C for 12 min and 4.5 mm thickness of biscuit (Figure 3).

The interaction of baking temperature and baking time was highly significant (P< 0.01) on water activity. As indicated in Figures 3A and 3B as baking time and temperature increases there is a decrement of water activity as expected. This is from the fact that application of heat results in the evaporation of the water molecules contributed for less aw value. Similar reports from Bojana [32], showed that increase of baking time and temperature on gluten free biscuits enriched with blueberry pomace with a decreased in aw from 0.6 to 0.326. Similar results also

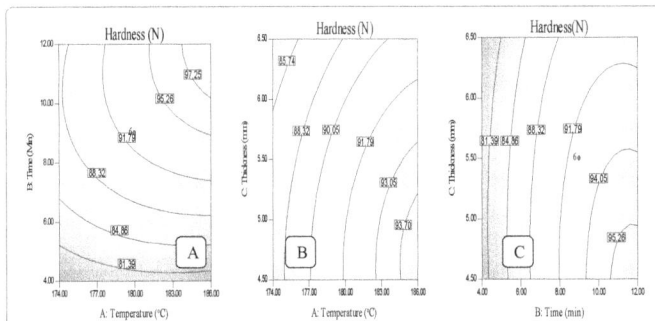

Figure 4: Effects of baking temperature and time (A) baking temperature and thickness (B) on rehydration ratio of biscuits.

reported in Manohar [33], indicated that sample of cereals and fruit-containing biscuits showed a decrease in aw with an increase in baking temperature.

Rehydration Ratio: Rehydration ratio as indicated on Table 2 ranged from 12.10% to 16.15% for biscuits. Higher result was recorded from biscuit baked at 186°C for 12 min at 4.5 mm thickness and lowest value from biscuit baked at 174°C, 6 min, 6.5 mm thickness. Rehydration ratio shows highly significant effect (P<0.01) in linear, quadratic and interaction of model terms. Effect of time and thickness had not a significant effect on rehydration ratio (Figure 4) .

The interaction of baking temperature and time was found to have highly significant (P<0.01) effect (Table 3) on rehydration ratio of biscuit. As indicated by contour line in Figure 4A as baking time and temperature increases there is an increase in rehydration ratio. This may be due to the more moisture removed from samples resulted in the more increase in water holding capacity. Similar result is reported in Mitra et al. [34] an increase in rehydration ratio with an increase in baking temperature and time. The interaction of baking temperature and thickness also exhibited highly significant (P<0.01) effect on rehydration ratio of biscuit (Table 3). As indicated in Figure 4B with decrease in thickness and baking temperature close to 180°C increases rehydration ratio. This might be due to the thinner biscuit with optimum baking temperature the change in microstructure of the biscuit might be in a position to absorb more water during rehydration process.

Nutritional composition of Teff flour based biscuit

Teff flour from different varieties contains almost similar proximate composition [35]. Nutritional compositions of pure Teff flour and other ingredients used for making of biscuits are indicated in Table 2. Proximate composition results obtained for Keyetena Teff variety is close to what is reported in Lovis [36] and Bultosa [37]. Also changes in nutritional composition of biscuits baked at different conditions are indicated in the same table.

Moisture: Moisture is an important parameter in baked foods that significantly affects shelf life and growth of microbial contaminants [38]. Moisture contents for biscuits baked at different conditions resulted in a range of 4.20 (186°C, 12 min, 4.5 mm) to 6.98% (180°C, 4 min and 5.5 mm) (Table 2). Moisture content showed significant effect (P<0.05) in main, quadratic, and interaction of model terms (Table 3). It is apparent that low moisture content is due to relatively high temperature, long time and thinner biscuit thickness as indicated in Figures 5A-5C.

As indicated by contour line in Figure 5A as baking time and temperature increases there is a reduction of moisture content. A similar result was reported in Piergiovanni and Farris [39] in which baking temperature and time was negatively affecting the amaranth

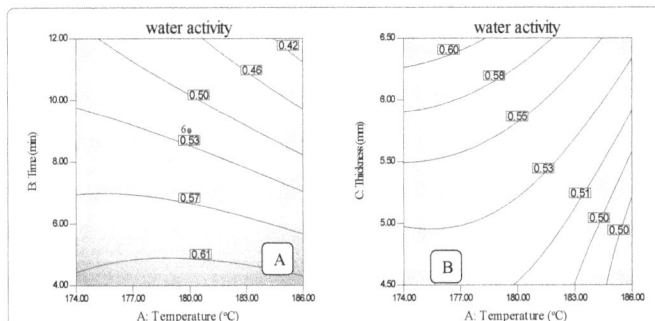

Figure 2: Effects of baking temperature and time (A) baking temperature and thickness (B) baking time and thickness (C) on hardness of biscuits.

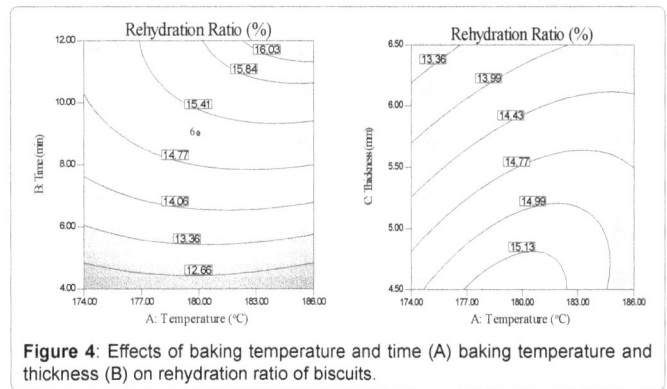

Figure 3: Effects of baking temperature and time (a) baking temperature and thickness (b) on water activity of biscuits.

cookie moisture content. Bojana [32] also showed that increase of baking time and temperature on gluten free biscuits enriched with blueberry pomace showed a decrease in moisture content from 10% to 5.2%. Similar result also reported in Patela et al. [40] and Kotoki and Deka [41], as bread baked at higher time-temperature combinations showed loss more water and the bread become harder and underweight. Figure 5B shows with an increase in baking temperature and a decrease in biscuit thickness, a decrease in moisture content. Baking time and thickness of biscuit have also a significant effect on moisture content as shown in Figure 5C. This might be associated with high rate of heat penetration on thinner biscuit than thicker one (Figure 5).

Ash: Ash content of Keyetena Teff flours was indicated on Table 2. This result is in agreement with Corke et al. [42] on review report indicated that, the ash level in Teff varieties varied from 2.66%-3.00% with typical value of 2.8%. Ciferri and Baldrati, [43] found that ash content of Teff flour ranged between 2.4% and 2.94% . The result for ash in this finding was comparable to the values indicated in Bultosa [13], which was reported in the range of 1.99% to 3.16%.

Results of ash for baked biscuits at different baking conditions are presented in Table 2 in range of 3.38 to 3.65. Ash content was not significantly affected by interaction of baking temperature, time, and thickness. According to Fenema [44] mineral elements, unlike vitamins and amino acids, cannot be destroyed by exposure to heat for a long time and this could be the reason that interaction of factors showed insignificant effects (P>0.05). However, increase in ash content of biscuits as compared to teff flour is associated with additional other biscuit ingredients which is in line with what is reported in Biniyam [25].

Crude protein: The statistical results regarding crude protein content of different biscuits baked at different conditions are presented in Table 2. The protein content of the biscuits significantly affected

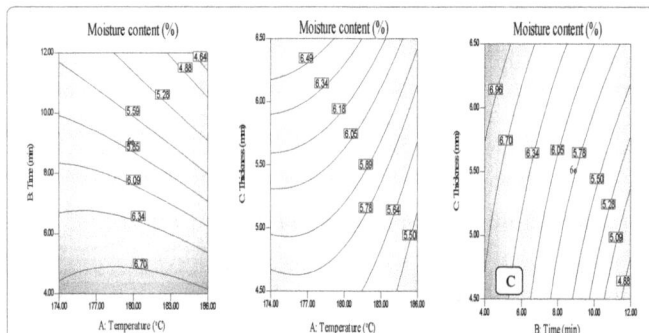

Figure 5: Effects of baking temperature and time (A) baking temperature and thickness (B) baking time and thickness (C) on moisture content of biscuits.

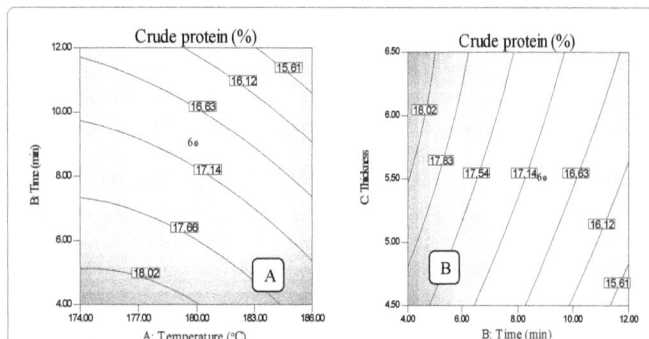

Figure 6: Effects of baking temperature and time (A) baking time and thickness (B) on moisture content of biscuits.

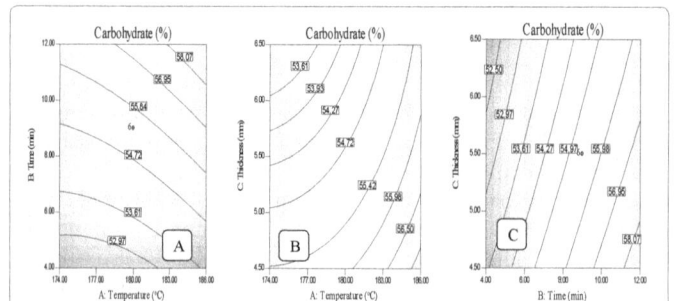

Figure 7: Effects of baking temperature and time (A) baking temperature and thickness (B) baking time and thickness (C) on carbohydrate content of biscuits.

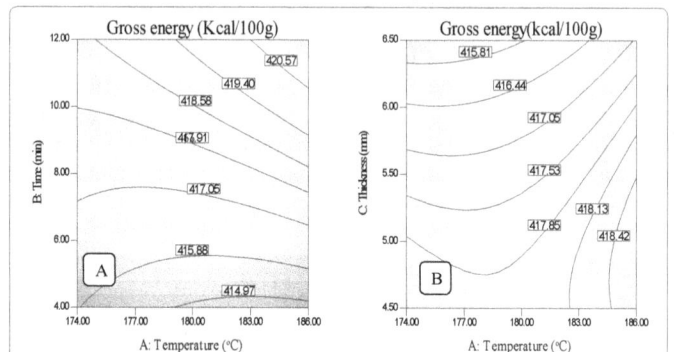

Figure 8: Effects of baking temperature and time (A) baking temperature and thickness (B) on gross energy of biscuits.

by main, quadratic and interaction of model terms at (P <0.05) as indicated on Table 3. The value varied from 14.59 to 18.14% this showed a significant increase as compared to Teff flour which might be due to effect of milk powder and mayonnaise added as an ingredient. Among the values, biscuits baked at 174°C, 6 min, and 6.5 mm thickness resulted in higher protein value than baked at 186°C, 12 min, and 4.5 mm thickness. This shows that, the higher the temperature, longer baking time and the thinner biscuit resulted in the reduction of crude protein content as indicated in Figures 6A and 6B.

Interaction effects of baking temperature and time as well as time and thickness have highly significant (P<0.01) effect on crude protein content. As indicated in Figure 7A as baking time and temperature increases crude protein content declines. This might be associated that, during baking, the physical and chemical properties of protein are altered due to the denaturation of protein where in the hydrogen bonds and non-polar hydrophobic interactions of the secondary and tertiary structures of proteins are disrupted by heat and the soluble amino acids leached out in the baking medium [45]. According to Hui et al. [46], the Maillard reaction, is also a responsible chemical reaction for the loss of protein due to reaction of free amino acids with reducing sugars (glucose and fructose) is favored at temperatures above 120°C. The extent of loss is increased by higher temperatures, longer baking time, and availability of larger amounts of reducing sugars [45].

Ether extract: Ether extract of flour is indicated in Table 3 with a value 2.93%. The ether extract value was comparable with those of Corke et al. [42], reported that the Ether extract content of Teff flour ranged from 2-3% with mean of 2.3%. Results presented in Table 2 indicated that the ether extract content of biscuits were not significantly (P<0.05) affected by interaction of model terms but linear terms had a significant effect (Table 3) and values ranged between 13.63 to 14.80 %. Usha [47] categorized biscuits in terms of their fat content as low

(7.5-15%), medium (15-27%), and high (more than 27%) fat biscuits. According to the author, Teff based biscuits baked under this study conditions are categorized in low fat biscuits. But when ingredient compositions are considered, similar to enhanced crude protein value of biscuits, addition of fat reach ingredient significantly enhances the ether extract of biscuits as compared to Teff flour as a major component. Higher ether extract measured for biscuits baked at relatively lower temperature and shorter time of baking, but when heating increased loss of ether extract observed. This finding was also in line with Fenema [44], high temperature heating trigger polymerization of fat molecules which contributing to its loss during baking, drying, and boiling. But when the effects of baking temperature and time were compared, baking time was found to be more effective on fat loss. This finding also in line with Biniyam [25] on fat content of cookies from wheat, quality protein maize and carrot composite flour which is ether extract was not significantly influenced by baking temperature.

Crude fiber: Fiber content of the Teff four is presented in Table 2. Relatively higher crude fiber content is reported for Keyetena Teff flour as compared to what are reported in Lovis [36] and Corke et al. [42], they reported that 3% and 3.8% respectively. However, the crude fiber of the final biscuit varied from 3.93 to 4.08 %. Higher fiber content was recorded from biscuit baked at 174°C for 6 min 6.5 mm thickness and lowest from 186°C for 12 min having 4.5 mm thickness. Fiber content significantly affected by baking temperature and biscuit thickness in both main and quadratic terms as indicated in Table 3. Lee et al. [46] and Bingham [48] reported that dietary fiber has a beneficial effect on bowel transit time, affects glucose and lipid metabolism, reduces the risk of colorectal cancer, and stimulates bacterial metabolic activity. Even though there is no report on incorporation of dietary fiber on gluten free biscuit, as indicated in this result, incorporation of Teff flour enhance fiber content of biscuits since other ingredients lack fiber.

Utilizable carbohydrate: Utilizable carbohydrates content of Keyetena Teff flours is presented in Table 2, 70.56%. This is lower than the values indicated in Lovis [35] (73.0%) and comparable with the value of 71.4% USDA [49]. Utilizable carbohydrate values of biscuits in this study varied from 52.4 to 60.25%. The decrease in carbohydrate could be associated with the addition of other biscuit ingredients high in protein and fat contents. Higher carbohydrate value obtained from biscuit baked at 186°C for 12 min having thickness of 4.5 and lowest value at 174°C for 6 min having 6.5 thicknesses (Figure 7).

Baking temperature and time significantly (P<0.01) affected carbohydrate content of biscuit Figure 7A with an increase in baking time and temperature carbohydrate content of biscuits increased. Since utilizable carbohydrate calculated by difference from other proximate compositions, the higher the other proximate compositions resulted in the lower carbohydrate value in similar baking condition. Furthermore, an increase in carbohydrate might also associate with starch degradation into dextrin and simple sugars [50]. Baking temperature and thickness has significant (P<0.05) effect on carbohydrate content of biscuit. As indicated in Figure 7B as baking temperature increased and biscuit thickness decreased there is an increase in carbohydrate content. Baking time and thickness effect was also significant (P<0.05) on carbohydrate content of biscuit as indicated in Figure 8.

Gross energy: Initially the gross energy was 346.61 kcal/100 g in Keyetena Teff flour. But an increase in gross energy was observed on biscuit in range of 414.22-422.15 kcal/100 g. The amount of calorie in the final product showed significant difference (P<0.05) in linear, quadratic and interaction model terms (Table 3). Higher calorie obtained from

biscuit baked at 186°C, 12 min, and 4.5 mm thickness and lower value for biscuit baked at 180°C, 4 min, and 5.5 mm thick.

Interaction effect of baking temperature and time was highly significant (P<0.01) on calorie value of biscuit. As indicated in Figure 8A as baking time and temperature increases there is an increase in energy content. Since the calorie of biscuit was calculated from value of protein, ether extract, and carbohydrate with their conversion factor, higher carbohydrate value played the role for increment of energy content as baking time and temperature increases. Baking temperature and thickness was found to have highly significant (P<0.01) effect on gross energy content of biscuit. As indicated in Figure 8B as baking temperature increases and biscuit thickness decreases there is increment of energy content in keyetena Teff biscuit. This could be explained in line with justification provided in above points (Figure 8).

Minerals content

Iron, Calcium and Zinc content of Keyetena Teff flour are 44.4, 110.95 and 3.95 mg/100 g (Table 2). Due to addition of ingredients the minerals content of biscuits increased as compared to Teff flour. Increase in iron content was small as compared to calcium and zinc. However, increase in calcium content in biscuits is almost double as compared to Teff flour content (Table 2). This indicates that, in addition to get functional benefits of the ingredients, the mineral content of Teff based biscuit can be improved significantly. When effect of baking temperature, time and thickness are considered they showed significant effects mainly in the linear terms (Table 3) and with little or no quadratic and interaction effects. This implies that, baking parameters have additive effect on mineral contents. Result if this work is also in agreement with Fenema [44], mineral elements, unlike vitamins and amino acids cannot be destroyed by exposure to heat for a long time.

Optimized processing variables for better biscuit quality

Numerical optimization conducted to select baking conditions for better physical, functional and nutritional value of biscuit. Based up on desired target values of physical, functional and nutritional variables, biscuits having 4.5 mm thickness and baked at 174°C for 9 minutes resulted in best baking condition for best combination of biscuit qualities. Optimized values of Keyetena based biscuits for physical parameters like difference in diameter, hardness, water activity, bulk density, and water rehydration capacity are close to 1.65 mm, 86.89 N, 0.52, 0.7 g/m³ and 19.6 ml respectively. Optimized proximate composition for moisture, ash, protein, ether extract, fiber, carbohydrate, and energy were; 5.81%, 3.65%, 17.8%, 14.3%, 4.04%, 55.15%, and 418.2 kcal/100 g respectively. For mineral elements iron 46.22, calcium 230.0, and zinc 6.18 mg/100 g were optimize results respectively.

Sensory evaluation

Sensory qualities are the main criteria that make the product to be liked or disliked by consumers [51]. Sensory evaluation was conducted after overall optimization of baking conditions for better physical, functional and nutritional quality parameters of biscuits (174°C, 9 min and 4.5 mm thickness). One of the major problems associated with production of gluten-free products is their inferior taste and/or structure. But the mean score value of biscuit samples for taste was 4.26 ± 0.09 (like moderately). This might be due to dominant effects of ingredients added (47.3%) in biscuit making. In agreement to this work, Gallagher et al. [52] also indicated that, Teff flour has a good taste than other gluten free cereals and hence allows an opportunity to produce gluten-free product with an attractive taste.

Color is one of the detrimental sensorial property significantly influences customers purchase decision. The color preference score ranged was 2.74 ± 0.19 which is in between dislike moderately to neither like nor dislike. The lower color value is might be due to dark color of biscuit because of red Teff color. This might be due to higher levels of pigmented material, such as tannins and polyphenols [28] in Keyetena Teff flour and Maillard reaction during high temperature baking. The aroma is one of the main sensorial properties for acceptability of biscuits. The average score value of panelist was 4.18 ± 0.02. This indicated that, baked biscuits were moderately liked by the panelist. This might be due to enhanced interaction and distinctive flavor of Teff flour together with ingredients interaction created during baking. Furthermore, aroma was mostly influenced by the reaction occurred between cross linking of Teff starch gelatinization and protein denaturation.

Biscuit lovers prefer to eat crispy biscuits since it is one of desirable textural properties. Average score value of crispness test of biscuit was 3.95 ± 0.10, which indicates that, panelist moderately liked the product. Hough and others [53] indicated that, dried or baked products loss their crispness when their water activity value greater than 0.5 ± 0.2. However, the water activity value for this study was above this value for better crispness. This might be associated with peculiar properties of Teff flour (high iron content and absence of gluten) for better crispness even at higher water activity value than wheat flour based biscuits [54-56].

The combination of the above sensorial parameters can be generalized or summarized by rating the overall acceptability of the product. Panelist's response to products is, of course, liking or acceptability to a certain degree. The overall acceptability of Teff based biscuits baked at optimized conditions was 3.98 ± 0.07. This implies that the biscuit was moderately liked by the panelists. Therefore, Teff based biscuit produced with the addition of required ingredients, not only good gluten free product in terms of other evaluated parameters, but also has good overall acceptability by the panelists.

Conclusion

The only effective treatment for celiac disease is the total lifelong avoidance of gluten ingestion or using a strict gluten-free but nutritious diet. Teff flour is an interesting gluten free raw material and considered as the major ingredient to produce gluten free biscuit. The study showed that, mainly higher baking temperature with longer baking time resulted in inferior biscuit quality. The sensory score lowered as baking temperature, time, and thickness increases. However, biscuits baked at relatively thickness of 4.5 mm, lower temperature of 174°C for 9 min, provided better quality product in terms of physical, functional, nutritional as well as sensory properties. The out comes from this study would be used to generate baseline information for subsequent use and studies to produce nutrient enriched gluten free products from Teff flour. Many gluten-free products may not meet the recommended daily intake for fiber and minerals. However, gluten free biscuit made from Teff based flour under recommended baking condition will mitigate such limitations and provide alternative snack food source for gluten intolerant individuals.

Acknowledgements

The authors thank Jimma University College of Agriculture and Veterinary Medicine of Ethiopia and PHMIL project of Global Affairs of Canada for financial support to conduct this research, Debrezeit Agricultural Research Center (DZARC) for supply of Teff grain.

References

1. Kebede L, WorkuS, Bultosa G, Yetneberek S (2010) Effect of extrusion operating conditions on the physical and sensory properties of Teff flour extrudates. Ethiopia J Applied Sci and Technol 1: 27-38.

2. Erica H, Dekking L, Blokland ML, Wuisman M, Zuijderduin W, et al. (2008) Teff in the diet of celiac patients in the netherlands. Scandinav J Gastro 43: 277-282.

3. Dekking LS, Winkelar YK, Koning F (2005) The Ethiopian cereal Teff in celiac disease. N Engl J Med 353: 1748-1749.

4. Boka B, Woldegiorgis AZ, Gulelat D Haki (2013) Antioxidant properties of ethiopian traditional bread (Injera) as affected by processing techniques and teff grain varieties. Center for Food Science and Nutrition, Addis Ababa University, Borderless Science 1: 7-24.

5. Bernardo D, Pena AS (2012) Developing strategies to improve the quality of life of patients with gluten intolerance in patients with coeliac disease. Eur J Intern Med 23: 6-8.

6. Haboubi NY, Taylor S, Jones S (2006) Coeliac disease and oats: a systematic review. Post grad Med J. 82: 672-678.

7. Claes H, Grant C, Grehn S, Grännö C, Hulten S, et al. (2002) Evidence of poor vitamin status in coeliac patients on a gluten-free diet for 10 Years. Alimentary Pharmacol Ther 16: 1333-1339.

8. Thompson T, Dennis M, Higgins LA, Lee AR, Sharrett MK (2005) Gluten-free diet survey: are Americans with coeliac disease consuming recommended amounts of fibre, iron, calcium, and grain foods. J Ther 16: 1333-1339.

9. Gallagher E (2002) Formulation and nutritional aspects of gluten-free cereal products and infant foods. Gluten-free cereal products and beverages, Academic Press, USA.

10. Emire SA, Tiruneh DD (2012) Optimization of formulation and process conditions of gluten-free bread from sorghum using response surface methodology. J Food Process Technol 3: 155.

11. AACC (2000) Approved methods of the american association of cereal chemists. AACC, Minnesota.

12. Adeleke RO, Odedeji JO (2010) Functional properties of wheat and sweet potato flour blends. Pak J Nutri 9: 535-538.

13. Ayo JA, Ayo VA, Nkama I, Adewori R (2007) Physicochemical, in-vitro digestibility and organoleptic evaluation of acha wheat biscuit supplemented with soybean flour. Niger Food J 25: 77-89.

14. Yu L, Ramaswamy HS, Joyce B (2009) Twin-screw extrusion of corn flour and soy protein isolate (SPI) blends: A response surface analysis. Food Bioprocess Technol 5: 485-497.

15. Nath A, Chattopadhyaya PK (2007) Optimization of oven toasting for improving crispness and other quality attributes of ready to eat potato-soy snack using response surface methodology. J Food Eng 80: 1282-1292.

16. Sourki H, Ghiafeh Davoodi M, Karimi M, Razavizadegan SH, Jahromi, et al. (2010) Staling and quality of Iranian flat bread stored at modified atmosphere in different packaging. World Academy of Science, Eng, and Technol 41: 5-29.

17. AOAC (2011) Association of official analytical chemists. Official methods of analysis. AOAC International, Washington, DC, USA.

18. AOAC (2005) Association of official analytical chemists. AOAC, Arlington, Virginia, USA.

19. Ponka R, Fokou E, Leke R, Fotso M, Souopgui J, et al. (2005) Methods of preparation and nutritional evaluation of dishes consumed in a malaria endemic zone in Cameroon (Ngali II). Afr J Biotechnol 4: 273-278.

20. Osborne DR, Voogt P (1978) The Analysis of Nutrients in Foods: LTD official methods 6.2 and 6.3. Academic press, Inc, London.

21. Meilgaard MC, Civille GV, Carr BT (2007) Sensory evaluation techniques. CRC Press, Boca Raton.

22. Khuri AI, Cornell JA (1987) Response surfaces, designs and analysis. Marcel Dekker Inc, New York.

23. Labuschagne MT, Coetzee MCB, Deventer CS (1996) Biscuit-making quality prediction using heritability estimates and correlations. J Sci Food Agri 70: 25-28.

24. Tesfaye B (2010) Development of cookies from wheat quality protein maize and carrot composite flour. Addis Ababa University, Technology, Ethiopia.

25. Odedeji J, oyeleke RO (2010) Functional properties of wheat and sweet potato flour blends. Pak J Nutri 9: 535-538.

26. Chevallier S, Della Valle G, Colonna P, Broyart B, Trystram G (2002) Structural and chemical modifications of short dough during baking. J Cereal Sci 35: 1-10.

27. Umeta M, Parker ML (1996) Microscopic studies of the major macro-components of seeds, dough and injera from Teff (Eragrostis teff). Ethiopian J Sci 19:141-148.

28. Singh S, Riar CS, Saxena DC (2008) Effect of incorporating sweet potato flour to wheat flour on the quality characteristics of cookies. Africa J Food Sci 20: 65-072.

29. Connick JR, Daigle DJ, Boyette CD, Williams KS, Vinyard BT (1996) Water activity and other factors that affect the viability of colletotrichum truncatum conidia in wheat flour kaolin granules (Pesta). J Biocontrol Sci and Technol 6: 277-284.

30. Young C, Linda S, Cauvain SP (2000) Bakery food manufacture and quality: water control and effects. Blackwell Science, Oxford.

31. Saric B (2014) The influence of baking time and temperature on characteristics of gluten free biscuits enriched with blueberry pomace. Food and Feed Res 41: 39-46.

32. Manohar RS, Rao PH (1999) Effects of water on the rheological characteristics of biscuits dough and quality of biscuits. Europe Food Res Technol 209: 281-285.

33. Mitra J, Shrivastava SL, Rao PS (2011) Vacuum dehydration kinetics of onion slices. Food Bioprod Process 89: 1-9.

34. EHNRI (1997) Food composition table for use in Ethiopia. Ethiopian Health and Nutrition Research Institute, Addis Ababa.

35. Lovis LJ (2003) Alternatives to wheat flour in baked goods. J Cereal Food World 48: 62-63.

36. Bultosa G (2007) Physicochemical characteristics of grain and flour in 13 Teff grain varieties. Appl Sci Res 3: 2042-2051.

37. ICMSF (2002) Microorganisms in food. Microbiological testing in food safety management. Kluwer Academic/Plenum, NY.

38. Piergiovanni L, Farris S (2008) Effects of ingredients and process conditions on 'Amaretti' cookies characteristics. Int J Food Sci Technol 43: 1395-1403.

39. Patela BK, Waniska RD, Seetharaman K (2005) Impact of different baking processes on bread firmness and starch properties in bread crumb. J Cereal Sci 42: 173-184.

40. Kotoki D, Deka SC (2010) Baking loss of bread with special emphasis on increasing water holding capacity. J Food Sci Technol 47: 128-131.

41. Corke H, Bultosa G, Taylor JRN (2004) Teff In Encyclopedia of grain science. Elsevier, UK 3: 281-290.

42. Ciferri R, Baldrati T (1939) Teff (Eragrostis Teff) PanifIcazion dell cereal.

43. Fennema RO (1996) Food chemistry. Marcel Dekker, Inc, USA.

44. Fellows JP (2000) Food process technology principle and practices. 2nd edn, Wood head Publishing Limited and CRC Press LLC.

45. Hui YH, Corke H, Leyn DI, Kit Nip W, Cross N (2006) Bakery products science and technology. Blackwell Publishing Ltd, UK.

46. Lee KW, Song KE, Lee HS (2006) The effects of Goami no. 2 rice, a natural fiber-rich rice, on body weight and lipid metabolism. Obesity 14: 423-430.

47. Usha B, Amarjit K (1995) Effect of ghee (butter oil) residue and additives on physical and sensory characteristics of biscuits. Chemie Mikrobiologie Technologie der Lebsensmittel 17: 151-155.

48. Bingham SA (2003) Dietary fiber in food and protection against cholesterol cancer in the European prospective investigation into cancer and nutritional (EPIC): an observational study. The Lancet 57: 9-14.

49. USDA Agricultural Research Service (2015) National nutrient database for standard reference release 27. Basic report: 20142, Teff, uncooked. USDA.

50. Guy RCE (2001) Raw material for extrusion Cooking. In: RCE Guy (edn) Extrusion cooking; Technologies and Application, Wood head publishing limited.

51. Falola AO, Olatidoye OP, Balogun IO, Opeifa AO (2011) Quality characteristics of biscuits produced from composite Flours of cassava and cucurbita Mixita seed. J Agri and Vet Sci 3: 1- 12.

52. Gallagher E, Gormley TR, Arendt EK (2004) Recent advances in the formulation of gluten-free cereal-based products. Trend Food Sci Technol 15: 143-152.

53. Hough G, Buera MD, Chirife J, Moro O (2001) Sensory texture of commercial biscuits as a function of water activity. J Texture Studies 32: 57-74.

54. AOAC (2000) Association of official analytical chemists, Official method of analysis, Washington, DC.

55. Mohamed HA, Elsoukkary MM, Doweidar MM, Atia AA (2004) Preparation characterizations and health effects of functional biscuits containing iso flavones. Minufiya J Agric Res 2: 425-434.

56. See J, Murray JA (2006) Gluten-free diet: the medical and nutrition management of celiac disease. Nutri Clin Pract 21:1-15.

28

Synthesis and Characterization of Carboxymethyl Cellulose from Sugarcane Bagasse

Saeid Alizadeh Asl, Mohammad Mousavi*and Mohsen Labbafi

Department of Food Science and Technology, Campus of Agricultural Engineering and Natural Resources, University of Tehran, Iran

Abstract

Various raw materials including plant biomass, bacteria, algae and the Tunicates (marine animals) have been used to produce cellulose. However, agricultural waste has rarely been utilized for this purpose. In this work, Sugarcane bagasse was used as raw material to produce cellulose. Cellulose was extracted from sugarcane bagasse through the elimination of lignin and hemicellulose. Cellulose was then converted to carboxymethyl cellulose (CMC_b) by using sodium monochloroacet (SMCA) and various sodium hydroxide (NaOH) concentrations. Fourier Transform Infrared Spectroscopy (FTIR) was applied to verify the effect of NaOH concentration on this property. The highest viscosity and degree of substitution (DS=0.78) were observed in 30 gr/100 ml NaOH of carboxymethylation. Maximum tensile strength of the films produced at these conditions was 37.34 Mpa. The addition of a various amount of glycerol (1 ml/ 100 ml, 2 ml/ 100 ml, 3 ml/ 100 ml) dramatically decreased the tensile strength. The highest level of water vapor permeability was also observed at the same NaOH concentration. Cellulose can be correctly extracted from sugarcane bagasse and converted to carboxymethyl cellulose. Based on the cellulose of the bagasse characteristic, proper amount of NaOH was found to get a high DS. CMC_b has considerable features for application on biodegradable coating materials.

Keywords: Sugarcane bagasse; Carboxymethyl cellulose; Edible film; Mechanical properties

Introduction

Nowadays huge quantities of plant wastes are produced globally. Although they are the most abundant and renewable resource for organic substances attainable today [1,2], still considerable amounts of these materials are not used in proper way. Thus, the conversion of plant wastes into valuable products can be helpful in reduction of the environmental problems [3]. Plant wastes consists of more than 90% (w/w) carbohydrate polymers which can be modified by both biochemical and chemical reactions to some products such as starch, cellulose, cotton linter, bagasse fiber etc. [4,5]. Sugarcane bagasse is a residue produced in large quantities every year by the sugar industries [6], and the most of this amount used as a fuel to supply the energy required for sugar mill [7]. Some reports reveal that bagasse is used as a raw material for industrial applications such as electricity generation, pulp and paper production, etc. New researches show the application of sugarcane bagasse to produce composites with bagasse linters [8]. However, the large quantities of sugar cane bagasse remain unused. Therefore, utilization of this huge agricultural waste for new applications has attracted growing interest because of their ecological and renewable characteristic. In general, each ton of sugarcane produces 280 kg bagasse. Sugar cane bagasse contains 40% to 50% cellulose (crystalline and amorphous structure), 25% to 35% hemicellulose (amorphous polymers usually composed of xylose, arabinose, galactose, glucose and mannose), 15% to 20% of lignin and the remainder lesser amounts of mineral, wax, and other compounds [9].

Cellulose is an important component for the application of new biomaterials obtained from agriculture wastes. Cellulose is usually found in the cell wall of plants and is generally associated with lignin and hemicellulose, which make it difficult to extract in pure form [10]. Cellulose is a high molecular weight and a linear homopolymer of repeating β-D-glucopyranosyl units joined by single oxygen atoms (acetal linkages) between the C-1 of one pyranose ring and the C-4 of the next ring, Because of their linearity and stereo regular nature, cellulose molecules associate over extended regions, forming polycrystalline, fibrous bundles. Large numbers of hydrogen bonds together hold crystalline regions. They are separated by, and connected to, amorphous regions. Cellulose is insoluble because, for to dissolve, most of these hydrogen bonds would have to be released at once. However, Cellulose through substitution can be converted into water-soluble gums [11,12]. Cellulose can be applied in various form, its original fibers (used in textile and paper) or its derivative forms such as methyl cellulose and carboxymethyl cellulose.

carboxymethyl cellulose is a linear, long-chain, water-soluble, anionic polysaccharide. Purified CMC is a white- to cream-colored, tasteless, odorless, powder [13,14]. Sodium carboxymethyl cellulose is formed when cellulose reacts with mono chloroacetic acid or its sodium salt under alkaline condition with presence of organic solvent, hydroxyl groups substituted by Sodium carboxymethyl groups in C2, C3 and C6 of glucose, which substitution slightly prevails at C2 position [15,16]. Carboxymethyl cellulose (CMC) is extensively used as a food gum. It has many applications in various industries such as food, pharmaceutics, detergent, lubricants, adhesives etc. [17-25]. However, Due to extensive use of CMC, many studies have been done to produce CMC from various resources such as durian rind [4], cotton linters [16], sugar beet pulp [26], cashew tree gum [27], Cavendish banana pseudo stem [28], sago waste [3], papaya peel [29], and Mimosa pigra peel [30]. Therefore, the purpose of this work was the production of carboxymethyl cellulose from bagasse (CMC_b) and study the effect of NaOH concentration on Characteristics of CMC synthesized from sugarcane bagasse and evaluation of this film.

*Corresponding author: Mohammad Mousavi, Department of Food Science and Technology, Campus of Agricultural Engineering and Natural Resources, University of Tehran, P.O. Box 4111, Karaj 31587-77871, Iran
E-mail: Mousavi@ut.ac.ir

Materials and Methods

Materials

Required quantities of sugarcane bagasse was collected from the farms of Khuzestan, Iran. Sodium hydroxide, potassium hydroxide and glacial acetic acid were prepared from Merck Chemical Co. (Darmstadt, Germany). Ethanol and methanol were provided from the local market. Isopropanol and sodium mono chloroacetate were purchased from Daejung Co. (South Korea). Sodium chlorite ($NaClO_2$) used in the experiment was from Sigma-Aldrich, USA.

Extraction of cellulose

First of all, bagasse was milled and de-pitched, then the fiber was cooked at 370°C in digester prior to bleaching with Sodium hypochlorite and Chlorine gas and washing with KCL 5%. After this process bagasse fiber with 70-80 of Degree of whiteness (Euro standard) was obtained. Additional bleaching on the samples was carried out in two steps; with sodium chlorite 3% and then 1% (1:10) at Ph=3.8-4 [31]. Elimination of hemicellulose from the pulp was then done with 10 g/100 ml KOH (1:20) at 80°C for 2 h (after 12 h remaining in room temperature) after every step for elimination residual, pulp was washed with distilled water. Alpha-cellulose content was measured by following Equation 1 according to the TAPPI T 203 cm-99 standard.

$$\alpha - cellulose \ (\%) \ = \frac{w_2 - w_1}{w_1} \times 100 \tag{1}$$

Synthesis of carboxymethyl cellulose (CMC_b) from cellulose of sugarcane bagasse

The synthesis of Carboxymethyl cellulose followed the procedure described by Rachtanapun et al. [29]. Nine grams of Cellulose powder from sugarcane bagasse, 30 ml of NaOH (20 g/100 ml, 30 g/100 ml, 40 g/100 ml, 50 g/100 ml) and 270 ml of solvent (isopropanol, due to its good ability in cellulose etherification based on Pushpamalar et al. [3] study was stirred in the beaker and let stand for 30 min at ambient temperature. Then 10.8 g of sodium mono chloroacetate was added and mechanically stirred for 90 min in a beaker and covered with aluminum foil and keep in 55°C for 180 min. During this time, the reaction continued and the slurry divided into two phases. The upper phase's discarded and sedimentary phases suspend in 70% methanol (100 ml) and neutralized using glacial acetic acid and then, filtered and washed five times with 70% ethanol (300 ml) to remove undesirable salts. Afterward, it was washed again with absolute methanol and filtered. The obtained CMC_b was dried at 55°C in an oven. The yield of CMC_b was calculated by the following Equation 2:

$$Yield of \ CMC_b \left(\%\right) \ = \frac{Weight \ of \ CMC}{Weight \ of \ cellulose} \times 100 \tag{2}$$

Degree of substitution (DS)

DS value shows the average amount of hydroxyl group that was replaced by (sodium) Carboxymethyl group in the cellulose structure at C2, C3 and C6. The DS value of CMC from bagasse was measured by USP XXIII method described for Croscarmellose sodium, this method included 2 steps titration and residue on ignition.

Film preparation

Some quantities of powdered CMC from bagasse was dissolved in 100 ml distilled water at 80°C and stirred. Mixing was done until a homogeneous solution obtained. 1 ml, 2 ml and 3 ml of glycerol were added to 100 ml of solution as a plasticizer. After obtaining a homogeneous solution, a film with a diameter of 10 cm (60 ml) was formed on glass plates by casting, then dried at 55°C within 24 hours.

Thickness

The thickness of the films was adjusted by the volume of film solution that cast on the plate (specific volume of solution poured on a plate with a fixed diameter) and measured by micrometers (model Mitutoyo, LIC.NO.689037, Japan). Measurements were done at ten different points on each sample and the average value was used to calculate mechanical properties.

Color

Colorimetry test was used to determine the color of samples with Chromameter CR-400 (japan). Three parameters which can be reached that in colorimeter is L (white=100, black=0), a (green=-60, red=+60) and b (blue=-60, yellow=+60). For each sample, at least three replications were done.

Mechanical properties

Measuring some mechanical properties of films was performed using tensile tester (M350-10CT, Testometric Co., Ltd., Rochdale, and Lancashire, England). Film samples were cute in 1 × 10 cm rectangles to be used as a test specimen, the samples preconditioned [32] at 53% RH for 48 hours in a desiccator containing magnesium nitrate. The initial grip separation and cross-head speed were set at 20 mm and 1 mm/min, respectively. Tensile strength and strain at break for specimen obtained using the curves of stress-strain. All measurement was done in triplicate.

Water vapor permeability

Water vapor permeability of the CMC_b films was performed based on modified ASTM (1995-method E96). According to this test, at first the films were cut into a circle (with 10 mm diameter), then placed on the glass cells with a specific diameter, and sealing completely with paraffin wax (anhydrous calcium chloride was poured into a glass cell that provides zero percent relative humidity). The glasses weight and was kept in a desiccator preserved at 75% RH with saturated sodium chloride at 25°C. Due to the Moisture differences on both side of the film (vapor pressure gradient), water vapor transmitted through the film specimen and absorbed by desiccant, evaluations were determined by measuring the weight gain. Changes in weight of the cells were recorded and plotted as a function of time. The slope of each line calculated using linear regression ($r^2 > 0.99$). Water vapor transmission rate (WVTR) (g/sm^2) calculated by Equation 3. And finally, the film permeability to water vapor (WVP) was calculated using the following equation (Equation 4):

$$WVTR = \frac{slope \left(g / s\right)}{surface area \left(m^2\right)} \tag{3}$$

$$WVP = \left[\frac{WVTR}{S \left(R_2 - R_1\right)}\right] \times D \tag{4}$$

Where S is saturation vapor pressure (Pa) at the test temperature (25°C); R_1 relative humidity at the desiccator; R_2 relative humidity inside the cell and D film thickness (m).

Viscosity

The viscosity of CMC samples was measured using rotational viscometer (LV model, Brookfield with ULA spindle). Samples were prepared by dissolved of 1 gr CMC in 100 ml distilled water followed by vigorous mixing, then the solution was standing for a while to remove air bubbles. The viscosity of samples was measured by the ASTM D1439-94 at different temperature (30°C, 40°C, and 50°C). All measurements were performed in triplicate.

FTIR

The Functional groups of carboxymethyl cellulose were investigated using infrared spectroscopy spectrum (EQUINOX 55, BRUKER Germany). Pellets were made by CMC_b with KBr. Transmission levels were measured for wave numbers of 4000-400 cm^{-1}.

Statistical analysis

The collected data were analyzed by ANOVA and compared by Duncan's multiple range test (p ≤ 0.05) with SPSS 18.

Results

α-cellulose

The alpha cellulose content was about 81% ± 2%, which indicates sufficient purification of the sample from all impurities.

Degree of substitution (DS)

The Degree of Substitution (DS) is the average number of hydroxyl groups in the cellulose structure substituted by carboxymethyl or sodium carboxymethyl groups at the carbon 2, 3 and 6. Each anhydroglucose (β-glucopyranose) unit has three reactive (hydroxyl) groups so theoretically DS value can be in the range from zero (cellulose itself) to three (fully substituted cellulose). In general, the DS of carboxymethyl cellulose obtained by alkalization reaction of cellulose with sodium mono chloroacetate was in the range of 0.4-1.3. CMC is fully soluble at DS above 0.4 and hydro affinity of CMC increases with increasing DS, while this polymer is swellable but insoluble below 0.4 [33]. The DS of CMC obtained in this work was in the expected range of 0.45- 0.78, as figured in Figure 1A. As presented in this figure, the DS of CMC_b increased by adding NaOH concentration up to 30%, which shows the efficient concentration of alkali reagent in carboxymethylation procedure. Carboxymethylation of cellulose takes place by three simultaneously reaction as shown in equations 5-7:

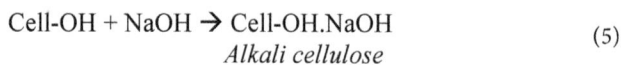

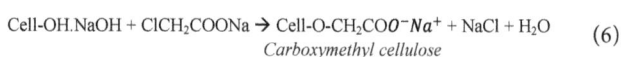

$$\text{Cell-OH} + \text{NaOH} \rightarrow \text{Cell-OH.NaOH} \tag{5}$$
$$\textit{Alkali cellulose}$$

$$\text{Cell-OH.NaOH} + \text{ClCH}_2\text{COONa} \rightarrow \text{Cell-O-CH}_2\text{COO}^-\textbf{\textit{Na}}^+ + \text{NaCl} + \text{H}_2\text{O} \tag{6}$$
$$\textit{Carboxymethyl cellulose}$$

$$\text{NaOH} + \text{Cl-CH}_2\text{COONa} \rightarrow \text{HO-CH}_2\text{COONa} + \text{NaCl} \tag{7}$$
$$\textit{Sodium glycolate}$$

According to Equation 5, Cellulose chains are swollen by Sodium hydroxide as an alkaline reagent (alkali cellulose), which provided the ability of substitution by sodium carboxymethyl groups in cellulose units. (Expose reactive site of anhydroglucose in compact cellulose chain to substitute by ether groups). The role of the solvent in the cellulose etherification is to provide miscibility and accessibility of the etherifying reagent (NaMCA) to the reaction centers of the cellulose chain rather than glycolate formation. Sodium mono chloroacetate (NaMCA) as an etherifying group participate in reaction and provide sodium carboxymethyl groups for substitution at C2, C3 and C6. As shown in Figure 1A, DS value augmented by adding NaOH concentration up to a maximum DS of 0.78 for 30% NaOH. At this level of NaOH concentration cellulose etherification (Equation 6) is

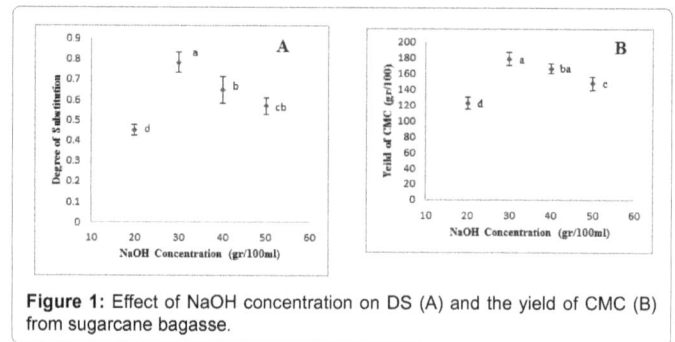

Figure 1: Effect of NaOH concentration on DS (A) and the yield of CMC (B) from sugarcane bagasse.

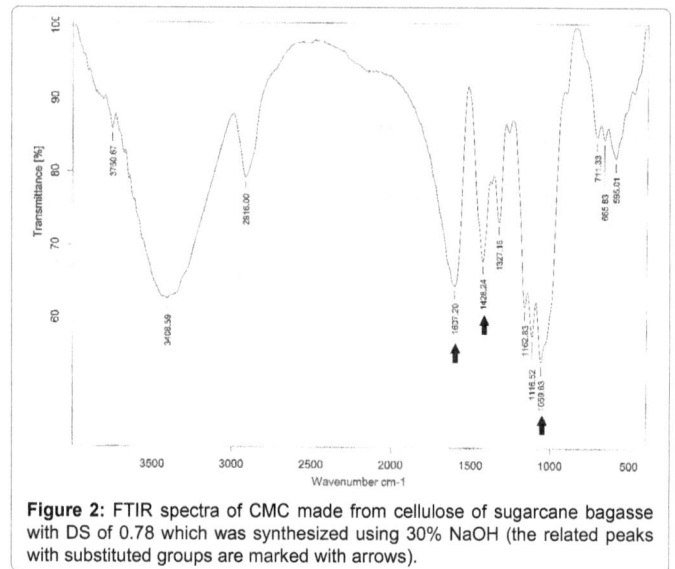

Figure 2: FTIR spectra of CMC made from cellulose of sugarcane bagasse with DS of 0.78 which was synthesized using 30% NaOH (the related peaks with substituted groups are marked with arrows).

predominating which produces CMC_b as a final product. Above 30% NaOH the DS value decreased. The reason for this observation is that an undesired side reaction happened which dominated the CMC_b production. Sodium glycolate was the product of such undesired reaction. With further increasing of NaOH concentration, more reduction in DS value was observed. It can be explained by degradation effect of high concentration of alkali reagent on CMC polymer chains. It must be noticed that DS value is affected by cellulose source. Crystallinity and regularity of cellulose structure according to its origin and considering the feasibility of substitution is more happened in an amorphous part of the cellulose structure [34,35]. Figure 1B shows the CMC_b yields from different experiments which indicated similarity with DS results. These results were in accordance with similar findings by several researchers.

FTIR

Fourier Transform Infrared Spectroscopy (FTIR) indicates chemical changes in the polymer structure. Cellulose and carboxymethyl have similar functional groups with same absorption bands in FTIR such as hydroxyl groups (-OH stretching) at 3200-3600 cm^{-1}, hydrocarbon groups (-CH$_2$ scissoring) at 1450 cm^{-1}, carbonyl groups (C=O stretching) at 1600 cm^{-1} and ether groups (-O-) at 1000-1200 cm^{-1}, also C-H stretching vibration at 3000 cm^{-1} [36]. Cellulose etherification with NaCMA causes the OH groups in cellulose replaced with CH$_2$COONa, which causes changes in the absorption spectrum of related bands. This lead OH groups to weaker peak and strengths or creates a new peak. However, from Figure 2, the differences can be observed in CMC_b

absorption bonds, at 1059, 1426 and 1607 absorption bands which are relevant to -O-, CH_2- and -COO, respectively. Pecsok [37] reported that the broadband at 1600-1640 cm^{-1} and 1400-1450 cm^{-1} is due to the carboxyl and its salts groups, which is confirmed substitution of carboxymethyl groups in cellulose structure. This peak doesn't exist in the FTIR spectra of cellulose from bagasse obtained in previous studies.

Viscosity

The viscosity of a solution is a measure of its resistance to gradual deformation by shear stress, which is due to intermolecular cohesive forces [37]. These forces are affected by some factors (CMC concentration, temperature, DS). Figure 3 shows the effect of NaOH concentration and temperature on viscosity of CMC_b solution. This figure indicates, as expected, the viscosity of samples decreases with increasing temperature. It can be explained by the fact that, during the heating to raising the temperature the energy of the molecules and molecular movement increased so intermolecular distances increase, which in turn, decreasing cohesive forces between molecules and causing lower viscosity. The viscosity of CMC_b samples increased with NaOH concentration (20% to 30%) at the constant temperature due to the presence of greater hydrophilic groups in polymer structure induced by greater DS, which gives more ability to the polymer to immobilize water in the aqueous system. As far as NaOH concentration raised above 30%, the viscosity fell down. This observation can be explained by decreasing of DS that provides less hydrophilic groups thereby ability of the polymer to bonding between water molecules reduced. Degradation effect of NaOH at a higher concentration on polymer chain also causes further declining in viscosity.

Water vapor permeability

Water Vapor Transmission Rate (WVTR) and Water Vapor Permeability (WVP) of the films obtained from CMC_b with different concentrations of sodium hydroxide is shown in Table 1. As tabulated,

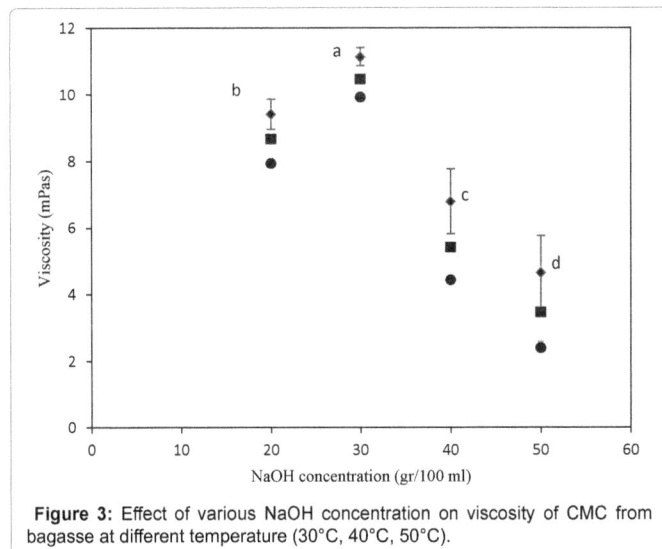

Figure 3: Effect of various NaOH concentration on viscosity of CMC from bagasse at different temperature (30°C, 40°C, 50°C).

Film type	WVP (g/msPa) × 10^{-10}	WVTR (g/s.m²) × 10^{-2}
CMC_b-20 gr/100 ml NaOH	6.9632 ± 0.472[c]	1.0477 ± 0.071[c]
CMC_b-30 gr/100 ml NaOH	11.0795 ± 0.217[a]	1.5025 ± 0.031[a]
CMC_b-40 gr/100 ml NaOH	9.4524 ± 0.121[b]	1.3510 ± 0.016[b]
CMC_b-50 gr/100 ml NaOH	9.5868 ± 0.094[b]	1.3270 ± 0.013[b]

Table 1: Water vapor transmission rate (WVTR) of CMC_b films synthesized with various NaOH concentrations at 25°C, 75% RH.

Films type	a*	b*	L*
CMCb-20% NaOH	-0.595 ± 0.12[a]	5.35 ± 1.87[b]	58.465 ± 10.98[a]
CMCb-30% NaOH	-0.685 ± 0.06[a]	4.195 ± 0.66[b]	59.755 ± 8.48[a]
CMCb-40% NaOH	-0.34 ± 0.03[a]	3.295 ± 0.62[cb]	42.795 ± 1.27[a]
CMCb-50% NaOH	-0.52 ± 0.04[a]	6.845 ± 0.33[ab]	57.24 ± 1.7[a]

Table 2: Color values of cellulose and CMC synthesized with various NaOH concentrations.

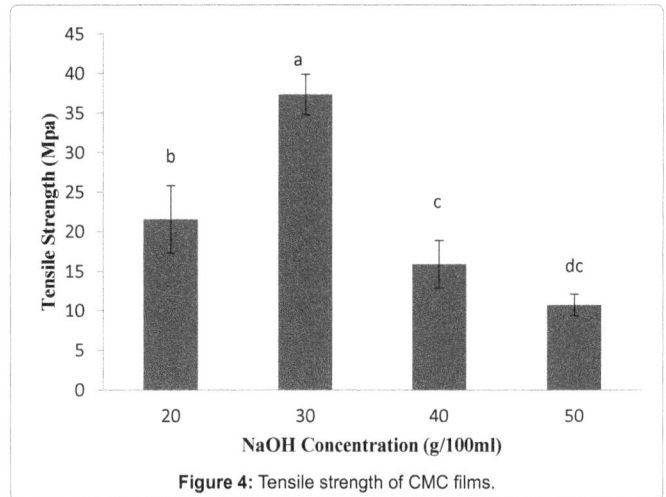

Figure 4: Tensile strength of CMC films.

when NaOH concentration increases, water vapor transmission rate significantly changes, and the highest WVP and WVTR values were observed for the films made with 30% NaOH. The Same trend was observed for DS of CMC_b meaning that higher DS values provided the polymer with greater hydrophilic groups which facilitate water molecules to easily pass through the films by dissolution–diffusion mechanism. It is obvious from different literatures that water vapor transmission rate through hydrophilic films depends on diffusivity and solubility of water molecules in the films Matrix [38,39]. In general, polymer polarities increased by conversion of cellulose to carboxymethyl cellulose also causes crystallinity reduction and changes in the granular morphology [40]. However, it can be observed from Table 2 that WVP of films with higher NaOH concentration (40%, 50%) reduced due to the reduction of hydrophilicity of films by declining polarity of the polymer.

Color

The result of color measurement of CMC_b films specimen is shown in Table 2. According to the values shown in this table, carboxymethyl cellulose from bagasse with different concentration of NaOH gives lighter, yellowish and slightly viridescent films. As can be seen, yellowness of samples decreased with increasing NaOH concentration up to 40%. It is probably due to the competitive reaction (Equations 2 and 3) to produce CMC or sodium glycolate. Beyond 40% NaOH yellowness increased which is most likely due to the intrinsic color of NaOH solution.

Mechanical properties

The results prove the effect of different concentrations of sodium hydroxide on mechanical properties of films. The tensile strength (TS) of film samples of carboxymethyl cellulose obtained under different conditions is given in the Figure 4. The TS of films increased with increasing NaOH concentration and maximum TS (37.34 Mpa) was observed in 30% NaOH, while TS decreased at higher concentration. With increasing the DS value due to the placement of more sodium carboxymethyl groups in cellulose structure, the polarity of the polymer chains increased which promoted more intermolecular bonds between the polymer chains. On the other hand, at higher concentration of NaOH due to the formation of sodium glycolate as a reaction byproduct during CMC production (decreasing in polymer chains ionic characteristic), TS decreased. TS reduction was intense with degradation effect of high concentration of NaOH on polymer chains. Decreasing in the CMC content at high NaOH concentration provides a reduction in intermolecular forces [41].

Discussion

The effect of various NaOH concentrations on the percent elongation at break (EB) of CMC_b films is depicted in Figure 5 As can be observed, the EB of CMC_b film increased with increasing NaOH concentration, beyond the 40% NaOH concentration the EB of films dropped. This increase can be attributed to the fact that at high concentration of NaOH polymer chains get more swelled that declined the crystallinity and increase the flexibility of cellulose structure. Under higher concentration of NaOH, due to hydrolysis reaction on polymer chains, the flexibility of CMC_b films decreased. The mechanical properties of CMC_b films after applying plasticizer were also studied. Glycerol, as an external plasticizer, was added to CMC_b film in various amounts to improve the mechanical properties before casting. This was applied only to CMC_b films with 30% NaOH (with maximum TS). The results (Figure 6) showed that adding glycerol concentration lowers the TS of CMC_b films. Glycerol lessens the internal hydrogen bonds between polymer chains which effect on film resistance under tension. On the other side, the EB of CMC_b films increased at higher glycerol content. Glycerol positions between polymer chains and prevent the formation of further hydrogen bonds and reduce the intermolecular forces, thereby the chains are able to move and flexibility of them increased. Similar results have been reported previously.

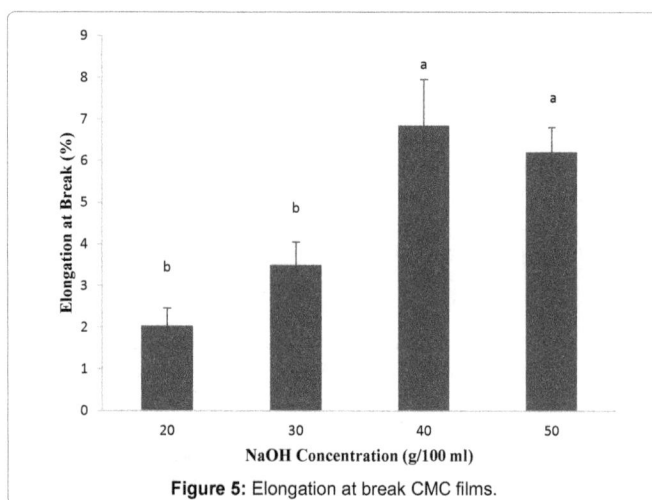

Figure 5: Elongation at break CMC films.

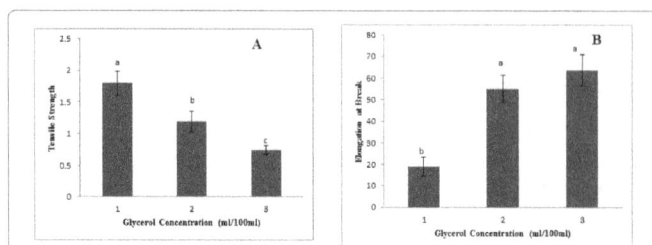

Figure 6: The TS (A) and The EB (B) of CMC_b-30% NaOH films with various amount of glycerol (1, 2 and 3%).

Conclusion

Sugarcane bagasse was properly used to extract cellulose and synthesis of carboxymethyl cellulose. The results showed maximum DS of CMC_b (0.78) at 30% of NaOH concentration. The viscosity, water vapor permeability, and mechanical properties are dependent to DS of CMC_b. Based on water vapor permeability and mechanical properties of CMC_b films, the possibility of its application on biodegradable coating materials is envisaged. Carboxymethylation of cellulose from bagasse was affected by various NaOH concentration, and crystallinity of cellulose sources is important for finding an appropriate concentration of NaOH. Carboxymethyl cellulose is one of most widely used cellulose derivatives in various industries like food, pharmaceutical, detergent, paper-making, oil drilling and Textile. Therefore, considering the methods that modified cellulose as cellulose derivatives, can help us to overcome environmental problems of agricultural waste (bagasse), which produced in large scale, and economical ways to use sugarcane bagasse.

Acknowledgements

The authors gratefully acknowledge the financial support from the Center of excellence for the application of novel technologies in functional foods and beverages of the University of Tehran.

References

1. Mandal A, Chakrabarty D (2011) Isolation of nanocellulose from waste sugarcane bagasse (SCB) and its characterization. J Carbpol 86: 1291-1299.

2. Shaikh HM, Pandare KV, Nair G, Varma AJ (2009) Utilization of sugarcane bagasse cellulose for producing cellulose acetates: Novel use of residual hemicellulose as plasticizer. J Carbpol 76: 23-29.

3. Pushpamalar V, Langford S, Ahmad M, Lim YY (2006) Optimization of reaction conditions for preparing carboxymethyl cellulose from sago waste. J Carbpol 64: 312-318.

4. Rachtanapun P, Luangkamin S, Tanprasert K, Suriyatem R (2012) Carboxymethyl cellulose film from durian rind. Food Sci Technol 48: 52-58.

5. Viera RG, Rodrigues Filho G, De Assunção RM, Meireles CdS, Vieira JG, et al. (2007) Synthesis and characterization of methylcellulose from sugar cane bagasse cellulose. J Carbpol 67: 182-189.

6. Chandel AK, Da Silva SS, Carvalho W, Singh OV (2012) Sugarcane bagasse and leaves: Foreseeable biomass of biofuel and bio□products. J Chem Technol Biotechnol 87: 11-20.

7. Sun J, Sun X, Zhao H, Sun R (2004) Isolation and characterization of cellulose from sugarcane bagasse. Polymer Degrad Stab 84: 331-339.

8. Loh Y, Sujan D, Rahman M, Das C (2013) Sugarcane bagasse: The future composite material: A literature review. Res Conserv Recycle 75: 14-22.

9. Sun J, Sun X, Sun R, Su Y (2004) Fractional extraction and structural characterization of sugarcane bagasse hemicelluloses. J Carbpol 56: 195-204.

10. Thomas G, Paquita E, Thomas J (2002) Cellulose ethers: Encyclopedia of polymer science and technology. Wiley online, New York, USA.

11. Damodaran S, Parkin KL, Fennema OR (2007) Fennema's food chemistry. CRC press, USA.

12. Yaşar F, Toğrul H, Arslan N (2007) Flow properties of cellulose and carboxymethyl cellulose from orange peel. J Food Eng 81: 187-199.

13. Hattori K, Abe E, Yoshida T, Cuculo JA (2004) New solvents for cellulose: II Ethylenediamine/thiocyanate salt system. Polymer J 36: 123-130.

14. Keller J (1984) Sodium carboxymethylcellulose (CMC). New York state agricultural experiment station special report, USA 53: 9-19.

15. Heinze T, Pfeiffer K (1999) Studies on the synthesis and characterization of carboxymethylcellulose. Die Angewandte Makromolekulare Chemie 266: 37-45.

16. Xiquan L, Tingzhu Q, Shaoqui Q (1990) Kinetics of the carboxymethylation of cellulose in the isopropyl alcohol system. Acta Polymerica 41: 220-222.

17. Boursier B, Bussiere G, Devos F, Hughette M (1985) Sugarless hard candy. Google Patents.

18. Charpentier D, Mocanu G, Carpov A, Chapelle S, Merle L, et al. (1997) New hydrophobically modified carboxymethylcellulose derivatives. J Carbpol 33: 177-186.

19. Gayrish G, Saychenko N, Kozlova Y, Melanichenko I, Liptuga N (1989) Carboxymethylcellulose urea resin blend adhesives. SU Patent 87: 4293505.

20. Koyama T (1988) Permanent waves of hair with carboxymethyl cellulose salts. Google Patents.

21. Lee M, Torras MF (1993) Pesticidal aqueous cellulose ether solutions. WO Patent: 9313657.

22. Leupin J, Gosselink E (1999) Laundry detergent compositions with cellulosic polymers as additives for improving appearance and integrity of laundered fabrics. WO Patent: 9914295.

23. Rachtanapun P, Thanakkasaranee S, Soonthornampai S (2008) Application of carboxymethylcellulose from papaya peel for mango (*Mangifera Indica* L.) 'Namdokmai'coating. Agricultural Sci J 39: 74-82.

24. Sánchez E, Sanz V, Bou E, Monfort E (1999) Carboxymethlycellulose used in ceramic glazes (Part III): Influence of CMC characteristics on glaze slip and consolidated glaze layer properties. CFI: Ceramic forum international.

25. Soper J (1991) Oily, free-flowing, microcapsules, comprising grafted and cross-linked gelatin: CMC wall material. US Patent: 89,401,189.

26. Toğrul H, Arslan N (2003) Production of carboxymethyl cellulose from sugar beet pulp cellulose and rheological behaviour of carboxymethyl cellulose. J Carbpol 54: 73-82.

27. Silva DA, De Paula RC, Feitosa JP, De Brito AC, Maciel JS, et al. (2004) Carboxymethylation of cashew tree exudate polysaccharide. J Carbpol 58: 163-171.

28. Adinugraha MP, Marseno DW (2005) Synthesis and characterization of sodium carboxymethylcellulose from cavendish banana pseudo stem (*Musa cavendishii* LAMBERT). J Carbpol 62: 164-169.

29. Rachtanapun P, Kumthai S, Yakee N, Uthaiyod R (2007) Production of carboxymethylcellulose (CMC) film from papaya peels and its mechanical properties. The Proceedings of the 45th Kasetsart University Annual Conference, Kasetsart University, Kasetsart, Thailand.

30. Rachtanapun P, Rattanapanone N (2011) Synthesis and characterization of carboxymethyl cellulose powder and films from Mimosa pigra. J Appl Polymer Sci 122: 3218-3226.

31. Mali S, Grossmann MVE, García MA, Martino MN, Zaritzky NE (2004) Barrier, mechanical and optical properties of plasticized yam starch films. J Carbpol 56: 129-135.

32. Yousefi H, Nishino T, Faezipour M, Ebrahimi G, Shakeri A (2011) Direct fabrication of all-cellulose nanocomposite from cellulose microfibers using ionic liquid-based nanowelding. Biomacromolecule 12: 4080-4085.

33. Waring M, Parsons D (2001) Physico-chemical characterization of carboxymethylated spun cellulose fibres. Biomaterial 22: 903-912.

34. Brown Jr RM (1992) Emerging technologies and future prospects for industrialization of microbially derived cellulose. Harnessing Biotechnology for the 21st Century.

35. Manguiat LS, Sabularse V, Sabularse DC (2001) Development of carboxymethylcellulose from nata de coco. Asian J Sci Technol Develop 18: 85-94.

36. Kondo T (1997) The assignment of IR absorption bands due to free hydroxyl groups in cellulose. Cellulose 4: 281-292.

37. Pecsok RL (1976) Modern methods of chemical analysis, John Wiley & Sons, USA.

38. El-Ghzaoui A, Trompette JL, Cassanas G, Bardet L, Fabregue E (2001) Comparative rheological behavior of some cellulosic ether derivatives. Langmuir 17: 1453-1456.

39. Gontard N, Guilbert S (1993) Biopackaging: Technology and properties of edible and/or biodegradable material of agricultural origin. Food Packag Preserv, Westport Connecticut, USA.

40. Li Y, Shoemaker CF, Ma J, Shen X, Zhong F (2008) Paste viscosity of rice starches of different amylose content and carboxymethylcellulose formed by dry heating and the physical properties of their films. Food Chem 109: 616-623.

41. Barai B, Singhal R, Kulkarni P (1997) Optimization of a process for preparing carboxymethyl cellulose from water hyacinth (*Eichornia crassipes*). J Carbpol 32: 229-231.

29

Post-harvest Shelf Life Extension and Nutritional Profile of Thompson Seedless Table Grapes Under Calcium Chloride and Modified Atmospheric Storage

Imlak M*, Randhawa MA, Hassan A, Ahmad N and Nadeem M

Food Safety Laboratory, National Institute of Food Science and Technology, University of Agriculture, Faisalabad, Pakistan

Abstract

Grapes are the fruits with highest rate of phenolic substances among all other fruits and vegetables but due to its enhanced susceptibility to *Botrytis cinerea*, its postharvest losses are also very high. In a country like Pakistan that has lesser advancement in post-harvest technological skills, a significant segment of this delicate fruit is lost. This study was conducted with an aim to minimize these losses by retaining healthy and attractive nutritional profile. Freshly harvested grapes after being sorted and graded were divided in three lots, two lots were immersed in aqueous solutions of 1% and 2% calcium chloride followed by modified atmosphere storage at 5% CO_2 level, and third lot was control sample that was simply dipped in tap water at ambient conditions, 80% relative humidity and $10 \pm 1°C$ temperature was kept same for all three lots. After being analyzed at harvest, stored grapes were then analyzed for total polyphenols, firmness, acidity, total sugars, total soluble solids and total viable count at 4th, 8th and 12th days of storage respectively. Overall results designated that grapes pretreated with 2% $CaCl_2$ stored at 5% CO_2 level retained maximum firmness, acidity and phenolic substances with minimum increase in soluble solid contents and significantly reduced incidence of browning caused by gray mold as compared to water washed control sample that was spoiled at the 8th day of storage. Grapes were stored for 12 days and the effect of storage days on differently pretreated samples was analyzed.

Keywords: $CaCl_2$; Modified atmosphere; Total phenolic compounds (TPC); Total viable count (TVC); Polypropylene perforated storage bags

Introduction

Grapes (*Vitis vinifera*) belong to the *Vitaceae* family and are amongst the largest cultivated global fruit crop due to their excessive use in winemaking and as table fruit [1] and are considered as highly perishable and delightful fruit with strong nutritional profile. Healthy berries contain fiber and folic acid that are helpful in lowering the bodyweight, blood cholesterol and chances of severe hypertension [2]. Berries contain polyphenols that are primarily confined to skin and involve anthocyanin and resveratrol that have documented biological activities as antiallergenic substances and minimizing the risks of atherosclerosis [3]. Food and Agriculture Organization of the United Nations reported that 75,866 km² of the world directly sanctified for grapes and practically one third of the global foodstuff manufactured for consumption is lost after being reaped [4]. In Pakistan, total grape production is 122,000 tons on an area of about 14,000 ha in 2008-09 that was recorded as the highest production of table grapes in the last nine successive years [5]. About 70% of the production of grapes in Pakistan is from Baluchistan Province with possible yield of almost 19 tons/ha only against the productive capacity of nearly 25 tons/ha, indicating poor infrastructure and postharvest technological skills [6].

Grapes experience many ripening changes during storage including biochemical and physical modifications in sugars, pH, acidity, total soluble solids, total polyphenols, and contents of vitamin C including minerals and different sensorial attributes. The ratio between sugars/acids is regarded to be the prime factor affecting taste and end quality of grapes and this ratio is highly exaggerated during prolonged storage [7]. Postharvest storage of fruits is highly susceptible to fungal infections such as *Botrytis, Aspergillus, Alternaria, Rhizopus* etc. that adversely affect the quality attributes of fruits [8]. Grey mold and stem browning caused by *Botrytis cinerea, is* a major biological hazard to horticultural produce as it can grow even below at -0.5°C and has the tendency to reproduce rapidly on berries skin [9]. To minimize the postharvest losses of grapes, several technologies have been introduced. The typical exercise is fumigation of the berries after being harvested with conventional sulfur dioxide gas (SO_2) but severe health issues of its residues specifically to the people that are allergic to sulfites have been reported [9]. Controlled and modified atmosphere strategies [10], use of various bio-control agents [11], different inorganic salt patterns [12], several nonchemical compounds [13] and natural antimicrobials [14] have been used as substitutes of SO_2.

Applications of calcium salts are a well experienced example to replace conservative fungicide sprays in controlling the postharvest antimicrobial decay of fruiting berries [15]. Calcium salt has been shown to play an eccentric part in upholding the cell wall of the fruit cells for its role as preservative and firming agent. Postharvest application of different salts pretreatments including calcium chloride [16], sodium carbonate and bicarbonates [17], calcium lactate [18] and various others are the promising salt treatments for maintaining the quality characteristics of fruits.

The positive impacts of calcium chloride have formerly been testified on wide ranges of horticultural produce including various fruits such as strawberries, grapes, mangoes, apples and different vegetables [19]. Akhtar et al. [20] illustrated that $CaCl_2$ treated loquats exhibited higher degree of firmness as compared to water washed control loquats. Yousefi et al. [16] reported that nutritional profile of $CaCl_2$ treated apricots such as Ca, Mg, N, and vitamin C were significantly higher than non-treated apricot fruits at termination of storage duration. Extended duration of fruits storage in modified atmosphere where

*Corresponding author: Imlak M, Food Safety Laboratory, National Institute of Food Science and Technology, University of Agriculture, Faisalabad, Pakistan E-mail: imlak.khalid@yahoo.com

increased CO_2 level was used proved useful in postponing the softening of skin of many fruits [21]. Al-Quarshi and Awad [22] reported that table grape vines sprayed with 1% $CaCl_2$ and 10% ethanol solution effectively reduced the incidences of decay. Furthermore, -Ca salt didn't show any negative influence on the overall worth of the fruit. Microbial infestation caused by Botrytis is a major cause of spoilage in fruits specifically grapes due to its soft texture and high perishability. Chervin et al. [23] reported reduction in grey mold incidence when berries were treated with ethanol and 1% calcium chloride (CC) during modified atmosphere storage. This study was conducted with the aim to prolong the shelf life of white seedless table grapes to cause an increase in export and processing quality of the fruit.

Materials and Methods

Procurement of raw material

Freshly harvested Thompson seedless table grapes (*Vitis vinifera*) were procured from botanical garden, University of Agriculture, Faisalabad, Pakistan. The berries were then brought to Food Safety Laboratory of National Institute of Food Science and Technology, UAF for further processing and storage.

Post-harvest treatments

After sorting and grading, healthy clusters were divided in three equal lots. 1st lot was simply immersed in tap water and named as control sample (T_0), 2nd lot was dipped in 1% $CaCl_2$ (T_1) and the 3rd lot in 2% $CaCl_2$ solution (T_2).

This dipping pretreatment to all samples at ambient conditions was given for 5 minutes followed by packing in polypropylene perforated storage bags and stored at modified atmospheric chambers (ICH260 Memmert, Germany). In all three storage chambers, temperature and relative humidity were attuned at $10 \pm 1°C$ and 80%, however the level of CO_2 for T_1 and T_2 inside the chambers was maintained at 5% and 0% for T_0 as shown in treatment plan (Table 1).

Total phenolic contents

Preparation of sample: Weighed amount (200 g) of samples were taken in glass bottles and the bottles were filled with methanol until a layer was formed above the sample. These samples were continuously shaken for 48 h with 3 h intervals at ambient temperature. After this, samples were filtered with filter paper and the extract obtained was concentrated to rotary evaporation for the removal of solvent from samples under vacuum. The distillation was stopped when the volume of the extract remains 1 mL. The solvent was further removed under purified gentle stream of N_2 gas. The sample was stored in freezer at -4°C till further analysis.

Determination of total polyphenols: The total phenolic compounds were estimated by Folin-Ciocalteu method by using an ultraviolet visible spectrophotometer (CECIL CE7200, UK) [24]. Extract mixture of 0.1 mL and 0.75 mL of Folin-Ciocalteu reagent that was previously diluted with deionized water was prepared in a test tube that was then allowed to stand at ambient conditions for 5 min. 0.75 mL of sodium carbonate 6% w/v was later added in the mixture. The absorbance of the samples was recorded from UV-Vis spectrophotometer at 275 nm. Gallic acid

was run as a standard along with the samples and its standard curve was used for the calculation of the total phenolic contents in the samples.

Total sugars and TSS

The total sugar contents (glucose, fructose and sucrose) were determined in terms of g/100 g by following the procedure (Method No. 967.21) as described in AOAC [25] and total soluble solids (°Brix) were measured using digital hand Refractometer (Carl Zeiss Jena-Germany) after being calibrated against sucrose.

Titratable acidity (%)

TA of the fruiting berries was determined by titrating diluted sample against 0.1 N NaOH with phenolphthalein indicator according to method described by AOAC [25].

Firmness (N/mm)

Firmness of the berries was measured by using texture measuring system fitted with needle probe. Berries were randomly selected from each treatment and placed at the base of the texture analyzer (Mod. TA-XT2, Surrey, UK). The force (N) required to penetrate the fruit surface up to a specific depth (mm) was recorded and expressed in terms of N/mm [26].

Total viable count (log CFU/g)

Fruits of known weight from treatments were washed in sterile distilled water using a shaker, with fruit to distilled water ratio as 1:9. The wash water was then further diluted using peptone water, up to 105 and plated in triplicate on potato dextrose agar prepared as per manufacturer's instructions and incubated at 25°C for 24 hours to count the microbial colonies [27].

Statistical analysis

The data was scrutinized statistically to determine the level of significance [28]. Experiment was a completely randomized design with factorial arrangement. Difference and comparison between the means were evaluated by LSD multiple range test at 5% level of significance.

Results and Discussion

The effects of calcium chloride and modified atmosphere storage showed a significant effect in maintaining the nutritional profile and physicochemical attributes of the berries as depicted (Table 2).

Total phenolic compounds (TPC)

Results: Present study investigations revealed that polyphenols of the berries were continuously decreasing throughout the storage irrespective of the given treatments and storage conditions; however the rate of this fall was highly dependent upon the given treatments (Figure 1). Maximum loss in TPC was observed in control sample. At harvest, TPC were recorded as 159.14 ± 6 mg GAE/100 g that reached to 116.2 and 94.6 at 4th and 8th day of storage, moreover significant lost was recorded at 12th day when the recorded value for trait reached 58.1 in T_0. Likewise, in T_1, a gradual decrease in phenolic contents was recorded that reached 99.2 at storage termination. Minimum fall in polyphenol count was recorded in T_2 under which grapes were treated with 2% $CaCl_2$ and stored at 5% CO_2 where value from harvest of 159.14 reached 124.2 at 12th day of storage indicating best and statistically significant results as compared to T_0 and T_1. This concludes that treated grapes with 2% $CaCl_2$ could adopt a sturdier defense mechanism which T_0 failed to adopt.

Treatments	$CaCl_2$ conc. (%)	CO_2 level (%)
T_0	0	0
T_1	1	5
T_2	2	5

Table 1: Treatment plan.

Storage Days	Treatments	TSS (^{0}Brix)	Total Sugars (g/100 g)	Titratable Acidity TA (%)	Firmness (N/mm)	Polyphenols (mg of GAE 100g^{-1})	TVC (log CFU g^1)
0	--	17.8 ± 0.9	14.12 ± 0.2	0.89 ± 0.03	2.74 ± 0.2	159.14 ± 6.0	2.73 ± 0.06
4	T_0	22.5 ± 1.3	16.56 ± 0.9	0.61 ± 0.08	2.12 ± 0.3	116.2 ± 5.1	3.2 ± 0.1
	T_1	19.0 ± 0.8	14.40 ± 0.2	0.83 ± 0.04	2.69 ± 0.2	138.3 ± 5.6	2.4 ± 0.1
	T_2	17.9 ± 1.0	14.21 ± 0.2	0.88 ± 0.07	2.73 ± 0.1	154.3 ± 9.1	2.3 ± 0.08
8	T_0	26.9 ± 2.0	19.10 ± 0.8	0.52 ± 0.04	1.5 ± 0.13	94.6 ± 4.8	3.9 ± 0.09
	T_1	19.9 ± 1.0	15.41 ± 0.3	0.80 ± 0.03	2.45 ± 0.1	119.7 ± 5.9	2.5 ± 0.07
	T_2	18.1 ± 1.0	14.50 ± 0.3	0.86 ± 0.05	2.7 ± 0.2	142.1 ± 6	2.4 ± 0.05
12	T_0	29.5 ± 2.9	20.40 ± 1.6	0.40 ± 0.04	0.7 ± 0.08	58.1 ± 4.9	4.95 ± 0.1
	T_1	23.6 ± 2.1	17.44 ± 0.2	0.68 ± 0.03	2.0 ± 0.09	99.2 ± 6.5	2.79 ± 0.2
	T_2	18.7 ± 0.5	14.59 ± 0.4	0.82 ± 0.08	2.66 ± 0.1	124.2 ± 5.5	2.6 ± 0.06

T_0= Water Washed (0% CO_2), T_1= 1% $CaCl_2$ (5% CO_2), T_2= 2% $CaCl_2$ (5% CO_2), Temperature (10 ± 1°C) and Relative Humidity (85%) was kept same in all. Given data are the mean values with 3 replications ± SD.

Table 2: Post-harvest quality attributes of Thompson seedless table grapes under different concentrations of $CaCl_2$ and modified atmosphere storage.

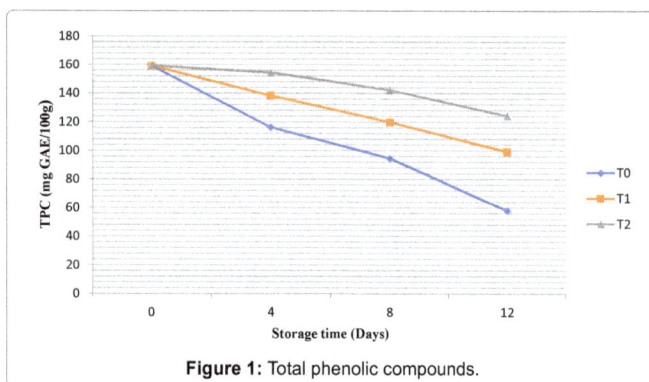

Figure 1: Total phenolic compounds.

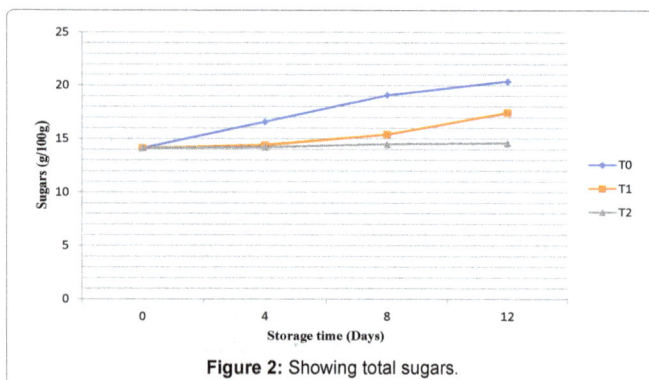

Figure 2: Showing total sugars.

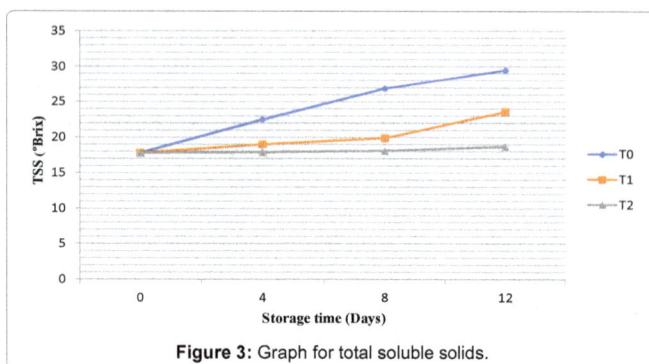

Figure 3: Graph for total soluble solids.

Discussion: Polyphenols are phytochemicals by nature and are present in highest amount in grapes than any other fruit; however, wine grapes have increased level of flavonoids, anthocyanin and phenolic contents as compared to table grapes [29]. This polyphenol composition also varies with berries skin color, cultivar, species, environmental and postharvest management skills [2]. Results of study in hand closely resemble with the investigations of Al-Quarshi and Awad [22] who reported that total polyphenols were significantly higher in $CaCl_2$ treated grapes as compared to control grapes and in promise with the fallouts of Pinheiro et al. [30] who explored total phenolic contents and anthocyanin in grape cultivars and illustrated that TPC of grapes were significantly decreasing with storage days irrespective of the given treatments.

Total sugars and total soluble solids (TSS)

Results: A gradual increase in total sugars composition and TSS was observed with progression in storage days. Amongst treatments, a similar behavior was shown by all treatments indicating a significant effect ($P<0.05$) of storage days on total sugars (Figure 2) and TSS (Figure 3), however this escalation was magnificently higher in T_0 as compared to $CaCl_2$ treated berries (T_1 and T_2).

Freshly procured grapes showed mean value of 14.1 g/100 g of total sugars at harvest that increased with storage period varying upon storage conditions and given treatments. Maximum increase in total sugars was observed in T_0 with mean values of 16.5, 19.1 and 20.4 at 4th, 8th and 12th day of storage respectively. Likewise, T_1 showed sugar contents of 14.1 at harvest to 17.44 at termination. Minimum nonconformity in total sugars was recorded in T_2 that had sugar contents of 14.59 at termination day indicating best positive results gained by 2% $CaCl_2$ given pretreatment followed by modified atmosphere storage at 5% CO_2 level.

Similarly, total soluble solids determined at harvest in fresh berries were 17.8 ± 0.9 °Brix with no noticeable changes in grapes stored at modified atmospheric conditions with 5% CO_2 level after being pretreated with $CaCl_2$ (1% and 2%). At the end of storage duration, TSS in $CaCl_2$ treated fruits were significantly lower and closer to the value at harvest. Contrary, a highly significant elevation in TSS level was found in T_0 where TSS were recorded as 22.5°, 26.9° and 29.5° at 4th, 8th and 12th days of storage respectively showing a greater degree of deviation in soluble solid contents. Likewise, T_1 also showed escalation in soluble solids reaching 19.0°, 19.9° and 23.6° during storage intervals. Minimum increase in total soluble solids was noticed in T_2 that stretched 17.8° to 18.7° from initiation to termination.

Discussion: A large percentage of the soluble solids in grapes are sugars, mainly glucose and fructose that are central sugars and are involved in cell respiration and synthesis and the third sugar is sucrose

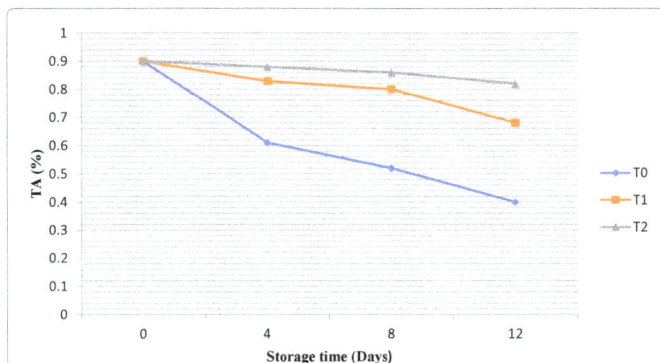

Figure 4: Showing graph for titratable acidity.

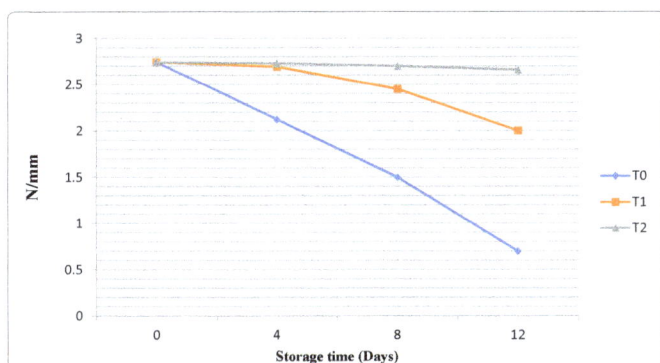

Figure 5: Showing firmness.

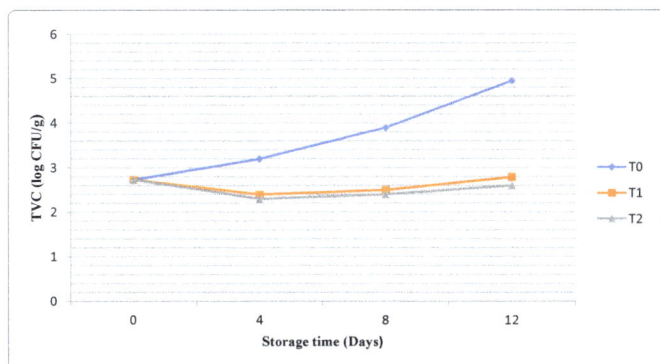

Figure 6: Total viable count.

that is non-reducing by nature and present relatively in smaller amounts with level not exceeding more than 1%. Prolonged storage resulted in greater deviation in sugar composition with reference to harvested value. This increment in soluble solids and sugars is attributed towards rapid conversion of complex starch molecules into simpler sugars [31]. Excess loss of water from the fruiting tissues may also be a valid reason behind this increment [18]. TSS of the fruits also increased during storage mainly due to glycogenesis and metabolism of fruiting tissues that becomes partially inactive due to changes in glucose and fructose. Findings of Samra [32] were in accordance with present study that SSC of fruiting berries were gradually increasing throughout the entire storage period irrespective of the given conditions and treatments.

Titratable acidity (TA)

Consumer preference of grapes relies on the peculiar taste that arises mainly from the organic acids present in the berries responsible for its titratable acidity (TA). The major organic acid present in table

grapes is tartaric acid that declined in all samples with progression in storage days.

Results: Freshly procured grapes at harvest showed a TA value of 0.89% that decreased with prolonged storage varying upon storage conditions and given treatments (Figure 4). Maximum loss of organic acids was recorded in T_0 that developed off flavors at the end of the storage showing TA values of 0.61, 0.52 and 0.40 (%) at 4th, 8th and 12th days of storage respectively. Relatively lower decrease in organic acids concentration was depicted by T_1 with TA of 0.89% at harvest to 0.68% at the end of storage period. Minimum loss in organic acids concentration from the given treatments specifically tartaric acid was observed in T_2 that depicted best results from TA of 0.89% at harvest to 0.82% at termination.

Discussion: Sugars/acids ratio is considered to be the main feature affecting taste of the berries and this ratio is highly affected during stretched storage periods [7]. In present investigation, TA of the berries was continuously decreasing throughout storage and this gradual decrease in acid level was physiologically attributed towards increase in membrane permeability of the fruit allowing the acids to respire in cell vacuoles resulting in transformation of acids into sugars [16]. These outcomes bear a resemblance with Sabir et al. [33] who exposed that TA levels in berries under modified atmosphere, apparently decreased. Randhawa et al. [34] also reported similar results with decreased level of acidity in citrus (non-climacteric) as storage duration was prolonged.

Firmness

Results: 2.74 ± 0.2 N/mm was the force required to the probe of texture analyzer to puncture the skin of the berries at harvest that gradually dropped in all treatments as storage period prolonged. However, $CaCl_2$ treated grapes stored in modified atmospheric chambers (T_1 and T_2) showed higher degree of firmness as compared to control sample (T_0) as shown (Figure 5). Higher degree of firmness was recorded in T_2 with mean values of 2.73, 2.70 and 2.66 followed by T_1 with 2.69, 2.45 and 2.0 (N/mm) at 4th, 8th and 12th days of storage respectively. Maximum deterioration and minimal degree of firmness was depicted by T_0 where force required to the probe of texture analyzer at 12th day of storage was only 0.7 N/mm indicating maximum quality degradation.

Discussion: Findings of present study are absolutely in accordance with that of Akhtar et al. [20] who treated loquats (non-climacteric) and Yousefi et al. [16] who treated apricots (climacteric) with different concentrations of $CaCl_2$ and reported that significantly higher degree of firmness was recorded in $CaCl_2$ treated fruits than untreated ones at the end of storage phase. Cell wall degrading enzymes (β-Glactocidase, Poly-Glacturonase and Pectinemethyl Esterase) may also be a reason for softening of fruit texture as mentioned by Pinzón-Gómez et al. [35].

Total viable count (TVC)

Results: TVC recorded at harvest was 2.73 ± 0.06 log CFU/g that significantly increased in T_0 while reduced microbial load was observed in $CaCl_2$ treated grapes stored at modified atmospheric chamber (T_1 and T_2) when storage phase terminated, as shown (Figure 6). Control grapes (T_0) had maximum yeast and mold count of 4.95 (log CFU/g) at 12th day of storage indicating severe mold growth caused by *Botrytis cinerea* turning the berries brown in color. No signs of browning were recorded in T_1 and T_2, however, higher the concentration of $CaCl_2$, lower the mold growth was observed. T_1 (1% $CaCl_2$) had microbial load of 2.79 at 12th day of storage that increased non-significantly from harvested value of 2.73. Likewise, in T_2 (2% $CaCl_2$), microbial load of

2.6 was recorded at termination that actually decreased from the values at harvest.

Discussion: Among all pathogens of different horticultural produce, *Botrytis cinerea* responsible for gray mold is amongst the list of world's top 10 most threatening fungal pathogenic microorganisms [36].

Increased CO_2 level has been proven to have antimicrobial and fungistatic effects that was the major reason behind lower microbial load in treated berries. Results of this study are in resemblance with the investigations of Romanazzi et al. [4] who reported that mere increased level of CO_2 limited the decay incidence but statistically highly significant results were observed when modified atmospheric conditions were provided. Findings of Nigro et al. [17] also endorses the current study status of minimal microbial load in grapes under calcium chloride dipping prior to storage.

Conclusion

Overall consequences of this study designate that modified atmospheric storage conditions (5% CO_2 level, 80% relative humidity and $10 \pm 1°C$ temperatures) in combination with 2% $CaCl_2$ pre-storage dipping, successfully escalated the shelf life of grapes up to 12 days' storage by preserving attractive nutritional and safety profile. Inexpensive $CaCl_2$ salt that is readily available and acceptable due to its non-toxic nature as compared to conventional fungicide sprays, efficiently minimized the incidence of pathogens on berries by retaining maximum firmness and freshness and modified atmosphere storage successfully delayed the postharvest ripening with mean values of quality attributes much closer to the ones at harvest as compared to tap water washed control sample.

Acknowledgement

The authors thank supervisors, National Institute of Food Science and Technology and Director General, NIFSAT for their help and encouragement during the course of work and preparation of this manuscript.

References

1. Fraige K, Pereira-Filho ER, Carrilh E (2014) Fingerprinting of anthocyanin from grapes produced in Brazil using HPLC-DAD-MS and exploratory analysis by principal component analysis. Food Chem 145: 395-403.

2. Yang J, Martinson TE, Liu RH (2009) Phytochemical profiles and antioxidant activities of wine grapes. Food Chem 116: 332-339.

3. Deedwania P, Singh V, Davidson MH (2009) Low and high-density lipoprotein cholesterol and increased cardiovascular disease risk: An analysis of statin clinical trials. Ameri J Cardiol 104: 3-9.

4. Romanazzi G, Smilanick JL, Feliziani E, Droby S (2016) Integrated management of post-harvest gray mold on fruit crops. Posthar Biol Technol 113: 69-76.

5. GOP (2010) Agriculture statistics of Pakistan. Ministry of Food, Agriculture and Livestock, Islamabad, Pakistan.

6. Khan AS, Ahmad N, Malik AU, Saleem BA, Rajwana IA (2011) Phenophysiological revelation of grapes germplasm grown in Faisalabad, Pakistan. Int J Agricul Biol 13: 391-395.

7. Sen F, Oksar R, Kesgin M (2016) Effects of shading and covering on 'Sultana Seedless' grape quality and storability. J Agricul Sci Technol 18: 245-254.

8. Pedreschi R, Lurie S, Hertog M, Nicolai B, Mes J, et al. (2013) Post-harvest proteomics and food security. Proteomics 13: 1772-1783.

9. Liu Q, Xi Z, Gao J, Meng Y, Lin S, et al. (2016) Effects of exogenous 24-epibrassinolide to control grey mold and maintain post-harvest quality of table grapes. Int J Food Sci Technol 51: 1236-1243.

10. Lichter A (2016) Rachis browning in table grapes. Austral J Grape Wine Resear.

11. Wang Y, Xu Z, Zhu P, Liu Y, Zhang Z, et al. (2010) Post-harvest biological control of melon pathogens using Bacillus subtilis EXWB1. J Plant Pathol.

12. Youssef K, Roberto SR (2014) Salt strategies to control Botrytis mold of 'Benitaka' table grapes and to maintain fruit quality during storage. Posthar Biol Technol 95: 95-102.

13. Ippolito A, Sanzani SM (2011) Control of post-harvest decay by the integration of pre and post-harvest application of nonchemical compounds. Acta Horticul 905: 135-143.

14. Aloui H, Khwaldia K, Licciardello F, Mazzaglia A, Muratore G, et al. (2014) Efficacy of the combined application of chitosan and Locust Bean Gum with different citrus essential oils to control post-harvest spoilage caused by Aspergillus flavus in dates. Int J Food Microbiol 170: 21-28.

15. Romanazzi G, Lichter A, Mlikota GF, Smilanick JL (2012) Natural and safe alternatives to conventional methods to control post-harvest gray mold of table grapes. Posthar Biol Technol 63: 141-147.

16. Yousefi S, Amiri ME, Mirabdulbaghi M (2015) Biochemical properties and fruit quality of 'Jahangiri' apricot fruit under calcium chloride treatment. Int J Agron Resear 48: 81-94.

17. Nigro F, Schena L, Ligorio A, Pentimone I, Ippolito A, et al. (2006) Control of table grape storage rots by preharvest applications of salts. Posthar Biol Technol 42: 142-149.

18. Javed MS, Randhawa MA, Butt MS, Nawaz H (2015) Effect of calcium lactate and modified atmosphere storage on biochemical characteristics of guava fruit. J Food Process Preserv.

19. Chepngeno J, Owino WO, Kinyuru J, Nenguwo N (2016) Effect of $CaCl_2$ and hydro-cooling on post-harvest quality of selected vegetables. J Food Resear 5: 23.

20. Akhtar A, Abbas NA, Hussain A (2010) Effect of calcium chloride treatments on quality characteristics of loquat fruit during storage. Pak J Bot 42: 181-188.

21. Gustavo HA, Juniorb LC, Antonio SF, Jose FD (2016) Quality of guava fruit stored in low O_2 controlled atmospheres is negatively affected by increasing levels of CO_2. Posthar Biol Technol 111: 62-68.

22. Al-Qurashi AD, Awad MA (2015) Effect of pre-harvest calcium chloride and ethanol spray on quality of ' El-Bayadi' table grapes during storage. Vitis J Grapevine Resear 52: 61.

23. Chervin C, Lavigne D, Westercamp P (2009) Reduction of gray mold development in table grapes by preharvest sprays with ethanol and calcium chloride. Posthar Biol Technol 54: 115-117.

24. Marinova D, Ribarova F, Atanassova MA (2005) Total phenolics and total flavonoids in Bulgarian fruits and vegetables. J Uni Chem Tech Metallur 40: 255-260.

25. AOAC (2006) Official methods of analysis. The Association Off Analyt Chem (18th edn.), Gaithersburg, Maryland, USA.

26. Tian SP, Jiang AL, Xu Y, Wang YS (2004) Responses of physiology and quality of sweet cherry fruit to different atmospheres in storage. Food Chem 87: 43-49.

27. Badosa E, Trias R, Pares D, Pla M, Montesinos E (2008) Microbiological quality of fresh fruit and vegetable products in Catalonia using normalized plate-counting methods and real time polymerase chain reaction. J Food Sci Agricul 88: 605-611.

28. Montgomery DC (2008) Design and analysis of experiments. John Wiley & Sons, USA.

29. Nile SH, Kim S, Ko EY, Park SW (2013) Polyphenolic contents and antioxidant properties of different grape cultivars. BioMed Resear Int.

30. Pinheiro ES, Costa JMC, Clemente E (2009) Total phenolics and total anthocyanin found in grape from Benitaka Cultivar (Vitis vinifera L.). J Food Technol 7: 78-83.

31. Gallo V, Mastrorilli P, Cafagna I, Nitti GI, Latronico M, et al. (2014) Effects of agronomical practices on chemical composition of table grapes evaluated by NMR spectroscopy. J Food Composit Analys 35: 44-52.

32. Samra BN (2015) Impact of post-harvest salicylic acid and jasmonic acid treatments on quality of "Crimson seedless" grapes during cold storage and shelf life. Int J 3: 483-490.

33. Sabir FK, Sabir A, Kara Z (2010) Effects of modified atmosphere packaging and ethanol treatments on quality of minimally processed table grapes during cold storage. Bulgarian J Agricul Sci 16: 678-686.

34. Randhawa MA, Rashid A, Saeed M, Javed MS, Khan AA, et al. (2014) Characterization of organic acids in juices of some Pakistani citrus species and their retention during refrigerated storage. J Anim Plant Sci 24: 211-215.

35. Pinzón-Gómez LP, Deaquiz YA, Álvarez-Herrera JG (2014) Post-harvest behavior of tamarillo treated with $CaCl_2$ under different storage temperatures. Agron Colomb 32: 238-245.

36. Dean R, Van Kan JA, Pretorius ZA, Hammond-Kosack KA, Di Pietro, et al. (2012) The top 10 fungal pathogens in molecular plant pathology. Molecul Plant Pathol 13: 414-430.

30

Osmotic Dehydration Characteristics of Pumpkin Slices using Ternary Osmotic Solution of Sucrose and Sodium Chloride

Mehnaza Manzoor[1], Shukla RN[2], Mishra AA[2], Afrin Fatima[2] and Nayik GA[3]*

[1]Department of Food Science and Technology, Sher-e-Kashmir University of Agricultural Sciences and Technology, Jammu, India
[2]Department of Food Process Engineering, Sam Higginbottom University of Agriculture, Technology and Sciences, UP, India
[3]Department of Food Engineering and Technology, Sant Longowal Institute of Engineering and Technology, Punjab, India

Abstract

In this study, drying characteristics of osmotically treated pumpkin slices were scrutinized at temperature within range of 30°C to 50°C and at the amalgamation of nine ternary solution (Sugar: Salt) concentration levels (30:5%, 30:10%, 30:15% w/w) (40:5%, 40:10%, 40:15% w/w) and (50:5%, 50:10%, 50:15% w/w). At eight time intervals (30 min, 60 min, 90 min, 120 min, 150 min, 180 min, 210 min and 240 min) moisture loss and solid gain were ascertained at all amalgamation. Sample to Solution ratio of 1:5 w/w was kept invariable from beginning to end of the experiments. The consequence of solution concentration and temperature was examined and it was established that preliminary water loss and solid gain are related to solution concentration and temperature. Both moisture loss and solid gain amplified non- linearly at dissimilar temperatures and at all concentrations. The investigational drying statistics for the pumpkin fruit was used to fit four thin layer drying models Parabolic, Hunderson and Pabis, Page and Logarithmic model. Non-linear regression assessment was used to check the statistical validness of models. The Parabolic model offered preeminent fit for all circumstances of drying, conferring utmost value of R² (0.999) and lowest RMSE values (0.004).

Keywords: Pumpkin; Osmotic dehydration; Solid gain; Moisture loss; Mathematical modelling; Microwave drying

Introduction

India being country with varied climate ensures accessibility of all varieties of fresh fruits and vegetables. It is second leading producer of fruits (81.258 million tons) and vegetables (162.19 million tons) in world, with 12.6% fruits and 14.0% vegetables production [1]. Only 4% of vegetable production and 2% of fruit production are being processed while 76% is being utilized in fresh form, out of entire vegetable and fruit production in India. The shortfall and wastage contribute 20% to 22%. Preservation of these vegetables can thus hamper these shortfalls and in the off-season make them promptly accessible at remunerative pay out. Proficient technique of preservation requires to be developed to preserve plant materials in order to obtain superior quality as they are seasonal. "Minimal processing" concept makes the foundation of all modern substitutive food preservation practice. This technique is used to attain products with elevated nutritional value and innate sensory characteristics, with minimum use of preservatives [2]. Pumpkin (*Cucurbita pepo*) contains 92% water and total solid content varying from 7% to 10% [3]. Thus requires to be preserved as being delicate in nature. The process of removal of water from food material mostly vegetables and fruits by drenching it in hypertonic solution of either sugar or salt or sometimes in amalgamation of both these solutes so as to partly dehydrate it by exclusion of moisture and at the same time mounting the solid content of sample is known as Osmotic dehydration process. This process is also known as "dewatering impregnation soaking process" (DISP) as dewatering occurs only after food material in soaked in osmotic solution which is accompanied by impregnation of osmotic solutes in food material from solution. During osmotic dehydration process two major counter-current flows happen at the same time: first is water flow out from food being dehydrated into the osmotic solution and second is simultaneous transport of solute from osmotic solution to food material being dehydrated [4]. Also, a third process: leaching of innate total soluble solutes such as organic acids, sugar, minerals, salts, and so on that trickle into osmotic solution from food being dehydrated [5]. Intermediate moisture foods having water activity varying from 0.65 to 0.90 are produced by osmotic dehydration

process, as it trims down the water activity of food material being dehydrated, between 0.95 and 0.90 [6]. The details about drying characteristics of foods during osmotic dehydration process and sketch of operational process are better interpreted by mathematical modeling of mass transfer. At present, no meticulous information on osmotic drying characteristics of pumpkin slices at a mixture of ternary solution concentration levels and at different temperature and osmotic time is accessible, although some text on drying of pumpkin is available [7,8]. In this study, the models that best depict the drying characteristics during the experimental conditions deemed is yet to be done. Drying characteristics of food material are evaluated by theoretical, semi-theoretical or solely empirical thin layer drying models. Various researchers have employed number of semi-theoretical drying models [9-11]. The paper aspires the experimental analysis and modeling on drying characteristics during osmotic dehydration of pumpkin slices.

Materials and Methods

Sample preparation

Fresh and ripe pumpkins purchased from the neighboring market (Allahabad, U.P) on daily basis were used as raw material. Prior to execute each set of experiments, the pumpkins were sorted out visually for color (light green), and no corporal damage. After sorting, the pumpkins were cleansed with tap water and then cut manually into slices of 3-4 mm thickness by very sharp and sterile knife. Finally, to

*****Corresponding author:** Nayik GA, Department of Food Engineering and Technology, Sant Longowal Institute of Engineering and Technology, Longowal, Punjab, India, E-mail: gulzarnaik@gmail.com

confiscate excess moisture the slices were blotted with absorbent paper. The average initial moisture content of fresh pumpkin samples was revealed by oven drying method [12] and it was found to be 94.211% (wet basis). The value was employed in estimation of moisture loss and solid gain.

Osmotic dehydration of pumpkin slices

Pumpkin slices were partly dehydrated via osmotic dehydration technique. The osmotic agents employed were sucrose and NaCl. Distilled water, commercial sucrose (30%, 40% and 50%), and table salt (5%, 10% and 15%) were used for formulating ternary osmotic solution. Eight 250 ml glass beakers were filled with 100 ml osmotic solution and 30 gm sample was dipped in each beaker. The beakers with sample were then put in water bath. One beaker at times was removed from the water bath after each 30-min interval from the commencement of osmosis. Slices were taken away and blotted tenderly with a blotting paper to take out the surface moisture and then weighed up on an electronic balance. Osmotically dehydrated sample was then employed for determination of moisture loss and solid gain.

Determination of moisture content

Hot air oven method suggested by Ranganna [13] for fruits and vegetables was utilized to calculate the initial and final moisture content of sample.

$$MC(\%)=\frac{(W+W_1)-W_2}{W}\times100 \qquad (1)$$

Where W= Net weight of sample taken (g), W_1 = Weight of petriplate (g), and W_2 = Weight of petriplate plus oven dried sample (g).

Determination of moisture loss and solid gain

The moisture loss (ML %) and solid gain (SG %) were determined by the equations given below.

$$ML(\%)=\frac{(M_0-M)}{W}\times100 \qquad (2)$$

Where M_o = Wt of initial moisture (g), M = Wt of final moisture (g), and W = Initial wt of sample (g).

$$SG(\%)=\frac{S-S_0}{W}\times100 \qquad (3)$$

Where S = Wt of final solid (g), S_o = Wt of initial solid (g), and W = Initial wt of sample (g)

Mathematical modeling

Four drying models were used to choose an appropriate model for illustrating the drying process of pumpkin slices. XLSTAT-2015 (Addinsoft, New York, USA) was used to fit drying models to the experimental data for each state of osmotic dehydration process.

Henderson and Pabis model

$$MR = a \exp(-k\,t) \qquad (4)$$

Page model

$$MR = \exp(-k\,t^n) \qquad (5)$$

Parabolic model

$$MR = a + bt + ct^2 \qquad (6)$$

Logarithmic model

$$MR = a \exp(-k\,t) + c \qquad (7)$$

Where M.R = Moisture loss ratio

$$M.R = (M-M_e)\,/\,(M_o-M_e)$$

M_o = Initial moisture content,

M = Moisture content after time t,

M_e = Equilibrium moisture content,

t = Time period, min,

and a, b, n and k are constants.

Results and Discussion

Effect of osmotic dehydration process parameters on moisture loss

Moisture loss from the pumpkin slices versus different process parameters is shown in Figure 1. From Figure 1, it could be depicted that for all process conditions moisture loss increases non-linearly with time, being quicker in beginning of dehydration process. The rate then decreases; because of declining chemical potential gradient of water as moisture keeps moving from sample to solution. Also, elevated turgor pressure gradient produced during initial period of osmosis trigger structural deformation resulting in mass transfer resistance for water. Similar results have been conveyed for osmotic dehydration of apples by Derossi et al. [14]. After 240 min time moisture loss at all process conditions varies between 44.058% to 47.361% (w.b).

In this study, increasing solution concentration at all process temperatures increases moisture loss, because NaCl being ionizable in water has higher water activity lowering power, which along with sucrose increases chemical potential gradient for water and with increase in their concentration moisture loss increases. The results are in agreement of Ozen et al. [15] on behalf of osmotic dehydration of red paparika and green pepper in sucrose-salt combination solution. At 50: 15% (sugar: salt) solution concentration moisture loss is found to be utmost.

Temperature showed prominent effect on moisture loss. It is understandable from Figure 1 that with increase in temperature of osmotic solution moisture loss by pumpkin slices increases. At 50°C temperature, higher moisture loss is perceived at all process conditions. This occurs because at elevated temperature viscosity of osmotic solution decreases which consecutively decreases solution resistance to mass so moisture loss occurs.

Effect of osmotic dehydration process parameters on solid gain

Solid gain from the pumpkin slices at different ternary solution concentration and solution temperatures versus osmosis time is shown in Figure 2. Solid gain also presented non-linearly relation with time at all process conditions. Mass transfer mostly occurs at beginning of osmotic process. At beginning rate of solid gain increases quickly with time than the latter; because of more solids present primarily in solution which then get utilized by sample with time resulting in reduced chemical potential gradient for process. Also as immersion time increases, sucrose comprising more molecular weight than NaCl leads to development of solid barrier at superficial surface of sample, which makes solid gain more intricate thus lowering rate of solid gain.

Solution composition shows substantial influence on solid gain. Solid gain over entire osmotic process Increases with increase in osmotic solution concentration. Maximum value of solid gain is seen

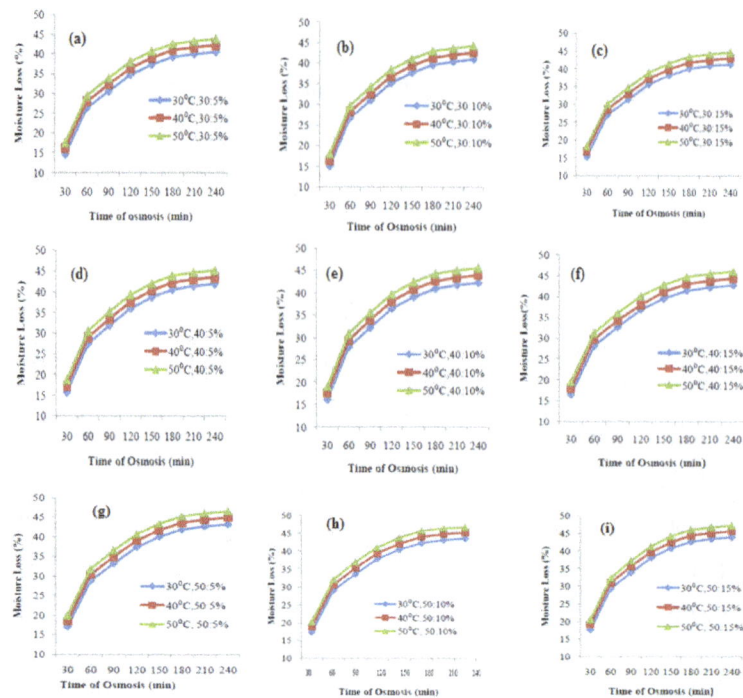

Figure 1: Effect of osmotic solution temperature on moisture loss of pumpkin slices during osmotic dehydration at (a) 30: 5% (b) 30: 10% (c) 30%: 15%(d) 40: 5% (e) 40: 10% (f) 40%: 15% (g) 50: 5% (h) 50: 10% (i) 50: 15% osmotic solution concentration.

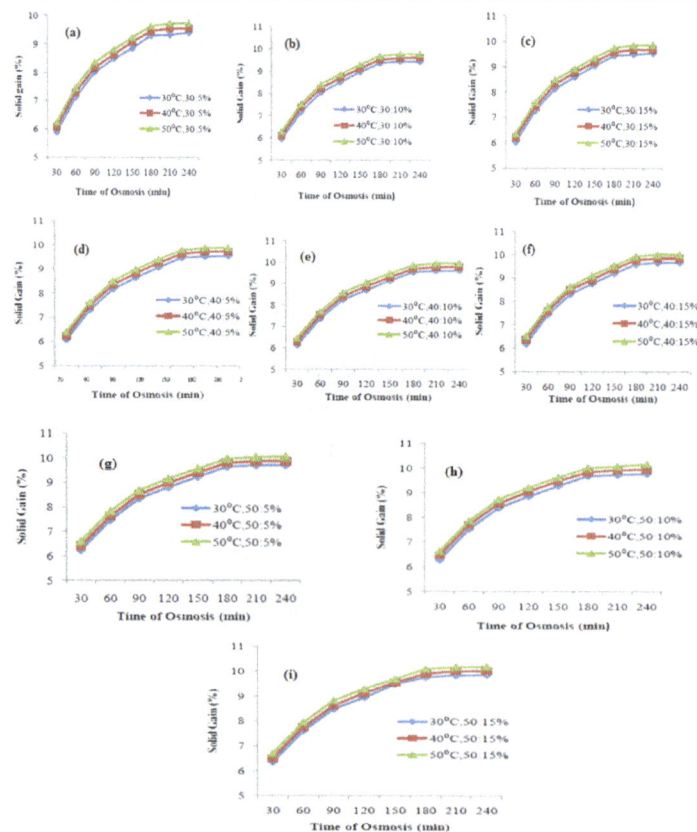

Figure 2: Effect of osmotic solution temperature on solid gain of pumpkin slices during osmotic dehydration at (a) 30: 5% (b) 30: 10% (c) 30%: 15%(d) 40: 5% (e) 40: 10% (f) 40%: 15% (g) 50: 5% (h) 50: 10% (i) 50: 15% osmotic solution concentration.

Model	R^2	RMSE
Henderson and Pabis	0.983	0.022
Page	0.854	0.569
Logarithmic	0.569	0.112
Parabolic	0.999	0.004

Table 1: R^2 and RMSE values of Henderson and Pabis, Page, Logarithmic and Parabolic models for the osmotic dehydration data.

in sample osmosed for 240 min at 50:15% osmotic agents' solution concentration at all process temperatures and its value varies from 6.375 to 10.196%. Temperature also shows immense effect on solid gain as is evident from Figure 2. Increase in solid gain due to increase in solution temperature occurs because of increase in cell membrane permeability to solutes due to swelling and plasticizing of cell membrane. Also, higher temperature reduces viscosity of solution making solute transfer effortless. High temperature of 50°C shows elevated solid gain.

Fitting of drying models

Henderson and Pabis, Page, Logarithmic, and Parabolic models were analyzed by fitting the experimental data using XLSTAT-2015. The R^2 and RMSE values for each of the tested models are given in Table 1 and it is clear from this table that the Parabolic model gives the best values in terms of highest R^2 and lowest RMSE.

Conclusion

From this study it can be concluded that non-linear relation exists between moisture loss and solid gain with osmotic dehydration process parameters. Mass transfers during osmotic dehydration process were mostly influenced by osmotic solution temperature and concentration pursued by immersion time. Parabolic model presents the best fit with utmost values for the coefficient of determination (0.999) and lowest RMSE value followed by Henderson and Pabis model and then by page model. Logarithmic model least illustrated the dehydration process for all temperature and concentration combinations.

Acknowledgments

The author one acknowledges Department of Food Process Engineering, SHIATS for making facilities available and Dr. R. N. Shukla for his valuable suggestions and timely guidance during entire period of research.

References

1. FAO (2014) Indian horticulture database. FAO, Paris.

2. Rzaca M, Witrowa-Rajchert D, Tyewicz U, Dalla Rosa M (2009) Mass exchange in osmotic dehydration process of kiwi fruit. Zywnosc: Nauka Technologia Jakosc, Zywn Nauk Technol J 6: 140-149.

3. Arevalo-Pinedo A, Murr FEX (2006) Kinetics of vacuum drying of Pumpkin (*Cucurbita maxima*): Modelling with Shrinkage. J Food Eng 76: 562-576

4. Torregiani D (1993) Osmotic dehydration in fruits and vegetable processing. Food Res Int 26: 59-68.

5. Dixon GM, Jen JJ (1997) Changes of sugars and acids of osmovac dried slices. J Food Sci 42: 11-26.

6. Yadav AK, Singh SV (2014) Osmotic dehydration of fruits and vegetables: A review. J Food Sci Technol 51: 1654-1673

7. Alibas I (2007) Microwave air and combined microwave air drying of pumpkin slices. LWT-Food Sci Technol 40: 1445-1451.

8. Sacilik K (2007) Effect of drying methods on thin layer drying characteristics of hull less seed pumpkin (*Cucurbita pepo* L.). J Food Eng 79: 23-30.

9. Sharma GP, Prasad S (2004) Effective moisture diffusivity of Garlic cloves undergoing microwave convective drying. J Food Eng 65: 609-617.

10. Simal S, Femenia A, Garau MC, Rosello C (2005) Use of exponential, page and diffusion model to simulate the drying kinetics of kiwi fruit. J Food Eng 66: 323-328.

11. Togrul IT, Pehlivan D (2004) Modelling of thin layer drying kinetics of some fruits under open air Sun drying process. J Food Eng 65: 413-425.

12. AOAC (1990) Official methods of analysis. (15th edn), Association of official analytical chemists, Arlington.

13. Ranganna S (2001) Handbook of analysis and quality control of fruits and vegetables product. 3rd edn, Tata McGraw-Hill publishing house, New Delhi, India.

14. Derossi A, De Pilla T, Severivi C, Mc Carthy MJ (2008) Mass transfer during osmotic dehydration of apple. J Food Eng 86: 519-528.

15. Ozen BF, Dock LL, Ozdemir M, Floros JD (2002) Processing factors affecting the osmotic dehydration of diced green pepper. Int J Food Sci Technol 4: 537-542.

31

Stability and Functionality of Grape Pomace Used as a Nutritive Additive During Extrusion Process

Bibi S, Kowalski RJ, Zhang S, Ganjyal GM and Zhu MJ*

School of Food Science, Washington State University, Pullman, WA 99164, USA

Abstract

Grape pomace (GP) is a major byproduct of wine and juice industry, rich in polyphenolics with demonstrated health benefits. Extrusion processing for development of healthy and quality GP supplemented cornstarch snack foods was evaluated using response surface methodology. The retainability of polyphenolic content and antioxidant activity after extrusion processing were further assessed. The processing variables were feed moisture (16, 20, and 24 ± 0.2% w.b.), screw speed (150, 200, and 250 rpm), and the level of GP supplementation (0, 5, and 10% w/w). Extrudates with 5% GP and 16 ± 0.2% feed moisture had a high overall expansion ratio of 3.83 ± 0.14, and overall low density (0.11 ± 0.00 g/cm³). Total polyphenolic content (TPC) of the extrudates (5% GP, and 16% feed moisture) extruded at 150 and 250 rpm retained up to 74.1% and 78.57% respectively, while TPC was retained at 95% when extruded under 200 rpm with 10% GP and 16% feed moisture. Additionally, the total antioxidant activity and 2,2-diphenyl-1-picrylhydrazyl scavenging activity of the 5% GP extrudates retained 98% after extrusion processing. Moreover, polyphenolic extract of 5% GP extrudates suppressed reactive oxygen species in Caco-2 cells induced by hydrogen peroxide. In conclusion, GP incorporation in cornstarch extrudates improved both the physicochemical quality as well as nutritional value of products. Our study indicates that GP can be effectively incorporated into extruded foods by providing enhanced nutritional value without losing the expansion characteristics.

Keywords: Grape pomace; Extrudates; Expansion; Polyphenolics; Antioxidant activity; Reactive oxidative species

Introduction

Grape is the 4th largest fruit crop in the world with a production of 67 million tons per year [1]. Grapes are generally cultivated for the wine production with about 80% of the global grape production being used in the wine industry [2]. After the juicing process of grapes, huge amounts of the grape pomace (GP), consisting of seeds, skin and stems is obtained as the byproduct. GP is currently used as animal feed because of its protein value, as a fertilizer, and as a source for extraction of bioactive compounds [3,4]. GP is a rich source of polyphenolic compounds including gallic acid, catechin, epicatechin, and proanthocyanins among others [5]. GP also contains substantial amount of non-digestible fiber (60%-70%), essential fatty acids (13%-19%), proteins (11%), as well as non-phenolic tocopherols, beta-carotene, and minerals [6,7]. The polyphenolics from GP are known for their antioxidant and anti-inflammatory effects [8,9], which demonstrate preventive effects on cancer [10], cardiovascular diseases [11], and inflammatory bowel disease [12]. The GP polyphenolics have been used in food technology as antioxidants to inhibit the oxidation of fish lipids, frozen fish muscle, and fish oil [13]. It is also used as a source of dietary fiber and antioxidant to enhance functionality and shelf life of yogurt and salad dressing [14], and ingredient in baking products [15].

As another alternative usage, GP has been used in extrusion cooking [16], by mixing with barley flour at levels of 2%-10% (w/w). However, using response surface methodology, product responses including bulk density, expansion, texture, and color were greatly affected by inclusion of GP and temperature of the extruder [16]. The GP barley flour extrudates with 2% GP (at 160°C, 200 rpm) had higher preferences for overall product quality, while increasing GP proportion above 6% reduced expansion [16]. Extrusion processing also reduced the total phenolic content (TPC) of the GP barley flour extrudates compared to the raw material [17]. However, both of these studies lack a reasonable explanation of the positive interactions between GP and barley flour, which can greatly affect the textural properties

and is evident from our recent findings on carrot and cherry pomace incorporation in cornstarch extrudates [18,19]. Furthermore, feed moisture can also affect the textural as well as nutritional quality of the extruded products. Extrusion of GP with white sorghum (at ratio of 30:70 w/w, barrel temperature of 170°C and screw speed of 200 rpm) with a moisture content of 45% resulted in 120% increase in monomer contents of the polyphenolics in GP extrudates [20]. All these suggest that extruded products can serve as a great vehicle for the delivery of beneficial nutrients of the GP, as extrusion is a high shear, high pressure, and short time processing technology [21], proven to help improve digestibility and nutrient bioavailability [20,22], and to have distinct textural properties [23]. However, both the nutritional and textural quality of the extrudates is highly affected by the operational process including screw speed, feed moisture and raw material composition used [24], which remain unclear for GP cornstarch extrudates.

Expansion, an important quality parameter of the extruded puffed products, is greatly affected by the feed moisture content and composition of feed [18,19,25]. Cornstarch, widely used in extrusion cooking, undergoes gelatinization and mechanical degradation during extrusion, and expands well, while other raw ingredients such as proteins, added sugars, lipids, fiber and moisture generally tend to reduce expansion [24,26]. The cornstarch expanded products are usually energy dense, with high glycemic index and low nutrients (vitamins, minerals, proteins and fiber). Nutritional value of these products can be improved via incorporation of bioactive compounds and fibers from

---

*Corresponding author: Mei-Jun Zhu, School of Food Science, Washington State University, Pullman, WA 99163, USA, E-mail: meijun.zhu@wsu.edu

legumes, fruits, and vegetables [23]. Studies on the incorporation of pomace into extruded products showed that increased pomace levels beyond a certain level had negative effect on the expansion quality [16,17,20,27,28]. However, recent studies indicated that carrot and cherry pomace incorporation into cornstarch extrudates resulted in better expansion and nutritional quality [18,19]. The objective of the current study was to examine the impact of GP incorporation on the quality of the cornstarch extrudates during extrusion processing and to produce a nutrient enhanced extruded product with good quality attributes.

Materials and Methods

Feed material

Cornstarch (native dent cornstarch with 23% amylose) was obtained from Tate & Lyle Hoffman Estates, IL, USA). White GP was received from Woodward Canyon Winery (Lowden, WA, USA) that mainly contained Chardonnay grape skins, seeds, as well as some stems. The GP was freeze-dried in VirTis freeze drier (Vertis Comp. Gardiner, NY, USA) and ground using a cyclone mill (Model# 3010-060, UDY Corp. Fort Collins, CO, USA). Particle size of the white GP was determined through the U.S.A standard test sieves of 75, 125, 150, 212 and 250 μm (Model# 78-700, Field master, Science First, Yulee, FL, USA). The GP bulk density was determined by measuring weight and volume of the GP content using standard graduated cylinder, and expressed in grams per cubic centimeter [18]. The moisture (oven drying), ash, protein content (Kjeldahl, protein factor: 6.25), and total sugar (using Fehling's reagents) were determined by the official methods (number 968.21, 945.18, and 974.06, respectively) of AOAC [29]. Total dietary fiber content was determined using Sigma total dietary assay kit (TDF-A100 Sigma, St. Louis, MO, USA).

Feed mixture preparation

The cornstarch and GP (0, 5 and 10% w/w) were mixed in a Hobart mixer (Model #A-200, Hobart, OH, USA) and equilibrated at three moisture levels: 16, 20, and 24 ± 0.2 (% w.b.) by adding the calculated amount of water. The feed mixtures were then stored in airtight plastic containers at 4°C overnight.

Processing conditions of extruder and process response

A 20 mm co-rotating twin-screw extruder (Model# TSE 20/40, CW Brabender, S. Hackensack, NJ, USA) with a length of 400 mm and the length to diameter (L/D) ratio of 20:1, was used for all the extrusion experiments. The temperature profile of the four independent zones of the extruder were kept constant at 50°C, 100°C, 140°C, and 140°C. A cylindrical die with a diameter of 3 mm was used. The feed rate of the material was fixed at 3.25 kg/h using a calibrated twin-screw volumetric feeder (Model# 15-37-000, CW Brabender Technologie, NJ, USA). The screw speeds were varied from 150 rpm to 250 rpm. These extrusion parameters were selected based on the literature data and our previous extrusion research experiments [19].

The premixed cornstarch and GP combinations were extruded as per the experimental design. After 5 min, when the extruder attained the stable conditions of pressure, torque and output flow, extrudates were collected. The collected extrudates were dried in a convection oven (Model# 414004-568, VWR International, LLC, PA, USA) at 45°C for 18 h. The extrudates had final average moisture of 4%-6% (w.b.) and were stored in airtight plastic bags at 4°C until analysis. During processing, continuous data from the extruder, including die pressure, motor torque, and zone temperature, were recorded using

Data Acquisition System for ATR and Intelli-Torque (CW Brabender, S. Hackensack, NJ, USA). The average of 10 data points, being taken at 20 second intervals during stable conditions, was used to calculate the specific mechanical energy (SME), motor torque, and die pressure according to previously reported method [30].

Product response

Expansion ratio (ER): Radial expansion was determined by measuring the diameter of 10 randomly chosen extrudates from one processing condition with a calipers (Mitutoyo America Corp., Aurora, IL, USA) according to the method reported previously [25]. For each experimental trial, two data points per single extrudate, with a total of 10 random extrudates, were taken and the mean diameter was calculated. ER was determined by dividing the mean diameter of the extrudates by the die diameter (3 mm).

Unit density: Unit density of the extrudates was determined in triplicate through displacement of 1.0 mm diameter glass beads (General Laboratory Supply, Pasadena, TX, USA) according to the method reported previously [25]. Glass beads (30 ml) were filled in a 50 ml graduated cylinder, and the displacement of the beads by the extruded sample (2.0 g) was recorded. The density was equal to the sample mass divided by the sample volume.

Water absorption and water solubility index: Water absorption index (WAI) and water solubility index (WSI) were determined by the previously described procedure [31]. Briefly, a portion of milled extrudates (2.5 g) was taken in 50 ml vial containing 30 ml of 30°C distilled water. Samples were mixed for 30 min in a 30°C water bath and then centrifuged at 3,000 × g speed for 10 min. The supernatant was removed and dried overnight. The weight of wet precipitate was recorded after 10 min for WAI. WAI was calculated as the ratio of the mass of the wet precipitate to the mass of the original dry weight. WSI was expressed as percentage of the overnight dried solid weight of the collected supernatant to the original sample weight.

Hardness: Hardness of the extrudates was determined by the previously reported method [32]. Briefly, texture analyzer TA.XT2i (Texture Technologies, Scarsdale, NY, USA) with a single-blade and a 25 kg load cell with a test speed of 1 mm/s was used. The maximum force required to break an extrudate was recorded. Ten extrudates were analyzed per treatment and the mean peak force was reported.

Color: Color measurements were performed on milled extruded samples from all the trials, using a CM-5 spectrophotometer (Konica Minolta, NJ, USA) according to the method reported previously [28]. The color was recorded using a Hunter color scale as the mean of three L*, a*, and b* readings, where L* indicates lightness, a* indicates redness, and b* indicates yellowness. A standard white and black calibration plate was used to equilibrate the spectrophotometer prior to color measurements.

Total polyphenolic compounds extraction and analysis: Total polyphenolics were extracted from raw material before processing and from extrudates after processing as previously reported [33]. Briefly, samples were defatted twice with n-hexane at 70°C for 20 min. Total polyphenols were extracted from 1 g of GP using 10 ml of 80% ethanol that contained 1% formic acid (v/v) for 12 h, and then centrifuged at 10,000 × g for 15 min. The supernatants were collected and residue samples were re-extracted once under the same conditions. The TPC was determined by modified Folin–Ciocalteu assay in a 96-well plate format [34]. Briefly, 200 μl of the extracts were mixed with 12.5 μl of Folin-Ciocalteu reagent and then 37.5 μl 20% Na_2CO_3 in a 96-well plate. The plate was incubated at room temperature for 2 h when the absorbance was read at 760 nm on Synergy™ H1 microplate reader (BioTek, Winooski, VT, US). Gallic acid was used to generate the standard curve. The TPC was expressed as milligrams of gallic acid equivalents per gram of dried weight (mg GAE/ g DW). TPC data was subjected to General Linear Model of Statistical Analysis System (2000) and was expressed as mean ± standard error of mean. A significant difference was considered as $P \leq 0.05$.

Total anti-oxidant activity and 2,2-diphenyl-1-picrylhydrazyl (DPPH) free radical scavenging analysis: Assay was performed in a 96-well plate using previously published method with some modifications [35]. Briefly, 200 µl of DPPH solution (60 µM) (Sigma, St. Louis, MO, USA) was added to wells of 96-well plate containing 50 µl of diluted extracts of the raw mix and GP extrudates or gallic acid standard solution. After incubation for 90 min at room temperature, the DPPH scavenging activity was measured at 517 nm absorbance on Synergy™ H1 microplate reader (BioTek, Winooski, VT, US). The total antioxidant activity was expressed as µg GAE/g DW based on the gallic acid standard calibration curve. The DPPH radical scavenging activity expressed as percentage of inhibition was calculated using the formula:

$$Inhibition(\%) = \frac{A_{control} - A_{sample}}{A_{control}} \times 100$$

where,

A is the absorbances

$A_{control}$ is reading without samples.

Effects of GP extrudates extracts on intracellular reactive oxygen species (ROS) production in CACO-2 human colonic epithelial cells: Caco-2 cell line was obtained from American Type Culture Collection (Manassas, VA, USA). The cells were grown in in Dulbecco's Modified Eagle's medium complete media: (DMEM, Sigma, St. Louis, MO, USA) supplemented with 10% fetal bovine serum (FBS, Sigma) and 1% penicillin–streptomycin (Sigma) at 37°C with 5% CO_2 in a humidified incubator. Intracellular ROS levels were measured using a cell-permeable fluorescent probe, 2,7-dichlorofluorescein diacetate (DCFH-DA) (Millipore, MA, USA) as previously described with some modifications [36]. Briefly, 100 l of Caco-2 cells (5×10^5 cell/ml) were seeded in 96-well plate, and cultured in DMEM media with 10% FBS at 37°C with 5% CO_2 for 12 h. Then, cells were pre-treated with 10 l of the polyphenolic extracts in DMEM media for 12 h. After washing with PBS once, the cells were incubated with 100 µl of 10 µM fresh DCFH-DA in PBS for 30 min. Then the cells were washed with PBS once, and incubated with 100 µl of PBS or 0.5 M H_2O_2 for 30 min. Fluorescence of each well was measured at an excitation wavelength of 485 nm and an emission wavelength of 530 nm on Synergy™ H1 microplate reader (BioTek, Winooski, VT, USA). All the values were normalized with the control (no extract no H_2O_2).

Experimental design and statistical analysis

Response surface methodology was used to investigate the effects of extrusion processing conditions on the process responses (SME, motor torque, and die pressure) and product responses (ER, unit density, hardness, WAI, WSI, and color) of the extrudates. The independent variables were feed moisture (16, 20, and 24 ± 0.2% w.b.), screw speed (150 rpm, 200 rpm, and 250 rpm), and the GP level (0%, 5%, and 10%). A Box-Behnken design was used to determine the experimental conditions (Table 1). Data was subjected to regression analysis and a second-order polynomial regression model:

Moisture (%)	Corn Starch (%)	GP (%)	Screw Speed (rpm)
16	100	0	200
16	100	0	200
16	95	5	150
16	95	5	150
16	95	5	250
16	95	5	250
16	90	10	200
16	90	10	200
20	100	0	150
20	100	0	150
20	100	0	250
20	100	0	250
20	95	5	200
20	95	5	200
20	95	5	200
20	95	5	200
20	95	5	200
20	95	5	200
20	90	10	150
20	90	10	150
20	90	10	250
20	90	10	250
24	100	0	200
24	100	0	200
24	95	5	150
24	95	5	150
24	95	5	250
24	95	5	250
24	90	10	200
24	90	10	200

Table 1: Extrusion parameters and experimental design.

$$Y = \beta_0 + \sum_{i=1}^{3} \beta_i X_i + \sum_{i=1}^{3} \sum_{j=1}^{3} \beta_{ij} X_i X_j \qquad (1)$$

was used to fit the data for each response. Terms β_0, β_i, β_{ij} where $i = j$, and β_{ij} where $i \neq j$ are the coefficients for intercept, linear, quadratic, and interactive effects respectively. The term Y is the response, and X_i and X_j are the independent variables of moisture, screw speed, and GP content. The responses were plotted as a function of two independent variables, keeping the other one constant at the middle level using Origin software (version 9.0, Origin Lab, Northhampton, MA, USA). A significant difference for a coefficient was considered as $P \leq 0.05$.

Results and Discussion

GP is a potential nutritive adjunct ingredient for extrudates

The GP had a moisture content of 13.96 ± 0.14 g/100 g, ash content of 3.79 ± 0.36 g/100 g, protein content of 7.15 ± 0.12 g/100 g, total sugar content of 29.31 ± 0.63 g/100 g, total dietary fiber content of 39.24 ± 1.40 g/100 g, and crude fats content of 6.55 ± 0.5 g/100 g. The bulk density of the GP was 0.48 ± 0.01 g/cm³. The particle size distribution in GP was 44.65% > 250 µm, 19.15% = 250-212 µm, 26.34 % = 212-150 µm, 2.27% = 150-125 µm, 2.27% = 125-175 µm, and 0.85% < 75 µm. GP is rich in dietary fiber and other nutrients, and its combination in cornstarch expanded products can enhance the nutritional quality and market value of GP incorporated products. The percentage of fiber, sugar, protein and fat in the GP can interact with cornstarch forming complexes that can greatly affect the textural properties of the extrudates [37]. Though fiber was previously considered as inert material that reduces expansion [38,39], recent studies indicated that carrot and cherry pomace in cornstarch can play an active role in the expansion of extrudates that is potentially due to the interaction between pomace fiber particles and starch [18,19].

Process response

The process response parameters (SME, motor torque, and die pressure) indicate energy input to the materials in the extrusion process, which are affected by the extrusion processing inputs such as material composition (%GP, moisture), and screw speed [21]. Regression results obtained for SME, motor torque, and pressure are shown in Table 2. Feed moisture had a negative linear effect ($P\leq0.01$) on the die pressure. Screw speed had a positive linear effect ($P\leq0.01$) on SME but no effect on motor torque or die pressure was observed. Feed moisture, GP, and screw speed had negative quadratic effects on the SME. The interactive effects of moisture and GP were significant for SME ($P\leq0.01$), motor torque ($P\leq0.01$), and die pressure ($P\leq0.05$). The calculated SME ranged from 160.88 ± 4.79 to 511.21 ± 11.08 kJ/kg,

Level	SME	Motor Torque	Die Pressure	Expansion	Density	WSI	WAI	Hardness	Color
C	-346.038	30.603*	5672.985**	8.307**	-0.016	72.249	-19.018*	82.825	96.667**
M	22.078	-0.443	-449.617**	-0.281	-0.035	-5.612	1.950**	-1.498	-0.104
G	-8.350	-0.230	9.890	0.222*	-0.083**	5.021*	-0.065	-5.451*	-0.139
SS	6.545**	0.037	-2.137	-0.016	0.005	0.122	0.043	-0.388	-0.030*
M × M	-0.996*	-0.041	9.440**	0.004	0.003	0.171	-0.041**	0.070	0.002
G × G	-1.151**	-0.051**	0.485	-0.001	0.002**	-0.050	-0.024**	0.140*	0.021**
SS × SS	-0.010**	0.000*	-0.004	0.000	0.000	0.001	0.000	0.001	0.000*
M × G	1.312**	0.057**	-1.173*	-0.012**	0.004**	-0.152	0.009	0.124	-0.012**
M × SS	-0.071	0.003	0.151**	0.000	0.000**	-0.014	0.000	-0.003	0.000
G × SS	-0.020	-0.001	0.060	0.000	0.000*	-0.008	0.001	0.002	0.000
R²	0.961	0.953	0.984	0.891	0.831	0.795	0.886	0.759	0.981
F	55.062	44.643	138.441	18.159	10.930	8.602	17.206	6.991	116.450
Sig. F	0.000	0.000	0.000	0.000	0.000	0.000	0.000	0.000	0.000

SME: Specific mechanical energy; WAI: Water absorption index; WSI: Water solubility index; C: Model constant; M, G and SS: Linear effects of moisture content, grape pomace, and screw speed, respectively; M × G: Interaction of moisture content and grape pomace; M × SS: Interaction of moisture content and screw speed; M × M, G × G; SS × SS: Quadratic effects of moisture content, grape pomace, and screw speed, respectively.
* and ** indicate significant at $P\leq0.05$ and 0.01 respectively.

Table 2: Results of regression analysis for cornstarch grape pomace extrudates.

Moisture (%)	GP (%)	Screw Speed (rpm)	SME (kJ/kg)	Motor Torque (Nm)	Die Pressure (psi)
16	0	200	404.6 ± 4.51	17.56 ± 0.20	814.00 ± 4.57
			441.00 ± 6.79	19.14 ± 0.29	816.10 ± 5.28
	5	150	374.47 ± 10.5	21.67 ± 0.16	843.50 ± 17.39
			375.16 ± 8.92	21.71 ± 0.52	859.70 ± 12.2
		250	511.21 ± 11.08	17.75 ± 0.38	730.90 ± 16.86
			500.56 ± 9.05	17.38 ± 0.31	700.90 ± 12.12
	10	200	409.20 ± 4.14	17.76 ± 0.18	852.50 ± 5.74
			382.47 ± 7.08	16.6 ± 0.31	858.00 ± 7.71
20	0	150	251.26 ± 2.8	14.54 ± 0.16	462.00 ± 2.61
			233.80 ± 2.42	13.53 ± 0.14	460.50 ± 2.56
		250	376.14 ± 13.59	13.06 ± 0.47	402.80 ± 3.34
			346.47 ± 2.98	12.03 ± 0.10	400.40 ± 3.99
	5	200	365.19 ± 4.38	15.85 ± 0.20	464.90 ± 6.22
			352.29 ± 6.02	15.29 ± 0.26	464.90 ± 5.50
			359.66 ± 4.34	15.61 ± 0.19	470.20 ± 4.94
			402.08 ± 8.18	17.49 ⊥ 0.30	461.90 ± 6.84
			358.51 ± 10.75	15.56 ± 0.47	471.60 ± 6.06
			357.59 ± 6.10	15.52 ± 0.26	468.00 ± 3.74
	10	150	284.78 ± 5.42	16.48 ± 0.31	502.60 ± 2.74
			262.66 ± 3.47	15.2 ± 0.20	507.30 ± 3.27
		250	397.16 ± 11.60	13.79 ± 0.40	504.10 ± 2.54
			347.34 ± 7.29	12.06 ± 0.25	507.3 ± 4.63
24	0	200	197.46 ± 3.86	8.57 ± 0.17	453.00 ± 2.70
			191.24 ± 2.4	8.3 ± 0.10	450.70 ± 2.68
	5	150	160.88 ± 4.79	9.31 ± 0.28	437.40 ± 2.19
			185.07 ± 3.40	10.71 ± 0.20	439.20 ± 3.19
		250	238.76 ± 5.47	8.29 ± 0.19	423.40 ± 3.60
			256.04 ± 5.11	8.89 ± 0.18	423.30 ± 2.96
	10	200	250.45 ± 8.00	10.87 ± 0.35	400.30 ± 3.42
			294.23 ± 5.06	12.77 ± 0.22	396.10 ± 3.33

Table 3: Results values of process response (SME, motor torque and pressure).

Level	SME	Motor Torque	Pressure	Expansion	Density	Hardness	WSI	WAI	Color (L)
SME	1	.....	.....	.....	.....	.....	.....	.....	.....
Motor Torque	0.750**	1	.....	......	.....	.....	.....	.....	.....
Pressure	0.605*	0.753**	1	......	.....	.....	.....	.....	.....
Expansion	-0.024	0.028	0.050	1	.....	.....	.....	.....	.....
Density	-0.008	-0.148	-0.125	-0.672*	1	.....	.....	.....	.....
Hardness	0.184	0.084	0.463	-0.421	0.474	1	.....	.....	....
WSI	-0.149	-0.119	-0.142	0.749**	-0.681*	-0.583*	1	.....	....
WAI	0.036	0.075	0.049	-0.766**	0.487	0.246	-0.766**	1	.....
Color (L)	0.210	0.163	0.443	0.377	-0.068	0.497	-0.019	-0.396	1

SME: Specific mechanical energy; WAI: Water absorption index; WSI: Water solubility index; L: Lightness.
* and ** indicate significant at $P \leq 0.05$, and 0.01, respectively.

Table 4: Cross-correlation values for cornstarch grape pomace extrudates.

and measured torque values ranged from 8.29 ± 0.19 to 21.71 ± 0.52 Nm, while the mean calculated values for pressure were 396.10 ± 3.33 to 859.70 ± 12.20 psi (Table 3).

The feed moisture and GP levels affected all the process response parameters (Table 2). The values of SME, motor torque, and pressure were lower at higher feed moisture and GP levels (Table 3). The feed moisture and pomace affect the melt viscosity of the material being processed [18,19]. The main reason for lower values of the process parameters at elevated feed moisture levels is reduction in the viscosity of the melt inside the extruder. Low viscosity melt faces less resistance in the screws and thus will be easily pushed out through the extruder with minimum energy input. It is likely that the inclusion of the GP disrupted the starch melt and contributed to the decrease in viscosity by acting as lubricant inside the extruder, which is seen in other studies [18,19]. A lower melt viscosity due to GP could account for the lower SME, motor torque and pressure observed in our study. This trend was also in line with reduced expansion of the extrudates at a higher level of GP (10%) and moisture content (24%). Further, the SME integrates torque, screw speed, and pressure. The SME, motor torque, and pressure, values were low at high moisture content via regression analysis (Table 2). These results were coincided with the correlation results, which showed that SME, motor torque and pressure positively correlated to each other (Table 4). Increase in screw speed increased SME while the motor torque decreased (Table 2). At GP level of 5%, the SME increased with the increasing screw speed and the decreasing feed moisture level (Figure 1). Generally, with increased screw speed, the SME increases. It is because increasing screw speed applies increased shear to the material being processed, resulting in greater SME input and decrease in motor torque. This is associated with decreased viscosity due to degradation of the starch in the material being processed [18,19,30].

Product response

Expansion ratio: The ER of cornstarch extrudates with GP ranged from 2.06 ± 0.12 to 3.83 ± 0.14. ER was found to be most dependent on GP and feed moisture levels, as shown by the regression analysis (Table 2). GP had a positive linear effect ($P \leq 0.05$) on the ER, along with a negative interactive effect ($P \leq 0.01$) in combination with moisture. It was observed in the regression response surface plot (Figure 2A) that increased feed moisture led to significant reduction in ER. As explained earlier, the high moisture level in the feed led to a reduction in the SME that could cause reduced physicochemical transformation of starch, and hence reduced the ER. Moreover, the post expansion collapse of cells at the die exit due to decrease in temperature, and migration of additional moisture from inside to the outside in expanding starch reduced the ER. Previous researchers also found an inverse relation

between the ER and feed moisture [25,39], which are in line with our observations.

Addition of GP (5%) at low feed moisture (16%) resulted in an increase in the ER compared to the control with 0% GP, while the expansion ratio decreased with the addition of GP (10%) at a high feed moisture (24%) (Figure 2A,2B). The expansion phenomenon of the extruded products is mainly based on the flashing off of the steam that expands the starch matrix, forming a porous product, along with the die swell property of the materials [24,25,40]. Traditionally, it was thought that addition of fiber reduces the ER of the final product [39]. However, the results suggest that instead of the GP being inert at low moistures, it is playing an active role either through the fiber, or the other components of the GP. It is possible that the fiber could have played an active role at 5% GP by being uniformly distributed in the starch matrix, which would allow maximum expansion, with fiber not disrupting the expanded starch structure. This trend is prominent in the carrot pomace and cherry pomace incorporated cornstarch expended products [18,19]. Other possibilities could be that sugar in addition to fiber in GP might have more specific effects. Sugar crystals act as filler at lower feed moisture and enhances expansion of the extrudates [41].

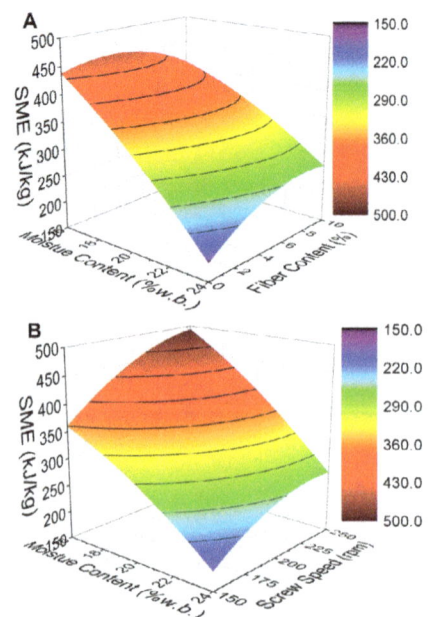

Figure 1: Specific mechanical energy (SME) response surface of cornstarch extrudates with different moisture content, grape pomace content, and screw speed. (A) Feed moisture vs grape pomace, and (B) Feed moisture vs screw speed.

Fiber in GP might have acted as a nucleating agent allowing cornstarch to trap more air cells by surrounding them firmly, while sugars in GP at low feed moisture might have enhanced the film forming effect by allowing bubble walls to be stabilized, resulting in larger porous structure. On the other hand, high GP and high moisture level has the tendency to disturb the starch matrix and the GP distribution during extrusion. This could have affected the expansion phenomenon such as collapsing of the cells due to aggregated GP in different spots of the starch [39] and plasticizing effect of sugars on starch [41] resulting in the reduction of the ER. Similarly, cherry pomace at 15% in cornstarch significantly reduces the ER compared to control [18]. Our findings are opposite to that of Altan and colleagues who found a decrease in the expansion with up to 6% GP of barley flour extrudates [16]. This difference could be due to difference in the raw material ingredients, as well as processing parameters, which is also evident from our previous studies on carrot and cherry pomace incorporation in cornstarch [18,19]. Overall, this suggests that there was positive active interaction between the GP constituents (sugars and fiber) and starch during extrusion that enhanced the expansion.

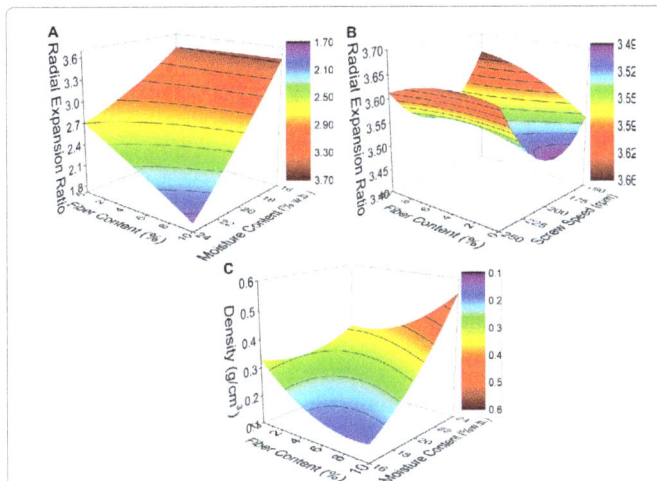

Figure 2: Product response characteristics of cornstarch extrudates processed with different moisture content, grape pomace content and screw speed. Expansion ratio (A & B), Unit density (C).

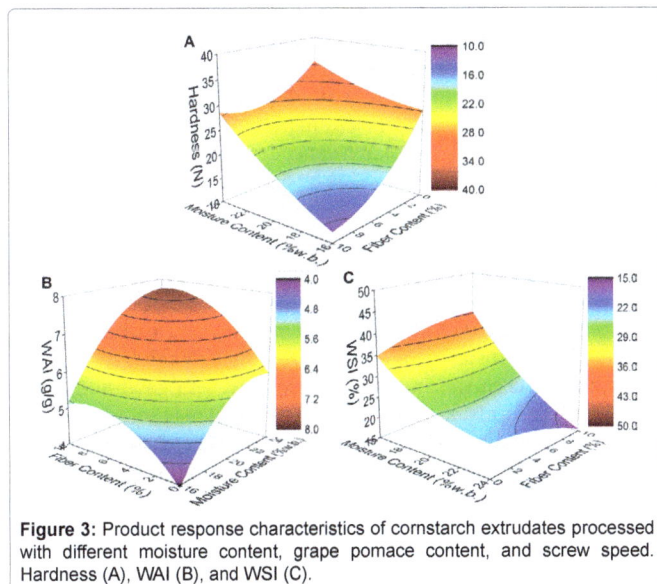

Figure 3: Product response characteristics of cornstarch extrudates processed with different moisture content, grape pomace content, and screw speed. Hardness (A), WAI (B), and WSI (C).

Unit density: Unit density is inversely proportional to expansion. Unit density values ranged from 0.11 ± 0.00 to 0.58 ± 0.13 g/cm³ for all extrudates. The regression analysis (Table 2) indicated that the linear and quadratic effect of GP significantly affected the unit density of the extrudates. The individual two-way interactive effects of all the variables: moisture, GP, and screw speed positively affected the unit density of the extrudates ($P \leq 0.01$). Increase in GP resulted in low unit density of the extrudates at a constant screw speed, while unit density of the extrudates decreased with decrease in moisture content (Figure 2C). This trend of the unit density can be attributed to the same factors that increased the ER at low feed moisture. Also with the inclusion of GP, the starting flour mix became less dense as GP unit density (0.48 ± 0.01 g/cm³) was significantly lower than that of cornstarch (0.78 ± 0.02 g/cm³). Further, the ER was negatively correlated with unit density (Table 4). Lower density and higher expansion are considered as favorable characteristics of extruded products. The GP levels effect on the unit density, are opposite to those found for extruded product of barley flour with GP [16], and corn flour with pineapple pomace [28], which could be due to the unique physical properties of GP and cornstarch.

Hardness: Hardness is a measure of the amount of force applied to break the sample. Lower the hardness the crispier and crunchier the extruded product is typically. GP had a significant negative linear, and significant positive quadratic effect on hardness (Table 2). The response surface plot for the GP and moisture content at fixed screw speed (200 rpm) is shown in Figure 3A. For low feed moisture (16%), increase in the GP led to lower hardness than the control samples. This low hardness, is anticipated from the fact that low density and high expansion correlates with low hardness [32,42]. These results are in line with unit density, which had a positive correlation with hardness (Table 1). Similar results are found in barley flour extrudates with GP [16] and the chick pea flour extruded snack foods [42]. The decrease in feed moisture resulted in an increased melt viscosity and SME, while the GP incorporation would result in uniform distribution of GP in cornstarch, forming more air cells with a final porous structure, which could decrease hardness. The lower hardness with GP inclusion (at 5%) can provide a crispier texture than only cornstarch extrudates when extruded under low moisture condition, in addition to health beneficial attributes.

Water absorption index (WAI) and water solubility index (WSI)

WAI measures the volume occupied by starch after swelling in excess water, which can indicate the starch integrity in water after the extrusion processing [43]. Feed moisture had positive linear, and negative quadratic effect while GP had positive quadratic effect on the WAI (Table 2). According to the response surface plots (Figure 3B) with increased feed moisture and GP, the WAI was lowered (screw speed 200 rpm). A similar trend was observed for GP (5% and 10%) levels that with an increase in moisture and screw speed heightened the WAI. At each feed moisture level (16, 20 or 24 ± 0.2%), the increase in screw speed resulted in lower WAI values, while the inclusion of GP resulted in higher WAI values. As increased screw speed results in an increased SME, that would ultimately result in more breakdown of the starch. This suggests that high screw speed renders more soluble products [17]. GP contains fiber, carbohydrates, and proteins components that can provide more hydrophilic forces to compete for water than the starch [16], which could be another explanation to the higher WAI found in extrudates with GP. WAI is an indicator of the hydrophilic groups and their gel-forming capability within the starch matrix [43]. High WAI values indicate that the extrudates can hold water with lesser solubility. This is a valuable property in processing breakfast cereal products, which helps to increase bowl life.

WSI measures soluble polysaccharides liberated from the starch after extrusion process, indicating molecular components degradation during extrusion [43]. WSI values ranged from 14.23 ± 3.85 to 72.99 ± 2.57% for all the extrudates. GP exhibited positive linear effect on WSI (Table 2). At screw speed of 200 rpm, WSI had inverse relation with GP and feed moisture as shown by the response surface plot (Figure 3C). At 16% moisture level, WSI exhibited a direct trend with increased screw speed, and an inverse trend with 5% GP inclusion. These results are in line with the previously reported WSI range [17]. The lower WSI of GP extrudates was consistent with their higher WAI, which was further proved by the negative correlation (-0.76) between WSI and WAI (Table 1). Lower values of WSI are often associated with the less dextrinization of starch during extrusion. Addition of GP provides insoluble fiber and other hydrophilic groups that can interact with starch to reduce the overall gelatinization, and hold water instead of being solubilized [27,43]. In fact cherry pomace in swelled starch granules trapes water and lowers WSI [18]. However, soluble sugars within the GP could also lead to an increase in WSI since they are more dextrinized components. The overall WSI values are likely a balance between the WSI from the starch components and from the GP components. This is likely why WSI is seen to increase with GP inclusion at low moistures, but decrease at high moistures. The lower moistures favor more mechanical breakdown in the starch leading to dextrinization as opposed to at higher moisture. It is also relevant from the high SME as it enhanced the starch breakdown and hence increased WSI. The low WSI value is favorable for the extruded product to maintain the structure; indicating GP can be utilized in the development of direct expanded breakfast cereals with crispier textures and longer bowl life.

Color

The regression analysis for the lightness (L) is shown in Table 2. GP level had a prominent effect on color parameters (L, *a*, *b*) with positive linear effect and quadratic effects on *a*, and *b* values (P≤0.01). Screw speed had a linear and quadratic effect on L and *b* values (P≤0.01). In addition, the L value was also affected by interactive effect of feed moisture and GP level (P≤0.01). Reduction in lightness with increasing GP level was observed at screw speed of 200 rpm. These findings are in agreement with those of Altan and colleagues who also found a reduction in lightness with GP inclusion [16]. The low L value with increased GP could be due to the browning Maillard reaction and caramelization due to the presence of more simple sugars and proteins in the GP compared to cornstarch. Increased *a* and *b* values could also be due to the yellowish pigments in the GP and its heat degraded products.

Total polyphenolic content of GP extrudates extracts

Based on the quality analysis of cornstarch extrudates with GP, the extrudates at low feed moisture (16%) conditions were further chosen for TPC quantification. TPC of the dried GP was 58.15 ± 5.21 mg GAE/g DW. TPC of the raw material mixes at 16% feed moisture with 0%, 5% and 10% GP were beyond detection level, 1.12 ± 0.03, and 1.89 ± 0.06 mg GAE/g DW, respectively. The loss of polyphenolics depended on the extrusion process combinations. TPC retention in the extrudates with 5% GP was 74.10% (150 rpm), and 78.57% (250 rpm), while TPC in products with 10% GP under 200 rpm had no significant loss (Figure 4A). Results indicated that most of the polyphenolics retained in extrusion processing and thus could contribute to the nutritional

quality of the extrudates. It has been reported that heat (baking) above 180°C negatively affects the TPC of the grape seed flour [15]. The difference may be attributed to the short residence time usually less than 45 S in extrusion processing [44] as opposed to significantly longer baking times. In support, recently, we found that extrusion technology has a very minimal effect on the loss of water soluble TPC in cherry pomace incorporated in cornstarch, with no significant decrease of TPC in the extruded product [18].

Effect of extrusion on the total antioxidant activities of GP extrudates

Based on the percent retention of TPC in extrudates with 5% GP, we further accessed the possible influence of extrusion processing on GP functionality. We evaluated the antioxidant activity of the 5% GP level before and after extrusion process. The antioxidant activity of the 5% GP at 16% feed moisture before extrusion was 22.15 ± 0.03 µg GAE/g DW, after extrusion at 150 rpm was 21.80 ± 0.21 µg GAE/g DW, and at 250 rpm was 21.31 ± 0.38 µg GAE/g DW. The antioxidant activity and DPPH free radical scavenging activity of extrudates with 5% GP at 150 rpm maintain the same as the raw material before extrusion with a 98% retention (Figure 4B,4C). Our findings showed that functionality of the GP bioactives could be maintained if the extrusion processing was conducted at an optimal condition, and hence GP incorporation enhances the nutritional value of extrudates. Previously, the baking process of grape seed flour significantly reduced its antioxidant activity [15]. The difference in time duration of exposure to heat which is less than 45s in extrusion processing [44] can explain the retention of antioxidant activity of the GP extrudates. Based on the physicochemical quality of the extruded products with GP, the addition of 5% GP level in cornstarch processed at 16% moisture level and 150 rpm screw speed is promising to produce healthy extrudates.

Effect of GP extrudates extracts on intracellular ROS production in Caco-2 human colonic epithelial cells

Furthermore, we examined the protective effect of GP extrudates against ROS production in human colonic epithelial cells using poplyphenolic extract prepared from GP raw mixes and extrudates. Hydrogen peroxide (H$_2$O$_2$) treatment induced enhanced ROS production in Caco-2 cells, which was mitigated by the GP polyphenolic extract treatment (Figure 4D). In fact, GP polyphenolics quenched ROS production in Caco-2 cells bringing it lower than the control with no detrimental effect of extrusion cooking. GP polyphenolics possess antioxidant as well as anti-infla mmatory effects with a well-established health beneficial effect on intestine [9,12,45,46]. These finding suggest that GP can be used as adjunctive ingredient in the production of healthy extrusion based cornstarch expended products.

Conclusion

Grape pomace incorporation in combination with feed moisture greatly affected the expansion quality of cornstarch extrudates. The 5% grape pomace level at the 16% feed moisture and 150 rpm resulted in enhanced expansion with substantial retention of TPC, total antioxidant activity in the cornstarch extrudates. This research can be applied in the making of corn starch expanded products with good nutritional value, which will enhance usage of this byproduct maximizing revenue of expending grape and wine industry. Further research on the sensory characteristics and physical and chemical interactions of GP with cornstarch need to be investigated.

Acknowledgement

This work was financially supported by Washington State University new

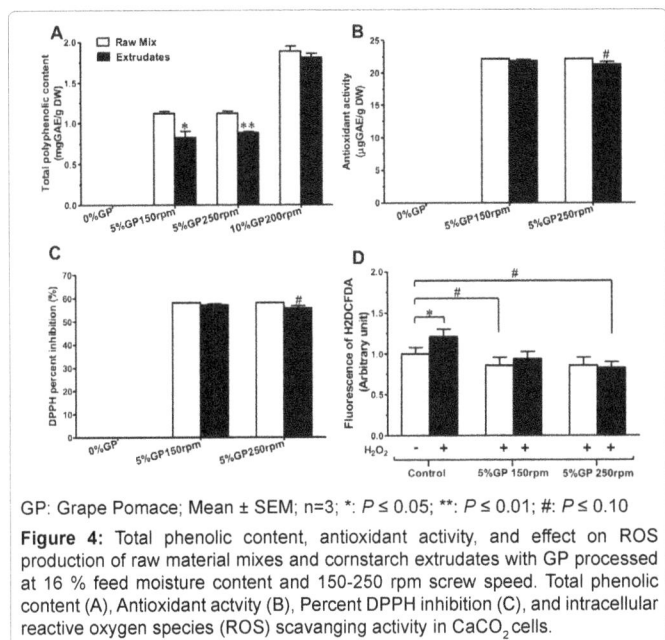

GP: Grape Pomace; Mean ± SEM; n=3; *: $P \leq 0.05$; **: $P \leq 0.01$; #: $P \leq 0.10$

Figure 4: Total phenolic content, antioxidant activity, and effect on ROS production of raw material mixes and cornstarch extrudates with GP processed at 16 % feed moisture content and 150-250 rpm screw speed. Total phenolic content (A), Antioxidant actvity (B), Percent DPPH inhibition (C), and intracellular reactive oxygen species (ROS) scavanging activity in $CaCO_2$ cells.

faculty seed grant (10A-3057-9906) to Dr. Mei-Jun Zhu. We thank the Woodward Canyon Winery (Lowden, WA) for providing the grape pomace and the Tate & Lyle (Decatur, IL, US) for providing cornstarch for the experiments. We also thank to Frank Younce and Bhim Thapa at Washington State University for their help during the project.

References

1. FAOSTAT (2012) Agricultural production domain online. Food and Agricultural Organization of the United Nations.

2. Kammerer D, Gajdos Kljusuric J, Carle R, Schieber A (2005) Recovery of anthocyanins from grape pomace extracts (Vitis vinifera L. cv. Cabernet Mitos) using a polymeric adsorber resin. Eur Food Res Technol 220: 431-437.

3. Woodard F (2001) Industrial waste treatment handbook. Butterworth-Heinemann, USA.

4. Aghsaghali MA, Sis MN, Mansouri H, Razeghi ME, Safaei AR, et al. (2011) Estimation of the nutritive value of tomato pomace for ruminant using in vitro gas production technique. Africa J Biotechnol 10: 6251-6256.

5. Fontana AR, Antoniolli A, Bottini R (2013) Grape pomace as a sustainable source of bioactive compounds: extraction, characterization, and biotechnological applications of phenolics. J Agri Food Chem 61: 8987-9003.

6. Bravi M, Spinoglio F, Verdone N, Adami M, Aliboni A, et al. (2007) Improving the extraction of α-tocopherol-enriched oil from grape seeds by supercritical CO_2. Optimisation of the extraction conditions. J Food Eng 78: 488-493.

7. Deng Q, Penner MH, Zhao Y (2011) Chemical composition of dietary fiber and polyphenols of five different varieties of wine grape pomace skins. Food Res Int 44: 2712-2720.

8. Rockenbach II, Rodrigues E, Gonzaga LV, Caliari V, Genovese MI, et al. (2011) Phenolic compounds content and antioxidant activity in pomace from selected red grapes (Vitis vinifera L. and Vitis labrusca L.) widely produced in Brazil. Food Chem 127: 174-179.

9. Bibi S, Kang Y, Yang G, Zhu M (2016) Grape seed extract improves small intestinal health through suppressing inflammation and regulating alkaline phosphatase in IL-10-deficient mice. J Funct Foods 20: 245-252.

10. Zhou K, Raffoul JJ (2012) Potential anticancer properties of grape antioxidants. J Oncol.

11. Feringa HH, Laskey DA, Dickson JE, Coleman CI (2011) The effect of grape seed extract on cardiovascular risk markers: a meta-analysis of randomized controlled trials. J America Diet Asso 111: 1173-1181.

12. Wang H, Xue Y, Zhang H, Huang Y, Yang G, et al. (2013) Dietary grape seed extract ameliorates symptoms of inflammatory bowel disease in IL10-deficient mice. Mol Nutr Food Res 57: 2253-2257.

13. Pazos M, Gallardo JM, Torres JL, Medina I (2005) Activity of grape polyphenols as inhibitors of the oxidation of fish lipids and frozen fish muscle. Food Chem 92: 547-557.

14. Tseng A, Zhao Y (2013) Wine grape pomace as antioxidant dietary fibre for enhancing nutritional value and improving storability of yogurt and salad dressing. Food Chem 138: 356-365.

15. Ross CF, Hoye CJ, Fernandez-Plotka VC (2011) Influence of heating on the polyphenolic content and antioxidant activity of grape seed flour. J Food Sci 76: 884-890.

16. Altan A, McCarthy KL, Maskan M (2008) Twin-screw extrusion of barley–grape pomace blends: Extrudate characteristics and determination of optimum processing conditions. J Food Eng 89: 24-32.

17. Altan A, McCarthy KL, Maskan M (2009) Effect of extrusion cooking on functional properties and in vitro starch digestibility of barley-based extrudates from fruit and vegetable by-products. J Food Sci 74: 77-86.

18. Wang S, Kowalski RJ, Kang Y, Kiszonas AM, Zhu M-J, et al. (2017) Impacts of the Particle Sizes and Levels of Inclusions of Cherry Pomace on the Physical and Structural Properties of Direct Expanded Corn Starch. Food Bioprocess Technol.

19. Kaisangsri N, Kowalski RJ, Wijesekara I, Kerdchoechuen O, Laohakunjit N, et al. (2016) Carrot pomace enhances the expansion and nutritional quality of corn starch extrudates. LWT - Food Sci Technol 68: 391-399.

20. Khanal RC, Howard LR, Prior RL (2009) Procyanidin content of grape seed and pomace, and total anthocyanin content of grape pomace as affected by extrusion processing. J Food Sci 74: 174-182.

21. Ganjyal GM, Hanna MA, Jones DD (2003) Modelling selected properties of extruded waxy maize cross-linked starches with neural networks. J Food Sci 68: 1384-1388.

22. Gu L, House SE, Rooney LW, Prior RL (2008) Sorghum extrusion increases bioavailability of catechins in weanling pigs. J Agri Food Chem 56: 1283-1288.

23. Brennan MA, Derbyshire E, Tiwari BK, Brennan CS (2013) Ready-to-eat snack products: the role of extrusion technology in developing consumer acceptable and nutritious snacks. Int J Food Sci Technol 48: 893-902.

24. Chinnaswamy R, Hanna MA (1988) Relationship between amylose content and extrusion-expansion properties of corn starches. Cereal Chem J 65: 138-143.

25. Ganjyal GM, Hanna MA (2006) Role of blowing agents in expansion of high-amylose starch acetate during extrusion. Cereal Chem J 83: 577-583.

26. Shevkani K, Kaur A, Singh G, Singh B, Singh N (2013) Composition, rheological and extrusion behaviour of fractions produced by three successive reduction dry milling of corn. Food Bioprocess Technol 7: 1414-1423.

27. Kumar N, Sarkar BC, Sharma HK (2010) Development and characterization of extruded product using carrot pomace and rice flour. Int J Food Eng 6: 1-24.

28. Selani MM, Brazaca SG, Dos Santos Dias CT, Ratnayake WS, Flores RA, et al. (2014) Characterisation and potential application of pineapple pomace in an extruded product for fibre enhancement. Food Chem 163: 23-30.

29. AOAC (2003) Official methods of analysis of AOAC. Association of official analytical chemists, Gaithersburg, USA.

30. Godavarti S, Karwe MV (1997) Determination of specific mechanical energy distribution on a twin-screw extruder. J Agri Eng Res 67: 277-287.

31. Anderson R, Conway H, Pfeifer V, Griffin E (1969) Gelatinization of corn grits by roll-and extrusion-cooking. Cereal Sci Tod 14: 4.

32. Kowalski RJ, Morris CF, Ganjyal GM (2015) Waxy soft white wheat: Extrusion characteristics and thermal and rheological properties. Cereal Chem J 92: 145-153.

33. Zhang S, Zhu M (2015) Characterization of polyphenolics in grape pomace extracts using ESI Q-TOF MS/MS. J Food Sci Nutri 1: 001.

34. Singleton VL, Rossi JJA (1965) Colorimetry of total phenolics with phosphomolybdic-phosphotungstic acid reagents. America J Enology Viticultur 16: 144-158.

35. Masuda T, Yonemori S, Oyama Y, Takeda Y, Tanaka T, et al. (1999) Evaluation of the Antioxidant Activity of Environmental Plants: Activity of the Leaf Extracts from Seashore Plants. J Agri Food Chem 47: 1749-1754.

36. Bellion P, Olk M, Will F, Dietrich H, Baum M, et al. (2009) Formation of hydrogen

peroxide in cell culture media by apple polyphenols and its effect on antioxidant biomarkers in the colon cell line HT-29. Mol Nutr Food Res 53: 1226-1236.

37. Seth D, Badwaik LS, Ganapathy V (2015) Effect of feed composition, moisture content and extrusion temperature on extrudate characteristics of yam-corn-rice based snack food. J Food Sci Technol 52: 1830-1838.

38. Camire ME, King CC (1991) Protein and fiber supplementation effects on extruded cornmeal snack quality. J Food Sci 56: 760-763.

39. Ganjyal GM, Reddy N, YQ Y, Hanna MA (2004) Biodegradable packaging foams of starch acetate blended with corn stalk fibers. J Appl Polymer Sci 93: 2627-2633.

40. Launay B, Lisch JM (1983) Twin-screw extrusion cooking of starches: Flow behaviour of starch pastes, expansion and mechanical properties of extrudates. J Food Eng 2: 259-280.

41. Carvalho CW, Mitchell JR (2000) Effect of sugar on the extrusion of maize grits and wheat flour. Int J Food Sci Technol 35: 569-576.

42. Meng X, Threinen D, Hansen M, Driedger D (2010) Effects of extrusion conditions on system parameters and physical properties of a chickpea flour-based snack. Food Res Int 43: 650-658.

43. Gonzalo de Gutierrez MV, Gomez MH (1987) A model for the extrusion of a corn: Soybean blend. Arch Latinoam Nutr 37: 494-502.

44. Ganjyal G, Hanna M (2002) A review on residence time distribution (RTD) in food extruders and study on the potential of neural networks in RTD modeling. J Food Sci 67: 1996-2002.

45. Yang G, Wang H, Kang Y, Zhu M (2014) Grape seed extract improves epithelial structure and suppresses inflammation in ileum of IL-10-deficient mice. Food Funct 5: 2558-2563.

46. Yang G, Xue Y, Zhang H, Du M, Zhu M (2015) Favourable effects of grape seed extract on intestinal epithelial differentiation and barrier function in IL10-deficient mice. Br J Nutr 114: 15-23.

32

Physicochemical and Sensory Evaluation of Dhakki Dates Candy

Zeeshan M[1]*, Saleem SA[1], Ayub M[2], Shah M[2] and Jan Z[2]

[1]*Food Technology Section, Agricultural Research Institute, D.I. Khan, K.P, Pakistan*
[2]*University of Agriculture, Peshawar, K.P, Pakistan*

Abstract

The experiment was conducted to develop candy from Dhakki dates picked at Khalal stage. Physicochemical and sensory characteristics like moisture, pH, TSS, color, flavor, texture and overall acceptability were studied for total period of six months. Candy was prepared from 5 different sugar concentrations, i.e., T0 (control), T1 (20%), T2 (40%), T3 (60%), and T4 (70%). Among them, best treatment was identified on the basis of overall acceptability. Candy prepared from T3 (60%) proved to be best but the candy prepared from T2 (40%) was equally good. The least acceptable was the candy of T1 followed by T4. Sensorial properties, moisture and pH decreased while TSS increased during six months of storage. Candy packed in HDPE bags can be kept safely up to six months.

Keywords: Physicochemical; Sensory properties; Sucrose; Khalal stage

Introduction

Date palm (*Phoenix dactylifera L.*) is an important fruit crop known from centauries as a high-energy food and desert. The nutritional value of date is high since it contains sugar, minerals and vitamins [1]. Sugars contain more than 70% of dry weight and are an important nutrient in dates. Dates play an important role in both mental and sexual activities. They also provide protection against age related problems which may be contributed to their high content of antioxidant compounds [2].

Carbohydrates, vitamins and minerals rich dates are good not only for humans but can also be used as a feed supplement for livestocks. (Zohary and Hopf [3]; Al-Shahib and Marshal [4]; Hassan et al. [5].

Asia stands first among region with 60 million date palms mostly grown in Saudi Arabia, Bahrain, Oman, Pakistan, Iran, Iraq and Yemen followed by African region with 32.5 million date palms. Mexico and USA have 600,000 date palms followed by Europe and Australia with 320,000 and 30,000 date palms respectively [6].

Pakistan was the 5th largest date producer with an annual production of 600,000 t in 2012 which may be contributed to the increase in cultivation of area of date palms in 2012 [7]. Many varieties are grown in Pakistan and some cultivars are grown in specific areas, like Dhakki in Dera Ismail Khan, Begum Jhangi in Panjgur and Aseel in Khairpur, all have the capability to compete with the world dates [8].

Among the local varieties in Dera Ismail Khan Dhakki is very popular for its jumbo size and weight with pleasure taste and texture [9]. Consumer demand for Dhakki dates is increasing rapidly in Dera Ismail Khan [10]

But the problem with Dhakki date in Dera Ismail Khan is that, monsoon rains occurs from July to September, which coincide with the date ripening season in the area. Date palm is mainly sensitive to rain from khalal (maturity) to rutab and Tamar (ripening). Rain can easily deteriorate dates within 2 days, at late khalal stage crakes in epicarp can appear even in hours. At early khalal stage, rain does not cause any damage and even may have a beneficial effect by washing dirt and dust [11].

To overcome these problems many studies have been carried out on date palm by-products like date pickles, jams, date syrup, ice creams, chocolates and date paste. (Al-Hooti et al. [12]; Hamad et al. [13]; Khatchadourain et al. [14]

Yousif et al. [15] conducted his research on the possibility of using date paste as a replacer for caramel or sugar paste in preparing candy bars. Processing conditions, nutritive value and organoleptic properties of the prepared date bars as well as their storability were evaluated. The results indicated that the prepared date bars either plain or chocolate coated had good acceptability, possessed a high nutritive value and could be stored for more than 5 months under refrigeration (5°C) without affecting their qualities.

Candy or confectionary is popular food item. Most of fruit candies available in market are imported. The availability of dates in substantial quantities in Dera Ismail Khan justify their use in various products and processing plain which can be used as a replacer for dates when the season is off.

Therefore, the objectives of the present study were to develop date candy from dhakki variety using different sugar concentration and to assess its quality in order to make the product available through the year, to generate income and improve marketability of the dates in a proper way for the benefits of farmers.

Materials and Methods

The experiment was conducted in month of August 2016, at Agriculture Research Institute Dera Ismail Khan. Dhakki dates at khalal stage were picked from the orchard of Agriculture Research Institute Dera Ismail khan and brought to laboratory of Food Technology Section.

Preparation of dhakki date candy

The dates were steeped in 0% sugar syrup for one day and then syrup was drained and dates were dried. Sugar was added to the syrup until TSS reached 20% and dates were steeped for 24 hrs. After 24 hr's dates

---

*Corresponding author: Zeeshan M, Food Technology Section, Agricultural Research Institute, D.I. Khan, K.P, Pakistan, E-mail: zeeshanfst07@gmail.com

Treatment	Storage Interval							% Dec	Mean
	0	30	60	90	120	150	180		
T0	16.14	15.02	14.23	13.67	13.04	12.35	12.00	25.65	13.7d
T1	15.18	15.00	14.56	14.12	13.78	13.13	12.45	17.98	14b
T2	16.77	16.12	15.44	14.89	14.10	13.77	13.33	20.51	14.9a
T3	14.16	13.68	13.12	12.78	12.22	12.00	11.89	16.03	12.8e
T4	14.87	14.45	14.12	13.78	13.44	13.14	12.89	13.32	13.8c
Mean	15.4a	14.8b	14.2c	13.8d	13.3e	12.8f	12.5g	--	--

a-g Means followed by different letters are significant (P≤0.05).

Table 1: Effect of storage period and treatments on (%) moisture of date candy.

Treatment	Storage Interval							% Dec	Mean
	0	30	60	90	120	150	180		
T0	5.96	5.67	5.14	4.67	4.35	4.11	4.00	32.89	4.81c
T1	5.46	5.35	5.13	4.97	4.66	4.37	4.14	24.13	4.84bc
T2	5.67	5.35	5.16	4.03	3.88	3.76	3.45	39.15	4.4d
T3	5.70	5.37	5.21	5.09	4.79	4.56	4.30	24.56	5a
T4	5.40	5.26	5.08	4.91	4.76	4.49	4.28	20.74	4.83b
Mean	5.63a	5.4b	5.1c	4.7d	4.4e	4.2f	4g	--	--

a-g Means followed by different letters are significant (P≤0.05).

Table 2: Effect of storage period and treatment on pH of date candy.

Treatment	Storage Interval							% Inc	Mean
	0	30	60	90	120	150	180		
T0	45.14	45.34	45.61	45.92	46.13	46.37	46.59	3.11	47.6e
T1	71.12	71.79	72.36	73.12	73.79	74.37	74.97	5.14	71.7d
T2	72.18	73.20	73.76	74.11	74.57	74.97	75.34	4.19	73.6b
T3	72.25	72.78	73.23	73.47	73.88	74.35	74.80	3.41	73.5c
T4	73.12	73.57	73.98	74.25	74.56	74.88	75.65	3.34	74.2a
Mean	66.6g	67.2f	67.7e	68.2d	68.5c	69b	69.5a		

a-g Means followed by different letters are significant (P≤0.05).

Table 3: Effect of storage period and treatments on TSS of date candy.

Treatment	Storage Interval							% Dec	Mean
	0	30	60	90	120	150	180		
T0	8	5	2	1	1	1	1	87.50	2.7c
T1	9	8.7	8.5	8	7.3	6	5	44.44	7.5b
T2	9	8.8	8.4	8	7.5	7.2	7	22.22	7.9a
T3	9	8.5	8.2	8	7.7	7.5	7.1	21.11	8.0a
T4	8.6	8.2	7.8	7.4	7	6.6	6	30.23	7.3b
Mean	8.7a	7.8b	6.9c	6.4d	6.1e	5.6f	5.2g	--	--

a-g Means followed by different letters are significant (P≤0.05).

Table 4: Effect of storage period and treatments on color of date candy.

were removed from syrup and dried. Syrup having 20% total soluble solids was boiled till TSS reached 40%, the dates were kept in that 40% syrup for one day. This process continued till the TSS of syrup reached 70%. At 70% TSS the dates were kept for about 48 hrs. At each level of TSS i.e., 0%, 20%, 40%, 60% and 70% the syrup was drained from dates and dried in mechanical dehydrator at $60^{\circ}C$ till moisture content of <16% is achieved. In the preparation of candy osmotic dehydration step prior to drying was used as described by Ramamurthey et al. [16]. The drying time requirement was similarly followed as described by Islam and Flink [17,18].

Treatments

T_0= Control

T_1 = Dates + 20% sugar syrup+ 0.1% potassium metabisulphite

T_2= Dates + 40% sugar syrup+ 0.1% potassium metabisulphite

T_3= Dates + 60% sugar syrup+ 0.1% potassium metabisulphite

T_4= Dates + 70% sugar syrup+ 0.1% potassium metabisulphite

Storage

Prepared date candy of different treatments was then wrapped in polyethylene bags and kept in cool airtight boxes for storage and further analysis.

Chemical analysis

The date candies were analyzed for moisture, pH and total soluble solids according the official standard method (AOAC 2003).

Sensory analysis of date candy

For statistical analysis samples were evaluated for moisture, pH, TSS, and sensory evaluation for color, flavor, texture and overall acceptability was performed by panel of 9 members. The samples were presented to 9 members. The members were asked to rate the different composition presented to them on a 9-point hedonic scale with the ratings of: 9 = Like extremely; 8 = Like very much; 7 = Like moderately; 6 = Like slightly; 5 = Neither like nor dislike; 4 = Dislike slightly; 3 = Dislike moderately; 2 = Dislike very much; and 1 = Dislike extremely. The result was analyzed by statistical software (statistics).

Results and Discussion

Moisture

The results of changes in moisture content of dhakki date candy during storage are presented in Table 1. A significant decrease was observed in moisture content during total period of storage. The moisture content decreased significantly (p<0.05) from 15.4 to 12.5 during total storage interval. For treatment, the highest mean was observed in T2 (14.9) while minimum was recorded in T3 (12.8).

The decrease in moisture content may be due to evaporation during storage. Variation in loss of moisture can due to variation in treatment. Chavan et al. [18] reported a decreasing trend in moisture content of osmo dried banana slices during six-month storage.

pH

The mean value of pH decreased from 5.63 to 4 during total period of storage. Highest mean value for treatment was observed in T1 (4.84) while the minimum value was observed in T2 (4.4) as presented in Table 2. During storage, highest percent decrease was observed in T2 (39.15) while lowest fall was recorded in T4 (20.74).

The decrease during total period of storage might be due to increase in acidity and can also be contributed to some other chemical reactions. Natalia et al. [19] also observed a decreasing trend in pH during storage while studying apple leather.

TSS

The mean value for TSS increased from 66.6 to 69.5 during total period of storage (Table 3). Highest mean value for treatment was recorded in T4 (74.2) while minimum score was recorded in T0 (47.6). Highest percent increase was recorded in T1 (5.14) while minimum increase was observed in T0 (3.11).

The increase in Total soluble solids might be due to the conversion of starch and other insoluble carbohydrates into sugar and may also be due to the loss of moisture that tends to increase TSS. A similar increasing trend was observed by Phimpharian et al. [20] reported an increase in TSS (from 82.42-86.9) while studying apple leather.

Color

The mean score of judges for color significantly (p<0.05) decreased from 8.7 to 5.2 during storage (Table 4). For treatment, maximum mean score was observed in T3 (8.0) while minimum mean score was observed in T0 (2.7).

Color of T3 was comparatively attractive during storage period at ambient temperature. Similar results have been reported by Durrani et al. [21] in development and quality evaluation of honey based carrot candy that osmotic drying had a protective effect upon the color and flavor of fully dried fruits.

Flavor

Flavor is a vital quality factor that determines the consumer attraction to the product. The results pertaining to the response of flavor on the storage interval of the candies prepared from dhakki dates are presented in Table 5.

The mean score of judges for flavor significantly (p<0.05) decreased from 8.47 to 5.4 during total period of storage. For treatment, maximum mean score was observed in T1 (7.7) while minimum mean score was observed in T0 (1.7). These results were in harmony with the observation of Dermesonlouoglou et al. [22] who reported 54.55% decrease in flavor scores of osmo- dehydrofrozen tomatoes, during 12 months' storage.

Texture

Originally the mean score of juries for texture of date candy from

Treatment	Storage Interval							% Dec	Mean
	0	30	60	90	120	150	180		
T0	6.5	1.0	1.0	1.0	1.0	1.0	1.0	84.62	1.7c
T1	9.0	8.5	8.0	7.6	7.3	7.0	6.5	27.78	7.7ab
T2	8.8	8.5	8.1	7.7	7.2	7.0	6.8	22.73	7.6a
T3	9.0	8.8	8.3	7.6	7.1	6.8	6.5	27.78	7.4ab
T4	9.0	8.5	8.1	7.6	7.2	6.8	6.2	31.11	7.6b
Mean	8.47a	7.06b	6.7c	6.3d	5.9e	5.7f	5.4g	--	--

a-g Means followed by different letters are significant (P≤0.05).

Table 5: Effect of storage period and treatments on flavor of date candy.

Treatment	Storage Interval							% Dec	Mean
	0	30	60	90	120	150	180		
T0	9.0	2.0	1.0	1.0	1.0	1.0	1.0	88.89	2.2e
T1	9.0	8.3	8.0	7.6	7.3	7.0	6.5	27.78	7.6b
T2	9.0	8.5	8.0	7.6	7.2	6.7	6.1	32.22	7.5c
T3	9.0	8.7	8.3	8.0	7.6	7.1	6.7	25.56	7.9a
T4	9.0	8.3	8.0	7.4	7.1	6.5	6.1	32.22	7.4d
Mean	9a	7.1b	6.6c	6.3d	6e	5.6f	5.2g	--	--

a-g Means followed by different letters are significant (P≤0.05).

Table 6: Effect of storage period and treatments on texture of date candy.

Treatment	Storage Interval							% Dec	Mean
	0	30	60	90	120	150	180		
T0	6.0	2.0	1.0	1.0	1.0	1.0	1.0	83.33	1.8d
T1	9.0	8.5	8.1	7.7	7.4	6.6	6.0	33.33	7.6b
T2	9.0	8.6	8.1	7.7	7.3	6.9	6.6	26.67	7.7a
T3	9.0	8.6	8.2	7.8	7.3	6.9	6.5	27.78	7.8a
T4	8.8	8.3	7.9	7.4	7.1	6.6	6.1	30.68	7.4c
Mean	8.3a	7.2b	6.6c	6.3d	6e	5.6f	5.2g	--	--

a-g Means followed by different letters are significant (P≤0.05).

Table 7: Effect of storage period and treatments on overall acceptability of date candy.

T0 0 to T4 was 9 for all, which was progressively reduced to1, 6.5, 6.1, 6.7, and 6.1 correspondingly during the total period of storage.

The mean score of judges for texture significantly (p<0.05) decreased from 9.00 to 5.2 during storage as presented in Table 6. For treatments, maximum mean score was observed in T3 (7.9), while minimum was recorded in T0 (2.2). The highest percent decrease was observed in T0 (88.89) while minimum was observed in T3 (25.56) Similar results were observed by Muhammad et al. [23] during storage of pear glaces.

Overall acceptability

The results on changes in overall acceptability of dhakki date candy during storage are presented in Table 7. A mean overall acceptability score of all treatments was found to gradually decrease from 8.3 to 5.2 during storage which may be due to reduction in score of color, flavor, and texture of date candy. For treatment, maximum mean score was observed in T2 and T3 both having value of 7.7 while minimum score was observed in T0 (1.8).

Chavan et al. [18] reported that a gradual decrease in overall acceptability score from 8.40 to 7.80 during six-month storage may be due to reduction in score of colour and appearance texture, taste of osmo-dried banana slices [24,25].

Conclusion

The different sugar solutions significantly affected the quality of date candy. The best quality date candy was that prepared with 60% (T3) sugar solution, followed by T2. The color, flavor and texture of date candy with 60% and 40% sugar solutions were preferred by the judges because of unique and sweat flavor. The color was excellent; taste was sweat, and a tender texture. From the results, it can be concluded that there is possibility of for utilization of surplus dates in candy making.

References

1. El-Shaarawy MI (1971) Intakes of phosphorus and calcium through excessive consumption of certain food items. In Phosphorus and Calcium Intakes by Dutch Diets. Utrecht Univ Holl.

2. Ashraf JF (2007) Antioxidant content of dates. In Proceedings of 4th Symposium on Date Palm. Al-Hassa, Saudi Arabia: 418.

3. Zohary D, Hopf M (2000) Domestication of plants in the old world: the origin and spread of cultivated plants in West Asia, Europe and the Nile valley (3rd edn). Oxford University Press, New York, USA.

4. Al-Shahib W, Marshall RJ (2003) The fruit of the date palm: Its possible use as the best food for the future. Inter J Food Sci Nutri 54: 247-259.

5. Hassan S, Bakhsh K, Gill ZA, Maqbool A, Ahmad W (2006) Economics of growing date palm in Punjab, Pakistan. Inter. J. Agri. Bio 8: 788-792.

6. Zaid A (2001) The world date production: a challenging case study. 2nd International conference on date palms, Al-Ain, UAE: 902-915.

7. FAO (2014) Food and Agriculture Organization of the United Nations. Food and agricultural commodities production for Pakistan for 2012.

8. PHDEB (2008) Pakistan Horticulture Development and Export Board. Dates Marketing Strategy.

9. Baloch AK (1999) Enhancement of post-harvest quality and stability of Dhakki dates using advanced technology. A publication of Pakistan Science Foundation, Islamabad.

10. Abul-Soad AA (2010) Date palm in Pakistan, current status and prospective. USAID Firms project: 9-11.

11. Zaid A, De Wet PF (2002) Date palm cultivation. FAO UN Plant Production and Protection.

12. Al-Hooti S, Sidhu JS, Qabazard H (1997) Physico-chemical characteristics of five date fruit cultivars grown in the United Arab Emirates. Plant Foods for Human Nutrition 50: 101-113.

13. Hamad AM, Mustafa AI, AI-Kahtani MS (1983) Possibility of utilizing date syrup as a sweetening and flavoring agent in ice cream making. Proc. of the 1ˢᵗ Symp. on Date Palm, King Faisl Univ AI-Hassa, Saudi Arabia, March 23-25: 544-550.

14. Khatchadourian HA, Sawaya WN, Khalil J, Safi WM, Mashadi AA (1983) Utilization of dates (*Phoenix dactylifera L*). grown in the Kingdom of Saudi Arabia in various date products. Proceedings of the 1ˢᵗ Symposium on Date Palm, K.F. University, AI Hassa, Saudi Arabia 2: 504-518.

15. Yousif AK, Alshaawan AF, Mininah MZ, Eltaisan SM (1987) Processing of date preserve and date jelly. Date Palm J 5: 73-86.

16. Ramamurthy MS, Bongi RC, Ward R, Banoyopaifihay C (1978) Osmotic dehydration of fruits possible alternative to freeze-drying. Indian Food Pack 32: 108-112.

17. Islam MN, Flink JN (1982) Osmotic concentration and its effect on air drying behavior. Inter J Food Sci Tech. 1365-2621.

18. AOAC (2000) Official methods of analysis. (17th edn) Association of Official Analytical chemists, Arlington, VA, USA.

19. Larmond E (1977) Method for sensory evaluation of food. Canada Department of Agriculture Publication.

20. Chavan UD, Prabukhanolkar AE, Pawar VD (2010) Preparation of osmotic dehydrated ripe banana slices J Food Sci and Tech 47: 380-386.

21. Natalia A, Ruiz Q, Demarchi SM, Massolo JF, Rodoni LM, et al. (2012) Evaluation of quality during storage of apple leather. LWT. Food Sci and tech 47: 485-492.

22. Phimpharin C, Jangchud A, Jangchud K, Kyoon H (2011) Physicochemical and sensory optimization of pineapple leather snack as affected by glucose syrup and pectin concentration. Inter J Food Sci and Tech 46: 972-981.

23. Durrani AM, Srivastava PK, Verma S (2011) Development and quality evaluation of honey based carrot candy. J Food Sci Tech 48: 502-505.

24. Dermesonlouoglou EK, Giannakourou M, Taoukis PS (2006) Kinetics modeling of the degradation of quality of osmo-dehydrofrozen tomatoes during storage. J Food Chem 103: 998-993.

25. Muhammad N, Shah AS, Riaz A, Hashim MM, Mahmood Z, et al. (2007) Prepration and evaluation of pear glace at different stages of maturity. Sarhad J Agric 23: 305-308.

33

Temperature Dependence of Bulk Viscosity in Edible Oils using Acoustic Spectroscopy

Sunandita Ghosh*, Melvin Holmes and Malcolm Povey

Department of Food Science and Nutrition, University of Leeds, Leeds, UK

Abstract

When ultrasound waves are applied to a compressible Newtonian fluid, bulk viscosity plays an important parameter to cause attenuation. Ultrasound spectroscopy is an important technique to characterise and determine the physico-chemical properties of many food components because it is a non-invasive, non-destructive, easy and accurate technique. The aim of this study was to find the bulk viscosity of three brands of sunflower and extra-virgin olive oil by using the Navier's-Stoke equation across a temperature range of 5°C to 40°C and to test the hypothesis that there is a significant difference in the value of bulk viscosity between the different brands of sunflower and olive oil used. The value of bulk viscosity was not found to be constant over the operating frequency range of 12-100 MHz, which suggested edible oils are non-Newtonian fluids. Also, no significant statistical difference of bulk viscosity values was found between different brands of the same oil ($p \geq 0.05$). This shows bulk viscosity is not affected by small compositional variations. Acoustic spectroscopy is increasingly being used to characterise food materials. More studies on bulk viscosity must be employed in order to be able to utilise this technology to its full strength.

Keywords: Bulk viscosity; Acoustic spectroscopy; Attenuation

Introduction

Edible oils occupy an important position in the human diet because of its nutritive value and also because of its organoleptic and rheological properties [1]. Acoustic spectroscopy is increasingly being used to characterise oils and fats due to its many advantages. Ultrasound waves are longitudinal sound waves of frequency of 20 KHz or more [2]. Ultrasonic analysis is a useful technique to characterize and determine many physico-chemical properties of oil and other food component mainly because it is a non-invasive, non-destructive, easy and accurate technique. It can be used 'on-line or off-line' and also works for opaque food objects where characterising food on visual methods can be difficult [1].

Compressible fluids are fluids whose density changes when high-pressure gradient is applied. When force is applied to these fluids, they flow in the form of transverse-pressure waves. The velocity of propagation of these pressure waves in compressible fluid is known as velocity of sound [3]. Newtonian fluids are ones whose shear stress is proportional to shear strain. Compressible Newtonian liquids exhibit two types of viscosity: shear and bulk viscosity. Shear viscosity is the resistance to the change in shape under shear stress and the bulk or volume viscosity is the resistance to change in volume under an applied pressure [4]. Bulk viscosity is also termed as volume viscosity, second viscosity coefficient, expansion coefficient of viscosity and coefficient of bulk viscosity. Bulk viscosity is significant if the compression or expansion in the fluids proceeds so rapidly that it takes longer time than the duration of change in volume to restore the thermodynamic equilibrium like with the absorption or dispersion of sound waves [5]. The experimental values obtained for ultrasonic absorption is often found to be much larger than the values obtained from classical Stroke's equation where only shear viscosity is considered [4]. This increase in the absorption can be attributed to the bulk viscosity and thermal conduction. Fluid molecules have translation, rotational and vibrational degrees of freedom. The translational motion is due to the dynamic viscosity and the rotational and vibrational motion is due to bulk viscosity. Therefore, to know the effect of vibrational and rotational energy on fluids obtaining the bulk viscosity data is very important [5,6]. There is also a certain bulk viscosity effect in fluid flow for fluids with large Reynold's number when the ratio of bulk to shear viscosity is of the order of the square root of Reynold's number [7]. However, given the importance of bulk viscosity little research has been conducted to characterise it and for many fluids it is unknown or inaccurately known particularly across different temperatures. Bulk viscosity is observed when sound particularly ultrasonic waves travels through fluids. Hence, ultrasound waves can also be used to measure the bulk viscosity of fluid.

In a study by Dukhin and Goetz [6], three methods were used to find the bulk viscosity: Brillouin spectroscopy, laser gradient spectroscopy and acoustic spectroscopy. It was found that the acoustic spectroscopy gave the most precise results for bulk viscosity; as with Brillouin spectroscopy there were 'high errors due to the difficulty in measuring the Brillouin linewidth' and with laser gradient spectroscopy complications arrived due to the fitting of the laser gradient with five adjustable parameters. Acoustic spectroscopy gives value for the speed of sound which can be used to measure compressibility and it is also the only method where multi-frequency measurements, in the range of 1-100 MHz can be taken. This is important to find the nature of the fluid, if it's a Newtonian or non-Newtonian fluid. If the fluid is Newtonian than the calculated bulk viscosity will be independent of the frequency changes [6].

The Navier-Stokes equation is important to study physical fluid dynamics. The general Navier-Stokes equation is written as:

$$\frac{\partial v}{\partial t} + (\mathbf{v}\nabla)\mathbf{v} = -\frac{1}{\rho}\nabla P + \eta\nabla^2\mathbf{v} + \frac{1}{\rho}\mathbf{F} \qquad (1)$$

Where ρ (kg.m^{-3}) is the density, t (s) is time, $\mathbf{v}$ is the velocity vector,

---

*Corresponding author: Sunandita Ghosh, Department of Food Science and Nutrition, University of Leeds, Leeds, UK, E-mail: sunandita.ghosh11@gmail.com

P (Pa) is pressure, η (Pa.s) is shear viscosity and F (N) is the body force term as such forces act on the volume of a fluid particle.

Bulk viscosity is an important term in the Navier-Stokes equation for a Newtonian compressible liquid [6].

$$\rho\left[\frac{\partial v}{\partial t}+\left(\mathbf{v}\nabla\right)\mathbf{v}\right]=-grad\,P+\eta\Delta\mathbf{v}+\left(\mu+\frac{4}{3}\eta\right)grad\,div\,\mathbf{v} \quad (2)$$

μ (Pa.s) is the bulk viscosity. For an incompressible liquid, the last term on the right hand side may be neglected as this term accounts for compressibility.

grad div $\mathbf{v}=0$

Thus, the bulk viscosity term has no contribution for incompressible fluids. Therefore, for incompressible fluids the Navier-Stokes equation can be written as:

$$\rho\left[\frac{\partial v}{\partial t}+\left(\mathbf{v}\nabla\right)\mathbf{v}\right]=-grad\,P+\eta\Delta\mathbf{v} \quad (3)$$

Thus the effect of bulk viscosity is not very significant for incompressible fluids and for ideal monoatomic gas for which $\mu=0$ [5].

When a wave propagates through a viscous and thermally non-conductive fluid then the general solution obtained from the Navier-Stokes equation with respect to attenuation is [6]:

$$2\left(\frac{\alpha_{long}v}{\omega}\right)^2=\frac{1}{\sqrt{1+t^2\omega^2}}-\frac{1}{\sqrt{1-t^2\omega^2}} \quad (4)$$

Where α_{long} (Np.m^{-1}) is ultrasound attenuation coefficient, v (ms^{-1}) is the velocity of sound, ω is the ultrasound frequency, t (s) is the viscous relaxation time and takes into account both bulk and shear viscosity and is given by:

$$t=\frac{1}{\rho v^2}\left(\frac{4}{3}\eta+\mu\right) \quad (5)$$

If attenuation is plotted as a function of frequency a normal distribution curve is obtained and at the critical frequency the maximum value obtained is approximately equal to the viscous relaxation time. The critical frequency is around 1000 GHz around, this high ultrasound range is difficult to achieve in real instruments but for low frequency this is achievable. The low frequency asymptotic function is given by [8]:

$$\alpha_{long}=\frac{\eta\omega^2}{2\rho v^3}\left[\frac{4}{3}+\frac{\mu}{\eta}+\frac{(\gamma-1)\tau}{\eta C_p}\right] \quad (6)$$

Where γ is the ratio of specific heats, τ (w.m^{-1}.K^{-1}) is the thermal conductivity and C_p (J.K^{-1}) is the specific heat at constant pressure, v (m.s^{-1}) is velocity, T(°C) is temperature, β (K^{-1}) is the bulk compressibility and ω is the angular frequency $\omega=2\pi f$, f (Hz) is the frequency of acoustic wave and i is the imaginary number. From equation (6) can be expressed as equation (7) to calculate bulk viscosity.

$$\mu=\frac{2\alpha\rho v^3}{\omega^2}-\frac{4\eta}{3}-\frac{(\gamma-1)\tau}{C_p} \quad (7)$$

α (Np.m) is the attenuation coefficient. The contribution of thermal conduction to bulk viscosity is dependent on $(\gamma-1)$ · For liquids the ratio of specific heats is close to one and for gases it is greater than one as liquids are less compressible as compared to gases. Therefore, there is not much contribution to the bulk viscosity from the thermal properties of the material and the thermal term can be neglected [6].

Hence equation (7) can be re-written as:

$$\mu=\frac{2\alpha\rho v^3}{\omega^2}-\frac{4\eta}{3} \quad (8)$$

The temperature dependence of physical parameters of edible oils is demonstrated by the following model equations [9]:

$$c=c_0+c_1 T \quad (9)$$

$$\rho=\rho_0+\rho_1 T \quad (10)$$

$$\eta=\eta_0\exp\left[-\frac{\eta_1}{k(T+273.13)}\right] \quad (11)$$

Where c (m.s^{-1}) is velocity, ρ (kg.m^{-3}) is density, η (kj.mol^{-1}) is viscosity, k is Boltzmann, T is temperature (°C). The subscripted terms are constants. For sunflower oil, the values of the constants have been reported as: ρ_0 (kg.m^{-3}) is 933.76, ρ_1 (kg.m^{-3}, °C^{-1}) is -0.61 for 20 to 80°C; ln η_0 is -13.83, η_1 (kJ.mol^{-1}) is 27.17 for 25°C to 50°C; c_0 (m.s^{-1}) is 1538 and c_1 (m.s^{-1}, °C^{-1}) is -3.28 for 5°C to 70°C. For olive oil c_0 (m.s^{-1}) is 1528.9 and c_1 (m.s^{-1}.°C^{-1}) is -3.23 for 20°C to 70°C. All the three parameters velocity, density and shear viscosity are temperature dependent (equations 9-11) and plays an important role in finding bulk viscosity. Also, Coupland and McClements [9] emphasised that these bulk properties of oils depend upon their chemical composition. Therefore, there is expected to be a dependence of bulk viscosity on temperature for edible oils which will be discussed in this work (Figure 1).

The aim of this work was to calculate the bulk viscosity of sunflower oil and extra-virgin olive oil across a temperature range of 5°C to 40°C and to test the hypothesis that there is a significant difference in the value of bulk viscosity between the different brands of sunflower (Tesco, Morrisons, Floras) and extra-virgin olive (Tesco, Morrisons, Sierra mágina) oil. Even for the same type of oil there are differences in the physico-chemical properties as the composition of food oil varies significantly depending on the geographical source, processing parameters (like distillation), storage time (as crystallisation or oxidation might take place) [9]. The length of fatty acid chain has an effect on the viscosity of the oil [10]. Hence, three different brands were investigated to find if these differences have any significant effect on bulk viscosity. The experimental procedure of conducting the study is mentioned in the next section. The results obtained from this work is illustrated in the Results and Discussion part. First, the justification of using the frequency squared equation (8) for finding the bulk viscosity was given using the graph of log attenuation v/s log frequency (Figure 2). Second, all the bulk properties like velocity, density, shear viscosity and bulk viscosity was tabulated in Table 1 and the dependence of bulk viscosity on the frequency was studied (Figures 3 and 4). Third,

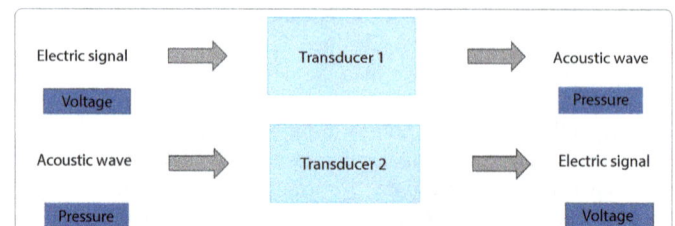

Figure 1: The first type of transducer emits ultrasound wave on application of electrical voltage which is picked up by the sample. The second type of transducer detects the ultrasound wave emitted by the transducer and converts it into electrical voltage which is measured as attenuation.

a comparative study between the bulk and shear viscosity was made (Table 2 and Figure 5). Fourth, the hypothesis was tested to find out if

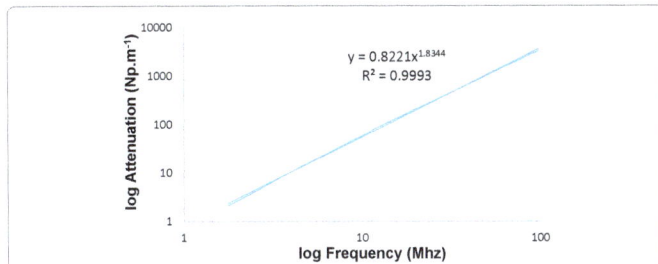

Figure 2: Plot of log Attenuation v/s log Frequency of Tesco sunflower oil at 25°C showing a best fit polynomial.

Temperature	Density	Velocity	Shear viscosity	Bulk viscosity		
T (°C)	ρ (kg m^{-3})	v (m s^{-1})	$\eta \times 10^{-2}$ Pa.s	$\mu \times 10^{-2}$ Pa.s		CV
				Mean*	SD	
6	928.71	1515.69	5.22	5.79	1.89	0.33
10	926.01	1501.93	4.75	4.51	1.59	0.35
15	922.57	1485.51	4.15	3.57	1.43	0.22
20	919.15	1469.02	3.55	2.91	1.21	0.41
25	915.74	1452.04	2.37	3.33	0.99	0.3
30	912.35	1435.36	2.1	2.77	0.86	0.31
35	908.97	1419.2	2.06	2.07	0.71	0.2
40	905.6	1403.98	1.42	2.36	0.6	0.25

Standard Deviation (SD)
Coefficient of Variation (CV)
*Mean was taken for the bulk viscosity in the frequency range of 12 MHz-100MHz Ultrasound velocity readings was obtained from the Ultrasizer. Density was measured by Anton Paar DMA 4500 M density-meter and shear viscosity by Anton Paar MCR 302 rheometer.

Table 1: Density, velocity, shear viscosity and mean bulk viscosity of tesco sunflower oil at the selected temperatures.

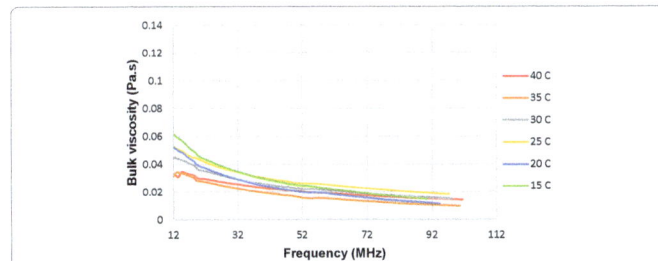

Figure 3: Plot of bulk viscosity v/s frequency of Tesco sunflower oil at each selected temperature. A decrease in the value of bulk viscosity is seen with the increase in temperature, showing frequency dependence and possible non-Newtonian behaviour.

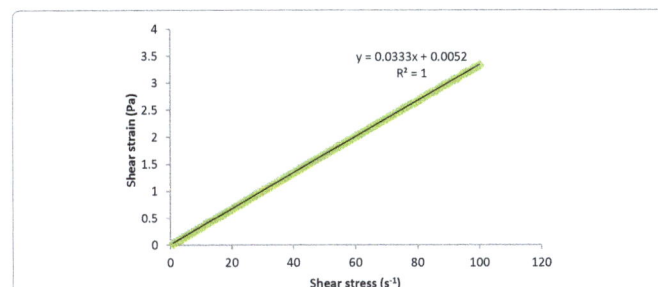

Figure 4: Plot of shear stress v/s shear strain, giving a linear relationship between stress and strain suggesting Newtonian behaviour of edible oils.

Temperature	Bulk viscosity		Shear viscosity	Ratio ($\frac{\mu}{\eta}$)
T°C	$\mu \times 10^{-2}$ Pa.s		$\eta \times 10^{-2}$ Pa.s	
	Mean	SD		
6	5.79	1.89	1.89	1.11
10	4.51	1.59	1.59	0.95
15	3.57	1.43	1.43	0.86
20	2.91	1.21	1.21	0.82
25	3.33	0.99	0.99	1.41
30	2.77	0.86	0.86	1.32
35	2.07	0.71	0.71	1
40	2.36	0.6	0.6	1.66

Table 2: The bulk and shear viscosity values of Tesco sunflower oil at the selected temperatures and the ratio between bulk viscosity to shear viscosity.

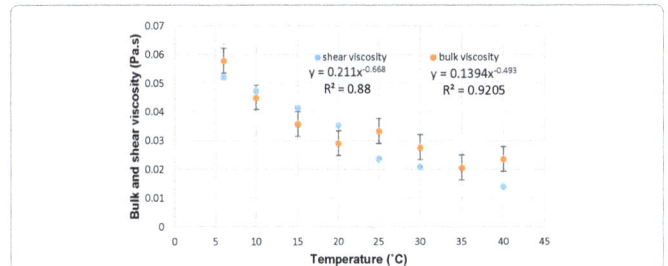

Figure 5: Plot of bulk and shear viscosity v/s temperature of Tesco sunflower oil showing a decrease in the values with the increase in the temperature.

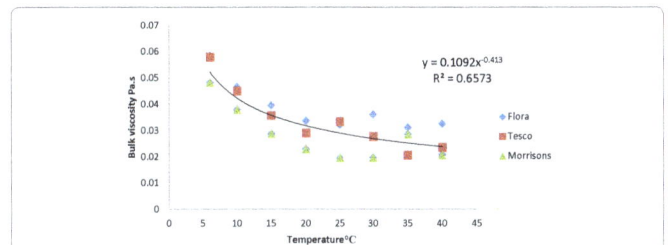

Figure 6: A plot of bulk viscosity v/s temperature of all the three brands (Tesco, Morrisons, Flora) of sunflower oil. As there is no significant difference between the brands an averaged polynomial best fit is obtained from this plot to know the temperature dependence of bulk viscosity for sunflower oil.

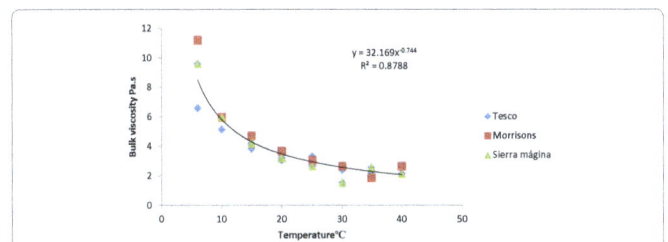

Figure 7: A plot of bulk viscosity v/s temperature of all the three brands (Tesco, Morrisons, Sierra magina) of extra-virgin olive oil. As there is no significant difference between the brands an averaged polynomial best fit is obtained from this plot to know the temperature dependence of bulk viscosity for extra-virgin olive oil.

there is a significant difference in the bulk viscosity values between the different brands of edible oils. Lastly, the temperature dependence of bulk viscosity was investigated (Figures 6 and 7).

Materials and Methods

Three brands of sunflower oil and three brands of extra-virgin olive oil have been used to evaluate potential variance of bulk viscosity with different brands of the same oil. The samples used for sunflower oil were Morrisons sunflower oil, Tesco sunflower oil and Floras sunflower oil. The samples used for olive oil was Morrisons extra virgin olive oil, Tesco extra virgin olive oil and Sierra mágina extra virgin olive oil. All the samples were locally purchased from Leeds supermarkets during July, 2015. Ultrasonic waves have been employed in this work to find the bulk viscosity. The attenuation coefficient, velocity of the sound wave after passing through the material across a range of frequencies is obtained from Ultrasizer MSV by Malvern Ltd. Density of the samples was measured by Anton Paar DMA 4500 M Density-meter. Anton Paar MCR 302 (Modular Compact Rheometer) was used to find the shear viscosity of the samples. All these parameters were measured to calculate the bulk viscosity using equation (8).

Ultrasizer MSV by Malvern Ltd was used to determine attenuation coefficient and velocity of sound as it passes through the oil samples at different selected temperatures. This is an acoustic spectroscopy instrument for liquids and emulsions operating in the frequency range of 1-100 MHz. Transducers are devices that convert energy from one form to other. This device makes use of two such transducers where one emits ultrasound waves on the application of voltage into the sample while the other detects it and converts into the corresponding voltage. Two pairs of such transducers are used. One pair operates in the low frequency range and the other in the high frequency range. On each run 50 measurements are taken covering the frequency range of 1 to 100 MHz. As sound waves travels through a medium attenuation is caused due to dissipation of energy in the form of shear viscosity, bulk viscosity, thermal conductivity and molecular relaxations. Ultrasizer measures this attenuation. This instrument needs 500 ml of sample to measure which is a measure drawback for limited sample volumes. It is connected to an external Huber Ministat temperature control unit which operates in the range of 5°C to 50°C. While conducting the experiment a stirrer constantly agitated the sample in order to reduce the thermal variation in the bulk sample, the speed of the stirrer can be adjusted. Care was taken so that no air bubbles were formed as these bubbles causes excess attenuation [11]. 10 repeat measurements were taken at each selected frequency and the mean was calculated to take into consideration any uncertainties or variations due to measurement.

The density measurements for the samples were accomplished by using Anton Paar DMA 4500 M Density-meter across a temperature range of 5°C to 40°C. The measurements by this instrument is based on the oscillating U-tube method. The thermal control is provided by two integrated Pt100 platinum thermometers together with Peltier elements. Viscosity related errors are automatically corrected over the full range of sample viscosities by measuring the damping effect of the viscous sample followed by a mathematical correction of the density value. Error while measuring the shear viscosity may arise due to 'sample under filling, uncertainties in the gap size, viscous heating effects, wall-slip errors, edge failure and radial migration.'

Anton Paar MCR 302 was used to take the shear viscosity readings. This rheometer is driven by air bearing supported EC (Electrically Commutated) motor technology. This ensures accuracy over a wide viscosity range. It is a digital instrument using digital signal processing technology. It makes use of patented normal force sensor. It also makes use of some patented features for convenience and to increase the efficiency. The measuring system used for our measurements was CP50-2 (Conical plate with diameter 50 mm and angle 2°). The temperature is controlled by a water bath. The shear viscosity values from 25°C to 40°C were measured and values outside the measurement range were extrapolated.

Thermal properties of the materials to be utilised in calculating the bulk viscosity were obtained from literature [6]. The ratio of specific heats, $\gamma = C_p/C_v$ was found to be almost equal to unity. Hence, the term due to the thermal property term in equation (9) was neglected. The bulk viscosity value was calculated from equation (10) using the values of mean attenuation, density, shear viscosity and the velocity of ultrasound as measured in experimental procedures.

Statistical analysis

The mean, standard deviation and coefficient of variance of the bulk viscosity were calculated at each temperature of the selected temperature range. T-test of two samples assuming equal variances between each brand and also single factor Annova was performed for both sunflower and olive oil to evaluate the significant difference in the value of bulk viscosity between different brands of sunflower oil and olive oil. Microsoft Excel (XLS) was used for the statistical analysis performed.

Results

Tesco sunflower oil has been taken as an example to show all the calculations and graphs. The experimental calculations for the other samples have been provided in the appendix for clarity and brevity.

A graph of log attenuation v/s log frequency has been plotted as shown in Figure 2 and the best fit polynomial was obtained in the frequency of the form f^δ, (where δ is the exponent). δ is found to be almost equal to 2 ($\delta = 1.8344$). A linear relationship between log attenuation and log frequency was seen (as regression coefficient > 0.999) for all the samples at each temperature.

The bulk parameters (density, shear viscosity, velocity) measured and the bulk viscosity calculated for Tesco sunflower oil across a temperature range of 6°C to 40°C is summarised in Table 1. The mean bulk viscosity and standard deviation has been calculated across the range of 12 MHz-100 MHz. A graph of bulk viscosity against frequency was plotted to illustrate the dependence of bulk viscosity on the frequency (Figure 3). Frequency below 12 MHz has not been included, as at this frequency range the attenuation value is too small due to molecular relaxations. The attenuation values obtained for 10 repeats of the same sample were averaged at each frequency. Repeated measurements of the same sample have been taken to increase the confidence level of calculating an accurate averaged value as exact value is not attainable at each time. These attenuation values were put into equation (10) along with the other parameters to calculate the bulk viscosity at each frequency. The standard deviation (SD) and coefficient of variation (CV) which is the ratio of the standard deviation by the mean were calculated. To find out about the nature of the fluid (Newtonian or non-Newtonian), a shear stress against strain diagram was obtained for Tesco sunflower oil (Figure 4). This shows an excellent linear relationship ($R^2 = 1$) between stress and strain for all the samples at the selected temperatures.

The ratio of bulk viscosity to shear viscosity was calculated in order to compare their values (Table 2). A graph of bulk viscosity and shear viscosity has been plotted to find their dependence on temperature (Figure 5). Statistical analysis performed on the different brands to find the existence of significant difference in the bulk viscosity value has been summarised in Table 3. The p-value obtained from single factor Anova between the different brands were 0.81($p \geq 0.05$) for olive oil, 0.17($p \geq 0.05$) for sunflower oil. This suggests there is no statistical significant difference in the bulk viscosity value between the different brands of the same oil. This is further supported by the p-values (≥ 0.05) obtained from performing t-test of two samples assuming equal

Sunflower oil	Samples	T-M	T-F	M-F
	p-value	0.33	0.05	0.40
Olive oil	Samples	T-M	T-S	M-S
	p-value	0.51	0.77	0.74

T: Tesco; M: Morrisons; F: Flora; S: Sierra mágina

Table 3: p-values from t-test of two samples assuming equal variances between each brand of sunflower oil and extra virgin olive oil.

Temperature (°C)	Tesco		Morrisons		Flora	
	Mean × 10⁻² Pa.s	SD × 10⁻² Pa.s	Mean × 10⁻² Pa.s	SD × 10⁻² Pa.s	Mean × 10⁻² Pa.s	SD × 10⁻² Pa.s
6	5.79	1.89	4.83	1.87	5.84	1.83
10	4.51	1.59	3.8	1.65	4.67	1.61
15	3.57	1.43	2.89	1.44	3.97	1.42
20	2.91	1.21	2.3	1.18	3.39	1.19
25	3.33	0.99	1.98	0.98	3.22	0.98
30	2.77	0.86	1.97	0.87	3.62	0.83
35	2.07	0.71	2.87	0.78	3.12	0.72
40	2.36	0.6	2.07	0.6	3.26	0.61

*Standard deviation SD
*Mean was taken for the bulk viscosity in the frequency range of 12 MHz-100MHz. the lowest temperature is decided by the operating limit of the instrument

Table 4: The bulk viscosity values of three different brands of sunflower oil at the selected temperature range.

Temperature (°C)	Tesco		Morrisons		Flora	
	Mean × 10⁻² Pa.s	SD × 10⁻² Pa.s	Mean × 10⁻² Pa.s	SD × 10⁻² Pa.s	Mean × 10⁻² Pa.s	SD × 10⁻² Pa.s
8	6.62	2.25	11.22	3.8	9.64	3.35
10	5.17	1.9	5.98	1.82	6.01	2.07
15	3.87	1.54	4.72	1.48	4.22	1.63
20	3.09	1.35	3.69	1.34	3.24	1.44
25	3.34	1.23	3.07	1.19	2.68	1.22
30	2.44	1.01	2.67	1.02	1.56	0.88
35	2.2	0.85	1.88	0.84	2.58	0.85
40	2.74	0.73	2.66	0.71	2.16	0.74

Standard deviation SD
*Mean was taken for the bulk viscosity in the frequency range of 12 MHz-100MHz. The lowest temperature is decided by the operating limit of the instrument

Table 5: The bulk viscosity values of three different brands of olive oil at the selected temperature range.

Temperature T(°C)	Density ρ (kg m⁻³)	Velocity v (m s⁻¹)	Shear viscosity (η × 10⁻² Pa.s)
6	928.96	1516.2	5.82
10	926.29	1501.93	5.29
15	922.84	1485.02	4.64
20	919.41	1469.02	3.99
25	915.82	1451.75	3.35
30	912.51	1434.15	2.7
35	909.16	1451.75	1.77
40	905.82	1403.29	1.61

Table 6: Measured parameters: density, velocity, shear viscosity of Morrisons sunflower oil at the selected temperatures.

Temperature T(°C)	Density ρ (kg m⁻³)	Velocity v (m s⁻¹)	Shear viscosity (η × 10⁻² Pa.s)
6	927.92	1513.26	4.93
10	925.9	1502.03	4.53
15	922.45	1485.02	3.87
20	919.03	1467.22	3.21
25	915.59	1451.6	2.48
30	912.22	1435.59	1.49
35	908.84	1419.59	1.33
40	905.48	1403.52	0.76

Table 7: Measured parameters: density, velocity, shear viscosity of Flora's sunflower oil at the selected temperatures.

Temperature T (°C)	Density ρ (kg m⁻³)	Velocity v (m s⁻¹)	Shear viscosity (η × 10⁻² Pa.s)
8	922.41	1505.02	6.26
10	920.48	1495.16	5.83
15	917.11	1477.55	5.12
20	913.2	1462.53	4.41
25	910.33	1445.98	3.26
30	906.95	1429.56	3
35	903.58	1413.53	2.53
40	900.21	1397.39	1.56

Table 8: Measured parameters: density, velocity, shear viscosity of Tesco extra-virgin olive oil at the selected temperatures.

Temperature T(°C)	Density ρ (kg m⁻³)	Velocity v (m s⁻¹)	Shear viscosity (η × 10⁻² Pa.s)
6	920.88	1502.07	5.26
10	919.52	1492.69	5.04
15	916.07	1479.57	4.5
20	912.65	1462.53	3.95
25	909.23	1444.25	3.35
30	905.84	1429.1	2.86
35	902.44	1412.98	2.76
40	899.06	1397.4	1.61

Table 9: Measured parameters: density, velocity, shear viscosity of Morrisons extra-virgin olive oil at the selected temperatures.

variances between each brand (Table 3). As there is no significant difference in bulk viscosity between the brands, a generalised temperature dependence model of bulk viscosity was established by taking the best fit from the average plot of the three brands. The bulk viscosity values calculated for all the samples have been listed in Tables 4 and 5.

Discussion

Justification of the use of frequency squared equation

Since the exponent term δ was almost equal to 2 (δ=1.8344) (Figure 2), hence the use of frequency squared equation (equation 10) to find the bulk viscosity was justified. Attenuation is the result of both classical mechanisms (shear and bulk viscosity, thermal contributions) as well as due to molecular relaxations. The reason for δ being less than 2 is due to the occurrence of molecular relaxations which has not been accounted for in the equation (Tables 6-9). One of the most important reasons for molecular relaxations maybe due to the molecular rearrangements that occur during the compression of oil in the ultrasonic field [12]. There might be some error as excess attenuation due to the formation of air bubbles in the sample inside the Ultrasizer. Also, any thermal fluctuations during the measurement might result in some error as attenuation, density, velocity are dependent on temperature as explained earlier.

Dependence of bulk viscosity on frequency

The higher the value of CV the more dispersed is the data. CV for the mean bulk viscosity across the frequency range was found to be ≥ 0.05 at each temperature (Table 1). This indicates that across the frequency range there is a variation in the bulk viscosity value which cannot be ignored. This suggests that the bulk viscosity values calculated for all

Temperature T (°C)	Density ρ (kg m⁻³)	Velocity v (m s⁻¹)	Shear viscosity (η × 10⁻² Pa.s)
6	922.41	1506.8	6.05
10	920.48	1494.07	5.67
15	917.11	1479.5	5.03
20	913.2	1461.05	4.4
25	910.33	1445.72	3.77
30	906.95	1429.1	3.2
35	903.58	1412.98	2.27
40	900.21	1397.4	2.05

Table 10: Measured parameters: density, velocity, shear viscosity of Sienna mágina extra-virgin olive oil at the selected temperatures.

the selected temperatures are frequency dependent and not constant. Also, from Figure 3 it is seen that the bulk viscosity decreases with the increase in frequency for all temperatures, showing bulk viscosity is frequency dependent. This indicates the non-Newtonian behaviour for sunflower oil and olive oil over the selected temperature and frequency range, as for a fluid to be Newtonian the bulk viscosity must be constant for a selected frequency range [6]. However, from the stress and strain diagram obtained from Figure 4, edible oils are showing Newtonian nature as the shear stress is proportional to the shear strain.

Comparison of bulk and shear viscosity

The ratio of bulk viscosity to shear viscosity was found to be around 1 at all the selected temperatures (Table 2). This shows for edible oils the bulk viscosity is almost equal to its shear viscosity. However, the contribution of bulk viscosity to sound propagation due to non-Newtonian fluid seems to be quite less than due to Newtonian fluid as the bulk viscosity of water (a Newtonian fluid) was reported to be almost three times larger than its shear viscosity [13]. A plot of bulk and shear viscosity against temperature (Figure 5) shows there is a decrease in both the values with the increase in temperature indicating both are temperature dependent.

Test of hypothesis

The hypothesis tested that there is a significant difference between the different brands of the same edible oil is rejected based on the results obtained from the statistical analysis ($p \geq 0.05$) (Table 3). This shows bulk viscosity of edible oils do not seem to be much affected by small compositional differences. This is an important finding for future bulk viscosity studies as only one brand can be used to represent a class of oil, saving both time and resources (Table 10).

Temperature dependence of bulk viscosity

A decrease in the bulk viscosity is seen as the temperature increases (Figures 6 and 7).

The temperature dependent bulk viscosity model has been established as:

$$\mu = \mu_1[T \exp(A)]$$

Where μ (Pa.s) is the bulk viscosity, μ_1 (Pa.s.°C⁻¹) T is temperature (°C), A is constant, μ_1 is 0.1092 and A is -0.413 for sunflower oil (Figure

6) and μ_1 is 32.169 and A is -0.744 for extra-virgin olive oil (Figure 7). The findings from this study cannot be generalised for edible oils as the sample size was not adequate as only two types of oil: sunflower oil and extra-virgin olive oil was investigated and only three brands from each were taken into consideration. Further studies should be conducted with larger sample size like with other commonly used edible oils. For the different brands simple random sampling must be undertaken so that the samples well represents the entire class.

Conclusion

The temperature dependence of bulk viscosity in edible oil using acoustic spectroscopy was established using a model equation. The mean bulk viscosity decreases with the increase in temperature. The value of bulk viscosity is not constant over the frequency range, it decreases with increase in frequency. Therefore, in terms of bulk viscosity edible oils are non-Newtonian fluids. There is no significant statistical difference of bulk viscosity value between different brands of the same oil. This suggests that future studies with only one variety of oil will be enough. Even though other physical properties of edible oils have been extensively studied, little research has been done on bulk viscosity. More research should be undertaken to check the reproducibility of these results and validate the data.

Acknowledgement

I would like to thank the entire School of Food Science, University of Leeds and my supervisor Dr. Melvin Holmes for his constant support and guidance.

References

1. McClements DJ, Povey MJW (1992) Ultrasonic analysis of edible fats and oils. Ultrasonics 30: 383-388.

2. Leighton TG (2007) What is ultrasound? Prog Biophys Mol Biol 93: 3-83.

3. Morrison FA (2004) Compressible fluids. Michigan Technological University, USA pp: 94-98.

4. Hirai N, Eyring H (1958) Bulk viscosity of liquids. J Appl Phys 29: 810.

5. Baidakov VG, Protsenko SP (2014) Metastable Lennard-Jones fluids: III. Bulk viscosity. J Chem Phys 141: 114503.

6. Dukhin AS, Goetz PJ (2009) Bulk viscosity and compressibility measurement using acoustic spectroscopy. J Chem Phys 130: 124519.

7. Cramer MS, Bahmani F (2014) Effect of large bulk viscosity on large-Reynolds-number flows. J Fluid Mechanic 751: 142-163.

8. Gladwell N, Javanaud C, Peers KE, Rahalkar RR (1985) Ultrasonic behavior of edible oils: Correlation with rheology. J Am Oil Chem Soc 62: 1231-1236.

9. Coupland JN, McClements DJ (1997) Physical properties of liquid edible oils. J Am Oil Chem Soc 74: 1559-1564.

10. Rodrigues Jr. JDA, Cardoso FDP, Lachter ER, Estevao LRM, Lima E, et al. (2006) Correlating chemical structure and physical properties of vegetable oil esters. J Am Oil Chem Soc 83: 353-357.

11. Dalen J, Lo A (1981) The influence of wind-induced bubbles on echo integration surveys. J Acou Soc Am 69: 1653-1659.

12. Chanamai R, McClements DJ (1998) Ultrasonic attenuation of edible oils. J Am Oil Chem Soc 75: 1447-1448.

13. Holmes M, Parker NG, Povey MJW (2011) Temperature dependence of bulk viscosity in water using acoustic spectroscopy. J Phys Conf Ser.

34

The Effect of Soybean Extracts on Serum Lipid Profile and the Accumulation of Free Cholesterol and Cholesteryl Ester in the Aorta, Carotid Artery and Iliac Artery-Experimental Study

Uyar SI[1]*, Ugur Ozdemır[2], Ilter MS[3], Muhittin Akyıldız[4] and Sivrikoz NO[5]

[1]Cardiovascular Surgery Department, Medical Faculty, Sifa University, Izmir, Turkey
[2]Anesthesiology Department, Medical Faculty, Sifa University, Izmir, Turkey
[3]Department of Food Engineering, Engineering and Natural Science Faculty, Gumushane University, Gumushane, Turkey
[4]Biochemistry Department, Medical Faculty, Sifa University, Izmir, Turkey
[5]Pathology Department, Medical Faculty, Sifa University, Izmir, Turkey

Abstract

Background: We evaluated the effect of soybean extracts on serum lipoprotein profile and cholesterol accumulation on the arterial walls.

Objective and design: Sixty-four female Sprague-Dawley rats were randomly divided into eight groups. Soybean extracts were given to the rats via oral gavage every day for eight weeks, after which serum was collected. In the thoracic aorta, left carotid artery, and right iliac artery, we measured the lipoprotein fractions in the serum and the accumulation of free cholesterol and cholesteryl ester, which are predictors of subclinical atherosclerosis.

Results: After eight weeks of a continuous soybean diet, only two groups showed a lipid-lowering effect (n-hexane extract for 200 mg/kg dose and ethyl acetate extract for 200 mg/kg dose). We found lower free cholesterol and cholesteryl ester accumulation in the aortas and iliac arterial walls only in these two groups.

Conclusion: The results indicated that soybean extract intake leads to weight change and may influence lipid metabolism. The positive effects of the soybean diet involved not only serum lipids but also aortic wall cholesterol accumulation.

Keywords: Cholesterol; Soybean extracts; Experimental model; Cholesterol accumulation; Arterial wall

Introduction

It has been known for many years that dietary soybeans have positive health benefits, particular for atherosclerosis-related diseases. Many experimental studies have shown that a diet high in soy protein reduces hyperlipoproteinemia and atherosclerosis [1-3]. However, the role of individual soy components and the mechanisms underlying the beneficial effects of soybeans are not fully understood. The increased amount of research over the past 20 to 25 years has resulted in the theory that soy protein consumption may improve cardiovascular health [2,4]. Soy protein extract has been commonly used in experimental studies. The major components of soy protein extract are protein/peptide fractions. The estrogen-like compounds in soybeans are called isoflavones, which are also present in many legumes and grains [5]. The aim of the present study is to evaluate the effects of soybean extracts on the serum lipoprotein profile and atherosclerosis in the aortic, iliac, and carotid arterial walls of female Sprague-Dawley rats.

Materials and Methods

Animals: Sixty-four female Sprague-Dawley rats (Afyon Kocatepe University in Turkey) were randomly assigned to eight groups of eight rats each. The animals were two months old and weighed 200 ± 15 grams. During the experiment, they were kept in a standard controlled environment (12 hours light-dark cycle, temperature 23°C ± 2°C, and relative humidity 60% ± 5%) in individual metabolic cages for approximately 15 days in order to adapt to the surroundings. They were given free access to standard feed and water. During the study period, the gratings were cleaned regularly every two days, and feed and water containers were kept filled. The total study period was eight weeks. Soybean extracts were given to each group via oral gavage every day for eight weeks. The daily dose was adjusted to 1 ml in volume,

and the soybean extract density was 100 or 200 mg/kg/dose. Variations in body weight and food intake were checked daily during the eight-week period. All procedures were conducted in compliance with the Republic of Turkey's state laws. The Institutional Animal Care and Use Committee and the ethical committee of experimental study of Afyon Kocatepe University approved all procedures involving the animals. The procedures were performed in accordance with the Guide for the Care and Use of Laboratory Animals. The experimental procedure was approved by the Research Ethics Committee (Res. No. 49533702/341/2013) [6].

Soybean extracts of n-hexane, ethyl acetate, and ethanol were prepared as described by Carrao-Panizzi et al. [7]. The collected soybeans were gradually extracted with n-hexane, ethyl acetate, and ethanol. Soybeans weighing 25 grams were extracted two times by using 500 ml of n-hexane, ethyl acetate, and ethanol at room temperature. The samples were then shaken for 48 hours. The combined extracts were concentrated under low pressure at 40°C in a Rotavapor˙ and then dehydrated under low pressure in a glass vacuum-controlled desiccator dryer. After extraction, the extract was transferred to an Eppendorff tube and stored at 5°C.

*Corresponding author: Ihsan Sami Uyar, Cardiovascular Surgery Department, Medical Faculty, Sifa University, Izmir, Turkey, E-mail: ihsansami@hotmail.com

Solvent removal process

This process was conducted using an ultimate vacuum system. To achieve optimal distillation conditions, the distillation energy supplied by the heating bath was removed by the condenser. All procedures were done according to the D20 5°C principle, which is mainly known as the solvent-removing method. The vacuum was adjusted to the rotary evaporation system appropriately for each solvent with an operating bath temperature of 60°C to yield a solvent vapor temperature of 40°C, which was subsequently condensed at 20°C.

At the end of the eight-week period, after 12 hours of fasting, the animals were anesthetized with ketamine HCl (40 mg/kg) and xylazine (8 mg/kg), and then 1 ml of blood was collected via heart puncture. The blood was collected into tubes containing SST gel separator II (BD Vacutainer° B, Franklin Lakes, NJ, USA), and the serum was promptly separated from the cells via centrifugation at 4500 RPM for 15 minutes at 4°C. Serum was collected and stored at -20°C for subsequent biochemical analysis. The rats were then euthanized via cervical dislocation. The carotid artery, aorta, and iliac artery, were promptly removed and placed in a 10% neutral buffered formalin solution for subsequent examination.

Measurement of serum lipid and lipoprotein

Serum lipoproteins were separated as described by Haug et al. [8] and isolated lipoprotein fractions were used for the enzymatic determination of cholesterol. Total serum cholesterol (TC), high-density lipoprotein cholesterol (HDL-C), non-high-density lipoprotein cholesterol (non HDL-C) and serum triglycerides (TG) were determined in the Clinical Chemistry and Hormone Laboratory of the Department of Biochemistry at Sifa University Hospital. Non HDL-C was calculated by subtracting the HDL-C value from the TC value. The TC, non-HDL-C, HDL-C, and TG concentrations in the serum were measured enzymatically by using commercially available reagent kits (HDL and LDL/VLDL Quantification Colorimetric/Fluorometric Kit, BioVision Incorporated, Milpitas Boulevard, Milpitas California, USA). We measured the lipoprotein fractions at 570 wavelengths by using an Epoch Micro-Volume Spectrophotometer System (Bio Tec Instruments, Inc., Tigan Street Winooski, VT, USA).

Histochemical examination

The levels of free cholesterol and cholesteryl ester concentrations in the aorta, carotid, and iliac arteries levels were analyzed. The analysis was performed as described in the commercial kit prospectus (Cholesterol Colorimetric Assay Kit Cell Biolabs Inc., San Diego, USA). The aorta, iliac, and carotid arteries were placed on the platform of a dissecting microscope, and the adventitia was carefully dissected and removed.

Measurement of free cholesterol and cholesteryl ester concentrations

We used 10 mg of tissue that was extracted with 200 μL of a mixture of chloroform and isopropanol in a micro-homogenizer. The extracts were centrifuged for 10 minutes at 15,000 g. Air drying was performed at 50°C to remove the chloroform. Samples were then placed in a vacuum for 30 minutes to remove any trace amounts of organic solvent. The dried lipids were dissolved in 200 μL of 1 X Assay Diluent and then treated by sonicating and vortexing until the solution was homogeneous. After sample preparation, we assayed at 570 wavelengths by using an Epoch Micro-Volume Spectrophotometer System (Bio Tec Instruments, Inc., Tigan Street Winooski, VT, USA), using a commercial kit (Cholesterol Colorimetric Assay Kit Cell Biolabs Inc., San Diego, USA).

Statistical analysis

All data were assessed by a one-way ANOVA and a Tukey's test in order to detect the main effects of diet type on serum lipoproteins and atherosclerosis. Multiple linear regressions were used to assess the relationship between the effects of the treatment on serum lipoproteins and the effects on atherosclerosis. We selected covariates for the analysis of covariance. The analyses were conducted using R Statistical Analysis V. 3.0.2 Statistical Software. The results were presented as the mean value ± standard deviation. Differences were considered significant at $p < 0.05$.

Results

The initial ages and body weights of the all animals were similar. The experimental groups and diets are shown in Table 1. Some statistical differences ($p < 0.05$) in the body weight gain and growth of the animals showed after the eight-week experiment. The effects of soybean extracts on the growth and weight gain of the study animals are shown in Table 2. The average daily gains (ADGs) and feed/gain ratios of the rats in Groups 4, 6, and 8 were higher than those in other groups (P < 0.05). Comparisons among Groups 3, 5, and 7 and the

	N	Diet	Doses
Group 1	8	Commercial diet	Negative Control
Group 2	8	0.5% Carboxymethylcellulose	Positive Control
Group 3	8	N-hexane extract	100 mg/kg
Group 4	8	N-hexane extract	200 mg/kg
Group 5	8	Ethyl acetate extract	100 mg/kg
Group 6	8	Ethyl acetate extract	200 mg/kg
Group 7	8	Ethanol extract	100 mg/kg
Group 8	8	Ethanol extract	200 mg/kg

Table 1: The experimental groups and diets.

Groups/ Parameters	Group 1 (n=8)	Group 2 (n=8)	Group 3 (n=8)	Group 4 (n=8)	Group 5 (n=8)	Group 6 (n=8)	Group 7 (n=8)	Group 8 (n=8)
Age (week)	8	7	8	8	7	8	8	8
Initial Body weight, g	208.9 ± 1.4	207.4 ± 1.6	206.6 ± 2.0	207.2 ± 2.7	207.7 ± 2.8	207.1 ± 1.8	207.9 ± 2.7	207.7 ± 2.2
Final Body weight, g	216.6 ± 1.9	215.9 ± 1.4	213.8 ± 2.3	226.9 ± 4.2*	215.3 ± 2.2	222.8 ± 25.1*	214.2 ± 2.9	223.1 ± 9.8*
Body Weight gain g/8week	7.7 ± 1.4	8.5 ± .6	7.2 ± 1.2	19.7 ± 1.2*	7.6 ± 2.6	15.7 ± 1.5*	6.3 ± 25.9	15.4 ± 1.4*
Food consumption	481.5 ± 19.1	492.3 ± 11.4	482.8 ± 18.8	491.4 ± 15.2	493.2 ± 11.1	492.9 ± 15.2	494.3 ± 12.9	489.1 ± 11.5

Each value is presented as the mean value ± standard deviation; g, gram.
*Mean value is significantly different when compared to control groups (p <0.05).

Table 2: Baseline values, total food intake, and body weight gain of rats fed on the soybean for 8 weeks.

mmol/l	Group 1	Group 2	Group 3	Group 4	Group 5	Group 6	Group 7	Group 8
T Chol	1.63 ± 0.05	1.59 ± 0.04	1.59 ± 0.04	1.49 ± 0.02*	1.62 ± 0.02	1.40 ± 0.03*	1.62 ± 0.03	1.59 ± 0.03
Non-HDL Chol	0.93 ± 0.03	0.95 ± 0.02	0.99 ± 0.29	0.77 ± 0.03*	1.1 ± 0.16	0.78 ± 0.03*	1.06 ± 0.13	0.89 ± 0.5
HDL-Chol	0.66 ± 0.04	0.67 ± 0.03	0.67 ± 0.04	1.02 ± 0.07*	0.68 ± 0.4	1.14 ± 0.12*	0.65 ± 0.02	0.66 ± 0.02
Triglyceride	0.67 ± 0.04	0.70 ± 0.03	0.68 ± 0.02	0.47 ± 0.03*	0.69 ± 0.06	0.50 ± 0.04*	0.70 ± 0.05	0.68 ± 0.03
Atherogenic Index	1.40 ± 0.8	1.42 ± 0.08	1.47 ± 0.07	0.75 ± 0.04*	1.61 ± 0.18	0.68 ± 0.9*	1.63 ± 0.19	1.34 ± 0.14

Values are represented as mean value ± standard deviation and mmol/L for each group.
T Chol, total cholesterol; HDL-Chol, high density lipoprotein; non-HDL-Chol, difference between T Chol and HDL- Chol; atherogenic index, non-HDL- Chol /HDL- Chol.
*Mean value is significantly different when compared to control groups (p <0.05).

Table 3: Serum total cholesterol, HDL-cholesterol, non-hdl-cholesterol, and triglyceride concentrations in rats fed on the soybean for 8 weeks.

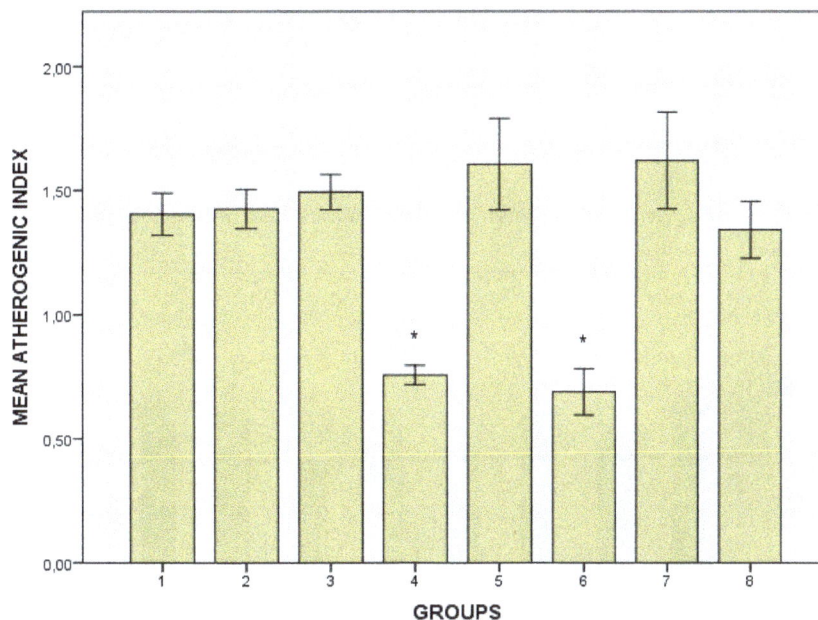

Figure 1: The atherogenic index values were significantly lower in Group 4 and 6 when compared to other groups. (*Mean value is significantly different when compared to other groups (p < 0.05).

control groups, there was no significant difference in terms of growth performance (P > 0.05).

After eight weeks of treatment, the distributions of cholesterol and triacylglycerol among the lipoprotein fractions was determined in mmol/L. The effects of diets supplemented with soybean extracts on serum lipoprotein profile are shown in Table 3. The serum lipoprotein levels were the same in Groups 1 and 2 (the negative and positive control groups, respectively). Between Groups 1 and 2, there were many differences in terms of all parameters. Lower serum TC, TG, and non-HDL-C levels were observed in Groups 4 and 6. When they were compared to other groups, this difference was significant (P < 0.01). The HDL-C concentrations of Groups 4 and 6 were higher than those of other groups (P < 0.05). Serum HDL-C levels in Groups 4 and 6 were 1.02 ± 0.07 and 1.14 ± 0.12 mmol/L, respectively, whereas those of the controls were 0.66 ± 0.04 and 0.67 ± 0.03 mmol/L, respectively. However, serum TC, TG, non-HDL-C and HDL-C levels were not significantly different in Groups 3, 5, 7, and 8 (P > 0.05). The control groups did not have higher serum cholesterol levels than Groups 3, 5, 7, and 8 did (P > 0.05). The differences between the control groups and Groups 3, 5, 7, and 8 were not significant. The highest HDL-C level was seen in Group 6, whereas the HDL-C level was lower in Group 7 than in the control groups. In this group, the TC and TG levels were similar to those in the control groups. The atherogenic index values were significantly lower in Groups 4 and 6. compared to other groups (Figure 1).

We measured the accumulation of free cholesterol and cholesteryl ester, which are *predictors of subclinical atherosclerosis,* in the thoracic aorta, the left carotid artery, and the right iliac artery. The atherogenic response, as measured by aortic cholesteryl ester concentration, is considered the primary indicator of the beneficial or detrimental effects of any diet supplementation on atherosclerosis. The analyses were performed in miligram/gram (mg/g) on cholesterol and cholesteryl ester. Table 4 shows the results of the effects of diets supplemented with soybean extracts on the arterial wall. Aortic and iliac artery triglyceride, free cholesterol, and cholesteryl ester concentrations were significantly lower in Groups 4 and 6 (P < 0.05). The lowest triglyceride values were found in Groups 4 and 6, which differed greatly from those found in the control groups (Table 4). Comparison showed no significant differences between other groups. The measurement results for the cholesteryl ester and free cholesterol concentrations in Groups 4 and 6 were significantly lower (P < 0.05) than those in Groups 1 and 2 (control groups). In addition, the aortic cholesteryl ester values in Groups 3, 5, 7, and 8 were similar to those in the control groups (Figures 2 and 3).

The aortic measurements yielded similar results for the iliac arteries. The greatest accumulation of both forms of cholesterol in the iliac artery was in the control groups. The cholesteryl ester values were the lowest in Groups 4 and 6. Interestingly, the results of the carotid artery measurements were different from those of the aorta and iliac artery. The carotid artery measurements showed no differences among

Values	Group 1	Group 2	Group 3	Group 4	Group 5	Group 6	Group 7	Group 8
Aortic Values, mg/g tissue								
TG	4.50 ± 0.87	4.58 ± 0.87	4.41 ± 0.96	3.7 ± 1.05	4.45 ± 1.08	3.58 ± 0.95	4.39 ± 0.88	4.47 ± 1.02
TC	1.92 ± 0.16	1.81 ± 0.15	1.72 ± 0.26	1.11 ± 0.23	1.83 ± 0.32	1.21 ± 0.23	1.89 ± 0.22	1.95 ± 0.24
FC	1.62 ± 0.07	1.78 ± 0.07	1.69 ± 0.07	1.19 ± 0.08	1.71 ± 0.07	1.21 ± 0.13	1.79 ± 0.11	1.69 ± 0.11
CE	0.25 ± 0.03	0.27 ± 0.04	0.21 ± 0.03	0.11 ± 0.02	0.24 ± 0.04	0.11 ± 0.01	0.19 ± 0.02	0.26 ± 0.03
Iliac Artery Values, mg/g tissue								
TG	4.58 ± 0.07	4.50 ± 0.35	3.99 ± 0.42	3.58 ± 0.59	4.51 ± 0.20	3.51 ± 0.16	4.79 ± 0.58	4.51 ± 0.19
TC	1.92 ± 0.18	1.93 ± 0.27	1.81 ± 0.23	1.03 ± 0.26	1.72 ± 0.26	1.01 ± 0.17	1.83 ± 0.22	1.82 ± 0.14
FC	1.67 ± 0.09	1.69 ± 0.12	1.62 ± 0.14	1.10 ± 0.08	1.50 ± 0.11	1.04 ± 0.06	1.79 ± 0.12	1.70 ± 0.09
CE	0.22 ± 0.02	0.27 ± 0.03	0.20 ± 0.04	0.02 ± 0.01	0.20 ± 0.03	0.07 ± 0.01	0.21 ± 0.03	0.30 ± 0.07
Carotid Artery Values, mg/g tissue								
TG	3.51 ± 0.10	4.01 ± 0.52	3.58 ± 0.10	3.48 ± 0.05	3.63 ± 0.09	3.69 ± 0.58	3.58 ± .0.34	3.60 ± .0.52
TC	1.12 ± 0.12	1.20 ± 0.29	1.02 ± 0.16	1.22 ± 0.16	1.13 ± 0.06	1.15 ± 0.16	1.02 ± 0.12	1.03 ± 0.11
FC	1.01 ± 0.03	1.01 ± 0.08	1.02 ± 0.03	1.11 ± 0.06	1.11 ± 0.05	1.21 ± 0.08	1.12 ± 0.07	1.01 ± 0.03
CE	0.20 ± 0.05	0.13 ± 0.05	0.09 ± 0.03	0.12 ± 0.02	0.09 ± 0.02	0.12 ± 0.03	0.12 ± 0.02	0.13 ± 0.02

Values are represented as the mean value ± standard deviation; g, gram; mg, milligram; TG, triglycerides; TC, total cholesterol; FC, free cholesterol; CE, cholesteryl ester. The bold digits, different from control groups (p < 0.05).

Table 4: Aortic, Iliac and Carotid arterial values of triglyceride, total cholesterol, free cholesterol and cholesteryl ester of rats fed on the soybean for 8 weeks (mg/g tissue).

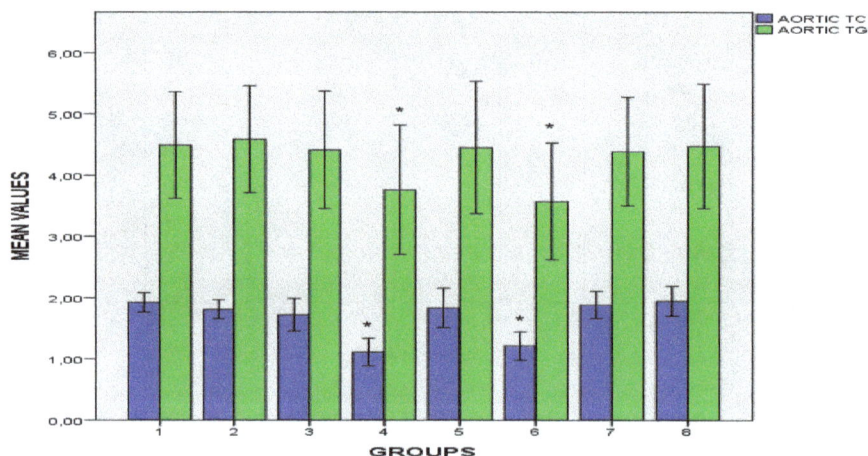

Figure 2: The mean values of total cholesterol and triglyceride in the thoracic aortic wall. (*Mean value is significantly different when compared to other groups (p < 0.05). TC: Total cholesterol; TG: Triglycerides. (mg/g tissue). Mean Values are represented as the value ± standard deviation and mg/g tissue).

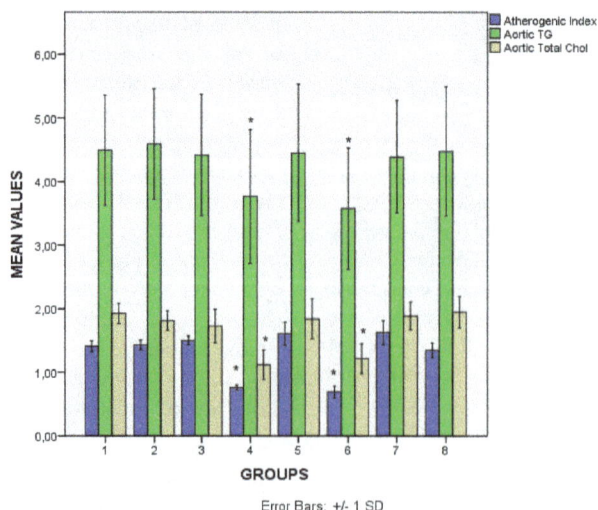

Figure 3: The mean values of total cholesterol and triglyceride in the thoracic aortic wall and the atherogenic index values were significantly lower in Group 4 and 6 when compared to other groups. (*Mean value is significantly different when compared to control groups (p <0.05). Atherogenic index, non-HDL- Chol /HDL- Chol; TG: Triglycerides; T Chol: Total cholesterol. Values are represented as mean value ± standard deviation and mg/g tissue for each group).

all groups. The results obtained from all study rats were similar, and there was no statistically significant difference between the control and study groups.

Discussion

The beneficial effects of soybean protein on serum lipid and lipoprotein concentrations and thus on cardiovascular diseases have been well-documented. Many studies have reported a correlation between the oral intake of soybean protein and serum lipid profile [8-10]. The increased number of studies performed over the past 10 to 12 years have provided evidence that soybean consumption has a positive effect on serum lipid profile and that it might protect against the accumulation of cholesterol on the vascular walls and thus improve cardiovascular health. Consequently, it may also inhibit the early progression of coronary artery atherosclerosis [11,12]. Several studies have investigated the soybean's effects on serum lipid and lipoprotein metabolism. These studies showed that soy protein intake was positively associated with HDL-C and negatively associated with total cholesterol, non-HDL-C, and TG [13-15]. Clarkson et al. [2] reported that soybean usage in human subjects caused a reduction in LDL-C by approximately 13%, a reduction in serum TG by approximately 10%, and an increase in HDL-C by approximately 2%. It has been shown that cholesterol accumulation in the vascular wall is associated with structural and functional changes in the vessels. Experimental data suggested that the alteration of cellular phospholipid metabolism is an adaptive response to prevent the ratio of cellular-free cholesterol and phospholipids from reaching cytotoxic levels in the macrophage foam cells of atherosclerotic lesions [16].

In this study, we showed that after eight weeks of continuous soybean diet consumption, only two groups displayed a lipid-lowering effect (n hexane extract at 200 mg/kg dose and ethyl acetate extract at 200 mg/kg dose). Our results showed that soybean extract intake led to weight change; thus, it may be considered a nutritional supplement and could influence lipid metabolism. In addition, we found that only Groups 4 and 6 had less free cholesterol and cholesteryl ester accumulation in the aorta and iliac arterial wall. The histochemical and biochemical analyses of other groups were similar to those of the control groups. However, we found no effects of the lower doses. Interestingly, the results of the carotid arterial wall analyses were similar in all groups; there were no significant differences among the groups. The results indicated that carotid artery atherosclerosis could not be significantly reduced by a soybean diet, given these doses and this period of study. These findings were similar to the literature [1,4,13]. In this respect, the carotid artery differed from the aorta and the iliac arteries. These findings showed that the positive effects of a soybean diet appear not only in lowered serum lipids but also less cholesterol accumulation in the aortic wall.

Many reports in the literature have indicated that soybean extracts may prevent coronary heart disease. Our findings suggested that the favorable cardiovascular effects of soybean may be enhanced with a 200 mg/kg dose of ethyl acetate extract and n hexane extract. We did not investigate higher doses. Nevertheless, in light of these findings, we can say that doses of ethanol extracts below 200 mg/kg did not affect the blood lipid profile and did not prevent free cholesterol and cholesteryl ester accumulation on the aortic wall. The aortic cholesteryl ester values in all groups were similar to those of the control groups, except for Groups 4 and 6. We interpret these findings to indicate that the ethyl acetate or n hexane extracts of the soybean diet may act to prevent atherosclerosis in a rat model at 200 mg/kg doses.

However, the components of the soybean diet that are responsible for this effect and the mechanisms involved remain uncertain. Recent research has focused primarily on efforts to identify the components of soybean protein that are responsible for its beneficial effects on the cardiovascular system. However, some experimental studies have shown that the isoflavones contained in soybeans and many soy-based products are responsible for these effects [4,5]. Conversely, various studies reported that isoflavone-free soy protein preparations reduce serum cholesterol [17]. How do soybeans exert these effects on the aortic wall and blood lipid profile? The pathways of the effects remain unclear. Adams et al. [1] concluded that the consumption of peptides from purified soybean beta-conglycinin has an inhibitory effect on the development of atherosclerosis, which greatly exceeds the effect of whole-isoflavone soy protein isolate and does not depend on low-density lipoprotein cholesterol (LDL-C) receptors or effects on serum lipoproteins. Cavallini et al. [18] emphasized that atherosclerosis is a chronic immune inflammatory disease, and they attributed the biological effects of soybeans to their antioxidant and anti-inflammatory effects. Because the initiation and progression of atherosclerosis are known to involve the oxidation of serum lipoproteins in the arterial intima, cell proliferation, and localized inflammatory reaction, these represent potential pathways by which soy peptides could directly inhibit atherosclerosis [19].

Some authors [3] determined that an undigested, insoluble high-molecular-weight Fraction (HMF), which bound bile acids, increased the excretion of acidic and neutral steroids and caused a marked reduction in serum and hepatic cholesterol concentrations in rats. Wang et al. [20] showed that the increased consumption of HMF of soybean by human subjects gave rise to bile acid secretion, raised serum HDL-C concentration, and decreased serum LDL-C concentration. Higher HDL-C and lower LDL-C concentrations have a preventive effect on cardiovascular diseases. Adams et al. [1] observed that soy protein isolate had a potent atheroinhibitory effect in female mice but that it had no effect in male rats. Conversely, experimental studies have demonstrated the inhibitory effects of dietary soy protein isolate on atherosclerosis in both male and female rats [16,17]. Therefore, we investigated only female rats. Some investigations in the literature reported that serum total cholesterol and lipoprotein cholesterol concentrations were unaffected by dietary soy protein isolate in animals and humans [21,22]. In contrast, in this study, we found a positive effect on blood lipid profile, the aorta, and the iliac arterial wall. However, we did not observe an atheroinhibitory effect on the carotid artery wall. In addition, we hypothesize that an effective dose of soybeans must be administered; otherwise, there is no atheroprotective effect [23]. This is supported by the findings of Clarkson et al. [24], who found that serum isoflavone concentrations must be in the optimal range for the inhibition of atherosclerosis to occur.

Conclusion

The present study supports the lipid-lowering and the atheropreventative effects of soybean extracts. We observed significantly decreased serum TC, non-HDL-C, and TG levels only in Groups 6 and 8 (P < 0.05). The results obtained for other groups were similar. The reason for this finding was probably that the dose was too low or that the duration of experiment was too short. After eight weeks of a soybean diet, rats had decreased levels of free cholesterol and cholesteryl ester accumulation in the aorta and the iliac arteries, but not in the carotid arteries. In the light of these findings, we can say that the soybean diet has a positive effect on the cardiovascular system; however, further long-term studies are needed.

References

1. Adams MR, Golden DL, Franke AA, Potter SM, Smith HS, et al. (2004) Dietary soy betaconglycinin (7S globulin) inhibits atherosclerosis in mice. J Nutr 134: 511-516.

2. Clarkson TB (2002) Soy, soy phytoestrogens and cardiovascular disease. J Nutr 132: 566- 569S. Review.

3. Høie LH, Morgenstern EC, Gruenwald J, Graubaum HJ, Busch R, et al. (2005) A double-blind placebo-controlled clinical trial compares the cholesterol-lowering effects of two different soy protein preparations in hypercholesterolemic subjects. Eur J Nutr 44: 65-71.

4. Das L, Bhaumik E, Raychaudhuri U, Chakraborty R (2012) Role of nutraceuticals in human health. J Food Sci Technol 49: 173-183.

5. Vidyavati HG, Manjunatha H, Hemavathy J, Srinivasan K (2010) Hypolipidemic and antioxidant efficacy of dehydrated onion in experimental rats. J Food Sci Technol 47: 55-60.

6. National Research Council (1985) Guide for the care and use of laboratory animals. Washington, DC: National Academy of Sciences.

7. Carrao-Panizzi MC, Favoni SP, Kikuchi A (2002) Extraction time for soybean isoflavone determination. Braz Arch Biol Technol 45: 515-518.

8. Haug A, Hostmark AT (1987) Lipoprotein lipases, lipoproteins and tissue lipids in rats fed fish oil or coconut oil. J Nutr 117: 1011-1017.

9. Rudel L, Kelley K, Sawyer JK, Shah R, Wilson MD, et al. (1998) Dietary monounsaturated fatty acids promote aortic atherosclerosis in LDL receptor-null, human apoB100-overexpressing transgenic rats. Arterioscler. Thromb Vasc Biol 18: 1818-1827.

10. Mizushige T, Mizushige K, Miyatake A, Kishida T, Ebihara K, et al. (2007) Inhibitory effects of soy isoflavones on cardiovascular collagen accumulation in rats. J Nutr Sci Vitaminol 53: 48-52.

11. Lovati M, Manzoni C, Gianazza E, Arnoldi A, Kurowska E, et al. (2000) Soy protein peptides regulate cholesterol homeostasis in Hep G2 cells. J. Nutr 130:2543-2549.

12. Gianazza E, Eberini I, Arnoldi A, Wait R, Sirtori CR, et al. (2003) A proteomic investigation of isolated soy proteins with variable effects in experimental and clinical studies. J. Nutr 133: 9-14.

13. Fassini PG, Noda RW, Ferreira ES, Silva MA, Neves VA, et al. (2011) Soybean glycinin improves HDL-C and suppresses the effects of rosuvastatin on hypercholesterolemic rats. Lipids Health Dis.

14. Hermansen K, Hansen B, Jacobsen R, Clausen P, Dalgaard M, et al. (2005) Effects of soy supplementation on blood lipids and arterial function in hypercholesterolaemic subjects. Eur J Clin Nutr 59: 843-850.

15. Ferreira ES, Silva MA, Demonte A, Neves VA (2010) β-Conglycinin (7S) and glycinin (11S) exert a hypocholesterolemic effect comparable to that of fenofibrate in rats fed a high-cholesterol diet. J Functional Foods 2: 275-228.

16. Gatica LV, Vega VA, Zirulnik F, Oliveros LB, Gimenez MS, et al. (2006) Alterations in the lipid metabolism of rat aorta: effects of vitamin a deficiency. J Vasc Res 43: 602-610.

17. Fukui K, Tachibana N, Wanezaki S, Tsuzaki S, Takamatsu K, et al. (2002) Isoflavone-free soy protein prepared by column chromatography reduces serum cholesterol in rats. J Agric Food Chem 50: 5717-5721.

18. Cavallini DC, Suzuki JY, Abdalla DS, Vendramini RC, Pauly-Silveira ND, et al. (2011) Influence of a probiotic soy product on fecal microbiota and its association with cardiovascular risk factors in an animal model. Lipids Health Dis 10: 126-130.

19. Chen G, Luo YC, Ji BP, Li B, Su W, et al. (2011) Hypocholesterolemic effects of Auricularia auricula ethanol extract in ICR mice fed a cholesterol-enriched diet. J Food Sci Technol 48: 692-698.

20. Wang MF, Yamamoto S, Chung HM, ChungSY, Miyatini S, et al. (1995) Antihypercholesteroemic effect of undigested fraction of soybean protein in young female volunteers. J Nutr Sci Vitaminol 41: 187-195.

21. Puska P, Korpelainen V, Høie LH, Skovlund E, Lahti T, et al. (2002) Soy in hypercholesterolaemia: a double-blind, placebo-controlled trial. Eur J Clin Nutr 56: 352-357.

22. Rebholz CM, Reynolds K, Wofford MR, Chen J, Kelly TN, et al. (2013) Effect of soybean protein on novel cardiovascular disease risk factors: a randomized controlled trial. Eur J Clin Nutr 67: 58-63.

23. Adams MR, Golden DL, Register TC, Anthony MS, Hodgin JB, et al. (2002) The atheroprotective effect of dietary soy isoflavones in apolipoprotein E −/− mice requires the presence of estrogen receptor-α. Arterioscler. Thromb Vasc Biol 22: 1859-1864.

24. Clarkson TB, Anthony MS, Smith M, Wilson L, Barnes S, et al. (2002) A paradoxical association between plasma isoflavone concentrations on a soy-containing diet, and both plasma lipoproteins and atherosclerosis. J Nutr 132: 583-584.

35

The Design of Double Screw Threads Soymilk Stone Mill

Pengyun Xu*, Xiaoshun Zhao and Haiyong Jiang

M & E College, Agriculture University of Hebei, Baoding, P.R.China

Abstract

In order to solve the problem of traditional stone mill, such as bulky structure, low efficiency, easy to be blocked, this paper design and implement a new conical double-screw threads stone mill refiner, rotating stone mill body is conical with two right-hand threads. As the diameter increases, its depth decreases. This structure can improve mechanical efficiency. The experiments showed that the machine's low-speed layer-by-layer uniform crushing grinding soybeans, the soymilk less susceptible to high temperature damage, is conducive to a variety of nutrients reserved.

Keywords: Soymilk; Stone mill; Conical mill; Refiner

Introduction

Amid the improvement of living conditions in China, the Chinese have more requirements for foods nutrition and variety. The grinded soymilk and flour porridge mixed by grains, such as rice, corn and bean, are nutritious and delicious. Having more than 2,000-year history and being still in use today, the stone mill is the traditional grinding device. But presently the stone mill refiner available has large size and high cost, high operation temperature, offering bad taste. Moreover, the grinding wheel has high rotate speed and short service life. Security accidents occur from time to time caused by different quality standards of grinding wheels, poor adhesive strength, and poisonous components in the binder [1].

The conical stone mill refiner, a kind of ecological stone mill device, is developed through researching the processing technique of the traditional stone mill and performing modern improvements. It delivers power by motors and may retain nutrition of agricultural products and effectiveness of Traditional Chinese Medicine.

It takes the granite in the Taihang Mountains area as the raw material of the stone mill. There are excellent granite resources in the Taihang Mountains area in China, such as Fengzhen, Hunyuan, and Fuping. Its products export to overseas countries and enjoy worldwide high reputation. The granite has excellent physical performance, namely, high hardness, abrasion resistance, low water absorption rate, compressive strength 100 ~ 127 MPa, and Shore hardness 78 ~ 80. Historically in the Taihang Mountains area, the granite is the traditional material to make the stone mill, which is still in use nowadays [2]. It has formed a complete system to manufacture the stone mill. There is no regional disease record of local residents caused by foods made by the granite stone mill. Therefore, this kind of granite is unlikely to do harm to human body. Based on its essential performance, considering that it is applicable regionally, it may develop resources in the Taihang Mountains area, which will be advantageous to promote the local economy.

Principles and Features of the Soymilk Stone Mill

As the main part of the processing, the stone mill designs include stone materials selection and determination of grinding marks.

The selection of stone materials

It is traditional to make the stone mill in various parts of China in history. To take regional materials may avoid long term conveyance.

Generally, it selects fine materials to manufacture stone mills, such as the sandstone, the bluestone, and the granite. As the crystallization formed by gradual magma condensate from deep underground, the granite is known as the King of the Rock. It is hard and durable in use, acid-resistant, alkali-resistant, and anti-weathering. The granite appearance is shown in Figure 1.

There are excellent granite resources in the Taihang Mountains area in China, such as Fengzhen, Hunyuan, and Fuping. Based on its prominent performance, considering that it is applicable regionally, the granite in the Taihang Mountains area is used as the raw material of the stone mill [3]. So that local residents will be richer than before through making stone mills, which can also develop resources in the Taihang Mountains area. The granite is illustrated in Figure 2.

The design of grinding marks

The earliest grinding marks were shown in millstones in the Spring and Autumn and Warring States Periods of China. The history

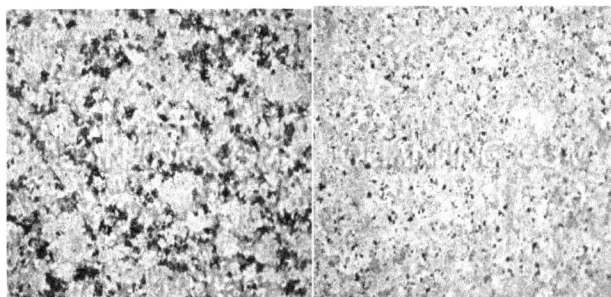

Figure 1: Different types of granite.

*Corresponding author: Pengyun Xu, M & E College, Agriculture University of Hebei, Baoding 071001, P.R.China, E-mail: pengyun99@qq.com

of stone mill forms change can be divided into three periods, with corresponding forms of grinding marks. During the initial forming stage from Warring States to the Western Han Period, the stone mills had irregular shapes and the corresponding grinding marks showed miscellaneous forms. Pit was the most popular grinding mark, having the shapes of rectangle, circle, triangle and shuttle. There were scattered pits distributed at the upper and lower parts of the millstone. The second phase for stone mill developments was from the Eastern Han Period to Three Kingdoms Period. Thanks to manufacture technique developments at that time, the round stone mill was the domain form in this stage. Grinding gears had diversified developments during this period, having the radial divisional sector types, including four-section, six- section and eight- section [4]. The third developing phase, Sui and Tang Period, was the mature state of the stone mill, when the most popular grinding mark was eight-section sector type, also ten-section sector type, which was still in use nowadays.

The stone mill can be used to process soymilk, flour, sesame oil, and tea, also used as a kind of handiwork. Different functions of the stone mill require different grinding marks. Figure 3 shows stone mills for flour and soymilk. Comparing the stone mill for flour with that for soymilk, it shows that their grinding marks are similar and characteristic respectively. They all have divisions in the millstones and are scattered anti-clockwise relative to the center and its edge. What makes these marks different is that the grinding mark of the flour millstone is connecting from its center to its edge, while that of the soymilk stone mill is not connecting at 2 cm from the its edge. The grinding lip, the disconnecting part on the millstone, refers that the grinding mark cannot carve to the edge of the millstone and it shall

reserve a circle of smooth blank. It depends on different processing objects. The flour stone mill aims to process wheat to powders, whose particle diameter is greater than 100 nm. But the soymilk stone mill aims to successfully get the protein out of the grinded soya beans. As a kind of colloid, its particle diameter is smaller, ranging from 1-100 nm. The grinding lip of the soymilk stone mill is meant to smash larger particles into pieces by a dull knife so that the colloid in the refined granule will be extracted.

The Structure Design of Double Screw Threads Stone Mill

As Figure 3 illustrated, the traditional stone mill is composed of the upper and the lower flat cylindrical millstones that are made of carved boulders. The top of the upper millstone is a concave that is low in the center and high around. The lower millstone is a little bit thicker that the upper one. There are grinding marks on the surface where the upper and lower millstones cooperate. On the soya bean stone mill there is only one feed inlet.

The upper millstone makes low-speed rotation movement, which can be driven by manpower and animal power. Grinding marks of the upper and lower millstones occlude and interweave each other. Their grooves become shallower and shallower from the center to the edge, which will be advantageous for the grinded fines to flow to the outside. Generally, directions of grinding marks are anticlockwise, in accordance with the habit of using the strength [2-4].

In this paper it purposes a new modernized electric conical stone mill, in order to ensure effective processing area in limited spaces and to make the process smooth. The structure is illustrated in Figure 4. The stone mill is composed of two millstones, (1) the inner one and (2) the outer one. The inner one is a cone that its small end is at the top, with double right-turn thread grinding marks. The outer stone mill is a hollow column. Two millstones cooperatively compose the finished surface (inside discharge hopper). The conical stone mill, the gap between two stone mills becomes smaller and smaller from top to bottom. The soaked soya beans are a little bit large so that the great gap on the finished surface will be convenient for soya beans to be put and grinded here. The grinded soya beans fall into gaps under the functions of flow, friction and squeezing for further accurate grinding. By taking the conical millstone it is possible to increase its valid finished surface area within limited space and to lower the weight of the stone mill. By adjusting the position of two stone mills it may adjust the gap of the finish surface, so that the juice grain fineness can pass through 100 mesh screen. Nutrition can be fully released after several grinding processing. It may avoid the bean pulp (bean dregs and skins) to be mixed into the soymilk, which will affect the taste.

The outer stone mill rotates with side type motor. Structures are shown in Figure 5A.

The whole set of structure is placed in a sealed enclosure, shown in Figure 5B.

Pour beans and water from the funnel. Soymilk flow out from the bottom left.

In order to ensure that the soybean into the wedge gap, should meet the following conditions:

$$L < \frac{(D - d/2)}{2} \tag{1}$$

Where,

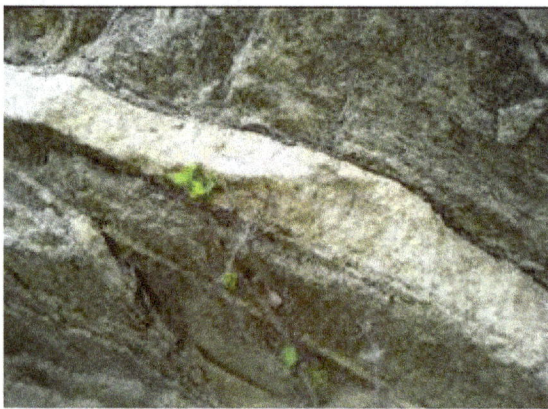

Figure 2: Taihang Mountain granite.

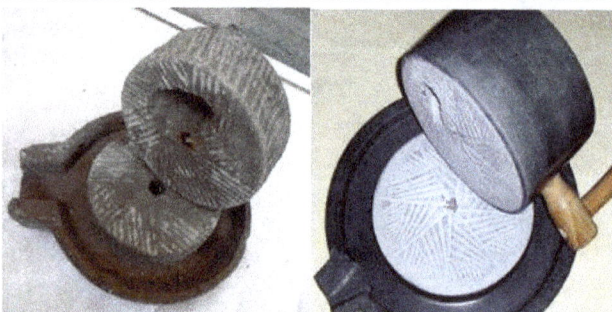

A   B

Figure 3: Stone mills for (A) Flour and (B) Soymilk.

L = soaked soybean width, about 7 mm

D = the inner stone mill top diameter, mm

D = the inner stone mill bottom diameter, mm

Then the wedge angle between the stone mills:

$$a = \arctan \frac{D-d}{2H} \qquad (2)$$

H = the inner stone mill height, mm; When H is too large, easy to plug; when H is too small, the efficiency is low.

Due to high speed will cause temperature rise, affecting the quality and taste of soy milk, so, the bottom diameter must meet the following conditions:

$$D \leq \frac{60 \times 1000 \times V_{max}}{n \times p}$$

V_{max} = Maximum line speed, m/s

N = rotation rate, r/min

Through the experiment contrasts, it suggests that when it is processing the soymilk, the linear speed of the grinding slice 9.16 m/s is proper. This millstone diameter is 150 mm and the setting speed is 80-350 r/min in operation. The allowed maximum rotating speed is 960 r/min.

A lot of experiments show that the following parameters are the best, shown in Table 1.

Results and Discussion

The double threads can make the stone body symmetrical structure, smooth operation, speed up feed, and is not easy to be blocked. The right screw is used to conform to the artificial force, and the artificial intervention is convenient when necessary. Pitch of screws is 30 mm. The thread depth gradually becomes shallow and completely

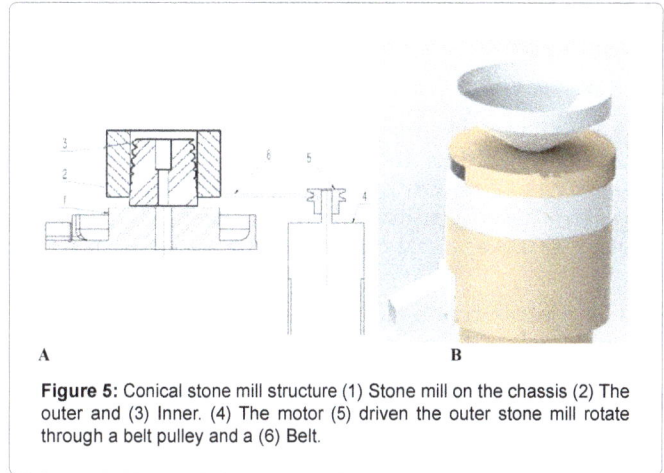

Figure 4: Conical stone mill.

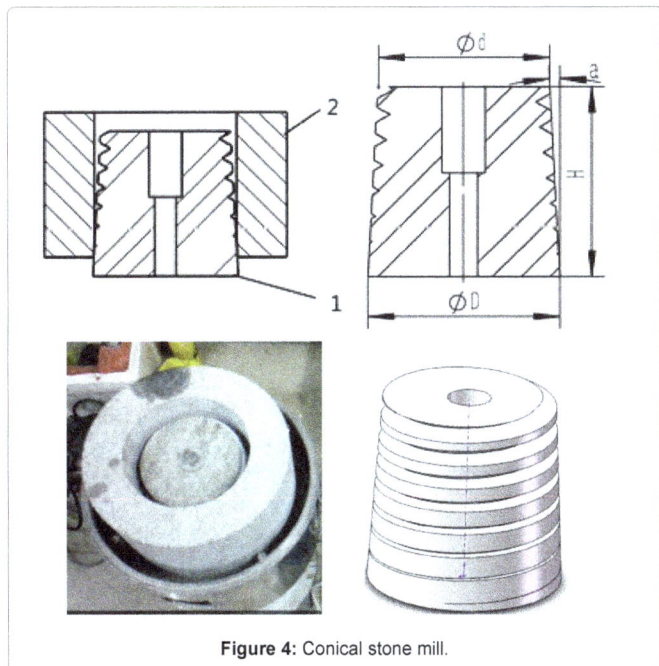

Figure 5: Conical stone mill structure (1) Stone mill on the chassis (2) The outer and (3) Inner. (4) The motor (5) driven the outer stone mill rotate through a belt pulley and a (6) Belt.

D	d	H	α	V_max	Pitch
150 mm	130 mm	190 mm	3°	400 r/min	30 mm

Table 1: Parameters of the conical stone mill.

disappeared at the distance 20 mm from the bottom. The design parameters of conical stone mill, outer diameter are 150 mm, height is 130 mm, taper is 3 degrees, and working surface area is about 43977 mm². But the traditional 150 mm diameter stone working surface area is about 13266 mm², only 1/3 of the conical stone mill.

Summary

By modernization technology and devices, considering features of mechanical juice and the traditional stone mill, it improves the design of the traditional stone mill. It proposed a kind of conical stone mill refiner, which increases grinding area. It grinds for several times for accurate grinding, in order to avoid gross powder. Therefore, the nutrition can be easily mixed with water. It will be homogenized and emulsified in a better way, which makes it easier to digest. The gaps of the finishing surface can be adjusted to meet requirements of different processing grains. This machine has stable reliable structure, being easy to install and maintain. It does not need to separate juice and residue, all of which promote its efficiency.

Acknowledgment

The authors thank the Science research project of Hebei Province, the Science and Technology Department of Hebei province program (11231005D-2), and Baoding science and technology research and development plan (12ZN013) for support.

References

1. Pengfei Qi, Min Liu (2012) The design of automatic mill. Journal of Agricultural Mechanization Research 34: 157-160.

2. Pengyun Xu, Zehe Wang, Jin Jin (2013)Three layers of stone grinding machine design. Journal of Agricultural Mechanization Research 34: 107-109.

3. Zhang Yuzhong, Wang Xiaolong, Zhang Maolong (2012) Research on the energy consumption and the characteristics of soybean milk in different wet crushing devices. Food industry technology 33: 97-100.

4. Chen Minxin (2011) The stone form and aesthetic culture value. Art panorama 4: 66-67.

PERMISSIONS

LIST OF CONTRIBUTORS

Aisha Idris Ali and Genitha Immanuel
Department of Food Process Engineering, Sam Higginbottom University of Agriculture Technology and Sciences, Allahabad, UP, India

Belay Binitu Worku
Department of Food Process Engineering and Postharvest Technology, Ambo University, Ambo, Ethiopia

Ashagrie Zewdu Woldegiorgis
Centre for Food Science and Nutrition, Addis Ababa University, Addis Ababa, Ethiopia

Habtamu Fekadu Gemeda
Centre for Food Science and Nutrition, Addis Ababa University, Addis Ababa, Ethiopia
Department of Food Technology and Process Engineering, Wollega University, P.O.Box: 395, Nekemte, Ethiopia

Kumari PV
IICPT, Thanjavur, Tamilnadu, India

Sangeetha N
Department of Food Science and Technology, Pondicherry University, Puducherry, India

Jude-Ojei BS
Department of Nutrition and Dietetics, Rufus Giwa Polytechnic, Owo Ondo State, Nigeria

Lola A and Ilemobayo Seun
Department of Food Science and Technology, Rufus Giwa Polytechnic, Owo Ondo State, Nigeria

Ajayi IO
Department of Science Laboratory Technology, Rufus Giwa Polytechnic, Owo Ondo State, Nigeria

Ali M
School of Food Science and Biotechnology, Key Laboratory of Fruits and Vegetables, Zhejiang Gongshang University, Hangzhou, China

Khan MR, Rakha A, Khalil AA, Lillah K and Murtaza G
National Institute of Food Science and Technology, Fruits and Vegetables Processing Laboratory, University of Agriculture Faisalabad, Pakistan

Ibrahim Khan, Rehman AU and Qazi IM
The University of Agriculture Peshawar, Khyber Pakhtunkhwa, Pakistan

Khan SH
Gomal University of D.I. Khan, Pakistan

Arsalan khan and Shah FN
Agricultural Research Institute ARI Tarnab Peshawar, Khyber Pakhtunkhwa, Pakistan

Rehman TU
Abdul Wali Khan University, Mardan, Pakistan

Insha Zahoor and Khan MA
Department of Post-Harvest Engineering and Technology, Faculty of Agricultural Sciences, Aligarh Muslim University, Aligarh, India

Anane MA and Immanuel G
Department of Food Processing Engineering, Sam Higginbottom Institute of Agriculture, Technology and Sciences (SHIATS), Allahabad, UP, India

Ragava SC, Loganathan M, Vidhyalakshmi R and Vimalin HJ
Department of Microbiology, Hindusthan College of Arts and Sciences, Coimbatore, Tamil Nadu, India

Abdel-Moemin AR
Department of Nutrition and Food Science, Faculty of Home Economics, Helwan University, Bolak, Cairo, Egypt

Kalse SB, Swami SB, Sawant AA and Thakor NJ
Department of Agricultural Process Engineering, College of Agricultural Engineering and Technology, Dr. BS Konkan Krishi Vidyapeeth, Ratnagir, India

Hafiya Malik and Dar BN
Department of Food Technology, Islamic University of Science and Technology, Awantipora Pulwama, Jammu & Kasmir, India

Gulzar Ahmad Nayik
Department of Food Engineering and Technology, Sant Longowal Institute of Engineering and Technology, Longowal, Punjab, India

Kumar S, Khadka M, Kohli D and Upadhaya S
Department of Food Technology, Uttaranchal University, Dehradun, India

Mishra R
Department of Agricultural science and Engineering, IFTM University, Moradabad, India

Menure Heiru
Department of Chemical Engineering, Dire Dawa University Institute of Technology, Dire Dawa, Ethiopia

Nassar KS and Shamsia SM
Department of Food, Dairy Science and Technology, University of Damanhour, Egypt

Attia IA
Department of Dairy Science and Technology, Alexandria University, Egypt

Sasikumar R, Vivek K, Chakaravarthy S and Deka SC
Department of Agri-Business Management and Food Technology, North-Eastern Hill University, Meghalaya, India

Famuwagun AA, Taiwo KA and Gbadamosi SO
Department of Food Science and Technology, Obafemi Awolowo University, Ile-Ife, Nigeria

Oyedele DJ
Faculty of Agriculture, Department of Soil and Land Management, Obafemi Awolowo University, Ile-Ife, Nigeria

Yui Sunano
School of Agricultural Sciences, Graduate School of Bio-agricultural Sciences, Nagoya University, Japan

Mawada E Yousif and Babiker E Mohamed
Department of Food Science and Technology, Faculty of Agriculture, University of Khartoum, Sudan

Elkhedir AE
Food Industry Department, Industrial Research and Consultancy Center (IRCC), Khartoum, Sudan

Mohammad H Rahman, Nasim Marzban, Theresa Schmidl, Saiful Hasan and Caroline Nandwa
Department of Organic Agricultural Sciences, Kassel University, Hesse, Germany

Kiranmai E, Uma Maheswari K and Vimala B
Department of Food Processing Engineering, Sam Higginbottom Institute of Agriculture, Technology and Sciences, Allahabad, UP, India

Samadder M, Someswararao CH and Das SK
Department of Agriculture and Food Engineering, Indian Institute of Technology, Kharagpur, West Bengal, India

Gbadamosi SO and Famuwagun AA
Department of Food Science and Technology, Obafemi Awolowo University, Ile-Ife, Nigeria

Bartolini Susanna and Leccese Annamaria
Institute of Life Science, Scuola Superiore Sant'Anna, Piazza Martiri della Libertà 33, Pisa, Italy

Viti Raffaella
Department of Agriculture, Food and Environment-Interdepartmental Research Center, Nutrafood 'Nutraceuticals and Food for Health', University of Pisa, Via del Borghetto, Pisa, Italy

Ashraf G, Sonkar C, Masih D and Shams R
Department of Food Process Engineering, Sam Higginbottom University of Agriculture, Technology and Sciences, Allahabad, UP, India

Oluwabukola Ojo D and Ndigwe Enujiugha V
Federal University of Technology, Akure, School of Agriculture and Agricultural Technology, Department of Food Science and Technology, Akure, Ondo State, Nigeria

Teshome E, Tola YB and Mohammed A
Department of Post-Harvest Management, Jimma University College of Agriculture and Veterinary Medicine, Jimma, Ethiopia

Saeid Alizadeh Asl, Mohammad Mousavi and Mohsen Labbafi
Department of Food Science and Technology, Campus of Agricultural Engineering and Natural Resources, University of Tehran, Iran

Imlak M, Randhawa MA, Hassan A, Ahmad N and Nadeem M
Food Safety Laboratory, National Institute of Food Science and Technology, University of Agriculture, Faisalabad, Pakistan

Mehnaza Manzoor
Department of Food Science and Technology, Sher-e-Kashmir University of Agricultural Sciences and Technology, Jammu, India

Shukla RN, Mishra AA and Afrin Fatima
Department of Food Process Engineering, Sam Higginbottom University of Agriculture, Technology and Sciences, UP, India

Nayik GA
Department of Food Engineering and Technology, Sant Longowal Institute of Engineering and Technology, Punjab, India

Bibi S, Kowalski RJ, Zhang S, Ganjyal GM and Zhu MJ
School of Food Science, Washington State University, Pullman, WA 99164, USA

Zeeshan M and Saleem SA
Food Technology Section, Agricultural Research Institute, D.I. Khan, K.P, Pakistan

Ayub M, Shah M and Jan Z
University of Agriculture, Peshawar, K.P, Pakistan

Sunandita Ghosh, Melvin Holmes and Malcolm Povey
Department of Food Science and Nutrition, University of Leeds, Leeds, UK

Uyar SI
Cardiovascular Surgery Department, Medical Faculty, Sifa University, Izmir, Turkey

Ugur Ozdemır
Anesthesiology Department, Medical Faculty, Sifa University, Izmir, Turkey

Ilter MS
Department of Food Engineering, Engineering and Natural Science Faculty, Gumushane University, Gumushane, Turkey

Muhittin Akyıldız
Biochemistry Department, Medical Faculty, Sifa University, Izmir, Turkey

Sivrikoz NO
Pathology Department, Medical Faculty, Sifa University, Izmir, Turkey

Pengyun Xu, Xiaoshun Zhao and Haiyong Jiang
M & E College, Agriculture University of Hebei, Baoding, P.R.China

Index